Advances on Methodological and Applied Aspects of Probability and Statistics

Advances on Methodological and Applied Aspects of Probability and Statistics

Edited by

N. Balakrishnan

McMaster University
Hamilton, Canada

CRC Press
Taylor & Francis Group
Boca Raton London New York

CRC Press is an imprint of the
Taylor & Francis Group, an **informa** business
A TAYLOR & FRANCIS BOOK

ADVANCES ON METHODOLOGICAL AND APPLIED ASPECTS OF PROBABILITY AND STATISTICS

CRC Press
Taylor & Francis Group
6000 Broken Sound Parkway NW, Suite 300
Boca Raton, FL 33487-2742

First issued in paperback 2020

ISBN 13: 978-0-367-57851-0 (pbk)
ISBN 13: 978-1-56032-980-0 (hbk)

Cover design by Ellen Seguin.

A CIP catalog record for this book is available from the British Library.

Library of Congress Cataloging-in-Publication Data is available from the publisher.

Visit the Taylor & Francis Web site at
http://www.taylorandfrancis.com

and the CRC Press Web site at
http://www.crcpress.com

CONTENTS

Part II   Models and Applications

Part III   Estimation and Testing

PREFACE

This is one of two volumes consisting of 33 invited papers presented at the *International Indian Statistical Association Conference* held during October 10–11, 1998, at McMaster University, Hamilton, Ontario, Canada. This Second International Conference of IISA was attended by about 240 participants and included around 170 talks on many different areas of Probability and Statistics. All the papers submitted for publication in this volume were refereed rigorously. The help offered in this regard by the members of the Editorial Board listed earlier and numerous referees is kindly acknowledged. This volume, which includes 33 of the invited papers presented at the conference, focuses on *Advances on Methodological and Applied Aspects of Probability and Statistics.*

For the benefit of the readers, this volume has been divided into nine parts as follows:

Part I   Applied Probability
Part II   Models and Applications
Part III   Estimation and Testing
Part IV   Robust Inference
Part V   Regression and Design
Part VI   Sample Size Methodology
Part VII   Applications to Industry
Part VIII   Applications to Ecology, Biology and Health
Part IX   Applications to Economics and Management

I sincerely hope that the readers of this volume will find the papers to be useful and of interest. I thank all the authors for submitting their papers for publication in this volume.

Special thanks go to Ms. Arnella Moore and Ms. Concetta Seminara-Kennedy (both of Gordon and Breach) and Ms. Stephanie Weidel (of Taylor & Francis) for supporting this project and also for helping with the production of this volume. My final thanks go to Mrs. Debbie Iscoe for her fine typesetting of the entire volume.

I hope the readers of this volume enjoy it as much as I did putting it together!

N. BALAKRISHNAN   MCMASTER UNIVERSITY
                                                 HAMILTON, ONTARIO, CANADA

LIST OF CONTRIBUTORS

Abraham, Bovas, IIQP, University of Waterloo, Waterloo, Ontario, Canada N2L 3G1
babraham@setosa.uwaterloo.ca

Agarwal, Manju, Department of Operations Research, University of Delhi, Delhi-110007, India

Aggarwala, Rita, Department of Mathematics and Statistics, University of Calgary, Calgary, Alberta, Canada T2N 1N4
rita@math.ucalgary.ca

Agrawal, Rehka, GE Corproate Research & Devleopment, Schenectady, NY 12065, U.S.A.
rekha.agrawal@corporate.ge.com

Balakrishnan, N., Department of Mathematics and Statistics, McMaster University, Hamilton, Ontario, Canada L8S 4K1
bala@mcmail.cis.mcmaster.ca

Basawa, Ishwar V., Department of Statistics, University of Georgia, Athens, GA 30602-1952, U.S.A.
ishwar@stat.uga.edu

Basu, Sanjib, Division of Statistics, Northern Illinois University, DeKalb, IL 60115, U.S.A.
basu@niu.edu

Basu, Sujit K., National Institute of Management, Calcutta 700027, India

Bera, Anil K., Department of Economics, University of Illinois at Urbana-Champaign, Champaign, IL 61820, U.S.A.
anil@fisher.econ.uiuc.edu

Bhat, U. Narayan, Department of Statistical Science, Southern Methodist University, Dallas, TX 75275-0240, U.S.A.
nbhat@mail.smu.edu

Bhattacharjee, Manish C., Center for Applied Mathematics & Statistics, Department of Mathematical Sciences, New Jersey Institute of Technology, Newark, NJ 07102-1982, U.S.A.
mabhat@chaos.njit.edu

Bhoj, Dinesh S., Department of Mathematical Sciences, Rutgers University, Camden, NJ 08102-1405, U.S.A.
dbhoj@crab.rutgers.edu

Billah, Md. Baki, Department of Statistics, University of Dhaka, Dhaka-1000, Bangladesh

Chaubey, Yogendra P., Department of Mathematics and Statistics, Concordia University, Montreal, Quebec, Canada H4B 1R6
chaubey@vax2.concordia.ca

Chaudhry, M. L., Department of Mathematics and Computer Science, Royal Military College of Canada, P.O. Box 17000, STN Forces, Kingston, Ontario, Canada K7K 7B4
chaudhry-ml@rmc.ca

Datta, Susmita, Department of Mathematics and Computer Science, Georgia State University, Atlanta, GA 30303-3083, U.S.A.
sdatta@cs.gsu.edu

Desu, M. M., Department of Statistics, State University of New York, Buffalo, NY 14214-3000, U.S.A.
desu@calcutta.med.buffalo.edu

Dhar, Sunil K., Department of Mathematical Sciences, New Jersey Institute of Technology, Newark, NJ 07102-1824, U.S.A.
sunidh@stat.njit.edu

Gadbury, Gary, Department of Mathematics, University of North Carolina at Greensboro, Greensboro, NC, U.S.A.

Gardiner, Joseph C., Department of Epidemiology, College of Human Medicine, Michigan State University, East Lansing, MI 48823, U.S.A.
gardine3@pilot.msu.edu

Geddes, R. R., Department of Economics, Fordham University, 441 East Fordham Road, Bronx, NY 10458-5158, U.S.A.

Ginebra, Josep, Departament d'Estadística, E.T.S.E.I.B., Universitat Politècnica de Catalunya, Avgda. Diagonal 647, 6ª planta, 08028 Barcelona, Spain
ginebra@eio.upc.es

González, Enrique, Departamento de Estadística, Universidad de La Laguna, 38271 La Laguna, Spain
egonzale@ull.es

Govindarajulu, Z., Department of Statistics, University of Kentucky, Lexington, KY 40506, U.S.A.
raju@ms.uky.edu

Gupta, U. C., Department of Mathematics, Indian Institute of Technology, Kharagpur 721 302, India
umesh@maths.iitkgp.ernet.in

Huzurbazar, Aparna V., Department of Mathematics and Statistics, University of New Mexico, Albuquerque, NM 87131-1141, U.S.A.
aparna@stat.unm.edu

Indurkhya, Alka, Department of Epidemiology, College of Human Medicine, Michigan State University, East Lansing, MI 48823, U.S.A.

Iyengar, Satish, Department of Statistics, University of Pittsburgh, Pittsburgh, PA 15260, U.S.A.
su@bacchus.stat.pitt.edu

Iyer, Hari, Department of Statistics, Colorado State University, Fort Collins, CO 80523, U.S.A.
hari@stat.colostat.edu

Jammalamadaka, S. Rao, Department of Statistics and Applied Probability, University of California, Santa Barbara, CA 93106, U.S.A.
rao@pstat.ucsb.edu

Kinateder, Kimberly K. K., Department of Mathematics and Statistics, Wright State University, Dayton, OH 45453, U.S.A.

Lu, Xuewen, Food Research Program, Sourthern Crop Protection and Food Research Centre, Agriculture and Agri-Food Canada, 43 McGilvray Street, Guelph, Ontario, Canada N1G 2W1
lux@em.agr.ca

Luo, Zhehui, Department of Epidemiology, College of Human Medicine, Michigan State University, East Lansing, MI 48823, U.S.A.

Mallick, Naresh C., Department of Economics and Finance, Alabama Agricultural and Mechanical University, Normal, AL, U.S.A.

Marchetti, Carol E., Department of Mathematics and Statistics, Rochester Institute of Technology, Rochester, NY 14623-5603, U.S.A.
cemsma@rit.edu

Mudholkar, Govind S., Department of Statistics, University of Rochester, Rochester, NY 14727, U.S.A.
govind@metro.bst.rochester.edu

Natarajan, Rajeshwari, Department of Statistics, University of Rochester, Rochester, NY 14727, U.S.A.
rajn@stat1.bst.rochester.edu

Patil, G. P., Department of Statistics, Pennsylvania State University, University Park, PA 16802, U.S.A.
gpp@stat.psu.edu

Paul, Sudhir R., Department of Mathematics and Statistics, University of Windsor, Windsor, Ontario, Canada N9B 3P4
smjp@uwindsor.ca

Prabhu, N. U., School of Operations Research and Industrial Engineering, Cornell University, Ithaca, NY 14853-3801, U.S.A.
questa@orie.cornell.edu

Rao, J. N. K., School of Mathematics and Statistics, Carleton University, Ottawa, Ontario, Canada K1S 5B6
jrao@math.carleton.ca

Saleh, A. K. Md. E., School of Mathematics and Statistics, Carleton University, Ottawa, Ontario, Canada K1S 5B6
esaleh@math.carleton.ca

Sarkar, Sanat K., Department of Statistics, Temple University, Philadelphia, PA 19122, U.S.A.
sanat@sbm.temple.edu

Sen, Kanwar, Department of Statistics, University of Delhi, Delhi-110007, India
ksen@ndb.vsnl.net.in

Serfling, Robert, Department of Mathematical Sciences, University of Texas at Dallas, Richardson, TX 75083-0688, U.S.A.
serfling@utdallas.edu

Singh, R. S., Department of Mathematics and Statistics, University of Guelph, Guelph, Ontario, Canada N1G 2W1
rsingh@msnet.mathstat.uoguelph.ca

Srivastava, Deo Kumar, Department of Biostatistics and Epidemiology, St. Jude Children's Research Hospital, 332 North Lauderdale St., Memphis, TN 38105-2794, U.S.A.
kumar.srivastava@stjude.org

Taillie, C., Department of Statistics, Pennsylvania State University, University Park, PA 16802, U.S.A.

Venkateswarlu, K., Department of Mathematics and Statistics, Concordia University, Montreal, Quebec, Canada H4B 1R6

Vinod, H. D., Department of Economics, Fordham University, 441 East Fordham Road, Bronx, NY 10458-5158, U.S.A.
vinod@murray.fordham.edu

Voss, Daniel T., Department of Mathematics and Statistics, Wright State University, Dayton, OH 45435, U.S.A.
dvoss@math.wright.edu

Wadhwa, Neerja, Card Services, GE Capital, Stamford, CT 06820, U.S.A.
pwadhwa@sprintmail.com

Wang, Weizhen, Department of Mathematics and Statistics, Wright State University, Dayton, OH 45435, U.S.A.

LIST OF TABLES

LIST OF FIGURES

Part I
Applied Probability

CHAPTER 1

FROM DAMS TO TELECOMMUNICATION – A SURVEY OF BASIC MODELS

N. U. PRABHU

Cornell University, Ithaca, NY

Abstract: In 1954 P.A.P. Moran formulated a simple discrete time model for a finite dam. This model was extended in several directions by J. Gani and the author during 1956–1963. The concepts underlying this model and the techniques used in its analysis are applicable in a wide variety of situations, as has already been demonstrated. Most recently, models for data communication systems have also been analyzed with these techniques. In this paper we survey some of this work.

Keywords and phrases: Buffer content, dam, data communication, idle time, input, fluid input, Lévy process, Markov chain, Markov-additive process, packets, Poisson arrivals, queues, subordinator, unmet demand, workload

1.1 INTRODUCTION

In 1954 P. A. P. Moran formulated a simple discrete time model for the finite dam. The basic components of this model are inputs that are independent and identically distributed random variables, a constant demand for water and the release policy "meet the demand if physically possible." During 1956–1963 J. Gani and the author extended this discrete time model to continuous time, where the input is described by a subordinator, the demand is at a unit rate and the release policy is the same as before. This continuous

3

time model has several applications, in particular, to single server queues with Poisson arrivals and first come, first served discipline or priority discipline of the static or dynamic type. Because these models have several common features in regard to the underlying concepts and techniques of analysis, the author proposed the term *stochastic storage processes* to describe the processes that arise from the family of such models and presented a unified theory of these processes [see Prabhu (1998)]. The most recent extension of this theory is to models for transmission of telecommunication data. Here the input of data is characterized as a Markov-additive process, the desired transmission (demand) rate depends on the Markov component of the input and the actual transmission (release) policy is to "meet the demand if physically possible." The resulting theory may be viewed as the Markov-modulated version of the theory of dams.

In this paper we survey some of this work, emphasizing only the modeling aspects in order to point out the common features of the models considered. For detailed results and recent references see Prabhu (1998). For historical references see Prabhu (1965).

In Section 1.2 we describe Moran's discrete time model for the finite dam. The continuous time dam model is described in Section 1.3, and its extension to the data communication model in Section 1.4.

1.2   MORAN'S MODEL FOR THE FINITE DAM

Moran's discrete time model for a dam (water reservoir) is the following. A dam of finite capacity is designed to meet the demand for electric power (expressed in terms of the volume of water required to produce it) or for water to be supplied to a city. The demand for water at time n is m ($< c$) and this demand is met "if physically possible," that is, to the extent that this quantity is available in the dam at time n. The dam is fed by inputs of water such that if X_{n+1} denotes the input during the time interval $(n, n + 1]$, then $\{X_n, n \geq 1\}$ is assumed to be a sequence of independent and identically distributed random variables. Because of this randomness the amount of water in the dam (the dam content) at time n is a random variable which we denote by Z_n ($n \geq 0$).

Since the capacity of the dam is finite there is a possibility of an overflow and the actual input during $(n, n + 1]$ is therefore

$$\eta_{n+1} = \min(c - Z_n, X_{n+1}) \quad (n \geq 0). \qquad (1.2.1)$$

The amount of water available for release at time $n + 1$ is then $Z_n + \eta_{n+1}$ and the release policy implies that

$$Z_{n+1} = Z_n + \eta_{n+1} - \min(m, Z_n + \eta_{n+1}).$$

The sequence $\{Z_n, n \geq 0\}$ satisfies the relation

$$Z_{n+1} = (Z_n + \eta_{n+1} - m)^+ \quad (n \geq 0). \tag{1.2.2}$$

To see how the dam operates subject to these assumptions we note that during a time interval $(0, n]$ there is a certain amount F_n of overflow from the dam, and an amount D_n of the total demand nm that is not met. Easy calculations show that

$$Z_n = Z_0 + (S_n - nm) - F_n + D_n \quad (n \geq 0) \tag{1.2.3}$$

where $S_n = X_1 + X_2 + \cdots + X_n (n \geq 1)$, $S_0 = 0$ and $S_n - nm$ is the net input (input minus demand) during $(0, n]$.

The assumption on the inputs X_n implies that $\{Z_n, n \geq 0\}$ is a time-homogeneous Markov chain on the state space $\mathbb{R}_+$. The problems of practical importance that arise in the analysis of the model are the derivation of (i) the steady state distribution of $\{Z_n\}$ and (ii) the distribution of the random variable

$$T(Z_0) = \min\{n \geq 1 : Z_n = 0\} \tag{1.2.4}$$

which is the duration of the wet period in the dam whose initial content is $Z_0 > 0$. Although these problems are standard in the theory of Markov chains, general solutions are not known because of the presence of the constant c $(< \infty)$ in (1.2.2). However, solutions are available for some important special cases of the input distributions [see Prabhu (1965)].

When $c = \infty$ (the case of the infinite dam) the equations (1.2.2) and (1.2.3) reduce to

$$Z_{n+1} = (Z_n + X_{n+1} - m)^+ \quad (n \geq 0) \tag{1.2.5}$$

and

$$Z_n = Z_0 + (S_n - mn) + D_n \quad (n \geq 0). \tag{1.2.6}$$

These lead to the expressions

$$Z_n = \max\{Z_0 + S_n - nm, S_n - nm - m_n\} \tag{1.2.7}$$

$$D_n = (Z_0 + m_n)^- \tag{1.2.8}$$

where m_n is the minimum functional of the random walk $\{S_n - nm, n \geq 0\}$, namely

$$m_n = \min_{0 \leq k \leq n} (S_k - km) \quad (n \geq 0). \tag{1.2.9}$$

The equation (1.2.5) arises in queueing theory, specifically for waiting times Z_n in the single server queue with constant interarrival times $(= m)$ and general service times $X_n(n \geq 1)$. The quantity D_n in (1.2.6) is the total idle period during (0,n], while the random variable $T(Z_0)$ defined by (1.2.4) is the number of customers served during the busy period initiated by a waiting time $Z_0 > 0$. Thus the results for the infinite dam are applicable to queueing theory.

1.3   A CONTINUOUS TIME MODEL FOR THE DAM

In developing a continuous time model for the dam we first assume that its capacity is ∞. For the input we postulate a nonnegative process with stationary independent increments, that is, a Lévy process $\{X(t), t \geq 0\}$ with nondecreasing sample functions (also called a subordinator) and zero drift. The demand for water occurs at a rate $d \circ Z(t)$, where $Z(t)$ is the dam content at time $t \geq 0$. As in the discrete time case, this demand is met "if physically possible". These assumptions lead to the integral equation

$$Z(t) = Z(0) + X(t) - \int_0^t d \circ Z(s)1_{\{Z(s)>0\}} ds. \tag{1.3.10}$$

We can rewrite this is

$$Z(t) = Z(0) + X(t) - \int_0^t d \circ Z(s)ds + \int_0^t d \circ Z(s)1_{\{Z(s)=0\}} ds. \tag{1.3.11}$$

Here on the right side of (1.3.11) the first integral represents the total demand during $(0, t]$ and the second integral is the part of this demand that is not met. The equation (1.3.11) is the continuous time analogue of (1.2.6).

The most extensively studied special case of (1.3.10) is the one with unit demand rate (that is, $d(x) \equiv 1$), which arises also in the queueing system $M/G/1$ and single server queues with Poisson arrivals and static or dynamic priorities. In the queue $M/G/1$, the input $X(t)$ of workload is a compound Poisson process, and $Z(t)$ represents the remaining workload (virtual waiting time) at time t. In dam models the special cases of input include the gamma process, stable process with exponent 1/2 and the inverse Gaussian process. The integral equation (1.3.11) reduces in the case of unit demand rate to

$$Z(t) = Z(0) + Y(t) - \int_0^t 1_{\{Z(s)=0\}} ds. \tag{1.3.12}$$

where $Y(t) = X(t) - t$ (the net input) and the integral

$$I(t) = \int_0^t 1_{\{Z(s)=0\}} ds \tag{1.3.13}$$

represents the amount of unmet demand (dry period in a dam or idle time in the queue $M/G/1$).

As formulated above, the integral equation (1.3.12) does not have a unique nonnegative solution. However, if we modify it by writing

$$Z(t) = Z(0) + Y(t) - \int_0^t 1_{\{Z(s) \leq 0\}} ds \qquad (1.3.14)$$

then the unique nonnegative solution of (1.3.14) is given by

$$Z(t) = \max\{Z(0) + Y(t), Y(t) - m(t)\} \qquad (1.3.15)$$

where $m(t)$ is the minimum functional

$$m(t) = \inf_{0 \leq s \leq t} Y(s). \qquad (1.3.16)$$

It follows from (1.3.14) that

$$I(t) = \int_0^t 1_{\{Z(s)=0\}} ds = [Z(0) + m(t)]^- \qquad (1.3.17)$$

on account of the nonnegativity of $Z(t)$. The results (1.3.15) and (1.3.17) are the continuous time analogues of (1.2.7) and (1.2.8) for the discrete time case.

Remarks.

1. When $Z(0) = 0$, the solution (1.3.15) reduces to

$$Z(t) = Y(t) - m(t). \qquad (1.3.18)$$

In current literature (1.3.18) is referred to as reflection mapping. This term does not give credit to the pioneering 1958 paper by E. Reich, who derived (1.3.15) for the virtual waiting time in $M/G/1$. Furthermore, the identification of the idle time with the minimum functional does not follow from the reflection mapping.

2. The joint distribution of $Z(t)$ and $I(t)$ can be obtained directly from (1.3.12). For the compound Poisson input the older technique of analysis is based on the forward Kolmogorov integro-differential equation for the distribution of $Z(t)$. □

1.4   A MODEL FOR DATA COMMUNICATION SYSTEMS

A buffer of infinite capacity receives inputs of data represented as a Markov-additive process $\{X(t), J(t), t \geq 0\}$ on the state space $\mathbb{R}_+ \times \mathcal{E}$ in which the additive component is a compound Poisson process. Specifically

$$X(t) = X_0(t) + \int_0^t a \circ J(s)ds. \tag{1.4.19}$$

Here $X_0(t)$ is a compound Poisson process in which the rate at which jumps occur as well as the jump sizes depend on the state of the Markov process J on a countable state space $\mathcal{E}$, these jumps representing the arrivals of packets. In addition X has a drift that occurs at a rate $a(j)$ when J is in state j, and the integral in (1.4.19) represents the amount of data that arrive in a fluid fashion. The desired transmission (demand) rate is $d(j)$ when J is in state j and the transmission (release) policy is to meet the demand "if physically possible." Let $Z(t)$ denote the buffer content at time $t \geq 0$. The above assumptions lead to the integral equation

$$Z(t) = Z_0(t) + X(t) - \int_0^t r \circ (Z(s), J(s))ds \tag{1.4.20}$$

where the release rate r is given by

$$\begin{aligned} r(x,j) &= d(j) \text{ if } x > 0 \\ &= \min(d(j), a(j)) \text{ if } x = 0. \end{aligned} \tag{1.4.21}$$

Comparison with (1.3.10) show that (1.4.20) is indeed an extension of the (now classical) dam model. The presence of J is to be understood with reference to specific models. We first consider two special cases.

A fluid model for data communication. If the arrival of data is only in a fluid fashion, then $X_0(t) \equiv 0$ and the integral equation (1.4.20) reduces to

$$Z(t) = Z(0) + \int_0^t x \circ J(s)^+ ds - \int_0^t x \circ J(s)^- 1_{\{Z(s)>0\}} ds \tag{1.4.22}$$

where $x(j)$ is the net input rate

$$x(j) = a(j) - d(j). \square \tag{1.4.23}$$

A model with packet arrivals. In the presence of packet arrivals we need to assume that the desired transmission rate $d(j)$ exceeds the rate of fluid arrival $a(j)$. The integral equation (1.4.20) then reduces to

$$Z(t) = Z(0) + X_0(t) - \int_0^t d_1 \circ J(s)1_{\{Z(s)>0\}}ds \qquad (1.4.24)$$

where $d_1(j) = d(j) - a(j) > 0$. $\qquad\qquad\qquad\qquad\qquad\qquad\qquad$ □

The integral equation that describes each of the above models is of the form

$$Z(t) = Z(0) + X(t) - \int_0^t r \circ (Z(s), J(s))ds \qquad (1.4.25)$$

where $\{X(t), J(t)\}$ is a Markov-additive process and

$$r(x, j) = d(j)1_{\{x>0\}}. \qquad (1.4.26)$$

Comparing (1.4.25) with the integral equation (1.3.10) we see that the data communication models described here are extensions of the continuous time dam model of Section (1.3). The unique nonnegative solution of (1.4.25), modified as in (1.3.14), is formally the same as (1.3.15), where the net input $Y(t)$ given by

$$Y(t) = X(t) - \int_0^t d \circ J(s)ds \qquad (1.4.27)$$

and it should be noted that $\{Y(t), J(t)\}$ is a Markov-additive process.

The following are two fluid models that have been investigated in the literature. The presence of the Markov component J will be clear from these models.

a. **A multiple source data handling system.** There are N sources of messages, which are "on" or "off" from time to time. A switch receives messages at a unit rate from each source and transmits them at a fixed maximum rate c ($1 \leq N < \infty, 0 < c < \infty$). Messages that are not transmitted are stored in a buffer of infinite capacity (see Figure 1.1). Denoting by $J(t)$ the number of "on" sources at time $t \geq 0$, we assume that $\{J(t), t \geq 0\}$ is a birth and death process on the state space $\{0, 1, 2, \ldots, N\}$. Of interest is the buffer content $Z(t)$. It is seen that $Z(t)$ satisfies the integral equation

$$Z(t) = Z(0) + \int_0^t J(s)ds - \int_0^t r \circ (Z(s), J(s))ds \qquad (1.4.28)$$

where

$$\begin{aligned} r(x, j) &= c \text{ if } x > 0 \\ &= \min(j, c) \text{ if } x = 0. \end{aligned} \qquad (1.4.29)$$

Clearly, this is a fluid model with $a(j) = j$ and $d(j) = c$. □

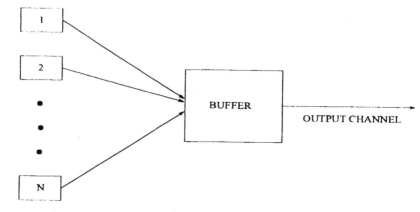

SOURCES ("ON" OR "OFF")

FIGURE 1.1 A buffer of infinite capacity for storage

b. An integrated circuit and packet switching multiplexer. A buffer of infinite capacity receives voice calls as well as data. There are $s+u$ output channels, of which u channels are reserved for data transmission, while the remaining s channels are shared by data and voice calls, with calls having preemptive priority over data and calls that find all s channels that serve them being lost (see Figure 1.2).

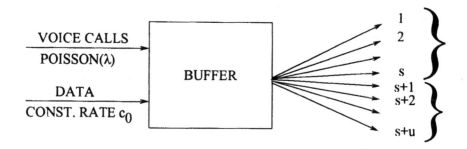

FIGURE 1.2 An integrated circuit and packet switching multiplexer

Voice calls arrive in a Poisson process and their service times have an exponential density. Data arrive continuously at a constant rate c_0 and are transmitted at a rate $c_1 (< c_0)$. At time $t \geq 0$, let $Z(t)$ denote the amount of data in the buffer and $J(t)$ the number of channels available for data transmission. It is clear that $s + u - J(t)$ represents the queue length in an $M/M/s$ loss system, and $Z(t)$ satisfies the integral equation

$$Z(t) = Z(0) + \int_0^t c_0 ds - \int_0^t r \circ (Z(s), J(s)) ds \qquad (1.4.30)$$

where
$$
\begin{aligned}
r(x,j) &= c_1 j \text{ if } x > 0 \\
&= \min(c_1 j, c_0) \text{ if } x = 0.
\end{aligned}
\qquad (1.4.31)
$$

This is a fluid model with $a(j) = c_0$ and $d(j) = c_1 j$.

Remarks.

1. Some authors take (1.3.18) as the starting point of their investigation of data communication models. Such an approach neglects the modeling aspects that are important in any area of applied probability. In particular it does not emphasize the role of Markov-additive inputs.

2. The forward Kolmogorov equation (in the matrix form) can be used to derive the joint distribution of $Z(t)$ and $J(t)$. However, as in the case of the dam model it is much more straightforward to derive the joint distribution of $Z(t), I(t)$ and $J(t)$ directly from (1.4.25), $I(t)$ being the amount of the unmet demand.

3. It is hoped that this brief survey has made it clear that all of the models described in Sections (1.3) and (1.4) are indeed *storage models*. The use of the term *fluid queue*, currently in fashion, is obviously based on lack of familiarity with earlier literature in this subject area. This term is both unnecessary and unpleasant, and the author hopes that discriminating researchers will not use it in the future.

REFERENCES

Prabhu, N. U. (1965). *Queues and Inventories: A Study of Their Basic Stochastic Processes*, John Wiley & Sons, New York.

Prabhu, N. U. (1980). *Stochastic Storage Processes*, Springer-Verlag, New York. (1998): 2nd Edition.

CHAPTER 2

MAXIMUM LIKELIHOOD ESTIMATION IN QUEUEING SYSTEMS

U. NARAYAN BHAT

Southern Methodist University, Dallas, TX

ISHWAR V. BASAWA

University of Georgia, Athens, GA

Abstract: This paper provides an overview of the literature on the use of the maximum likelihood method for estimating parameters in queueing models. Two cases, one when the system elements are fully observable and the second when only a limited amount of information is available are considered. The paper also includes some new results in later sections.

Keywords and phrases: Parameter estimation, maximum likelihood, GI/G/1 queue, M/G/1 queue, GI/M/1 queue, waiting time, queue length

2.1 INTRODUCTION

There are two key steps in the use of the method of maximum likelihood estimation (m.l.e.): constructing the likelihood function and deriving estimators that maximize the function. It was Clarke (1957) who first demonstrated that the likelihood function can be constructed for the underlying queue length process in the queueing system M/M/1 (Poisson arrivals, exponential service times and single servers) if one can describe its sample path as a realization of random events that can be described in terms of distributions. The general maximum likelihood theory for Markov processes, of which M/M/1 is a simple example, has been given by Billingsley (1961). Since then, researchers have explored ways of using this method to non-Markovian systems as well.

In stochastic models, many times factors such as system structure and cost may prevent full observation. In such cases, inference on system parameters will have to be made using other system characteristics. For instance, in a queueing system where embedded Markov chains can be identified, observations relative to those Markov chains can be used to estimate parameters. Goyal and Harris (1972) provides one of the first examples of this procedure.

In this paper we provide an overview of the use of maximum likelihood estimation in queueing systems under both cases of complete and incomplete information. In addition to describing some of the basic work on Markovian systems, we review research on non-Markovian systems when the processes are fully observable and when information only on certain characteristics is available. In the latter case some new results are also presented. The paper is arranged in eight sections. Parameter estimation in Markovian and non-Markovian systems is described in Sections 2.2 and 2.3 respectively. These procedures assume the availability of complete information on the system, although in continuous time, discrete state Markovian systems the set of sufficient statistics used is smaller than that we normally require for non-Markovian systems. Section 2.4 deals with estimation using the embedded Markov chains for the waiting time process and in Section 2.5, the procedure described in Section 2.4 is modified for system time (waiting time plus service time) instead of only waiting time. In Sections 2.6 and 2.7 the process considered is the number of customers in the system and the two sections deal with the queues M/G/1 and GI/M/1 respectively. Finally, Section 2.8 provides some concluding observations. Also Sections 2.5 and 2.7 include new results.

We do not plan to provide a long bibliography in this overview. Only those papers with major influence in the course of research are cited. For the general theory of inference on Markov processes Billingsley's book (1961) is an excellent reference. Basawa and Prakasa Rao (1980) and Karr (1991) provide the theory of inference on stochastic processes, in general. For inference on queues, Bhat *et al.* (1997) is a good reference which includes an extensive bibliography.

2.2   M.L.E. IN MARKOVIAN SYSTEMS

Any discussion of m.l.e. in Markovian queues has to start with the paper by Clarke (1957). Even though two earlier papers by Moran (1951, 1953) described a procedure to estimate the birth and death parameters in the simple birth-and-death process, it was Clarke who used the complete description of the sample path to construct the likelihood function.

Let the system be observed for a length of time t such that the time spent in a busy state is a preassigned value t_b. Let n_a, n_s, t_e represent the number of arrivals, number of service completions, and the time spent in the empty state, respectively, during $[0, t]$. Furthermore, let n_0 be the initial queue length. Also assume that the system is in the steady state. The likelihood function can be written as

$$L(\lambda, \mu) = \left(\frac{\lambda}{\mu}\right)^{n_0} \left(1 - \frac{\lambda}{\mu}\right) \lambda^{n_a} \mu^{n_s} e^{-(\lambda+\mu)t_b} e^{-\lambda t_e}, \qquad (2.2.1)$$

and the m.l.e.'s of λ and μ are found from the equations

$$\widehat{\lambda} = (\widehat{\mu} - \widehat{\lambda})(n_a + n_0 - \widehat{\lambda}t) \text{ and } \widehat{\lambda} = (\widehat{\lambda} - \widehat{\mu})(n_s - n_0 - \widehat{\mu}t_b). \quad (2.2.2)$$

Estimating $\widehat{\mu}$ from the second equation gives a quadratic in $\widehat{\lambda}$. Of the two solutions, any negative solution is rejected, and for the remaining values of $\widehat{\lambda}$, corresponding $\widehat{\mu}$ is obtained. Furthermore, any pair $(\widehat{\lambda}, \widehat{\mu})$ would be rejected for which $\widehat{\mu} \leq 0$ or $\widehat{\lambda}/\widehat{\mu} \geq 1$. If both solutions are valid, then the solution which maximizes the likelihood function is chosen.

For large $n_s - n_0$ Clarke gives a sample approximation for $\widehat{\lambda}$ and $\widehat{\mu}$ as

$$\widehat{\lambda} \simeq (n_a + n_0)/t, \ \widehat{\mu} \simeq (n_s - n_0)/t_b. \qquad (2.2.3)$$

The consistency of $\widehat{\lambda}$ and $\widehat{\mu}$ has been examined by Samaan and Tracy (1978) who could establish only a weak consistency for $\widehat{\lambda}$. If we ignore the initial queue size, the estimates of λ and μ are, respectively, n_a/t and n_s/t_b.

As noted by Cox (1965), specializing Billingsley's (1961) results, this procedure can be extended to the generalized birth-and-death models. The conditional likelihood function (ignoring the contribution of the initial state) is of the form

$$e^{-\Sigma(\lambda_i + \mu_i)t_i} \Pi \lambda_i^{n_{a_i}} \mu_i^{n_{s_i}}, \qquad (2.2.4)$$

where λ_i, μ_i are the rates of arrival and service compilations in state i, n_{a_i} and n_{s_i} are the numbers of arrivals and service completions in state i, and t_i is the total time spent in state i during the observation interval $(0, t]$.

For a finite state birth-death queue, ignoring the impact of the initial queue size, the m.l.e's of λ_i and μ_i are given by

$$\widehat{\lambda}_i = n_{a_i}/t_i \ (0 \leq i \leq M - 1), \qquad \widehat{\mu}_i = n_{s_i}/t_i \ (1 \leq i \leq M). \qquad (2.2.5)$$

The above results and similar estimates for parameters in M/M/s, M/M/∞, and machine interference problem have been given by Wolff (1965), where many details are provided. For an extension of these methods to a simple Markovian queueing network, commonly known as the Jackson network, see Thiruvaiyaru et al. (1991), where joint asymptotic normality of the estimators is also established. Also see, Beneš (1957) for a discussion of the set of sufficient statistics in similar problems, and Cox (1965), and Lilliefors (1966) for confidence intervals for estimates.

2.3   M.L.E. IN NON-MARKOVIAN SYSTEMS

In Markovian systems, due to the memoryless property of the exponential distribution data-collection gets simplified because of our ability to pool observations without losing information. In non-Markovian systems this is not the case and therefore the two cases, one with complete information and the second with incomplete information (which arises when the system cannot be observed fully), become relevant. In this section we cover two important papers by Basawa and Prabhu (1981, 1988) which assume the availability of complete information. Research on cases with incomplete information is discussed in later sections.

Basawa and Prabhu (1981) obtain the m.l.e.'s of parameters of the arrival and service time distributions with continuous densities $f(u; \theta)$ and $g(v; \phi)$, respectively. The sampling scheme is to observe the queue until the first n customers have departed from the system and the service times of these n customers, say $(v_1, v_2, \ldots, v_n)$. Let the nth departure epoch be D_n and observe the interarrival times of all customers who arrive during $(0, D_n]$, giving the interarrival sequence $(u_1, u_2, \ldots, u_{N_A})$, where $N_A = N_A(D_n) = \max\{k : u_1 + u_2 + \cdots + u_k \leq D_n\}$. Under this sampling scheme, the likelihood function is

$$L_n(f, g) = \left\{ \prod_{i=1}^{N_A} f(u_j; \theta) \right\} \left\{ \prod_{i=1}^{n} g(v_j; \phi) \right\} [1 - F(x_n; \theta)], \qquad (2.3.6)$$

where

$$x_n = x_n(D_n) = D_n - \sum_{j=1}^{N_A} u_j.$$

Since the factor $[1 - F(x_n; \theta)]$ causes difficulty in obtaining simple estimates, consider the alternative approximate likelihood function obtained by dropping the last terms in (2.3.6):

$$L_n^a(f, g) = \left\{ \prod_{i=1}^{N_A} f(u_j; \theta) \right\} \left\{ \prod_{i=1}^{n} g(v_j; \phi) \right\}. \tag{2.3.7}$$

If $\widehat{\theta}_n^a, \widehat{\phi}_n^a$ are the m.l.e.'s of θ and ϕ based on $L_n^a(f, g)$, they are solutions of the equations

$$\sum_{j=1}^{N_A} \frac{\partial}{\partial \theta} \log f(u_j; \theta) = 0, \ \sum_{j=1}^{n} \frac{\partial}{\partial \phi} \log g(v_j; \phi) = 0. \tag{2.3.8}$$

Basawa and Prabhu prove that $\widehat{\theta}_n^a, \widehat{\phi}_n^a$ are consistent estimators of θ and ϕ and that

$$\begin{bmatrix} \sqrt{n}(\widehat{\theta}_n^a - \theta) \\ \sqrt{n}(\widehat{\phi}_n^a - \phi) \end{bmatrix} \xrightarrow{D} N_2 \left\{ \begin{pmatrix} 0 \\ 0 \end{pmatrix}, \left\{ \begin{matrix} \sigma_\theta^2/\eta & 0 \\ 0 & \sigma_\phi^2 \end{matrix} \right\} \right\}, \tag{2.3.9}$$

where N_2 represents a bivariate normal density with

$$\sigma_\theta^2 = \left[E \left(\frac{\partial}{\partial \theta} \log f \right)^2 \right]^{-1}, \sigma_\phi^2 = \left[E \left(\frac{\partial}{\partial \phi} \log g \right)^2 \right]^{-1}, \tag{2.3.10}$$

$\eta = \max(1, \rho)$, and ρ being the traffic intensity.

Let $\widehat{\theta}_n$ and $\widehat{\phi}_n$ be the estimators based on the full likelihood function (2.3.6). It is seen that $\widehat{\phi}_n = \widehat{\phi}_n^a$, and $\widehat{\theta}_n$ differs from $\widehat{\theta}_n^a$, but it can be shown that $\widehat{\theta}_n$ and $\widehat{\theta}_n^a$ have the same limiting distributions whenever

$$\frac{1}{\sqrt{n}} \frac{\partial}{\partial \theta} \log(1 - F(x_n; \theta)) \xrightarrow{P} 0. \tag{2.3.11}$$

This condition is satisfied for Erlangian arrivals. For large samples, estimators of θ and ϕ can be determined from (2.3.8) at least numerically, if not in closed form. Using (2.3.9) confidence intervals for θ and ϕ can also be constructed. From a practical point of view, it is significant to note that the limit properties of these statistics are obtained without the assumption on the existence of equilibrium. Basawa and Prabhu also consider m.l.e.'s for arrival and service rates in the M/M/1 queue based on a sample function observed over a fixed interval $(0, t]$, as done by Wolff (1965), and obtain limit distributions of the m.l.e.'s without any restrictions on ρ.

In a subsequent paper, Basawa and Prabhu (1988) have provided a unified framework for the estimation problem described above where the observation period is $(0, T]$, with a suitable stopping time T. Four different stopping rules are considered. It is shown that the limit distribution does not depend on the particular stopping rule if a random norming is used. They assume that the interarrival and service time distributions belong to the class of non-negative exponential families. Basawa and Prabhu also derive similar results using a generalized linear model for interarrival and service time distributions.

An extension of these procedures to Jackson-type queueing networks with arrivals at each node following a renewal process and service times being arbitrary has been carried out by Thiruvaiyaru and Basawa (1996). As an illustration, the inter-arrival time and service time distributions are assumed to belong to two separate exponential families of distributions. Two sampling plans, one based on a realization over a fixed interval and the second with observations over a certain random interval are used.

2.4   M.L.E. FOR SINGLE SERVER QUEUES USING WAITING TIME DATA

In Sections 2.4–2.7 m.l.e. procedures are described when complete information on the systems under consideration is not available. This section uses waiting time data, Section 2.6 employs system time (waiting time plus service time) data and the following two sections use queue length data for estimation.

A maximum likelihood procedure for the estimation of parameters in a single server queueing system GI/G/1 was presented in a recent paper by Basawa, Bhat and Lund (1996) using information on waiting times $\{W_t\}$, $t = 1, 2, \ldots, n$ of n successive customers. Information is collected from each of n successive customers on the amount of time spent by them in waiting for service. Let W_t denote the waiting time of the tth customer. The waiting time process $\{W_t, t = 1, 2, \ldots\}$ satisfies the following well known equation:

$$W_{t+1} = \begin{cases} W_t + X_{t+1} & , \text{ if } W_t + X_{t+1} > 0 \\ 0 & , \text{ if } W_t + X_{t+1} \leq 0 \end{cases}$$
$$= \max(0, W_t + X_{t+1}) \tag{2.4.12}$$

where $X_t = V_{t-1} - U_t$, with V_t and U_t denoting, respectively, the service and inter-arrival times corresponding to the tth customer. It should be noted that $\{X_t\}$ is a sequence of independent and identically distributed random variables and X_{t+1} is independent of W_t. It is clear that $\{W_t\}$ is a Markov chain and its transition distribution function can be written as

$$P(W_{t+1} \leq y | W_t = w) = \begin{cases} F_x(y - w) & , \quad y \geq 0 \\ 0 & , \quad y < 0 \end{cases} \qquad (2.4.13)$$

where $F_x(\cdot)$ is the distribution of X_t. The transition distribution function has a discontinuity at 0. Define

$$\alpha(w) = 1 - F_x(-w). \qquad (2.4.14)$$

Then, for the transition density we have

$$p(y|w) = \begin{cases} 1 - \alpha(w) & y = 0 \\ f_x(y - w) & y > 0 \\ 0 & y < 0 \end{cases} \qquad (2.4.15)$$

Define the indicator function

$$Z_{t+1} = \begin{cases} 0 & \text{if } W_{t+1} = 0 \\ 1 & \text{if } W_{t+1} > 0 \end{cases} \qquad (2.4.16)$$

Using Z_{t+1}, for the transition density of W_t, we can write

$$p(W_{t+1}|W_t) = [1 - \alpha(W_t)]^{1-Z_{t+1}} [f_x(W_{t+1} - W_t)]^{Z_{t+1}}. \qquad (2.4.17)$$

The likelihood function based on the sample $(W_1, W_2, ..., W_n)$ is given by

$$L = p(W_1) \sum_{t=1}^{n-1} p(W_{t+1}|W_t). \qquad (2.4.18)$$

Let $\theta = (\theta_1, \theta_2, ..., \theta_r)'$ be the unknown parameter vector corresponding to the distribution of X_t. Basawa et al. (1996) show that estimates for θ can in fact be determined using the likelihood function (2.4.18) following the standard procedure. Basawa et al. also have established the consistency and the asymptotic normality of the estimators, and discussed issues pertaining to their efficiency.

2.5   M.L.E. USING SYSTEM TIME

The sampling plan used in the last section requires the knowledge of the amount of time customers spend in waiting for service. In practice, in many instances, it may not be as easy to determine the actual waiting time as it is to determine the total time spent by customers in the system; i.e., the waiting time plus service time. We shall call this characteristic system time.

Let Y_t be the system time corresponding to the tth customer. Based on its definition, we have

$$
\begin{aligned}
Y_{t+1} &= V_{t+1} + W_{t+1} \\
&= V_{t+1} + \text{Max}(0, W_t + V_t - U_{t+1}) \\
&= V_{t+1} + \text{Max}(0, Y_t - U_{t+1}) \quad (2.5.19)
\end{aligned}
$$

which can also be written in display form as

$$
Y_{t+1} = \begin{cases} Y_t - U_{t+1} + V_{t+1} & \text{if } Y_t - U_{t+1} > 0 \\ V_{t+1} & \text{if } Y_t - U_{t+1} \le 0 \end{cases} \quad (2.5.20)
$$

Incidentally, the continuous time analog $\{Y(t),\ t \ge 0\}$ of the process $\{Y_t,\ t = 1, 2, 3 \ldots\}$ was originally introduced by Prabhu (1964) in the context of queue GI/M/1. The process $Y(t)$ exhibits properties of duality with the virtual waiting time process $W(t)$ as defined by Takács (1955) and the graph of $Y(t)$ can be looked upon as a mirror image of the graph of $W(t)$ [see, Prabhu (1965, p. 102)].

Equation (2.3.9) shows that $\{Y_t\}$, $t = 1, 2, \ldots$ is a Markov process. We now proceed to derive the transition density corresponding to the Markov process $\{Y_t\}$. We have

$$
\begin{aligned}
P(Y_{t+1} &\le y_{t+1} | Y_t = y_t) \\
&= P(V_{t+1} \le y_{t+1}) P(U_{t+1} \ge y_t) \\
&\quad + \int_0^{y_t} P(V_{t+1} \le y_{t+1} - y_t + u) a(u) du, \quad (2.5.21)
\end{aligned}
$$

where $a(u)$ is the inter-arrival time density. The result in (2.5.21) follows readily from (2.5.20), considering the two possibilities: $U_{t+1} \ge Y_t$ and $U_{t+1} < Y_t$ and applying the addition law. The transition probability of Y_{t+1} given Y_t is then obtained by differentiating (2.5.21) with respect to y_{t+1}:

$$
\begin{aligned}
P(y_{t+1} | y_t) &= b(y_{t+1})(1 - A(y_t)) \\
&\quad + \int_0^{y_t} b(y_{t+1} - y_t + u) a(u) du, \quad (2.5.22)
\end{aligned}
$$

where $b(\cdot)$ and $A(\cdot)$ denote the density of service time V and the distribution function of inter-arrival time, U, respectively. The likelihood function based on $(Y_1, \ldots, Y_n)$ is then given by

$$L(\theta) = p(Y_1; \theta) \prod_{j=1}^{n-1} p(Y_{j+1}|Y_j; \theta), \qquad (2.5.23)$$

where θ is the parameter of interest, and $p(Y_1; \theta)$ is the initial density of Y_1. The ML estimator $\hat{\theta}$ is obtained as a solution of the equation

$$\frac{d \log L(\theta)}{d\theta} = 0. \qquad (2.5.24)$$

Since $\{W_t\}$ is an ergodic process (assuming that the traffic intensity $\rho < 1$), it follows that $\{Y_t\}$, $Y_t = W_t + V_t$, is also ergodic. The consistency and the asymptotic normality of the MLE, $\hat{\theta}$, can therefore be deduced as in Basawa et al. (1996).

2.6 M.L.E. IN M/G/1 USING QUEUE LENGTH DATA

In this and the next section, the sampling scheme used for collecting data includes only observing the number of customers in the system for a fixed length of time or some variation of it.

Consider the embedded Markov chain of the queue length in M/G/1, defined at departure epochs. Let Q_t be the number of customers in the system immediately after the tth departure. The process $\{Q_t, t = 0, 1, 2, \ldots\}$ is a Markov chain. Let $B(\cdot)$ be the service time distribution and the Poisson arrival rate be λ. If we denote by A_t, the number of arriving customers during the service period, we get the distribution of A_t as

$$P(A_t = j) = k_j = \int_0^\infty e^{-\lambda x} \frac{(\lambda x)^j}{j!} dB(x), \qquad j = 0, 1, 2, \ldots \quad (2.6.25)$$

It is well known that Q_t satisfies the relation

$$\begin{aligned}
Q_{t+1} &= \begin{cases} Q_t - 1 + A_{t+1} & \text{if } Q_t > 0 \\ A_{t+1} & \text{if } Q_t = 0 \end{cases} \\
&= \begin{cases} Q_t - 1 + A_{t+1} & \text{if } Q_t - 1 > 0 \\ A_{t+1} & \text{if } Q_t - 1 \leq 0 \end{cases} \qquad (2.6.26)
\end{aligned}$$

which is similar in structure to Eq. (2.5.20). For the transition probabilities of $\{Q_t\}$, we have

$$P(Q_{t+1} = i_{t+1}|Q_t = i_t) = \begin{cases} P(A_{t+1} = i_{t+1}) & \text{if } i_t = 0 \\ P(A_{t+1} = i_{t+1} - i_t + 1) & \text{if } i_t \geq 1. \end{cases} \quad (2.6.27)$$

Suppose the process is observed until the number of departures reaches a fixed value n. Now tracing the sample path of the process we may write down the likelihood function as

$$L(\theta) = p(Q_1; \theta) \prod_{t=1}^{n-1} p(Q_{t+1}|Q_t; \theta). \qquad (2.6.28)$$

Let n_{ij} be the number of transitions of Q_t from i to j on the sample path, and θ, the vector of parameters for which estimators are being sought. We get

$$\begin{aligned} \log L(\theta) &= \log P(Q_0 = i_0) \\ &+ \sum_{j=0}^{\infty} (n_{0j} + n_{1j}) \log k_j \\ &+ \sum_{i=2}^{\infty} \sum_{j=i-1}^{\infty} n_{ij} \log k_{j-i+1}. \end{aligned} \qquad (2.6.29)$$

Depending on the form of the service time distribution, an explicit expression for the likelihood function can be written down and maximized in the usual manner to determine maximum likelihood estimates. The same general formulation holds when the service times are dependent. Goyal and Harris (1972) consider two such systems: (i) service times are exponential but with different means when the queue size is 1 and when it is > 1, (ii) service times are exponential with means linearly dependent on the number of customers in the system ($\mu_t = t\mu$). They derive m.l.e.'s for utilization factors (arrival rate/service rate) in the case of these two systems when the effect of the initial queue length can or cannot be ignored. Depending on the complexity of likelihood functions to be maximized, some equations will have to be solved using numerical approximation methods.

Another approach to maximum likelihood estimation using embedded Markov chains is to observe only the number of arrivals during successive service periods. In particular, when the arrivals are Poisson and the service times are Erlangian, Harishchandra and Rao (1984) have constructed the likelihood function using the number of arrivals during successive service periods as the sample. In an M/E_k/1 queue, in which k is the shape parameter of the Erlangian distribution and ρ is the traffic intensity, let A_t denote the number of arrivals during the service of the $(t+1)$th customer. Then A_t has the negative binomial distribution given by

$$P(A_t = x) = f(x, \rho) = \binom{x + k - 1}{x} \left(\frac{\rho}{\rho + k}\right)^x \left(\frac{k}{\rho + k}\right)^k,$$
$$x = 0, 1, 2, \ldots . \qquad (2.6.30)$$

Suppose the system is observed only at departure epochs. Using equation (2.6.26), the queue length data can be easily converted into arrival data. Let $x_1, x_2, \ldots, x_n$ be the number of arrivals during the first n service times, respectively. The likelihood function for this sample is then

$$L(x_1, x_2, \ldots, x_n; \rho) = \prod_{i=1}^{n} \binom{x_i + k - 1}{x_i} \left(\frac{\rho}{\rho + k}\right)^{x_i} \left(\frac{k}{\rho + k}\right)^k. \qquad (2.6.31)$$

The maximum likelihood estimate of ρ is found to be $\hat{\rho} = \Sigma x_i / n$. This estimator is unbiased and consistent, since $E(\hat{\rho}) = \rho$ and $\text{Var}(\hat{\rho}) = \rho(\rho + k)/(kn)$. Furthermore, it turns out that $\hat{\rho}$ is also the minimum variance bound (MVB) estimator and therefore uniformly minimum variance unbiased estimator (UMVUE) of ρ. It can be shown that the probability distribution of X belongs to the one-parameter exponential family and hence $T = \Sigma x_i$ is a sufficient statistic for ρ. Finally, for large values of n,

$$\frac{1}{\sigma} \sqrt{n}(\hat{\rho} - \rho) \xrightarrow{D} N(0, 1), \qquad (2.6.32)$$

where

$$\sigma^2 = \left[E \left(\frac{\partial}{\partial \rho} \log f(x, \rho)\right)^2\right]^{-1} = \frac{\rho(\rho + k)}{k}. \qquad (2.6.33)$$

Even though, conceptually, estimating k using the likelihood function (2.6.31) is only a mathematical problem, due to the complexities of the expressions, the procedure does not become tractable. The results derived by Miller and Bhat (1997) overcome this problem by using a different approach.

Miller and Bhat use the number of customers served while the system has been busy for a specific length of time as the data element. In this formulation the service process, after eliminating idle times, resembles a renewal process. Consider the following two sampling plans for this renewal process.

Sampling Plan I: Assuming that the first observation period begins at time zero, observe the renewal process at time τ and record the number of renewals in $(0, \tau)$. To assure independent observations, the next observation period will begin when the next renewal occurs. Then after a period of τ time units, the number of renewals occurring in this second period is recorded. Wait until the next renewal occurs and the renewal epoch begins the following observation period, etc.

Sampling Plan II: Assuming the first observation period begins at time zero, observe the renewal process at time τ and record the number of renewals in $(0, \tau)$. Also record the time until the next renewal following time τ which will signal the start of a new observation period. Then after a period of τ time units, the number of renewals occurring in this second period is recorded. Record the time elapsed until the next renewal and the renewal epoch begins the following observations period, etc.

The second sampling plan uses the additional information on the waiting time to start the next observation.

Let N_1^τ, N_2^τ,.... denote the number of renewals (service completions) occurring in the observation periods, $1, 2, 3, \ldots$, respectively. In the second sampling plan the observations will be bivariate $\{(N_i^\tau, Y_i(\tau)), i = 1, 2, \ldots\}$ where $Y_i(\tau)$ is the excess life of the renewal period encountered at the ith observation. Using these observations, $\{N_i^\tau, i = 1, 2, \ldots\}$ with Sampling Plan I and $\{(N_i^\tau, Y_i(\tau))\}, i = 1, 2, \ldots\}$ with Sampling Plan II, Miller and Bhat construct likelihood functions which can be used to derive m.l.e. for k either assuming k to be continuous first and determining the best integer k from that result, or using the method of integer maximum likelihood estimation. As one would expect Sampling Plan II leads to better results in estimation.

2.7   M.L.E. IN GI/M/1 USING QUEUE LENGTH DATA

Consider the imbedded Markov chain $\{Q_t, t = 0, 1, 2, \ldots\}$ in a GI/M/1 queue in which arrivals from a renewal process and service times are exponential. Let Q_t represent the number of customers in the system just before the tth arrival. Let $A(\cdot)$ be the inter-arrival time distribution function and μ be the service rate so that the exponential service time density is given by $\mu e^{-\mu x} (x > 0)$. Define D_t as the number of potential departures during an inter-arrival period if an unlimited number of customers are available for service. The random variable D_t has the distribution

$$P(D_t = j) = \delta(j) = \int_0^\infty e^{-\mu x} \frac{(\mu x)^j}{j!} dA(x), \, j = 0, 1, 2, \ldots \quad (2.7.34)$$

It is well known that Q_t satisfies the relation

$$Q_{t+1} = \begin{cases} Q_{t+1} - D_t & , \quad textif\ Q_t + 1 - D_t > 0 \\ 0 & , \quad if\ Q_t + 1 - D_t \leq 0 \end{cases}. \qquad (2.7.35)$$

Let

$$X_{t+1} = 1 - D_t.$$

Then, (2.7.35) can be re-written in the form

$$Q_{t+1} = \begin{cases} Q_t + X_{t+1} & , \quad if\ Q_t + X_{t+1} > 0 \\ 0 & , \quad if\ Q_t + X_{t+1} \leq 0 \end{cases} \qquad (2.7.36)$$

which is similar in structure to Eq. (2.4.12).

From equation (2.7.36) we get

$$\begin{aligned} P(Q_{t+1}) = 0 | Q_t = i) &= P(i + X_{t+1} \leq 0) \\ &= P(D_t \geq i + 1) \\ &= \sum_{r=i+1}^{\infty} \delta(r) \\ &= 1 - \alpha(i) \qquad (2.7.37) \end{aligned}$$

where we have written $\sum_{r=0}^{i} \delta(r) = \alpha(i)$. Also

$$\begin{aligned} P(Q_{t+1} = j | Q_t = i) &= P(i + X_{t+1} = j) \\ &= P(D_t = i + 1 - j) \\ &= \delta(i + 1 - j), (j > 0). \qquad (2.7.38) \end{aligned}$$

Using the indicator function Z_t defined in (2.4.16), with W_t replaced by Q_t, we may write the transition probability as

$$p(Q_{t+1}|Q_t) = [1 - \alpha(Q_t)]^{1-Z_{t+1}} [\delta(Q_t + 1 - Q_{t+1})]^{Z_{t+1}} \qquad (2.7.39)$$

and the likelihood function as

$$L(\theta) = p(Q_1; \theta) \prod_{t=1}^{n-1} p(Q_{t+1}|Q_t; \theta). \qquad (2.7.40)$$

It should be noted that when estimating θ using maximization of (2.7.40), numerical methods maybe needed. For instance, when the inter-arrival time distribution is Erlangian with

$$dA(x) \;=\; e^{-k\lambda x}\frac{(k\lambda x)^{k-1}}{(k-1)!}k\lambda dx \qquad (x>0) \qquad (2.7.41)$$

$$\delta(r) \;=\; \int_0^\infty e^{-\mu x}\frac{(\mu x)^r}{r!}e^{-k\lambda x}\frac{(k\lambda x)^{k-1}}{(k-1)!}k\lambda dx$$

$$\;=\; \binom{r+k-1}{r}\left(\frac{\mu}{\mu+k\lambda}\right)^r\left(\frac{k\lambda}{\mu+k\lambda}\right)^k.$$

Even though $\delta(r)$ lends itself convenient for taking logarithms and differentiating, $\alpha(i)=\sum_{r=0}^i \delta(r)$ is not easily tractable in such operations. Then, direct maximization using numerical techniques is recommended.

If k is also an unknown parameter, methods using integer-maximum likelihood estimation will have to be incorporated in the process [see, Dahiya (1986) and Miller (1997)]. Another approach is to follow the procedure of Miller and Bhat (1977) described in Section 2.6. The arrival process is a renewal process and the estimation procedure proposed by Miller and Bhat gives m.l.e. for Erlang k of the arrival distribution.

In deriving Eq. (2.7.35), we note that D_t has been defined as the number of potential departures during an inter-arrival period. (It is the actual number when the system is busy throughout the period; otherwise it is the number of departures if there are an unlimited number of customers in the system). Consequently, the information available on $\{Q_t\}$ cannot be transformed into information on D_t completely as done for Eq. (2.6.30) in Section 2.6. Therefore, if one has to carry out inference based solely on queue length, the maximum likelihood method described above seems to be the best approach.

2.8 SOME OBSERVATIONS

From a review of research papers on the use of m.l.e. to estimate parameters of queueing models, it is clear that if one is interested in deriving simple readily usable results, a Markovian model is almost a necessity. Even when using information from an embedded chain in the queue M/G/1, the procedure leads to closed-form solutions only when the service time distribution is Erlangian. When likelihood function becomes complex, maximization can be accomplished only through numerical approximation methods. Therefore, in applications with non-Markovian models where easy numerical results are needed, regardless of the sophistication of the maximum likelihood procedure and the desirable properties possessed by the estimators resulting from it, we may not have any recourse but to use moment estimators.

However, with the increasing capability of computers one should be able to numerically maximize likelihood functions of increasing complexity. Alternatively, one could also use one-step maximum likelihood estimation starting with the moment estimator as the initial value [see, Lehmann (1983)].

REFERENCES

Basawa, I. V. and Bhat, B. R. (1992), Sequential inference for single server queues, In *Queueing and Related Models* (Eds., U. N. Bhat and I. V. Basawa), pp. 325–336, Oxford University Press, Oxford.

Basawa, I. V., Bhat, U. N. and Lund, R. (1996). Maximum likelihood estimation for single server queues from waiting time data, *Queueing Systems*, **24**, 155–167.

Basawa, I. V. and Prabhu, N. U. (1981). Estimation in single server queues, *Naval Research Logistics Quarterly*, **28**, 475–487.

Basawa, I. V. and Prabhu, N. U. (1988). Large sample inference from single server queues, *Queueing Systems*, **3**, 289–306.

Beneš, V. E. (1957). A sufficient set of statistics for a simple telephone exchange model, *Bell Systems Technical Journal*, **36**, 939–964.

Bhat, U. N., Miller, G. K. and Rao, S. S. (1997). Statistical analysis of queueing systems, *Frontiers in Queueing* (Ed., J. H. Dshalalow), Chapter 13, pp. 351–394.

Billingsley, P. (1961). *Statistical Inference for Markov Processes*, University Chicago Press, Chicago.

Clarke, A. B. (1957). Maximum likelihood estimate in a simple queue, *Annals of Mathematical Statistics*, **28**, 1036–1040.

Cox, D. R. (1965). Some problems of statistical analysis connected with congestion, *Proceedings of the Symposium on Congestion Theory* (Eds., W. L. Smith and W. B. Wilkinson), University of North Carolina at Chapel Hill, NC.

Dahiya, R. C. (1986). Integer-parameter estimation in discrete distributions, *Communications in Statistics—Theory and Methods*, **15**, 709–725.

Goyal, T. L. and Harris, C. M. (1972). Maximum likelihood estimation for queues with state dependent service, *Sankhyā, Series A*, **34**, 65–80.

Harishchandra, K. and Rao, S. S. (1984). Statistical inference about the traffic intensity parameter of $M/E_k/1$ and $E_k/M/1$ queues, *Report*, Indian Institute of Management, Bangalore, India.

Karr, A. F. (1991). *Point Processes and their Statistical Inference*, Second edition, Marcel Dekker, New York.

Lehmann, E. L. (1983). *Theory of Point Estimation*, John Wiley & Sons, New York.

Lilliefors, H. W. (1966). Some confidence intervals for queues, *Operations Research*, **14**, 723–727.

Miller, G. K. (1999). Maximum-likelihood estimation for the Erlang integer parameter, *Statistics & Probability Letters*, **43**, 335–341.

Miller, G. K. and Bhat, U. N. (1997). Estimation for renewal processes with unobservable gamma or Erlang inter-arrival times, *Journal of Statistical Planning and Inference*, **61**, 355–372.

Moran, P. A. P. (1951). Estimation methods for evolutive processes, *Journal of the Royal Statistical Society, Series B*, 141–146.

Moran, P. A. P. (1953). The estimation of the parameters of a birth and death process, *Journal of the Royal Statistical Society, Series B*, 241–245.

Prabhu, N. U. (1964). A waiting time process in the queue $GI/M/1$, *Acta Mathematica Acad. Sci. Hungary*, **15**, 363–371.

Prabhu, N. U. (1965). *Queues and Inventories*, John Wiley & Sons, New York.

Samaan, J. E. and Tracy, D. S. (1978). Properties of some estimators for the simple queue $M/M/1$, *Proceedings of the Third Symposium in Operations Research: Methods of Operations Research*, University of Mannheim, 377–387.

Takács, L. (1955). Investigation of waiting time problems by reduction to Markov processes, *Acta Mathematica Acad. Sci. Hungary*, **6**, 101–129.

Thiruvaiyaru, D. and Basawa, I. V. (1992). Empirical Bayes estimation for queueing systems and networks, *Queueing Systems*, **11**, 179–202.

Thiruvaiyaru, D. and Basawa, I. V. (1996), Maximum likelihood estimation for queueing networks, In *Stochastic Processes and Statistical Inference* (Eds., B. L. S. Prakasa Rao and B. R. Bhat), pp. 132–149, New Age International Publications, New Delhi.

Wolff, R. W. (1965), Problems of statistical inference for birth-and-death queueing models, *Operations Research*, **13**, 343–357.

CHAPTER 3

NUMERICAL EVALUATION OF STATE PROBABILITIES AT DIFFERENT EPOCHS IN MULTISERVER GI/Geom/m QUEUE

M. L. CHAUDHRY

Royal Military College of Canada, Kingston, Ontario, Canada

U. C. GUPTA

Indian Institute of Technology, Kharagpur, India

Abstract: In this paper, we analyze numerically a multiserver discrete-time queue with arbitrary interarrival and geometric service times. For completeness' sake, both early and late arrival models are considered. We first propose a way of evaluating arbitrary-epoch probabilities from those at prearrival epoch for the early arrival system. Then the results for the late arrival system with delayed access are derived. Outside observer's observation epoch probabilities are also discussed for both the models. Numerical results have been validated by computer simulation. It is hoped that the results obtained in this paper should be of interest to both specialists and practitioners of queueing theory.

Keywords and phrases: Queueing, discrete-time, multiserver

31

3.1 INTRODUCTION

The high-speed multi-access communication channels such as Broadband Integrated Services Digital Network (BISDN) are designed to support a wide range of services: transmission of video, voice and data signals. The Asynchronous Transfer Mode (ATM) is the first technology to merge video, voice and data into a common format and uses very short, fixed length packets called "cells." In all these systems, events (packet arrival and onward transmission of packets) occur only at a regularly spaced points in time. In the past, continuous-time queueing models have been used to evaluate performance measures of communication systems but due to recent changes in technology which is based on discrete time, they can only be used as approximations to real systems. In view of this, discrete time queueing models seem more appropriate. A detailed discussion and applications of discrete-time queues to telecommunication systems may be found in a recent book by Bruneel and Kim (1993).

Discrete-time queueing systems with a single server have been discussed extensively. However, very little seems to have been done on the corresponding multiserver queues. One of the earliest work in this direction was by Chan and Maa (1978). Using the imbedded Markov chain technique, they obtain only the distribution of number of customers in the system at a prearrival epoch. However, in many situations we need performance measures such as average queue length at other epochs, e.g. arbitrary and outside observer's observation epochs. In a recent paper, Chaudhry and Gupta (1997) develop relations among state probabilities at various epochs for two models: GI/Geom/m system with early arrivals (EAS) and GI/Geom/m system with late arrivals and delayed access (LAS-DA). They further show that, in the limiting case, these relations tend to the corresponding continuous-time results. It may be remarked here that using a recursive algorithm some results on the discrete-time GI/Geom/m/m queue have also been investigated by Chaudhry and Gupta (1999).

In this paper, we first consider GI/Geom/m queue with early arrival system (EAS). Some details on EAS may be found in Hunter (1983) or Chaudhry et al. (1996). The aim of this paper is to numerically evaluate the state probabilities at arbitrary epochs from those at prearrival epochs. Unfortunately, the direct substitution of prearrival epoch probabilities into equations (3.2.7) and (3.2.8) (see below) does not yield arbitrary epoch probabilities. Similar remarks also apply to the evaluation of outside observer's observation epoch probabilities though such probabilities can be easily obtained in the corresponding continuous-time multiserver GI/M/m queue, Takács (1962). It is shown later on how to resolve this in the discrete-time case.

Once the state probabilities at various epochs are evaluated for the GI/Geom/m queue with EAS, we derive similar results for the GI/Geom/m queue with LAS-DA. This is done by developing relation between prearrival epoch probabilities of EAS and LAS-DA. Subsequently, arbitrary epoch probabilities are evaluated for the LAS-DA GI/Geom/m queue using the relation developed by Chaudhry and Gupta (1997). Further, numerical results have also been validated by performing computer simulation experiments for both the EAS and LAS-DA systems.

3.2   MODEL AND SOLUTION: GI/Geom/m (EAS)

Though the GI/Geom/m queue with EAS is discussed in Chaudhry and Gupta (1997b), it is briefly described here again for the sake of completeness. We assume that the interarrival times are independent identically distributed (iid) random variables (rvs) having common probability mass function (pmf) $a_n = P(A=n)$, $n \geq 1$, probability generating function $A(z)$, and mean a. The transmission time S of each of the m servers is independent and geometrically distributed with distribution given by

$$P(S = n) = b_n = (1 - \mu)^{n-1}\mu, \quad 0 < \mu < 1, \quad n \geq 1.$$

Further, the probability that j customers are served given that there are i in the system is given by

$$c(j|i) = \binom{i}{j}\mu^j(1 - \mu)^{i-j}, \quad i=1,2,...,m, \quad j=0,1,...,i,$$

$$c(j|i) = \binom{m}{j}\mu^j(1 - \mu)^{m-j}, \quad i \geq m, \quad j=0,1,...,m,$$

with $c(0|0) = 1$ and $\binom{k}{r} = 0$, $r > k$ or $r < 0$.

Let the time axis be marked by $0, 1, 2, ..., t, ...$, and assume that the potential arrivals occur in $(t, t+)$ and the potential departures occur in $(t-, t)$. More specifically various time epochs at which events occur are depicted in Figure 3.1.

The state of the system just before a potential arrival is described by two variables: the number of customers in the system (N_t) and the remaining interarrival time for the next arrival (U_t). Let us define

$$Q_n(t, u) = P\{N_t = n, U_t = u\}, \quad u \geq 0$$
$$Q_n(u) = \lim_{t \to \infty} Q_n(t, u).$$

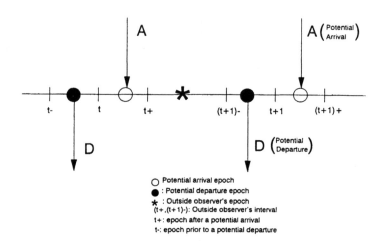

FIGURE 3.1 Various time epochs in early arrival system (EAS)

It follows that the marginal distribution Q_n is given by

$$Q_n = \sum_{u=0}^{\infty} Q_n(u).$$

In steady-state, assuming it exists ($\rho = 1/am\mu < 1$), we relate the states of the system at two consecutive epochs t and $t+1$, and get for $u \geq 1$

$$Q_0(u-1) = \sum_{j=0}^{m} Q_j(u)c(j|j) + a_u \sum_{j=0}^{m-1} Q_j(0)c(j+1|j+1)$$

$$(3.2.1)$$

and

$$Q_n(u-1) = \sum_{j=n}^{n+m} Q_j(u)c(j-n|j)$$

$$+ a_u \sum_{j=n-1}^{m+n-1} Q_j(0)c(j-n+1|j+1), \quad n \geq 1.$$

$$(3.2.2)$$

Let

$$Q_n^*(z) = \sum_{u=0}^{\infty} Q_n(u)z^u, \quad A(z) = \sum_{u=0}^{\infty} a_u z^u, \quad a_0 = 0.$$

Multiplying (3.2.1) and (3.2.2) by z^u, summing over $u = 1$ to ∞ and adjusting terms for $u = 0$, we obtain

$$[z - c(0|0)]Q_0^*(z) = \sum_{j=1}^{m} Q_j^*(z)c(j|j)$$

$$+ \sum_{j=0}^{m-1} Q_j(0)\{A(z)c(j+1|j+1) - c(j|j)\}$$

$$- Q_m(0)c(m|m) \qquad (3.2.3)$$

$$[z - c(0|n)]Q_n^*(z) = \sum_{j=n+1}^{n+m} Q_j^*(z)c(j-n|j) - \sum_{j=n}^{m+n} Q_j(0)c(j-n|j)$$

$$+ (z) \sum_{j=n-1}^{n-1+m} Q_j(0)c(j-n+1|j+15) \quad n \geq 1$$

$$(3.2.4)$$

Adding (3.2.3) and (3.2.4) over all possible values of n, we get, after simplification,

$$\sum_{n=0}^{\infty} Q_n^*(z) = \frac{A(z) - 1}{z - 1} \sum_{n=0}^{\infty} Q_n(0). \qquad (3.2.5)$$

Letting $z \to 1$ in (3.2.5) yields

$$\sum_{n=0}^{\infty} Q_n(0) = \frac{1}{a}. \qquad (3.2.6)$$

Equation (3.2.5) and (3.2.6) have intuitive and probabilistic interpretations. Whereas $\sum_{n=0}^{\infty} Q_n(0)$ represents the arrival rate of customers, $\sum_{n=0}^{\infty} Q_n^*(z)$ gives the transform of stationary residual interarrival time measured from a slot boundary. To obtain relations between distributions of numbers in system at prearrival epoch, $\{Q_n^-\}_0^{\infty}$, and arbitrary epoch, $\{Q_n\}_0^{\infty}$, we set $z = 1$ in (3.2.3) and (3.2.4) and use $Q_n^- = aQ_n(0)$. They are

$$\sum_{j=1}^{m} Q_j c(j|j) = \frac{1}{a}\left[\sum_{j=0}^{m} Q_j^- \{c(j|j) - c(j+1|j+1)\}\right] \qquad (3.2.7)$$

and

$$Q_n[1 - c(0|n)] - \sum_{j=n+1}^{n+m} Q_j c(j - n|j)$$

$$= \frac{1}{a}\left[\sum_{j=n-1}^{n+m} Q_j^-\{c(j - n + 1|j + 1) - c(j - n|j)\}\right], \quad n \geq 1$$

(3.2.8)

It appears from the above two expressions that once the distribution $\{Q_n^-\}_0^\infty$ is known, one can obtain $\{Q_n\}_0^\infty$. However, as stated earlier this is not straight forward. For details, see the next section.

The state probabilities $\{Q_n^-\}_0^\infty$ at a prearrival epoch can be obtained using Chan and Maa (1978). For easy reference and computational purposes, their main results have been reproduced below.

If the condition $\rho < 1$ is satisfied, then the limiting probability distribution, Q_n^-, is independent of the initial distribution and is given by

$$Q_n^- = \begin{cases} \sum_{r=n}^{m-1}(-1)^{r-n}\binom{r}{n}U_r & n = 0, 1, 2, ..., m - 1 \\ Dx^{n-m} & n \geq m \end{cases}$$

(3.2.9)

where

$$U_r = DB_r \sum_{j=r+1}^{m} \frac{H_j}{B_j(1 - A_j)}$$

$$D = 1\left/\left[\frac{1}{1 - x} + \sum_{j=1}^{m} \frac{H_j}{B_j(1 - A_j)}\right]\right.$$

$$H_j = \binom{m}{j} - \frac{(1 - \mu)^m}{x^{m+1}} \frac{(x - A_j)}{[(1 - \mu) + \mu x]^m - (1 - \mu)^j]} \sum_{r=1}^{m} \binom{m}{r}\left(\frac{\mu x}{1 - \mu}\right)^r$$

$$\times \sum_{k=m+1-r}^{m} x^k \binom{k}{j}$$

x is the solution of equation

$$x = A([(1 - \mu) + \mu x]^m) \equiv \sum_{k=1}^{\infty}[(1 - \mu) + \mu x]^{mk}a_k, \quad 0 < x < 1$$

and

$$B_j = \prod_{k=1}^{j} \frac{A_k}{1 - A_k}, \text{ and } A_j = \sum_{k=1}^{\infty}(1 - \mu)^{kj}a_k.$$

Because of the alternating signs in (3.2.9), the above expressions fail to give $\{Q_n^-\}_0^\infty$ completely for high values of m and ρ even if double precision is used. However, this problem can be resolved if we use MAPLE (1995) with extended precision. Further, we can also solve

$$\mathbf{P}\mathbf{q}^{-(i)} = \mathbf{q}^{-(i+1)}, \quad i \geq 0$$

iteratively, where $\mathbf{q}^{-(i)} = [Q_0^{-(i)}, Q_1^{-(i)}, ..., Q_{size}^{-(i)}]^T$ and $\mathbf{q}^{-(0)} = [1, 0, ..., 0]^T$. In this case, entries for the transition probability matrix (tpm) $\mathbf{P}$, see Chan and Maa (1978), are given below:

$$a_{ij} = \sum_{n=1}^{\infty} \binom{i+1}{j}((1-\mu)^n)^j(1-(1-\mu)^n)^{i+1-j}a_n,$$

$$0 \leq i < m \text{ and } 0 \leq j \leq i+1 \leq m$$

$$b_{ij} = \sum_{n=1}^{\infty}\left[\sum_{r=1}^{n}\sum_{s=j}^{m}\sum_{u=m-s+1}^{m}\binom{m(r-1)}{i+1-u-s}\mu^{i+1-u-s}\right.$$

$$\times (1-\mu)^{m(r-1)-(i+1-u-s)}\binom{m}{u}\mu^u(1-\mu)^{m-u}\binom{s}{j}$$

$$\left.\times (1-\mu)^{(n-r)j}(1-(1-\mu)^{n-r})^{s-j}\right]a_n,$$

$$j \leq m \leq i$$

$$c_{ij} = \sum_{n=1}^{\infty}\binom{nm}{i+1-j}\mu^{i+1-j}(1-\mu)^{mn-(i+1-j)}a_n,$$

$$i \geq m \text{ and } m+1 \leq j \leq i+1.$$

Also, for $i \geq 0$, the probability vector $\mathbf{q}^{-(i)}$ is iterated until the following condition is met

$$\sum_{n=0}^{size}|Q_n^{-(i+1)} - Q_n^{-(i)}| \leq 10^{-14}.$$

Though the iterative method is a bit slower, it is stable, even for inter-arrival distributions with infinite support such as geometric.

3.2.1   Evaluation of $\{Q_n\}_0^\infty$ from $\{Q_n^-\}_0^\infty$

In this section, we discuss the procedure for evaluating $\{Q_n\}_0^\infty$ from $\{Q_n^-\}_0^\infty$. First, we use (3.2.9) to get $\{Q_n^-\}_0^\infty$ as discussed earlier. Once prearrival

epoch probabilities are evaluated either by expression (3.2.9) or using iteration, we use (3.2.8) for the evaluation of $\{Q_n\}_1^\infty$. It may be pointed out
that equation (3.2.7) is not needed for the evaluation of the probabilities
$\{Q_n\}_0^\infty$. However, this equation can be used as a check on $\{Q_n^-\}_0^m$ and
$\{Q_n\}_1^m$. Let us denote the final prearrival epoch probabilities by the new
vector $\mathbf{q}^- = [Q_1^-, Q_2^-, ..., Q_{size}^-]^T$. Also, let $\mathbf{q} = [Q_1, Q_2, ..., Q_{size}]^T$ which
is yet to be evaluated. Re-arranging (3.2.8) by isolating the term Q_n yields

$$Q_n = \frac{1}{1 - c(0|n)}\left\{\sum_{j=n+1}^{n+m} Q_j c(j - n|j) - \frac{1}{a}\left[\sum_{j=n-1}^{n+m} Q_j^-\{c(j - n|j) - c(j - n + 1|j + 1)\}\right]\right\}, \qquad n \geq 1.$$

$$(3.2.10)$$

With $i \geq 0$ and using iteration on (3.2.10), the vector $\mathbf{q} = [Q_1, Q_2, ..., Q_{size}]^T$
can be obtained. Thus, we write

$$Q_n^{(i+1)} = \frac{1}{1 - c(0|n)}\left\{\sum_{j=n+1}^{n+m} Q_j^{(i)} c(j - n|j) - \frac{1}{a}\left[\sum_{j=n-1}^{n+m} Q_j^-\{c(j - n|j) - c(j - n + 1|j + 1)\}\right]\right\},$$

$$1 \leq n \leq (size - m)^+ \qquad (3.2.11)$$

and

$$Q_n^{(i+1)} = \frac{1}{1 - c(0|n)}\left\{\sum_{j=n+1}^{size} Q_j^{(i)} c(j - n|j) - \frac{1}{a}\left[\sum_{j=n-1}^{size} Q_j^-\{c(j - n|j) - c(j - n + 1|j + 1)\}\right]\right\},$$

$$(size - m)^+ < n \leq size \qquad (3.2.12)$$

where $Q_n^{(i)}$ are the probabilities evaluated from the i-th iteration with the
initial estimate $\mathbf{q}^{(0)} = [0, 0, ..., 0]^T$ and $(x)^+ = \max(0, x)$. Note that for
$size \leq m$ only (3.2.12) will be used. Further, note that Q_0 is not used in
(3.2.11) and (3.2.12), but can be obtained later using normalization. For
$i \geq 0$, the probability vector $\mathbf{q}^{(i)}$ is iterated until the following condition is

met

$$\sum_{n=1}^{size} |Q_n^{(i+1)} - Q_n^{(i)}| \leq 10^{-14}.$$

Finally, Q_0 is obtained from

$$Q_0 = 1 - \sum_{n=1}^{size} Q_n. \tag{3.2.13}$$

3.2.2   Outside observer's distribution

Since an outside observer's observation epoch falls in an interval after a potential arrival and before a potential departure, the probability Q_n^o that the outside observer sees n in the system can be obtained from Q_n using the following relation

$$Q_n = \sum_{j=n}^{n+m} Q_j^o c(j - n|j), \qquad n \geq 0 \tag{3.2.14}$$

which is obtained through probabilistic arguments. One may note that $c(i|j) = c(i|m)$ if $j \geq m$. Since we know $\{Q_n\}_0^\infty$, the vector $\{Q_n^o\}_0^\infty$ is obtained iteratively following the procedure discussed in the previous section.

3.3   GI/Geom/m (LAS-DA)

In this section, we obtain the state probabilities at various epochs in the case of LAS-DA by developing relations between state probabilities at pre-arrival epochs of EAS and LAS-DA. Again, for the sake of completeness, the GI/Geom/m queue with LAS-DA is described briefly and the relations are reproduced below. In this case, a potential customer arrives in $(t-, t)$ and a potential departure occurs in $(t, t+)$. More specifically, various time epochs at which events occur are depicted in Figure 3.2.

Here, the state of the system just before a potential arrival is described again by two variables: the number of customers in the system at $t-$ (N_{t-}) and the remaining interarrival time for the next arrival (U_{t-}). In what follows, the minus sign '$-$' after t is omitted for simplicity. Let us define

$$\begin{aligned} P_n(t, u) &= P\{N_t = n, U_t = u\}, \qquad u \geq 0 \\ P_n(u) &= \lim_{t \to \infty} P_n(t, u) \end{aligned}$$

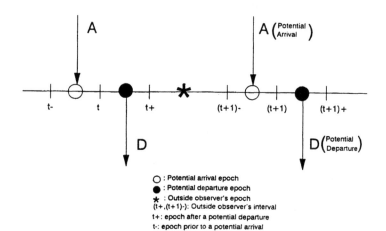

FIGURE 3.2 Various time epochs in late arrival system with delayed access
(LAS-DA)

It then follows that

$$P_n = \sum_{u=0}^{\infty} P_n(u).$$

We have, in steady state,

$$P_0(u-1) \;=\; \sum_{j=0}^{m} P_j(u)c(j|j) \tag{3.3.15}$$

and

$$P_n(u-1) \;=\; \sum_{j=n}^{n+m} P_j(u)c(j-n|j) + a_u \sum_{j=n-1}^{n+m-1} P_j(0)c(j-n+1|j),$$
$$n \geq 1 \tag{3.3.16}$$

Define

$$P_n^*(z) = \sum_{u=0}^{\infty} P_n(u)z^u, \quad A(z) = \sum_{u=0}^{\infty} a_u z^u, \quad a_0 = 0.$$

From (3.3.15) to (3.3.16) we obtain, respectively,

$$[z - c(0|0)]P_0^*(z) \;=\; \sum_{j=1}^{m} P_j^*(z)c(j|j) - \sum_{j=0}^{m} P_j(0)c(j|j) \qquad (3.3.17)$$

$$[z - c(0|n)]P_n^*(z) \;=\; \sum_{j=n+1}^{m+n} P_j^*(z)c(j-n|j)$$

$$+A(z) \sum_{j=n-1}^{n-1+m} P_j(0)c(j-n+1|j)$$

$$-\sum_{j=n}^{m+n} P_j(0)c(j-n|j) \quad n \geq 1. \qquad (3.3.18)$$

Adding equations (3.3.17) to (3.3.18) over all possible values of n, we get, after simplification,

$$(z-1)\sum_{n=0}^{\infty} P_n^*(z) = [A(z)-1]\sum_{n=0}^{\infty} P_n(0) \qquad (3.3.19)$$

Letting $z \to 1$ yields

$$\sum_{n=0}^{\infty} P_n(0) = \frac{1}{a}. \qquad (3.3.20)$$

Equations (3.3.19) and (3.3.20) may be interpreted as before. To obtain a relation between $\{P_n\}_n^{\infty}$ and $\{P_n^-\}_n^{\infty}$, we set $z = 1$ in (3.3.17) to (3.3.18) and use $P_n^- = aP_n(0)$. They are

$$\sum_{j=0}^{m} P_j^- c(j|j) = a\sum_{j=1}^{m} P_j c(j|j) \qquad (3.3.21)$$

$$P_n[1 - c(0|n)] - \sum_{j=n+1}^{n+m} P_j c(j-n|j)$$

$$= \frac{1}{a}\left[P_{n-1}^- c(0|n-1) + \sum_{j=n}^{n+m} P_j^- \{c(j-n+1|j) - c(j-n|j)\}\right],$$

$$n \geq 1. \qquad (3.3.22)$$

3.3.1 Evaluation of $\{P_n\}_0^\infty$ from $\{P_n^-\}_0^\infty$

It can be seen from (3.3.21) and (3.3.22) that to obtain P_n we first need P_n^- which can be obtained from the following relation between P_n^- and Q_n^-:

$$Q_n^- = \sum_{j=n}^{m+n} P_j^- c(j-n|j), \quad n \geq 0. \tag{3.3.23}$$

Since only departures occur between two arrivals for the two systems, the above relation between Q_n^- and P_n^- can be obtained by connecting probabilities at two prearrival epochs and using probabilistic arguments. Having known Q_n^-, P_n^- are obtained using iteration as discussed in Section 3.2.1.

Now once P_n^- is known we can obtain P_n using

$$P_n^{(i+1)}$$

$$= \frac{1}{1-c(0|n)}\left\{ \sum_{j=n+1}^{n+m} P_j^{(i)} c(j-n|j) \right.$$

$$\left. + \frac{1}{a}\left[P_{n-1}^- c(0|n-1) + \sum_{j=n}^{n+m} P_j^- \{c(j-n+1|j) - c(j-n|j)\} \right] \right\},$$

$$1 \leq n \leq (size - m)^+ \tag{3.3.24}$$

$$P_n^{(i+1)}$$

$$= \frac{1}{1-c(0|n)}\left\{ \sum_{j=n+1}^{size} P_j^{(i)} c(j-n|j) \right.$$

$$\left. + \frac{1}{a}\left[P_{n-1}^- c(0|n-1) + \sum_{j=n}^{size} P_j^- \{c(j-n+1|j) - c(j-n|j)\} \right] \right\},$$

$$(size - m)^+ < n \leq size \tag{3.3.25}$$

where $P_n^{(i)}$ are the probabilities calculated from the i-th iteration and $(x)^+ = \max(0, x)$. Note that for $size \leq m$ only (3.3.25) will be used. We use the initial estimate $\mathbf{p}^{(0)} = [0, 0, ..., 0]^T$. Further, note that P_0 is not used in (3.3.24) and (3.3.25), but can be obtained later using normalization.

3.3.2 Outside observer's distribution

In the case of LAS-DA, outside observer's observation epoch falls in a time interval after a potential departure and before a potential arrival, the probability that an outside observer sees n in the system, P_n^o is same as P_n, $\forall n \geq 0$.

3.4 NUMERICAL RESULTS

Numerical work has been performed using the procedures described in Sections 3.2.1 and 3.3.1. Further, numerical results have also been validated by performing computer simulation experiments. As expected, simulation took much more time to achieve desired level of accuracy. In other words, to get the same values which were obtained by an analytic method, one has to increase the number of trials which, in turn, increases time (in some cases it took 10 hours on a 486 PC). The results given in columns 2, 3, 4 and 5 of Table 3.1 for Geom/Geom/m queue have been obtained directly from the model equations derived independently, whereas the results in columns 6, 7, 8 and 9 were obtained from the procedure discussed in this paper and have been denoted by C&G. Finally, the results given in columns 10, 11, 12 and 13 were obtained using computer simulation method. It may be remarked here that since $P_n^o = Q_n^o = P_n$, no separate results have been reported for P_n^o and P_n. The results for D/Geom/m queue are obtained similarly and are given in Table 3.2. Numerical results for higher values of model parameters m and ρ are given in Table 3.3. For large values of m and/or ρ, numerical work can be done using the method discussed in this paper since Chan and Maa's procedure creates instability as stated earlier. Various measures such as average queue length at various epochs and average waiting time can be obtained in the usual way. One may note that since all the results reported here were rounded to four decimal places, the sum may not add to one in some cases.

A final remark may be in order. The method discussed here works even if we wish to get low probabilities such as $< 10^{-7}$. In this case, one only needs to increase the size of the vector $\mathbf{q}^-$ given in Section 3.2.1. In Tables 3.1, 3.2 and 3.3, the size was taken as 30, 30, 150, respectively.

Acknowledgements The authors are grateful to Mr. Haynes Lee, Research Assistant, RMC, for his help in the preparation of this paper. They are also thankful to ARP for supporting this research. The second author acknowledges with thanks the services provided by the Department of Mathematics and Computer Science, Royal Military College of Canada, Kingston, Ontario. The authors are grateful to an anonymous referee for his comments and suggestions which improved the paper significantly.

TABLE 3.1 Distributions of numbers in system, at various epochs, in the queueing system Geom/Geom/m with $\mu = 0.2$, $\lambda = 0.2$, $m = 5$, and $\rho = 0.2$

	Geom/Geom/m				C&G			
n	Q_n^-	Q_n	Q_n^o	P_n^-	Q_n^-	Q_n	Q_n^o	P_n^-
0	.4323	.4323	.3458	.3458	.4323	.4323	.3458	.3458
1	.3807	.3807	.3910	.3910	.3807	.3807	.3910	.3910
2	.1478	.1478	.1944	.1944	.1478	.1478	.1944	.1944
3	.0336	.0336	.0564	.0564	.0336	.0336	.0564	.0564
4	.0050	.0050	.0107	.0107	.0050	.0050	.0107	.0107
5	.0005	.0005	.0014	.0014	.0005	.0005	.0014	.0014
6	.0001	.0001	.0002	.0002	.0001	.0001	.0002	.0002
7	.0000	.0000	.0000	.0000	.0000	.0000	.0000	.0000

	simulation			
n	Q_n^-	Q_n	Q_n^o	P_n^-
0	.4323	.4323	.3458	.3458
1	.3807	.3807	.3910	.3910
2	.1478	.1478	.1944	.1944
3	.0336	.0336	.0564	.0564
4	..0050	.0050	.0107	.0107
5	..0005	.0005	.0014	.0014
6	.0001	.0001	.0002	.0002
7	..0000	.0000	.0000	.0000

TABLE 3.2 Distributions of numbers in system, at various epochs, in the queueing system D/Geom/m with $\mu = 0.2$, $a = 4$, $m = 5$, and $\rho = 0.25$

	C&G				simulation			
n	Q_n^-	Q_n	Q_n^o	P_n^-	Q_n^-	Q_n	Q_n^o	P_n^-
0	.4361	.2734	.1644	.3319	.4361	.2734	.1644	.3319
1	.4438	.4864	.4845	.4878	.4438	.4864	.4845	.4878
2	.1107	.2087	.2920	.1623	.1107	.2087	.2920	.1623
3	.0092	.0299	.0553	.0173	.0092	.0292	.0553	.0173
4	.0003	.0016	.0038	.0007	.0003	.0016	.0038	.0007
5	.0000	.0000	.0001	.0000	.0000	.0000	.0001	.0000

TABLE 3.3 Distributions of numbers in system, at various epochs, in the queueing system D/Geom/m with $\mu = 0.016666$, $a = 4$, $m = 20$, and $\rho = 0.75$

n	Q_n^-	Q_n	Q_n^o	P_n^-
≤ 4	.0000	.0000	.0000	.0000
5	.0001	.0002	.0001	.0001
6	.0005	.0007	.0003	.0005
7	.0018	.0027	.0013	.0020
8	.0058	.0081	.0044	.0064
9	.0150	.0199	.0120	.0163
10	.0323	.0406	.0271	.0347
11	.0587	.0701	.0513	.0623
12	.0911	.1033	.0828	.0953
13	.1216	.1311	.1147	.1253
14	.1409	.1445	.1376	.1429
15	.1425	.1391	.1439	.1421
16	.1266	.1177	.1320	.1240
17	.0993	.0880	.1067	.0955
18	.0690	.0584	.0765	.0652
19	.0428	.0346	.0487	.0397
20	.0239	.0187	.0279	.0218
21	.0129	.0101	.0151	.0118
22	.0070	.0055	.0082	.0064
23	.0038	.0030	.0044	.0035
24	.0020	.0016	.0024	.0019
25	.0011	.0009	.0013	.0010
26	.0006	.0005	.0007	.0005
27	.0003	.0003	.0004	.0003
28	.0002	.0001	.0002	.0002
29	.0001	.0001	.0001	.0001
30	.0001	.0000	.0001	.0000
≥ 31	.0000	.0000	.0000	.0000

REFERENCES

Bruneel, H. and Kim, B. G. (1993). *Discrete-Time Models for Communication Systems Including ATM*, Kluwer Academic Publishers, Boston.

Chan, W. C. and Maa, D. Y. (1978). The $GI/Geom/N$ queue in discrete time, *INFOR*, **16**(3), 232–252.

Chaudhry, M. L. and Gupta, U. C. (1999). Algorithmic discussions of distributions of numbers of busy channels for GI/Geom/m/m queues, *INFOR* (to appear).

Chaudhry, M. L. and Gupta, U. C. (1997). Relations among limiting distributions of the numbers of customers at different epochs in discrete- and continuous-time multiserver queues: GI/Geom/m and GI/M/m, Submitted for publication.

Chaudhry, M. L., Gupta, U. C. and Templeton, J. G. C. (1996). On the relations among the distributions at different epochs for discrete-time GI/Geom/1 queues, *Operations Research Letters*, **18**, 247–255.

Hunter, J. J. (1983). *Mathematical Techniques of Applied Probability, Volume II, Discrete Time Models: Techniques and Applications*, Academic Press, New York.

Maple (1995). *Maple V Release 4*, Waterloo Maple Software, 450 Philip Street, Waterloo, Ontario N2L 5J2, Canada.

Takács, L. (1962). *Introduction to the Theory of Queues*, Oxford University Press, New York.

CHAPTER 4

BUSY PERIOD ANALYSIS OF GI^{b}IM/1/N QUEUES — LATTICE PATH APPROACH

KANWAR SEN   MANJU AGARWAL

University of Delhi, Delhi, India

Abstract: Queuing theory literature reveals that steady state solutions are available for various types of infinite-space queuing models both Markovian and non-Markovian. Also that, while some work has been done to find transient solutions of finite/infinite Markovian queues, non-Markovian queues are not attempted much. However, transient solutions of non-Markovian finite bulk queues have not perhaps been attempted as yet. This study is an effort towards this direction and deals with the busy period analysis of GI^b/M/1/N queue. Via Lattice Paths Combinatorics (LPC), results are obtained in explicit computational form. The general interarrival time distribution is approximated by 2-phase Cox distribution C_2 that has Markovian property, amenable to Lattice Paths Combinatorics. The distribution C_2 covers a wide range of distributions that have square coefficient of variation lying in [1/2,). As such, the results obtained in this paper are applicable to a large class of real life situations. Some numerical results for the C_2^b/M/1/N model are also given.

Keywords and phrases: Lattice paths combinatorics, transient solutions, busy period density, non-Markovian queues, bulk queues

4.1 INTRODUCTION

Explicit closed form results for time dependent behaviour of non-Markovian bulk finite queues do not seem to be available in queuing literature though they are very much required in real life systems. This may be due to the inherent difficulties in analysing such systems. This paper aims at studying the queue $GI^b/M/1/N$ and provides busy period density in explicit closed form. For the purpose, lattice paths combinatorics (LPC) is used.

However, using LPC the transient analysis of finite queues GI/M/1/N and M/G/1/N have been carried out by Agarwal (2000) and Kanwar Sen (1999), respectively. In LPC analysis, the process is split up at suitable renewable epochs and thus can be represented by a LP. The general distributions involved have been approximated by 2-phase Cox distributions, C_2 [Cox (1955)], that have Markovian property amenable to LPC analysis. Same way, the busy period analysis of bulk queue $GI^b/M/1$ has also been carried out by Agarwal and Sen (1997). The results generalize those obtained by Sen and Gupta (1996a,b) for $M^b/M/1$.

The LPC method consists in providing transient solution through a discrete time analogue and a limiting process [Meisling (1958), Mohanty and Panny (1989) Bourn (1993)]. This method is found to be simple and elegant in studying Markovian queuing systems under different control policies, vacations and numerous other restrictions [see Kanwar Sen and Jain (1993), Kanwar Sen et al. (1993), Kanwar Sen and Gupta (1993,1994,1997)] as well as for non-Markovian queuing systems [Kanwar Sen and Agarwal (1997a,b,c, 1998)]. By using combinatorial methods involving LPs, transient solutions for M/M/1 queues have also been obtained by Mohanty and Panny (1990), Böhm (1993) and Böhm and Mohanty (1994a,b). However, whatever other transient solutions are available for non-Markovian queues, they are obtained by applying the much used so called top-to-bottom techniques [Böhm and Mohanty (1994a,b)] and thus are given either in terms of Laplace-Stieltjes transforms (LSTs) or other integral transforms [Takàs (1962), Benes (1963), Dalen and Natvig (1980), Neuts (1989), Takagi (1991,1993 a,b), Böhm (1993)]. As such, they are much complicated, intractable and hard to implement. This raises the question regarding the implementation of the models and as such one may have to be satisfied by getting their numerical solutions only [Grassman (1990)]. Lucantoni et al. (1994) and Logothetis et al. (1996) developed numerical computational algorithms for Batch Markovian Arrival Process (BMAP)/G/1 queues with infinite and finite waiting spaces, respectively, and took general distributions to be deterministic. Alfa (1982) considered time-inhomogeneous batch–server discrete time queuing model G/G/1/N for its transient behaviour. Mohanty (1991) studied the transient behaviour of a finite discrete time birth-death process. Recently, Mohanty (1996) surveyed briefly the work done on transient behaviour of discrete time queues.

The distribution C_2 consists of 2 independent exponential phases with arrival rate λ_j $(j = 1, 2)$ as shown below (Fig. 4.1). After phase 1 of arrival, the unit either enters phase 2 of arrival with probability α or joins the system for service with probability $\hat{a}(= 1 - \alpha)$.

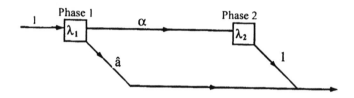

FIGURE 4.1 2-phase Cox distribution C_2

The distributions C_2 cover a wide range of distributions in terms of differing values of squared coefficient of variation, Marie (1978, 1980), Botta *et al.* (1987). As such the results obtained are applicable to a large class of real life situations.

4.2   THE GI^b/M/1/N MODEL

We assume that the system starts initially nonempty and has finite capacity N (assumed to be a multiple of b) including the one in service. The customers arrive in batches of size b and the service is done one by one. Interarrival time distribution of batches of customers is general, which is approximated by 2-phase Cox distribution C_2. Service time distribution is exponential. Therefore, as in Fig. 4.1,

λ_1: exponential interarrival rate in phase 1

λ_2: exponential interamval rate in phase 2

α: $P\{$a batch of b customers enters into phase 2 of arrival after completing phase 1 of arrival$\}$

$\hat{a}$: $P\{$a batch of customers joins the system for service after phase 1 of arrival$\}$ $(\alpha + \hat{a} = 1)$.

Let

μ: exponential service rate

i: number of batches of customers initially in the system.

4.3 LATTICE PATH APPROACH

To study the busy period distribution of the continuous queue $GI^b/M/1/N$ we first study the discretized system on segmenting the time interval, say, $(0, t]$ into a sequence of t/h (an integer) time slots, each of very small duration h (> 0). Obviously in a time slot only one of the following events takes place:

  (i) a batch of b customers joins the system (after either phase 1 or phase 2 of arrival)

 (ii) a customer departs from the system after getting service

(iii) a batch of b customers enters into phase 2 of arrival

(iv) none of these. This is termed as a stay.

Therefore, by Discretizing the system time, the sequence of events can be represented by a two dimensional LP representing, respectively, (see Fig. 4.3):

- an arrival of a batch after phase 1 by a horizontal step of length b units

- an arrival of a batch after phase 2 by a dotted horizontal step of length b units

- a departure by a vertical step of unit length

- entry of a batch into phase 2 of arrival by a diagonal step of length b units

- stay by a point.

It is obvious that, for the discretized model, system state at the end of any time slot is represented by a vertex (x, y) on a LP(x y and x i). To make understanding better, we first consider, as an example, a LP representing busy period of the server, Fig. 4.3. The server becomes free only at the vertex $B(y, y)$ when the LP touches the line $Y = X$ for the first time. Moreover, since no batch can join the system when it is full to its capacity N, obviously the LP cannot cross the line $Y = X - N$ and therefore is to lie between the lines $Y = X$ and $Y = X - N$. The points A_1, A_2, A_3, A_4, and A_5, where the LP touches the line $Y = X - N$, represent that the system is full to its capacity and hence would continue in this state until

  (i) either a departure, i.e. , a service completion takes place

(ii) or a batch enters phase 2 of arrival.

It may be noted that while at the point A_3 system state changes due to (ii) , it changes due to (i) at the other points A_1, A_2, A_4 and A_5.

Since for a fixed t, length of the busy period of the server, there can be more than one LP that touches the line $Y = X$ for the first time at the end of t/h time slots, to obtain busy period probability, we have to count the number of all possible LPs that lead the system to empty state and then associate the appropriate probabilities with the corresponding LPs and take their sum. Finally, on taking the limit as $h \to 0$, the desired continuous time transient results can be obtained [Meisling (1958), Mohanty and Panny (1989) and Böhm (1993)].

4.4 DISCRETIZED C_2^b/M/1/N MODEL

4.4.1 Transient Probabilities

According to the model assumptions, following *transitions* are possible in a time slot:

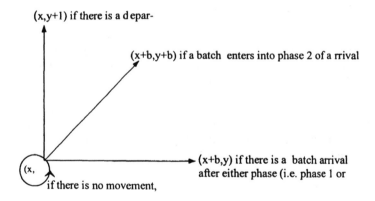

FIGURE 4.2 Possible transitions in a time

Therefore, we have the following *transition probabilities*:

(i) $P\{(x,y)(x+b,y)$ if arrival after phase $1, y < x < y + N - b + 1\} = \hat{a}\lambda_1 h + o(h)$

(ii) $P\{(x,y)(x+b,y)$ if arrival after phase $2, y < x < y + N - b + 1\} = \lambda_2 h + o(h)$

(iii) $P\{(x,y)(x+b,y+b)$, i.e., entry into phase 2 of arrival, $y < xy + N\} = \alpha\lambda_1 h + o(h)$

(iv) $P\{(x,y)(x,y+1)$, i.e., a departure, $y < xy + N\} = \mu h + o(h)$

(v) $P\{(x,y)(x,y)$ if a batch is in phase 1 of arrival, $y < x < y+N-b+l\} = 1 - (\lambda_1 + \mu)h + o(h)$

(vi) $P\{(x,y)(x,y)$ if a batch is in phase 2 of arrival, $y < zx < y + N - b + 1\} = 1 - (\lambda_2 + \mu)h + o(h)$

(vii) $P\{(x,y)(x,y)$ if a batch is in phase 1 of arrival, $y+N-b < xy+N\} = 1 - (\alpha\lambda_1 + \mu)h + o(h)$

(viii) $P\{(x,y)(x,y)$ if a batch is in phase 2 of arrival, $y+N-b < xy+N\} = 1 - \mu h + o(h)$.

Stays occurring in (v) to (viii) are called as *type 1, type 2, type 3* and *type 4*, respectively.

4.4.2   Counting of Lattice Paths

To see, in general, how to count the number of possible LPs, if in Fig. 4.3, all the diagonals are removed, then we have a LP having only horizontal steps (each of length b units), and vertical steps (each of unit length) as shown in Fig. 4.4. At this stage we define *Run* as:

Run: A sequence of consecutive horizontal (vertical) steps bounded on each side by a vertical (horizontal) step is called a horizontal (vertical) run. The sequence of horizontal steps starting from the origin followed by the first vertical as well as the sequence of vertical steps at the end following the last horizontal step are also called horizontal and vertical runs, respectively.

Now, Fig. 4.4, obtained from Fig. 4.3, does not contain any diagonals, therefore, it represents a very special case that all arrivals to the system take place after phase 1 on *ly*. But, since some arrivals (at the maximum all) could be after phase 2 as well, therefore, while counting the possible LPs, one has to think of all the different possibilities in which diagonals can be inserted into horizontal and vertical runs. Keeping in mind the distribution C_2, it is clear that we have to observe the following conditions:

(i) two or more consecutive diagonals should not occur,

(ii) in any horizontal run any number of diagonals may occur,

(iii) in any vertical run not more than one diagonal should occur,

(iv) the first horizontal step following a diagonal step has to be a dotted horizontal step,

(v) two or more consecutive dotted horizontal steps should not occur,

(vi) a dotted horizontal step should not be immediately preceded by a horizontal step,

(vii) from any vertex (x, y) such that $y + N - b < xy + N$, there should not be a horizontal step, only vertical steps or diagonal steps are possible.

The counting of LPs has to be done, therefore, keeping in view the above restrictions.

4.4.3 Notations

In a LP, let

$y - bk$: number of arrivals (including those initially in the system) as well as the number of departures in a busy period, obviously y will be a multiple of b (batch size),

r: number of horizontal runs and vertical runs, separately ($r \geq 1$),

bl_s: length of the sth horizontal run ($s = 1, 2, ..., r$),

L_s: length of the sth vertical run ($s = 1, 2, ..., r$),

$\boldsymbol{L}$: $(l_1, l_2, \ldots, l_r; L_1, L_2, \ldots, L_r)$,

k: total number of diagonals inserted in horizontal and vertical runs (k 0),

j: number of diagonals inserted, one each, in vertical runs, (0 j k),

$k - j$: number of diagonals inserted in horizontal runs,

$\boldsymbol{i}$: $(i_1, i_2, \ldots, i_s, \ldots, i_j)$, numbered vertical runs in which j diagonals are inserted (one each),

$\boldsymbol{L_i}$: $(L_{i_1}, L_{i_2}, \ldots, L_{i_s}, \ldots, L_{i_j})$, length of j vertical runs numbered i, in which the j diagonals are inserted,

$\boldsymbol{K_i}$: $(K_{i_1}, K_{i_2}, \ldots, K_{i_s}, \ldots, K_{i_j})$, respective distances from the lower end of the vertical runs at which j diagonals are inserted,

b_m: number of possible vertices where type m stays can occur ($m = 1, 2, 3, 4$),

c_m: number of type m stays ($m = 1, 2, 3, 4$),

$\boldsymbol{b}$: (b_1, b_2, b_3, b_4),

c: (c_1, c_2, c_3, c_4),

N: capacity of the system (assumed to be a multiple of b).

With the vertex (y, y) on the line $Y = X$, it is obvious that

\# of batches arrived during a busy period (number of horizontal steps each of length b) $= \frac{y}{b} - k - i$,

\# of departures during a busy period (number of vertical steps each of length 1) $= y - bk$,

\# of diagonals each of length b (number of batches entered into phase 2 of arrival) $= k$.

Therefore, the total number of transitions in a busy period is $= \frac{y}{b} - k - i + (y - bk) + k = \frac{b+1}{b}y - bk - i$.

Obviously, b_ms should satisfy the relation:

$$\sum_{m=1}^{4} b_m = \frac{b+1}{b}y - bk - i \quad \text{(excluding the end vertex } (y, y)\text{).}$$

To understand these notations we refer to Fig. 4.3,

$b = 4$, $i = l$, $N = 16$, $y = 68$, $r = 7$, $k = 6$, $j = 4$,

$L = (l_1, l_2, \ldots, l_7; L_1, L_2, \ldots, L_7) = (4, 1, 1, 1, 1, 2, 1; 4, 5, 3, 6, 8, 2, 16)$,

$i = (i_1, i_2, i_3, i_4) = (1, 4, 5, 7)$,

$L_i = (4, 6, 8, 16)$,

$K_i = (2, 0, 1, 6)$,

$b = (11, 23, 19, 7)$,

$$\sum_{m=1}^{4} b_m = \frac{b+1}{b}y - bk - i = 60.$$

The remaining 2 $(= k - j)$ diagonals are inserted in horizontal runs numbered 1 and 6.

To count the number of required LPs, therefore, we have

Theorem 4.4.1 *For fixed values of nonnegative integers i, b, y, N, k, j, r, L, i, K_i, b, c, let $LP^b_{(i,b,y,N,k,j,r,L,i,K_i,b,c)}$ denote the number of LPs from $A(bi, 0)$ to $B(y, y)$ remaining below the line $Y = X$ but not crossing the line $Y = X - N$, each comprising of k diagonals, of length b each, $\frac{y}{b} - k$ horizontal steps of length b each (including those from $(0,0)$ to $(bi, 0)$) and $y - bk$ vertical steps, such that*

(a) $\frac{y}{b} - k$ horizontal steps form r horizontal runs of lengths

$$bl_1, bl_2, \ldots, bl_r; l_1 \ i; l_2, l_3, \ldots, l_r > 0,$$

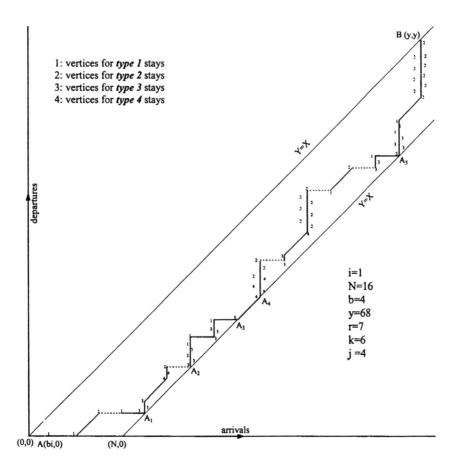

FIGURE 4.3 Busy period illustration

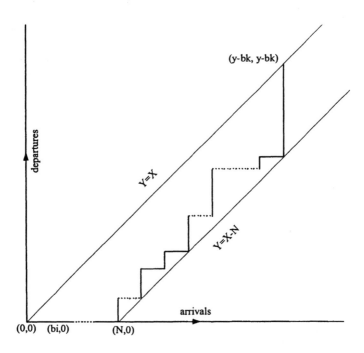

FIGURE 4.4 $C_2^b/M/1/N$ model. Lattice path ignoring the diagonals

Theorem 4.4.2 *(b)* $y - bk$ *vertical steps form* r *vertical runs of lengths* $L_1, L_2, \ldots, L_r$, *respectively,* $L_1, L_2, \ldots, L_r > 0$,
  (c) $\max(bi, L_1 + 1) \le bl_1 \le N$,

$$
\begin{aligned}
l \le l_u &\le \frac{N}{b} - 1, \qquad u = 2, 3, \ldots, r, \\
l \le L_u &\le N - 1, \qquad u = 1, 2, \ldots, r - 1, \\
b + 1 &\le L_r \le N, \\
b \sum_{u=1}^{v} l_u &> \sum_{u=1}^{v} L_u, \quad v = 2, 3, \ldots, r - 1, \quad and \\
b \sum_{u=1}^{r} l_u &= y - bk = \sum_{u=1}^{r} L_u,
\end{aligned}
$$

i.e.,

$$
\left\{ \boldsymbol{L} : (N \ge bl_1 \ge \max(bi, L_1 + 1)) \bigcap_{u=2}^{r} \left(l \le l_u \le \frac{N}{b} - 1 \right) \right.
$$

$$
\times \bigcap_{u=1}^{r-1} (1 \le L_u \le N - 1) \cap (b + 1 \le L_r \le N)
$$

$$
\left. \times \bigcap_{v=2}^{r-1} \left(b \sum_{u=1}^{v} l_u > \sum_{u=1}^{v} L_u \right) \cap \left(b \sum_{u=1}^{r} l_u = y - bk = \sum_{u=1}^{r} L_u \right) \right\},
$$

(These conditions ensure that the LP touches $Y = X$ *for the first time at* (y, y) *without crossing the line* $Y = X - N$),
  (d) j *diagonals are inserted one each into* j *vertical runs numbered* $\boldsymbol{i} = (i_1, i_2, \ldots, i_j)$, *respectively, of lengths* $\boldsymbol{L_i} = (L_{i_1}, L_{i_2}, \ldots, L_{i_j})$ *at distances* $\boldsymbol{K_i} = (K_{i_1}, K_{i_2}, \ldots, K_{i_j})$ *from bottom (including the vertices at both ends of the vertical runs except the vertex at the end of the last vertical run),*
  (e) the remaining $k - j$ *diagonals are inserted into horizontal runs; one or more diagonals can be inserted in any horizontal run except at the vertices at both ends of horizontal runs,*
  f) c_m *stays of type* m *can occur at* b_m *vertices (*$m = 1, 2, 3, 4$*), respectively.*
  Then for $r > 1$,

$$
LP^b_{i,y,N,k,j,r,\boldsymbol{L},\boldsymbol{i},\boldsymbol{K}_i,\boldsymbol{b},c)}
$$

$$
= \begin{cases} \binom{\frac{y}{b} - k - i - r}{k - j} \prod_{m=1}^{4} \binom{c_m + b_m - 1}{c_m}, & k > 0 \\ \binom{e_1 + d_1 - 1}{e_1} \binom{e_3 + d_3 - 1}{e_3}, & k = 0 \end{cases} \qquad (4.4.1)
$$

where, for $k > 0$,

$$b_1 = \left(\frac{b+1}{b}y - bk - i\right) - b_2 - b_3 - b_4, \qquad (4.4.2)$$

$$b_2 = \sum_{p \in (i_1, i_2, \ldots, i_j)} (L_p - K_p + 1) + (k - j)$$

$$- \sum_{p \in (i_1, i_2, \ldots, i_j)} V(b, N, K_{+p}, a_p) - \ddot{A}$$

$$\Delta = \begin{cases} 1, & \text{if } i_j = r \\ 0, & \text{otherwise ,} \end{cases} \qquad (4.4.3)$$

$$b_3 = \sum_{p \in (i_1, i_2, \ldots, i_j)} W(b, N, K_p, a_p) + \sum_{p \notin (i_1, i_2, \ldots, i_j)} U(b, N, a_p), \qquad (4.4.4)$$

$$b_4 = \sum_{p \in (i_1, i_2, \ldots, i_j)} V(b, N, K_p, a_p), \qquad (4.4.5)$$

$$a_p = b \sum_{s=1}^{p} l_s - \sum_{s=1}^{p-1} L_s, \qquad p = 1, 2, \ldots, r \qquad (4.4.6)$$

$$U(b, N, a_p) = \begin{cases} b - N + a_p, & \text{if } a_p = N - b + 1, \ldots, N, \\ 0, & \text{otherwise ,} \end{cases}$$
$$p \notin (i_1, i_2, \ldots, i_j), \qquad (4.4.7)$$

$$V(b, N, K_p, a_p)$$
$$= \begin{cases} b - K_p - N + a_p, & \text{if } a_p = N - b + 1, \ldots, N \\ & \text{and } K_p + N - a_p < b, \\ 0, & \text{otherwise,} \end{cases}$$
$$p \in (i_1, i_2, \ldots, i_j) \qquad (4.4.8)$$

$$W(b, N, K_p, a_p)$$
$$= \begin{cases} K_p + 1, & \text{if } a_p = N - b + 1, \ldots, N \\ & \text{and } K_p + N - a_p < b \\ b - N + a_p, & \text{if } a_p = N - b + 1, \ldots, N \\ & \text{and } K_p + N - a_p \geq b \\ 0, & \text{otherwise,} \end{cases}$$
$$p \in (i_1, i_2, \ldots, i_j) \qquad (4.4.9)$$

and for $k = 0$, defining

$$
\begin{aligned}
d_n &= \quad \textit{number of possible vertices for type n stays to occur } n = 1, 3, \\
e_n &= \quad \textit{number of type n stays, } n = 1, 3,
\end{aligned}
$$

we have

$$d_1 = \frac{b+1}{b}y - i - d_3, \qquad (4.4.10)$$

$$d_3 = \sum_{p=1}^{r} D(b, N, l_p, a_p) \qquad (4.4.11)$$

with

$$
D(b, N, l_p, a_p) = \begin{cases} b - N + a_p, & \textit{if } a_p = N - b + 1, \ldots, N, \\ 0, & \textit{otherwise} \end{cases}
$$
$$\textit{for } p = 1, 2, \ldots, r. \qquad (4.4.12)$$

PROOF To prove (4.4.1), it is obvious that fixed values of nonnegative integers $i, b, y, N, k, j, r, L, i, K_i$ would lead to only one unique LP with only j diagonals (each of length b) inserted, one each, in vertical runs. However, insertion of the remaining $k - j$ diagonals along the r horizontal runs as well as the distribution of different types of stays (c_ms and e_ns) into the corresponding vertices (b_ms. and d_ns) will generate the required number of LPs. Therefore the crux of the theorem is to compute the values of b_m ($m = 1, 2, 3, 4$) and d_n ($n = 1, 3$).

First we compute b_4, the number of possible vertices for *type 4* stays to occur. It is equal to the number of vertices of the type (x, y) ($x - Nyx - N + b - 1$) following the j diagonals inserted, one each, in j vertical runs numbered $i_1, i_2, \ldots, i_j$) of lengths $(L_{i_1}, L_{i_2}, \ldots, L_{i_j}$ at distances $K_{i_1}, K_{i_2}, \ldots, K_{i_j}$, respectively, from their lower end points (see Fig. 4.3). Therefore, by defining a_p, and the indicator function $V(b, N, K_p, a_p)$ as in (4.4.6) and (4.4.8), respectively, we get b_4 as given in (4.4.5).

For computing the value of b_3, it is observed that b_3 is the number of vertices of the type(x, y) ($x - Nyx - N + b - 1$) lying on the vertical runs following the r horizontal runs but preceding the diagonal steps, if any, inserted in vertical runs (see Fig. 4.3). Therefore, by defining the indicator functions $U(b, N, a_p)$ and $W(b, N, K_p, a_p)$ as in (4.4.7), and (4.4.9), respectively, we get b_3 as in (4.4.4).

b_2 clearly includes the end points of the $k - j$ diagonals inserted in the horizontal runs and the vertices of the type(x, y) ($x - N + b - 1 < y < x$) following the j diagonals inserted in vertical runs but preceding the subsequent horizontal steps, (see Fig. 4.3). In case $i_j = r$, then since there can not be any stays at (y, y), the value of b_2 is reduced by 1. This explains $\Delta = \begin{cases} 1, & \text{when } i_j = r \\ 0, & \text{otherwise.} \end{cases}$ Therefore b_2 is obtained as in (4.4.3).

Lastly b_1 follows from (A), i.e.,

$$b_1 = \left(\frac{b+1}{b}y - bk - i\right) - (b_2 + b_3 + b_4)$$

which is (4.4.2).

If $k = 0$ (see Fig. 4.4); then arrivals of all batches take place after phase 1 only and, only two types of stays are possible, i.e., *type 1* and *type 3*. The indicator function $D(b, N, l_p, a_p)$ which is defined in (4.4.12) on the lines of indicator function $U(b, N, a_p)$ in (4.4.7) is self explanatory. The values of d_1 and d_3 in (4.4.10)and (4.4.11), respectively, then follow easily.

As regards for the case $k > 0$, the insertion of $k - j$ diagonal steps, each of length b, into r horizontal runs, these can be inserted into any $k - j$ vertices out of the only $\frac{y}{b} - k - i - r$ vertices available along the horizontal runs in $\binom{\frac{y}{b}-k-i-r}{k-j}$ ways.

Finally, by identifying stays with balls and vertices with cells, and the n using the formula $\binom{e+f-1}{e}$ of the number of ways of distributing e similar balls into f cells [Feller (1985)], we get (4.4.1) for $k > 0$ and $k = 0$, respectively.   □

4.5   BUSY PERIOD PROBABILITY FOR THE DISCRETIZED $C_2^b/M/1/N$ MODEL

Theorem 4.5.1 *If $f_{i,N}^b(\frac{t}{h})$ denotes the probability that the busy period is of length $\frac{t}{h}$ time slots for the discretized $C_2^b/M/1/N$ system starting initially with bi units (i.e., with i batches), then*

$$f_{i,N}^b\left(\frac{t}{h}\right)$$

$$= \sum_{(R,R_2,R_3,R_1)} \binom{e_1 + d_1 - 1}{e_1}\binom{e_3 + d_3 - 1}{e_3}(\mu h)^y (\beta\lambda_1 h)^{\frac{y}{b}-i}$$

$$\times (1 - (\lambda_1 + \mu)h)^{e_1}(1 - (\alpha\lambda_1 + \mu)h)^{e_3}$$

$$+ \sum_{(R',R_2,\ldots,R_{10})} \binom{\frac{y}{b}-k-i-r}{k-j}\prod_{m=1}^{4}\binom{c_m + b_m - 1}{c_m}$$

$$\times (\mu h)^{y-bk}(\alpha\lambda_1 h)^k(\beta\lambda_1 h)^{\frac{y}{b}-2k-i+\Delta}(\lambda_2 h)^{k-\Delta}$$

$$\times(1 - (\lambda_1 + \mu)h)^{c_1}(1 - (\lambda_2 + \mu)h)^{c_2}$$
$$\times(1 - (\alpha\lambda_1 + \mu)h)^{c_3}(1 - \mu h)^{c_4} + o(h), \qquad (4.5.13)$$

where

$$e_3 = \frac{\tau}{h}, \qquad e_1 = \frac{t - \tau}{h} - \left(\frac{b+1}{b}y - i\right),$$

$$c_2 = \frac{t_2}{h}, \qquad c_3 = \frac{t_3}{h}, \qquad c_4 = \frac{t_4}{h},$$

$$c_1 = \frac{t - \sum_{m=2}^{4} t_m}{h} - \left(\frac{b+1}{b}y - bk - i\right)$$

and the summations are defined as

$R:$ $\quad \left\{ y : i \le \frac{y}{b} \le \left[\frac{1}{b+1}\left(\frac{t}{h} + i\right)\right]\right\}, \quad y$ *is a multiple of* b,

$R':$ $\quad \left\{ y : i \le \frac{y}{b} \le \left[\frac{1}{b+2}\left(\frac{2t}{n} - i(b-2) + b\right)\right]\right\},$

$R_1:$ $\quad \left\{ \frac{\tau}{h} : \frac{\tau}{h} = 0, 1, \ldots, \frac{t}{h} - \left(\frac{b+1}{b}y - i\right)\right\},$

$R_2:$ $\quad \{r : 1 \le r \le y - i + 1\},$

$R_3:$ $\quad \left\{ \boldsymbol{L} : (\max(bi, L_1 + 1) \le bl_1 \le N) \bigcap_{u=2}^{r} \left(1 \le l_u \le \frac{N}{b} - 1\right)\right.$

$\qquad \times \displaystyle\bigcap_{u=1}^{r-1}(1 \le L_u \le N - 1) \cap (b + 1 \le L_r \le N)$

$\qquad \left. \times \displaystyle\bigcap_{v=2}^{r-1}\left(b\sum_{u=1}^{v} l_u > \sum_{u=1}^{v} L_u\right) \cap \left(b\sum_{u=1}^{r} l_u = y - bk = \sum_{u=1}^{r} L_u\right)\right\},$

$R_4:$ $\quad \{k : 1 \le k \le c\},$

$\qquad c = \begin{cases} \frac{y}{b} - i - r + 1, & \text{if } i - r + 1 \le \frac{y}{b} \le i + 2r - 1 \\ r + \left[\frac{1}{2}\left(\frac{y}{b} - i - 2r + 1\right)\right], & \text{if } \frac{y}{b} \ge i + 2r - 1, \end{cases}$

$R_5:$ $\quad \left\{ j : \max\left(0, 2k + i + r - 1 - \frac{y}{b}\right) \le j \le \min(r, k)\right\},$

$R_6:$ $\quad \{\boldsymbol{i} : (i_1, i_2, \ldots, i_j) : 1 \le i_1 < i_2 < \cdots < i_j \le r\},$

$R_7:$ $\quad \left\{ \boldsymbol{K_i} : (K_{i_1}, K_{i_2}, \ldots, K_{i_j}), \ 0 \le K_{i_s} \le L_{i_s},\right.$

$\qquad \left. s = 1, 2, \ldots, j - 1, \ 0 \le K_{i_j} \le L_{i_j} - \Delta, \ \Delta = \begin{cases} 1, & \text{if } i_j = r \\ 0, & \text{otherwise} \end{cases}\right\},$

$$R_8: \quad \left\{ \frac{t_2}{h} : 0 \le \frac{t_2}{h} \le \frac{t}{h} - \left(\frac{b+1}{b} y - bk - i \right) \right\},$$

$$R_9: \quad \left\{ \frac{t_3}{h} : 0 \le \frac{t_3}{h} \le \frac{t - t_2}{h} - \left(\frac{b+1}{b} y - bk - i \right) \right\},$$

$$R_{10}: \quad \left\{ \frac{t_4}{h} : 0 \le \frac{t_4}{h} \le \frac{t - t_2 - t_3}{h} - \left(\frac{b+1}{b} y - bk - i \right) \right\},$$

PROOF For computing the $f_{i,N}^b(\frac{t}{n})$ when the total number of time slots is given to be $\frac{t}{h}$, we have to consider both the cases $k = 0$ and $k > 0$, and hence (4.5.13) consists of corresponding two terms. For the case $k = 0$, let $e_e = \frac{\tau}{h}$ be the number of time slots in which the system has type 3 stays. The total number of transitions in $\frac{t}{h}$ time slots is $\left(\frac{b+1}{b} y - i \right)$ therefore, the number of type 1 stays is

$$e_1 = \frac{t - \tau}{h} - \left(\frac{b+1}{b} y - i \right).$$

Using the transition probabilities given in Section 4.4.1, the probability of occurrence of:

(i) e_1 type 1 stays is $(1 - (\lambda_1 + \mu)h)^{e_1} + o(h)$

(ii) e_3 type 3 stays is $(1 - (\alpha \lambda_1 + \mu)h)^{e_3} + o(h)$

(iii) y departures is $(\mu h)^y + o(h)$

(iv) arrivals of $\frac{y}{b} - i$ batches is $(\beta \lambda_1 h)^{\frac{y}{b} - i} + o(h)$.

Thus multiplying the number of stipulated LPs from (2) by the above transition probabilities and then summing over (R, R_3, R_2, R_1) we get the first term in (4.5.13). For the case $k > 0$, let $c_m = \frac{t_m}{h}$ be the number of time slots out of the total $\frac{t}{h}$ time slots in which the system has type m stay $(m = 2, 3, 4)$. Since, the number of transitions in the remaining $\left(\frac{t - \sum_{m=2}^{4} t_m}{h} \right)$ time slots is $\left(\frac{b+1}{b} y - bk - i \right)$, we get the number of type 1 stays c_1, equal to

$$\frac{t - \sum_{m=2}^{4} t_m}{h} - \left(\frac{b+1}{b} y - bk - i \right).$$

Using the transition probabilities given in Section 4.4.1, we multiply by stipulated number of LPs given in (4.4.1) and then take the sum over $(R', R_2, \ldots, R_{10})$. This gives the second term in (4.5.13). $\qquad \square$

4.6 CONTINUOUS $C_2^b/M/1/N$ MODEL

On using a limiting process as h 0 in (4.5.13) [Meisling (1958), Mohanty and Panny (1989) and Böhm (1993)], we obtain the expression for the busy period density function as given in the following.

Theorem 4.6.1 *The probability density function of the busy period for $C_2^b 2b/M/1/N$ system starting initially with bi units (i.e., i batches) is given by*

$$f_{i,N}^b(t)$$

$$= e^{-(\lambda_1+\mu)t} \sum_{(R,R_2,R_3)} \sum_{p=0}^{\infty} \mu^y (\beta\lambda_1)^{\frac{y}{b}-i} t^{p+\frac{b+1}{b}y-i-1}$$

$$\times \frac{(\lambda_1(1-\alpha))^p \Gamma(p+d_3)}{p!\Gamma(d_3)\Gamma(p+\frac{b+1}{b}y-i)}$$

$$+e^{-(\lambda_1+\mu)t} \sum_{(R',R_2,R_3,R_4,R_5,R_6,R_7)} \sum_{p_1=0}^{\infty}\sum_{p_2=0}^{\infty}\sum_{p_3=0}^{\infty} \binom{\frac{y}{b}-k-i-r}{k-j}$$

$$\times \mu^{u-bk}(\alpha\lambda_1\lambda_2)^k(\beta\lambda_1)^{\frac{y}{b}-2k-i+\Delta}\lambda_1^{p_3}\lambda_2^{-\Delta}(\lambda_1(1-\alpha))^{p_2}(\lambda_1-\lambda_2)^{p_1}$$

$$\times \frac{t^{p_1+p_2+p_3+b_1+b_2+b_3+b_4-1}\Gamma(p_1+b_2)\Gamma(p_2+b_3)\Gamma(p_3+b_4)}{p_1!p_2!p_3!\Gamma(b_2)\Gamma(b_3)\Gamma(b_4)\Gamma(p_1+p_2+p_3+b_1+b_2+b_3+b_4)}.$$

$$(4.6.14)$$

PROOF On taking limit as h 0, (4.5.13) leads to

$$\lim_{h\to 0} f_{i,N}^b\left(\frac{t}{h}\right)$$

$$= f_{i,N}^b(t)dt$$

$$= e^{-(\ddot{e}_1+i)t} \sum_{(R,R_2,R_3)} \frac{i^y(\hat{a}\lambda_1)^{\frac{y}{b}-i}}{\tilde{A}(d_1)\tilde{A}(d_3)} \int_{\tau=0}^{t} \tau^{d_3-1}(t-\tau)^{d_1-1}d^{\ddot{e}_1\tau(t-\hat{a})}d\tau dt$$

$$+e^{-(\lambda_1+\mu)t} \sum_{(R',R_2,R_3,R_4,R_5,R_6,R_7)} \binom{\frac{y}{b}-k-i-r}{k-j}$$

$$\times \frac{\mu^{y-bk}(\alpha\lambda_1\lambda_2)^k(\beta\lambda_1)^{\frac{y}{b}-2k-i+\Delta}\lambda_2^{-\Delta}}{\Gamma(b_1)\Gamma(b_2)\Gamma(b_3)\Gamma(b_4)}$$

$$\times \int_{t_2=0}^{t}\int_{t_3=0}^{t-t_2}\int_{t_4=0}^{t-t_2-t_3} t_2^{b_2-1}t_3^{b_3-1}t_4^{b_4-1}(t-t_2-t_3-t_4)^{b_1-1}$$

$$\times e^{(\lambda_1-\lambda_2)t_2}e^{\lambda_1 t_3(1-\alpha)}e^{\lambda_1 t_4}dt_2 dt_3 dt_4 dt$$

$$(4.6.15)$$

which, on simplification yields (4.6.14).   □

4.7   PARTICULAR CASES

(i) $M^b/M/1/N$ model

Taking $\alpha = 0$, $\hat{a} = 1$, $\lambda_1 = \lambda$, $\lambda_2 = 0$, (4.6.14) yields the busy period density function for $M^b/M/1/N$ queue, i.e.,

$$f_{i,N}^b(t) = e^{-(\lambda_\mu)t} \sum_{(R,R_2,R_3)} \sum_{p=0}^{\infty} \frac{\Gamma(p+d_3)}{p!\Gamma(d_3)\Gamma(b+\frac{b+1}{b}y-i)}$$
$$\times \mu^y \lambda^{\frac{y}{b}-i+p} t^{p+\frac{b+1}{b}y-i-1} \tag{4.7.16}$$

Further, when N ($\Rightarrow d_3 = 0$, $p = 0$), (4.7.16) becomes

$$\lim_{n \to \infty} f_{i,N}^b(t) = f_i^b(t)$$

$$= e^{-(\lambda+\mu)t} \sum_{(R,R_2,R_3)} \frac{\lambda^{\frac{y}{b}-i}\mu^y t^{\frac{b+1}{b}y-i-1}}{\Gamma(\frac{b+1}{b}y-i)}$$

$$= e^{-(\lambda+\mu)t} \sum_{R} \frac{\lambda^{\frac{y}{b}-i}\mu^y t^{\frac{b+1}{b}y-i-1}}{\Gamma(\frac{b+1}{b}y-i)} \sum_{(R_2,R_3)} (l)$$

where $\sum_{(R_2,R_3)}(l) = \sum_{r=1}^{\frac{y}{b}-i+1} \sum_{R_3}(l) = \#$ of LPs from $(bi,0)$ to (y,y) touching $Y = X$ for the first time at (y,y) with horizontal steps each of length b and vertical steps, each of unit length. Therefore

$$\sum_{(R_2,R_3)} (l) = \frac{bi}{\frac{b+1}{b}y-i}\binom{\frac{b+1}{b}y-i}{\frac{y}{b}-i}, \tag{4.7.17}$$

[see Mohanty (1979)]. Then

$$f_i^b(t) = e^{-(\lambda+\mu)t} \sum_{y=bi}^{\infty} \frac{bi\lambda^{\frac{y}{b}-i}\mu^y t^{\frac{b+1}{b}y-i-1}}{(\frac{y}{b}-i)!y!}$$

$$= \frac{bi}{t}e^{-(\lambda+\mu)t}\left(\frac{\lambda}{mu}\right)^{-bi/2}\sum_{s=0}^{\infty}\frac{(\sqrt{\lambda\mu}t)^{(b+1)s+bi}}{s!b(s+i)!}\left(\frac{\lambda}{\mu}\right)^{-\frac{1}{2}(b-1)s},$$
$$\tag{4.7.18}$$

gives the busy period density of the $M^b/M/I$ queue.

For $b = 1$, (4.7.18) further reduces to

$$f_i(t) = \frac{i}{t}e^{-(\lambda+\mu)t}\left(\frac{\lambda}{\mu}\right)^{-i/2} I_i(2\sqrt{\lambda\mu}t), \qquad (4.7.19)$$

where

$$I_i(2a) = \sum_{s=0}^{\infty} \frac{a^{2s+i}}{s!(s_i)!}$$

is the modified Bessel function.

Eq. (4.7.19) gives the busy period density of the M/M/1 queue [see Saaty (1961) and Kanwar Sen and Jain (1993)].

(ii) C_2^b/M/1 model

Taking N ($\Rightarrow p_1 = p_2 = p_3 = 0$, $b_2 = b_4 = 0$, $p = 0$ and $d_3 = 0$) in (4.6.14) we get, as in case (i),

$$\begin{aligned}
f_i^b(t) &= \frac{bi}{t}e^{-(\ddot{e}_1+i)t}\left(\frac{\beta\lambda_1}{\mu}\right)^{-\frac{bi}{2}}\\
&\times \sum_{s=0}^{\infty} \frac{(\sqrt{\bar{a}\lambda_1\mu}t)^{(b+1)s+bi}}{s!b(s+i)!}\left(\frac{\beta\lambda_1}{\mu}\right)^{-\frac{(b-1)s}{2}}\\
&+ e^{-(\lambda_1+\mu)t}\sum_{(R',R_2,R_3,R_4,R_5,R_6,R_7)}\binom{\frac{y}{b}-k-i-r}{k-j}\\
&\times \frac{\mu^{y-bk}(\alpha\lambda_1\lambda_2)^k(\beta\lambda_1)^{\frac{y}{b}-2k-i+\Delta}\lambda_2^{-\Delta}}{\Gamma(b_1+b_2)}t^{b_1+b_2-1}. \qquad (4.7.20)
\end{aligned}$$

Eq. (4.7.20) gives the busy period density function for GI^b/M/1 system [Agarwal and Kanwar Sen (1997)].

4.8 NUMERICAL COMPUTATIONS AND COMMENTS

Numerical examples and graphs give insight into the effect of varying the parameter values. In view of this and even otherwise to test our results, numerical computations have been performed in Fortran 77 in double precision on PC Pentium-III for different sets of values of the parameters involved for busy period probability, $f_{i,N}^b(\frac{t}{h})$ given in (4.5.13), of the discretized model C_2^b/M/1/N. The computations could be performed in a short time since the program deals mainly with multiple summations. From illustration point of view some results are given in Tables 4.1–4.7 along with their corresponding graphs in Figs. 4.5–4.11.

It can be observed that in all the Tables, in general, for a given set of parameter values, the probabilities increase up to a certain value of busy period t/h, and then start decreasing, satisfying the expected normal pattern, thus justifying our results. Table 4.1 (Fig. 4.5) contains busy period probability for $b = 2, 3, 4$, when $h = 0.02$, $i = 1$, $N = 5$, $\alpha = 0.6$ ($\hat{\alpha} = 0.4$), $\lambda_1 = 3$, $\lambda_2 = 2$, $\mu = 5$. We note that probabilities decrease as b increases. Interestingly, it can also be noted that there are zeros at places when $b > t/h$ as it should. Besides, 1st row in each table also contains all zeros since for $b = 2$, busy period cannot tenninate in $t/h = 1$ time slot.

Table 4.2 (Fig. 4.6) gives behaviour of probabilities for different values of $\alpha = 0.0, 0.2, \ldots, 1.0$. The values for $\alpha = 0.0$ and $\alpha = 1.0$ correspond to busy period probabilities for $M^b/M/1/N$ and $E_2^b/M/1/N$ models, respectively. The probabilities are increasing w.r.t. α since when the probability of customers entering phase 2 of arrival increases the busy period should terminate early. But the reverse pattern should hold w.r.t. λ_1 as well as λ_2 which one can see in Tables 4.3 and 4.4 (Fig. 4.7 and 4.8). However, when repair rate μ increases, busy period terminates faster and so probabilities increase, as they should, see Table 4.5 (Fig. 4.9). In Table 4.6 (Fig. 4.10) we see behaviour of busy period probabilities w.r.t. i, the initial number of batches present in the system. Obviously when i increases, busy period probabilities decrease. In Table 4.7 (Fig. 4.11), the behaviour of busy period probabilities can be observed w.r.t. N. It can be noticed that probabilities remain equal for $N > t/h$ which is obvious otherwise too. Also it is noted in Table 4.7 that the differences in probabilities are very very small for N 5. Therefore the corresponding graphs in Fig. 4.11 have overlapped.

Further, in Table 4.4 it is interesting to note that λ_2 has no effect up to $t/h = 3$ when $i = 1$, $b = 2$ and $N = 5$. It is so since for $t/h = 2$, the probability is $(\mu h)^2$ and when $t/h = 3$, a fresh batch of size 2 can at most enter phase 2 of arrival but can not join the system. Hence λ_2 does not occur in the value for busy period probability. The expression for probability is $2(\mu h)^2(1 - (\lambda_1 + \mu)h) + 2(\mu h)^2(\alpha \lambda_1 h)$. Also in Tables 4.2, 4.3 and 4.4, row 2 (for $t/h = 2$) is constant containing the value 0.0 100 since this probability is $(\mu h)^2$ and is independent of α, λ_1 and λ_2 respectively. The numerical computations can similarly be done for other expressions.

Acknowledgements The authors thank the referees for their useful comments and suggestions, which led to improvements of the paper. The authors also thank Mr. A. K. Goswami, Research Scholar, Department of

Statistics, University of Delhi, Delhi, 10007 for his help in making the computations. Research Project Supported by Department of Science and Technology, Government of India, New Delhi, India, (DST/MS/049/96).

REFERENCES

Agarwal, M. (2000). Distribution of number served during a busy period of GI/M/1/N queues - Lattice path approach, To appear in the *Proceedings of the 4th International Conference on Lattice Paths Combinatorics and Applications*, University of Vienna, Austria, July 8 - 10, 1998.

Agarwal, M. and Kanwar Sen (1997). Lattice paths combinatorics applied to transient busy period analysis of bulk queues GI^b/M/1 communicated (presented at the *International Conference on Combinatorics, Information Theory and Statistics*, University of South Maine, Portland, July 18-20,1997).

Alfa, A. S. (1982). Time-inhomogeneous bulk server queue in discrete time: A transportation type problem, *Operations Research*, **30**, 650–658.

Benes, V. E.(1963). *General Stochastic Processes in the Theory of Queues*, Addison-Wesley: Reading, Massachusetts.

Böhm, W. M. (1993). *Markovian Queuing Systems in Discrete Time*, Frankfurt am main: Antonhain.

Böhm, W. M. and Mohanty, S. G. (1994a). On discrete time Markovian N-policy queue involving batches, *Sankhyā, Series A*, **56**, 1–20.

Böhm, W. M. and Mohanty, S. G. (1994b). Transient analysis of queues with heterogeneous arrivals, *Queueing Systems*, **18**, 27–45.

Botta, R. F., Harris, C. M. and Marchal, W. G. (1987). Characterization of generalized hyperexponential distribution functions, *Communications in Statistics—Stochastic Models*, **3**, 115–148.

Cox, D. R. (1955). A use of complex probabilities in the theory of stochastic processes, *Proceedings of the Cambridge Philosophical Society*, **51**, 313–319.

Dalen, G. and Natvig, B. (1980). On the transient waiting times for a GI/M/1 priority queue, *Journal of Applied Probability*, **17**, 227–234.

Feller, W. (1985). *An Introduction to Probability Theory and its Applications*, Volume 1, Fourth edition, John Wiley & Sons, New York.

Grassman, W. K.(1990). Computational methods in probability theory, In *Handbooks in O.R. & M.S.*, Volume 2 (Eds., D. P. Heyman and M. J. Sobel), pp. 199–255.

Kanwar Sen (2000). Lattice path approach to transient analysis of M/G/1/N non-Markovian queues using Cox distributions, To appear in the *Proceedings of the 4th International Conference on Lattice Paths Combinatorics and Applications*, University of Vienna, Vienna, Austria, July 8-10, 1998.

Kanwar Sen and Agarwal, M. (1997a). Transient busy period analysis of initially non-empty M/G/1 queue-Lattice path approach, In *Advances in Combinatorial Methods and Applications to Probability and Statistics* (Ed., N. Balakrishnan), pp. 301–315, Birkhäuser, Boston.

Kanwar Sen and Agarwal, M. (1997b). Lattice path approach to transient busy period analysis of initially non-empty GI/M/1 queues using Cox distribution, In *Statistical Methods in Quality and Reliability* (Eds., N. Unnikrishnan Nair and P. G. Sankaran), pp. 32–49, EPD, New Delhi.

Kanwar Sen and Agarwal, M. (1997c). Transient queue length distribution of GI/M/1 queues via lattice paths combinatorics using Cox distribution, Submitted for publication.

Kanwar Sen and Agarwal, M. (1998). Lattice path approach to transient queue length distribution of M/G/1 queues using Cox distribution, *Proceedings of the International Conference on Stochastic Processes* (Ed., A. Krishnamoorthy), pp. 35–55, Department of Mathematics, Cochin University of Science and Technology.

Kanwar Sen and Gupta, R. (1993). M/M/1 T-Policy queues with service control - a study of transient behaviour, In *Recent Development in Probability and Statistics*, Narosa Publishing House: New Delhi.

Kanwar Sen and Gupta, R. (1994). Transient analysis of thresh old T-policy M/M/1 queue with server control, *Sankhyā, Series A*, **56**, 39–51.

Kanwar Sen and Gupta, R. (1996a). Transient solution of $M^b/M/1$ system under threshold control policies, *Journal of Statisticall Research*, **30**, 109–120.

Kanwar Sen and Gupta, R. (1996b). Transient solution of M/M/1 queues with batch arrival - A new approach, *Statistica*, anno **LVI, n. 3**, 333–343.

Kanwar Sen and Gupta, R. (1997). Discrete time Markovian batch arrival queues with restricted vacations and non-exhaustive service, *Sankhyā, Series B* (to appear).

Kanwar Sen and Jain, J. L. (1993). Combinatorial approach to Markovian queuing models, *Journal of Statistical Planning and Inference*, **34**, 269–279.

Kanwar Sen, Jain, J. L. and Gupta, J. M. (1993). Lattice path approach to transient solution of M/M/1 queue with (0,k) control policy, *Journal of Statistical Planning and Inference*, **34**, 259–267.

Logothetis, D., Mainkar, V. and Trivedi, K. S. (1996). Transient analysis of non-Markovian queues via Markov regenerative processes, In *Probability Models and Statistics—J. Medhi Festschrift*, 109–131, New Age International Publishers, New Delhi.

Lucantoni, D. M., Choudhury, G. L. and Whitt, W. (1994). The transient BMAP/G/1 queue, *Communications in Statistics—Stochastic Models*, **10**, 145–182.

Marie, R. (1978). Methods iteratives de resolution de models mathematiques de systems informatiques, *R.A.I.R.0. Informatique/Computer Services*, **12**, 107–122.

Marie, R. (1980). Calculating equilibrium probabilities for $\lambda^{(n)}/C_k/1/N$ queues, ACM SIGMETRICS, *Conference on Measurement and Modeling of Computer Systems*, 117–125.

Meisling, T. (1958). Discrete time queuing theory, *Journal Operations Research Society of America*, **6**, 96–105.

Mohanty, S. G. (1979). *Lattice Path Counting and Applications*, Academic Press, New York.

Mohanty, S. G. (1991). On the transient behaviour of a finite discrete time birth death process, *Assam Statistics Review*, **5**, 1–7.

Mohanty, S. G. (1996). Transient behaviour of discrete time queues: A survey, *Journal of Orissa Mathematical Society*, **12-15**, 141–148.

Mohanty, S. G. and Panny, W. (1989). A discrete time analogue of the M/M/1 queue and the transient solution: An analytic approach, *Collqia Mathematica Societatis Jnos Bolyai*, **57**, *Limit Theorems In Probability and Statistics*, 417–424.

Mohanty, S. G. and Panny, W. (1990). A discrete time analogue of the M/M/l queue and the transient solution: A geometric approach, *Sankhyā, Series A*, **52**, 364–370.

Neuts, M. F. (1989). *Structured stochastic Matrices of M/G/1 Type and Their Applications*, McGraw-Hill, New York.

Saaty, T. L. (1961). *Elements of Queueing Theory with Applications*, McGraw-Hill, New York.

Takàcs, L. (1962). *Introduction to Theory of Queues*, Oxford University Press, New York.

Takagi, H. (1991). *Queueing Analysis, 1: Vacation and Priority System*, North-Holland, Amsterdam.

Takagi, H. (1993a). *Queueing Analysis, 2: Finite Systems*, North-Holland, Amsterdam.

Takagi, H. (1993b). *Queueing Analysis, 3: Discrete Time Systems*, North-Holland, Amsterdam.

TABLE 4.1 Busy period probabilities for different values of b when $h = 0.02$, $i = 1$, $N = 5$, $\alpha = 0.6$, $\hat{a} = 0.4$, $\lambda_1 = 3$, $\lambda_2 = 2$, $\mu = 5$

t/h \ b	2	3	4
1	0.000000000	0.000000000	0.000000000
2	0.010000000	0.000000000	0.000000000
3	0.017520000	0.001000000	0.000000000
4	0.023004000	0.002652000	0.000100000
5	0.026834304	0.004686048	0.000357600
6	0.029337095	0.006896461	0.000799027
7	0.030788515	0.009130058	0.001427942
8	0.031420796	0.011276176	0.002232394
9	0.031428172	0.013258230	0.003190242
10	0.030972387	0.015026660	0.004273319

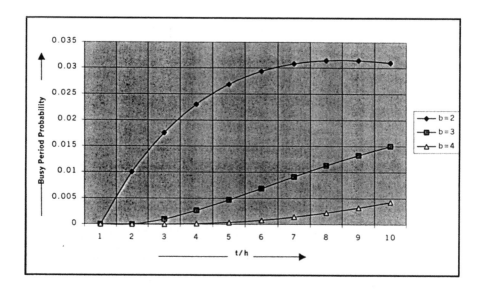

FIGURE 4.5

TABLE 4.2 Busy period probabilities for different values of α when
$h = 0.02$, $i = 1$, $b = 2$, $N = 5$, $\lambda_1 = 3$, $\lambda_2 = 2$, $\mu = 5$

t/h \ α	0.0	0.2	0.4	0.6	0.8	1.0
1	0.0000000000	0.0000000000	0.0000000000	0.0000000000	0.0000000000	0.0000000000
2	0.0100000000	0.0100000000	0.0100000000	0.0100000000	0.0100000000	0.0100000000
3	0.0168000000	0.0170400000	0.0172800000	0.0175200000	0.0177600000	0.0180000000
4	0.0211680000	0.0217800000	0.0223920000	0.0230040000	0.0236160000	0.0242280000
5	0.0237201600	0.0247582080	0.0257962560	0.0268343040	0.0278723520	0.0289104000
6	0.0249443280	0.0264086700	0.0278729260	0.0293370950	0.0308011780	0.0322651750
7	0.0252215605	0.0270776480	0.0289332990	0.0307885150	0.0326432990	0.0344976560
8	0.0248452821	0.0270384050	0.0292302360	0.0314207960	0.0336101010	0.0357981710
9	0.0240379712	0.0265042490	0.0289676280	0.0314281720	0.0338859430	0.0363410050
10	0.0229654226	0.0256398440	0.0283087770	0.0309723870	0.0336308340	0.0362842760

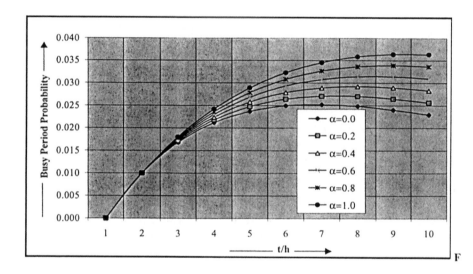

FIGURE 4.6

TABLE 4.3 Busy period probabilities for different values of λ_1 when $h = 0.02$, $i = 1$, $b = 2$, $N = 5$, $\alpha = 0.6$, $\hat{a} = 0.4$, $\lambda_2 = 2$, $\mu = 5$

t/h \ λ_1	1	3	5	7	9	11
1	0.000000000	0.000000000	0.000000000	0.000000000	0.000000000	0.000000000
2	0.010000000	0.010000000	0.010000000	0.010000000	0.010000000	0.010000000
3	0.017840000	0.017520000	0.017200000	0.016880000	0.016560000	0.016240000
4	0.023858400	0.023004000	0.022188000	0.021410400	0.020671200	0.019970400
5	0.028350464	0.026834304	0.025450240	0.024192128	0.023053824	0.022029184
6	0.031572703	0.029337095	0.027384536	0.025689378	0.024227204	0.022974826
7	0.033747039	0.030788515	0.028313731	0.026258730	0.024565544	0.023181898
8	0.035064715	0.031420796	0.028498286	0.026173600	0.024340072	0.022906321
9	0.035689912	0.031428172	0.028147559	0.025644040	0.023749828	0.022328260
10	0.035763072	0.030972387	0.027429585	0.024832361	0.022943635	0.021579597

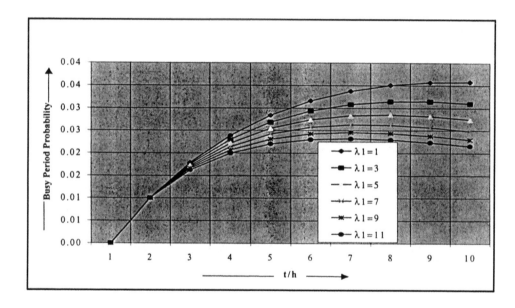

FIGURE 4.7

TABLE 4.4 Busy period probabilities for different values of λ_2 when $h = 0.02$, $i = 1$, $b = 2$, $N = 5$, $\alpha = 0.6$, $\hat{a} = 0.4$, $\lambda_1 = 3$, $\mu = 5$

λ_2 t/h	2	4	6	8	10
1	0.0000000000	0.0000000000	0.0000000000	0.0000000000	0.0000000000
2	0.0100000000	0.0100000000	0.0100000000	0.0100000000	0.0100000000
3	0.0175200000	0.0175200000	0.0175200000	0.0175200000	0.0175200000
4	0.0230040000	0.0229608000	0.0229176000	0.0228744000	0.0228312000
5	0.0268343040	0.0266891520	0.0265486080	0.0264126720	0.0262813440
6	0.0293370950	0.0290328990	0.0287477110	0.0284808400	0.0282315950
7	0.0307885150	0.0302798620	0.0298181810	0.0294001210	0.0290224650
8	0.0314207960	0.0306791260	0.0300275490	0.0294565990	0.0289575430
9	0.0314281720	0.0304438030	0.0296071370	0.0288978340	0.0282978730
10	0.0309723870	0.0297545880	0.0287539470	0.0279336710	0.0272624460

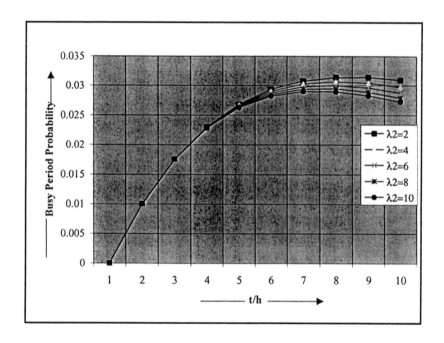

FIGURE 4.8

TABLE 4.5 Busy period probabilities for different values of μ when $h = 0.02$, $i = 1$, $N = 5$, $b = 2$, $\alpha = 0.6$, $\hat{a} = 0.4$, $\lambda_1 = 3$, $\lambda_2 = 2$

t/h \ μ	3	5	7	9	11
1	0.0000000000	0.0000000000	0.0000000000	0.0000000000	0.0000000000
2	0.0036000000	0.0100000000	0.0196000000	0.0324000000	0.0484000000
3	0.0065952000	0.0175200000	0.0327712000	0.0515808000	0.0731808000
4	0.0090555840	0.0230040000	0.0410612160	0.0615314880	0.0829033920
5	0.0110457446	0.0268343040	0.0457148520	0.0652408470	0.0835162520
6	0.0126249071	0.0293370950	0.0477192700	0.0649001100	0.0790200960
7	0.0138470761	0.0307885150	0.0478475750	0.0620836700	0.0720200840
8	0.0147612683	0.0314207960	0.0466985310	0.0579011640	0.0641560940
9	0.0154118003	0.0314281720	0.0447313790	0.0531204130	0.0564229230
10	0.0158386124	0.0309723870	0.0422955310	0.0482638430	0.0494000980

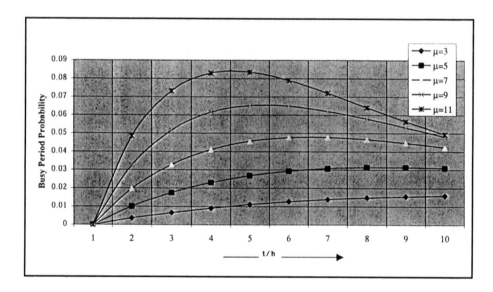

FIGURE 4.9

TABLE 4.6 Busy period probabilities for different values of i when $h = 0.02$, $b = 2$, $N = 5$, $\alpha = 0.6$, $\hat{a} = 0.4$, $\lambda_1 = 3$, $\lambda_2 = 2$, $\mu = 5$

i t/h	1	2
1	0.00000000000000000	0.00000000000000000
2	0.01000000000000000	0.00000000000000000
3	0.01752000001091510	0.00000000000000000
4	0.02300400001037120	0.00010000000000000
5	0.02683430411069560	0.00035280000000000
6	0.02933709531054610	0.00077742700000000
7	0.03078851467469500	0.00136975700000000
8	0.03142079570278680	0.00211075000000000
9	0.03142817232505390	0.00297270300000000
10	0.03097238703722280	0.00392389500000000

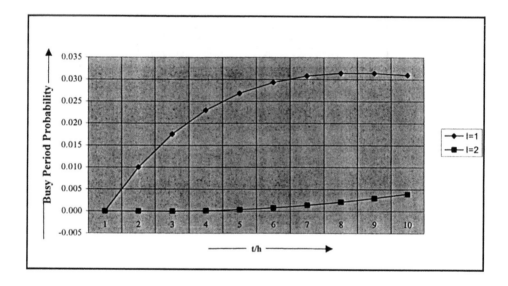

FIGURE 4.10

TABLE 4.7 Busy period probabilities for different values of N when $h = 0.02$, $i = 1$, $b = 2$, $\alpha = 0.6$, $\hat{a} = 0.4$, $\lambda_1 = 3$, $\lambda_2 = 2$, $\mu = 5$

N t/h	3	4	5	6	7	8
1	0.0000000000	0.0000000000	0.0000000000	0.0000000000	0.0000000000	0.0000000000
2	0.0100000000	0.0100000000	0.0100000000	0.0100000000	0.0100000000	0.0100000000
3	0.0177600000	0.0175200000	0.0175200000	0.0175200000	0.0175200000	0.0175200000
4	0.0236462400	0.0230040000	0.0230040000	0.0230040000	0.0230040000	0.0230040000
5	0.0279766235	0.0268319040	0.0268343040	0.0268343040	0.0268343040	0.0268343040
6	0.0310254925	0.0293265260	0.0293370950	0.0293370380	0.0293370380	0.0293370380
7	0.0330283407	0.0307605240	0.0307885150	0.0307882080	0.0307882080	0.0307882080
8	0.0341863036	0.0313629880	0.0314207960	0.0314198430	0.0314198430	0.0314198430
9	0.0346703364	0.0313255000	0.0314281720	0.0314259190	0.0314259230	0.0314259230
10	0.0346250342	0.0308076210	0.0309723870	0.0309678930	0.0309679090	0.0309679090

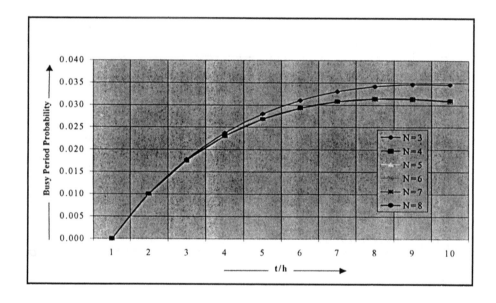

FIGURE 4.11

Part II
Models and Applications

CHAPTER 5

MEASURES FOR DISTRIBUTIONAL CLASSIFICATION AND MODEL SELECTION

GOVIND S. MUDHOLKAR   RAJESHWARI NATARAJAN

University of Rochester, Rochester, NY

Abstract: The use of conventional measures of skewness ($\sqrt{\beta_1}$) and kurtosis (β_2) for distributional classification and model selection is classical. In this paper, we review such existing measures and propose a new set (ξ_1, ξ_2) having some distinguishability advantages. In the (β_1, β_2)-chart the normal family appears as the limiting point of lines representing gamma, Type V, lognormal and the inverse Gaussian families. In the new (ξ_1, ξ_2)-chart the Gaussian, inverse Gaussian and gamma families appear as three distinct points. The asymptotic distributions of the estimates (J_1, J_2) of (ξ_1, ξ_2) for samples from some parent populations are derived and the implications of the distinguishability for the goodness-of-fit problems are discussed.

Keywords and phrases: Model selection, measure of skewness, measure of kurtosis, goodness-of-fit.

5.1 INTRODUCTION

Constructing a stochastic model for populations is an essential step prior to the statistical analysis of data from them. The space of all probability distributions being too large and complex, the choice for the model building is generally restricted to some well understood system of distributions such as those due to Pearson, Johnson, Burr, Perk etc.; see e.g. Johnson, Kotz and Balakrishnan (1994). The selection of a parametric family from the system, assuming generalities such as continuity and unimodality, involves use of common tools of exploratory data analysis and various measures of distributional classification.

The most commonly employed measures of distributional morphology, often used in tandem, are the classical cofficients of skewness and kurtosis, their variations, and the related charts discussed in Johnson, Kotz and Balakrishnan (1994), Elderton and Johnson (1969) or Ord (1972). More recently, Mudholkar and Natarajan (1998) while examining the remarkable similarity between the Gaussian and inverse Gaussian (IG) families, noted by many e.g., Folks and Chhikara (1978), Iyengar and Patwardhan (1988), introduced the coefficients δ_1 and δ_2 respectively of IG-skewness and IG-kurtosis. Mudholkar and Natarajan demonstrate several remarkable analogies between the measures $(\sqrt{\beta_1}, \beta_2)$ and (δ_1, δ_2), and between their respective sample versions. They propose (δ_1, δ_2)-chart for model selection and illustrate its use with some well known datasets. Ord (1967) proposes the ratio of the coefficient of skewness and coefficient of variation as a measure for classifying such distributions.

The purpose of this paper is to supplement Ord's (1967) measure, denoted here by ξ_1, with another ratio ξ_2 and consider the pair (ξ_1, ξ_2) for the purpose of classifying distributions. It is noted that in a chart based on (ξ_1, ξ_2), the Gaussian, inverse Gaussian and gamma appear as three distinct points.

The existing measures of distributional classification of distributions are briefly reviewed in Section 5.2. The new measures ξ_1 and ξ_2 are discussed and the uses of (ξ_1, ξ_2)-chart as of (β_1, β_2)-chart and of the sample estimates (J_1, J_2) are illustrated in Section 5.3. The asymptotic distributions of (J_1, J_2) for samples from Gaussian, inverse Gaussian and gamma populations are derived in Section 5.4. The final section is given to miscellaneous remarks including mainly applications to goodness-of-fit tests.

5.2   CURRENT MEASURES FOR DISTRIBUTIONAL MORPHOLOGY

The best known among the measures used for classifying statistical distributions are the conventional coefficients of skewness and kurtosis defined as $\sqrt{\beta_1} = \mu_3/\mu_2^{3/2}$ and $\beta_2 = \mu_4/\mu_2^2$, respectively where μ_j refers to the jth central moment. Even though there exist asymmetric distributions with $\sqrt{\beta_1} = 0$, (e.g. MacGillivray (1986) and Freimer et $al.$ (1988)) the coefficient of skewness is generally accepted as a measure of asymmetry. The meaning of the coefficient of kurtosis is somewhat amorphous; see Balanda and MacGillivray (1988) for a critical review in which they describe it "vaguely as the location- and scale-free movement of probability mass from the shoulders of a distribution into its center and tails...". MacGillivray (1986) and Balanda and MacGillivray (1988) discuss partial orderings of probability distributions in terms of $\sqrt{\beta_1}$ and β_2 respectively.

Many variations of $\sqrt{\beta_1}$, β_2 exist in literature. Well known among these are the measures based upon the quantiles. For example, an alternative

$$\gamma_U(F) = \frac{[F^{-1}(1-u) + F^{-1}(u) - 2F^{-1}(0.5)]}{[F^{-1}(1-u) - F^{-1}(u)]}, \ u \ \epsilon \ (0, \frac{1}{2}) \qquad (5.2.1)$$

to skewness appears in David and Johnson (1956), whereas Groeneveld and Meeden (1984) propose

$$\beta_2(\alpha, H) = \frac{H^{-1}(0.75 + \alpha) + H^{-1}(0.75 - \alpha) - 2H^{-1}(0.75)}{H^{-1}(0.75 + \alpha) - H^{-1}(0.75 - \alpha)}, \qquad (5.2.2)$$

as a measure of kurtosis.

Hosking (1990) has proposed and studied two ratios $\tau_3 = \lambda_3/\lambda_2$ and $\tau_4 = \lambda_4/\lambda_2$ of Sillitto's (1951) L-moments λ_r,

$$\lambda_r = r^{-1} \sum_{k=0}^{r-1} (-1)^k \left(\begin{array}{c} r-1 \\ k \end{array} \right) E\left(X_{r-k:r}\right), \ r = 1, 2, \ldots, \qquad (5.2.3)$$

where $X_{r-k:r}$ denotes the $(r-k)^{th}$ order statistic, as substitutes for $\sqrt{\beta_1}$ and β_2. The measures, L-skewness τ_3 and L-kurtosis τ_4 are believed to be less variable and more meaningful in the context of procedures such as Shapiro-Wilk test of normality; see Hosking (1992). More recently Mudholkar and Hutson (1998) have proposed LQ-moments, quantile analogs of L-moments, as

$$\zeta_r = r^{-1} \sum_{k=0}^{r-1} \left(\begin{array}{c} r-1 \\ k \end{array} \right) \tau_{p,\alpha}(X_{r-k:r}), \ r = 1, 2, \ldots, \qquad (5.2.4)$$

where $0 \leq \alpha \leq 1/2$, $0 \leq p \leq 1/2$, based on quick estimators such as Trimean, Gastwirth's estimator, and have developed measures $\eta_3 = \zeta_3/\zeta_2$ of LQ-skewness and $\eta_4 = \zeta_4/\zeta_2$ of LQ-kurtosis based upon them. The L-measures and LQ-measures behave similarly, except that LQ-moments always exist, whereas L-moments exist only for distributions with finite expectations.

Mudholkar and Natarajan (1998) define and discuss uses of two new measures δ_1 and δ_2 as IG-analogs of $\sqrt{\beta_1}$ and β_2 respectively. The coefficient δ_1 is based upon a suggestion in Mudholkar, Natarajan and Chaubey (1998) which contains $Z(IG)$-test analogous to Lin and Mudholkar's Z-test of normality against asymmetric alternatives. The coefficient δ_1 which appears as a parameter in the power function of the $Z(IG)$-test is proposed as a measure of IG-skewness, where a random variable X is said to be IG-symmetric about μ if it satisfies the countable equalities

$$E\left[\left(\frac{X}{\mu}\right)^{-r}\right] = E\left[\left(\frac{X}{\mu}\right)^{r+1}\right], \; r = 1, 2, \ldots \qquad (5.2.5)$$

To construct an analog of β_2, they compare the asymptotic distribution of the sample variance given by

$$\sqrt{n}\,(\log S^2 - \log \sigma^2) \xrightarrow{d} N(0,\,(\beta_2 - 1)), \qquad (5.2.6)$$

with that of IG parameter estimate $V = \lambda^{-1} = \sum_{i=1}^{n}(1/X_i - 1/\bar{X})$ to arrive at the coefficient δ_2 of IG-kurtosis. The measure δ_1 involves the first two positive moments and first negative moment, whereas δ_2 involves the first two positive and the first two negative moments. It is shown that the asymptotic distributions of the estimates d_1, d_2 of δ_1, δ_2 for IG-samples are exactly the same as those of $\sqrt{b_1}$ and b_2 for normal samples. Furthermore, they also offer, study, and illustrate the use of (δ_1, δ_2)-chart as the IG analog of the (β_1, β_2)-chart for distributional classification and data modelling.

Carver (1919) considered a difference equation

$$\frac{\Delta y_x}{\Delta x y_x} = \frac{a - x}{b_0 + b_1 x + b_2 x^2}, \qquad (5.2.7)$$

analogous to the well known differential equation underlying the Pearson system. When the lower threshold of the distribution is zero and $b_0 = 0$, the three constants a, b_1, b_2 may be associated with the first three moments of the distribution. In view of this Ord (1967) uses the two ratios $I = \mu_2/\mu_1'$ and $S = \mu_3/\mu_2$ to propose a measure $\omega = S/I$, to distinguish between distributions with positive support. We also note that Cox and Oakes (1984) discuss a use of the chart based on coefficient of variation γ and coefficient of skewness $\sqrt{\beta_1}$ in the context of survival analysis for model selection. We denote Ord's measure ω by

$$\xi_1 = \frac{\kappa_1 \kappa_3}{\kappa_2^2} = \omega = \frac{S}{I} = \sqrt{\beta_1}/\gamma, \qquad (5.2.8)$$

where κ_i refers to the ith cumulant.

5.3 (ξ_1, ξ_2) SYSTEM

If the coefficient b_0 in Equation (5.2.7) is non-zero, or the support of distribution is not strictly positive, then four moments or cumulants are necessary to describe the constants in the difference equation or corresponding differential equation. Hence, one may define another measure

$$\xi_2 = \frac{\kappa_4 \kappa_1{}^2}{\kappa_2{}^3} = \frac{\gamma_2}{\gamma^2}, \tag{5.3.9}$$

where γ_2 represents the coefficient of excess kurtosis, as a supplement to the index ξ_1 defined by Ord (1967). The two indices ξ_1 and ξ_2 can then be used in place of $\sqrt{\beta_1}$ and β_2 for distributional classification, and their sample versions J_1 and J_2 may be placed on the (ξ_1, ξ_2)-chart given in Figure 5.1 for model selection.

(ξ_1, ξ_2)-**Chart.** Several features of the (ξ_1, ξ_2)-chart are noteworthy. Since $\xi_1 = 0$ whenever $\beta_1 = 0$, all symmetric families are represented as the vertical line $\xi_1 = 0$, with an exception of the normal family. The normal family for which $\beta_1 = 0 = \gamma_2$, appears as the point $(0, 0)$ in the (ξ_1, ξ_2)-chart. As a matter of technicality, it may be noted that the point $(0, 3)$ in (β_1, β_2)-chart and the point $(0, 0)$ in the (ξ_1, ξ_2)-chart may also represent non-normal distributions for which $\sqrt{\beta_1} = 0 = \gamma_2$ but some higher order cumulants are non-zero. Both variance-ratio F and beta families occupy certain regions of the chart. The curves in Figure 5.1 corresponding to the variance-ratio $F(m, n)$, and the beta $B(m, n)$ distributions for fixed m and varying n and fixed n and varying m, give approximate location of these regions. Figure 5.1 also contains the curves corresponding to the Weibull and lognormal families and shows the position of the datasets discussed later in this section as points D1 and D2. Interestingly, the gamma and inverse Gaussian families which appear as lines in the (β_1, β_2)-chart, are represented in the (ξ_1, ξ_2)-chart by points $(2, 6)$ and $(3, 15)$ respectively. In (β_1, β_2)-chart, gamma and inverse Gaussian lines, both converge to the normal point $(0, 3)$ whereas in this chart they are distinct points and hence may be considered better discriminators between these distributions.

Model Selection. The consistent sample versions $J_1 = m_3 m_1 / (m_2^2)$ and $J_2 = [(m_4 - 3 m_2{}^2) m_1{}^2] / (m_2{}^3)$, obtained by plugging-in the sample moments, in conjunction with the (ξ_1, ξ_2)-chart, may be used to select a parametric model using data. We illustrate the process using two datasets D1 and D2. The values of the measures $\sqrt{b_1}$, b_2 and J_1, J_2 for these data appear in Table 5.1.

TABLE 5.1   Comparison of $(\sqrt{b_1}, b_2)$ and (J_1, J_2) for the datasets

Datset	$\sqrt{b_1}$	b_2	J_1	J_2
D1	1.00	4.60	1.77	4.5
D2	0.76	2.69	1.04	-0.58

TABLE 5.2   Rainfall (in mm) at Kyoto, Japan for the month of July from 1880–1960

Rainfall	Observed	Expected Weibull fit	Expected Gamma fit
0-50	5	4.83	3.78
50-100	9	11.05	12.27
100-150	12	13.86	15.45
150-200	18	13.84	14.24
200-250	17	11.93	11.27
250-300	6	9.16	8.14
300-350	5	6.38	5.53
350-400	4	4.07	3.60
400-above	4	4.83	5.72

Source: World Weather Records Smithsonian Institution, Miscellaneous Collections and U.S. Department of Commerce.

D1.   Rainfall Data. These data in Table 5.2, used in Mooley (1973) give the July rainfall (in millimeters) at Kyoto over a period of 80 years 1880-1960. Conventionally such meteorological data were analyzed by using normal fit. Mooley argued that for such data a gamma model would be more appropriate. In the (ξ_1, ξ_2)-chart, the point D1 $(1.77, 4.5)$ falls on the Weibull line but is very close to the gamma point. In order to compare these models we consider Pearson's chi-square statistics obtained from Table 5.2. The values $\chi^2(gamma) = 7.12$ and $\chi^2(weibull) = 5.57$; both with six degrees of freedom, and p-values of 0.31 and 0.473 respectively, suggests the preferability of the Weibull model over the gamma model.

D2.   Bus Motor Failures. These data, originally in Davis (1952), are reanalyzed in Mudholkar, Srivasatava and Freimer (1995) using the exponentiated Weibull family.

TABLE 5.3   Fifth bus motor failure

Mileage (Thousands of Miles)	0-20	20-40	40-60	60-80	80-up	Total
Observed Number	29	27	14	8	7	85

Source: Davis (1952), *Journal of the American Statistical Association,* **47**.

Table 5.3 gives the number of miles (in 1000's) between the fourth and fifth failures of the motors. The reanalysis shows appropriateness of a Weibull model for the data. The placement of D2 $(1.04, -0.58)$ in the (ξ_1, ξ_2)-chart confirms appropriateness of the model, whereas the choice of a parametric model in the (β_1, β_2)-chart is unclear.

5.4 ASYMPTOTIC DISTRIBUTIONS OF J_1, J_2

In addition to the model selection applications the empirical skewness and kurtosis coefficients $\sqrt{b_1}$ and b_2 have been widely used for testing goodness-of-fit hypotheses such as normality; see Chapter 7 by Bowman and Shenton in D'Agostino and Stephens (1986). In this section, with a similar goal in mind, we consider the asymptotic distributions of the coefficients J_1 and J_2 for samples from the normal, gamma and inverse Gaussian populations.

Gaussian Population. For a random sample of size n from an $N(\mu, \sigma)$ population, the asymptotic sampling distribution of $\sqrt{b_1}$ and b_2, as $n \to \infty$, is well known; see e.g. Kendall and Stuart (1969). Specifically, as $n \to \infty$,

$$\sqrt{n\, b_1} \xrightarrow{d} N(0, 6). \tag{5.4.10}$$

Hence, if the population mean $\mu \neq 0$, using (5.4.10) and an appeal to Slutsky's theorem we get

$$\sqrt{n}(J_1) \xrightarrow{d} N(0, \frac{6\mu^2}{\sigma^2}), \tag{5.4.11}$$

as $n \to \infty$. Similarly, from the sampling distribution of b_2,

$$\sqrt{n}(b_2 - 3) \xrightarrow{d} N(0, 24), \tag{5.4.12}$$

as $n \to \infty$, we see that

$$\sqrt{n}(J_2) \xrightarrow{d} N(0, \frac{24\mu^4}{\sigma^4}). \tag{5.4.13}$$

Furthermore, since $\sqrt{b_1}$ and b_2 are asymptotically independent, using the multivariate version of Slutsky's theorem, [e.g. Cramér (1946)] it is seen that J_1 and J_2 are also asymptotically independent. In other words, for a sample from normal population, as $n \to \infty$

$$\sqrt{n}\left[\begin{pmatrix} J_1 \\ J_2 \end{pmatrix} - \begin{pmatrix} 2 \\ 6 \end{pmatrix} \right] \xrightarrow{d} N\left[0, \begin{pmatrix} 6\mu^2/\sigma^2 & 0 \\ 0 & 24\mu^4/\sigma^4 \end{pmatrix} \right]. \tag{5.4.14}$$

Gamma Population. For a random sample of size n from a $G(\alpha)$ population with shape parameter α, as $n \to \infty$, the vector $(\bar{X}, m_2, m_3, m_4)'$ is asymptotically normally distributed with mean $(\mu, \mu_2, \mu_3, \mu_4)'$, and covariance matrix $(1/n) \sum = (\sigma_{ij})/n$, where, with the notation $\bar{X} = m_1$ and $m_i = E[(X - m_1)^i]$, $\sigma_{ii} = Var(m_i)$, $\sigma_{ij} = Cov(m_i, m_j)$ we have,

$$\sigma_{11} = \alpha, \ \sigma_{12} = 2\alpha,$$

$$\sigma_{13} = 6\alpha \ \ \sigma_{14} = 12\alpha\,(2 + \alpha),$$

$$\sigma_{22} = 2\alpha\,(3 + \alpha), \ \ \sigma_{23} = 12\alpha(2 + \alpha),$$

$$\sigma_{24} = 120\alpha + 108\alpha^2 + 15\alpha^3, \ \ \sigma_{33} = 120\alpha + 90\alpha^2 + 6\alpha^3,$$

$$\sigma_{34} = 720\alpha + 792\alpha^2 + 144\alpha^3, \ \ \sigma_{44} = 5040\alpha + 6888\alpha^2 + 2088\alpha^3 + 96\alpha^4.$$

Hence, by use of the multivariate version of Mann-Wald theorem, [Serfling (1980)] we see that, as $n \to \infty$, the vector $J' = (J_1, J_2)$ has a bivariate normal distribution with mean $(2, 6)$ and covariance matrix given by

$$Var(J_1) = \frac{20}{\alpha} + 26 + 6\,\alpha, \ \ Var(J_2) = 12\left(\frac{210}{\alpha} + 292 + 75\,\alpha - \alpha^2\right),$$

$$Cov(J_1, J_2) = 12\left(\frac{18}{\alpha} + 24 + 5\alpha\right). \tag{5.4.15}$$

Inverse Gaussian Population. For a random sample of size n from an $IG(\mu, \lambda)$ population, as $n \to \infty$, the vector $(\bar{X}, m_2, m_3, m_4)'$ is asymptotically normally distributed with mean $(\mu, \mu_2, \mu_3, \mu_4)'$, and covariance matrix $(1/n) \sum = (\sigma_{ij})/n$, where with the notation in gamma population case

$$\sigma_{11} = \frac{\mu}{\lambda}, \ \ \sigma_{23} = \frac{105\mu^9}{\lambda^4} + \frac{18\mu^8}{\lambda^3},$$

$$\sigma_{12} = \frac{3\mu^5}{\lambda^2}, \ \ \sigma_{24} = \frac{\mu^{11}}{\lambda^5} + \frac{264\mu^{10}}{\lambda^4} + \frac{15\mu^9}{\lambda^3},$$

$$\sigma_{13} = \frac{15\mu^7}{\lambda^3}, \ \ \sigma_{33} = \frac{945\mu^{11}}{\lambda^5} + \frac{216\mu^{10}}{\lambda^4} + \frac{6\mu^9}{\lambda^3},$$

$$\sigma_{14} = \frac{105\mu^9}{\lambda^4} + \frac{18\mu^8}{\lambda^3}, \ \ \sigma_{34} = \frac{10395\mu^{13}}{\lambda^6} + \frac{3440\mu^{12}}{\lambda^5} + \frac{216\mu^{11}}{\lambda^4},$$

$$\sigma_{22} = \frac{15\mu^7}{\lambda^3} + \frac{2\mu^6}{\lambda^2}, \ \ \sigma_{44} = \frac{135135\mu^{15}}{\lambda^7} + \frac{49230\mu^{14}}{\lambda^6} + \frac{5004\mu^{13}}{\lambda^5} + \frac{96\mu^{12}}{\lambda}.$$

Hence, by use of the multivariate version of Mann-Wald theorem, it follows that, as $n \to \infty$, the vector $J' = (J_1, J_2)$ has a bivariate normal distribution with mean $(3, 15)$ and covariance matrix given by

$$Var(J_1) = 6\phi + 72 + \frac{216}{\phi}, \quad Var(J_2) = 6\left(-2\phi^2 + 351\phi + 4380 + \frac{13260}{\phi}\right),$$

$$Cov(J_1, J_2) = 2\left(45\phi + 748 + \frac{1980}{\phi}\right), \tag{5.4.16}$$

where $\phi = \lambda/\mu$ denotes the shape parameter.

5.5 MISCELLANEOUS REMARKS

The classical (β_1, β_2)-chart provides the view of only one cross-section of the space of probability distributions. As noted earlier, in this cross-section the gamma, inverse Gaussian and type V families appear as lines converging to the $(0, 3)$ point representing the normal family. In the (ξ_1, ξ_2)-chart, however, the gamma and inverse Gaussian and normal families appear as distinct points. Hence, the sample estimates (J_1, J_2) may be useful in distinguishing between these families.

1. Testing Normality. Among various tests of the composite hypothesis of normality, those based on $(\sqrt{b_1}, b_2)$ are the oldest (Chapter 7 of D'Agostino and Stephens (1986)). The indices J_1 and J_2 can be similarly employed to construct an omnibus test of normality, and also tests directed at asymmetric alternatives or non-normal kurtosis alternatives. From the asymptotic distribution (5.4.14) it appears that, for a population with large coefficient of variation, such tests may have superior power properties. Furthermore, using an approach similar to that in Mudholkar, Marchetti and Lin (1998) J_1 and J_2 tests could be combined to detect restricted asymmetric and non-normal kurtosis alternatives.

2. Gamma Hypothesis. At present there do not exist reasonable goodness-of-fit tests for the composite gamma hypothesis. This is mainly because even the asymptotic null distributions of most goodness-of-fit test statistics involve the population parameters in both mean and variances. As shown in (5.4.15), the expectation of asymptotic distributions of (J_1, J_2) is parameter-free. Hence, the prospect of a gamma goodness-of-fit test based on J_1 and J_2 appears promising.

3. Inverse Gaussian hypothesis. Over the last few decades, the analytical simplicity of the inverse Gaussian inference procedures and the analogy of IG family with Gaussian family has intrigued the statistical community. The family is highly recommeded as a model for asymmetric data. The goodness-of-fit tests for the IG model are still very few, see Mudholkar, Natarajan and Chaubey (1998). Hence, use of (J_1, J_2) for testing goodness-of-fit of the composite IG hypothesis seems a reasonable project.

4. Confidence regions. The asymptotic joint distributions of the indices (J_1, J_2) may be used to construct confidence regions for (ξ_1, ξ_2). For large sample sizes, they may be useful in choosing between competing models suggested by the (ξ_1, ξ_2)-chart in Figure 5.1.

5. Variations of (ξ_1, ξ_2)**.** Several variations of the classical coefficients of skewness and kurtosis based on population quantiles, L-moments and LQ-moments are described in Section 5.2. Similar variations of the measures ξ_1 and ξ_2 are obviously feasible.

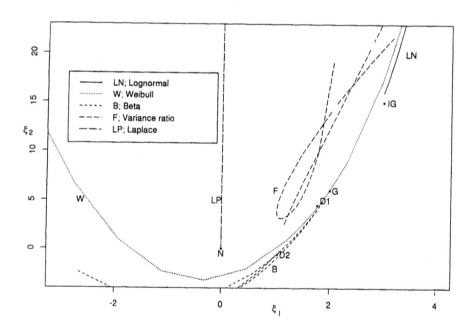

N: Normal Point, G: Gamma Point, IG: Inverse Gaussian Point, and D1, D2 locate datasets in Section 5.3

FIGURE 5.1 The (ξ_1, ξ_2)-chart

REFERENCES

Balanda, K. P. and MacGillivray, H. L. (1988). Kurtosis: A critical review, *The American Statistician*, **42**, 111–119.

Bartlett, M. S. (1937). Properties of sufficiency and statistical tests, *Proceedings in Royal Statistical Society, Series A*, **160**, 268–282.

Carver, H. C. (1919). On the graduation of frequency distributions, *Casualty Actuarial Society of America, Proceedings*, **6**, 52–72.

Chhikara, R. S. and Folks, J. L. (1989). *The Inverse Gaussian distribution*, Marcel Dekker, New York.

Cox, D. R. and Oakes, D. (1984). *Analysis of Survival Data*, Chapman & Hall, New York.

Cramér, H. (1946). *Mathematical Methods of Statistics*, Princeton University Press, Princeton.

D'Agostino, R. B. and Stephens, M. A. (1986). *Goodness-of-fit Techniques*, Marcel Dekker, New York.

Davis, D. J. (1952). An analysis of some failure data, *Journal of the American Statistical Association*, **47**, 113–150.

Elderton, W. P. and Johnson, N. L. (1969). *Systems of Frequency Curves*, Cambridge University Press.

Folks, J. L. and Chhikara, R. S. (1978). The inverse Gaussian distribution and its statistical application - A review, *Journal of the Royal Statistical Society, Series B*, **40**, 263–289.

Freimer, M., Mudholkar, G. S., Kollia, G., and Lin, C. T. (1988). A study of the generalized Tukey lambda family, *Communications in Statistics—Theory and Methods*, **17**, 3547–3567.

Groeneveld, R. A. and Meeden, G. (1984). Measuring skewness and kurtosis, *The Statistician*, **33**, 391–399.

Hosking, J. R. M. (1990). L-moments: Analysis and estimation of distributions using linear combinations of order statistics, *Journal of the Royal Statistical Society, Series B*, **52**, 105–124.

Hosking, J. R. M. (1992). Moments or L Moments? An example comparing two measures of distributional shape, *American Statistician*, **46**, 186–189.

Iyengar, S. and Patwardhan, G. (1988). Recent developments in the inverse Gaussian distribution, In *Handook of Statistics* (Eds., P. R. Krishnaiah and C. R. Rao), **7**, 479–490.

Johnson, N. L., Kotz, S. and Balakrishnan, N. (1994). *Continuous univariate distributions*, Second edition, John Wiley & Sons, New York.

Kendall, M. G. and Stuart, A. (1969). *The Advanced Theory of Statistics, Vol I*, Third Edition, Hafner, New York.

Lin, C. C. and Mudholkar, G. S. (1980). A simple test for normality against asymmetric alternatives, *Biometrika*, **67**, 455–461.

MacGillivray, H. L. (1986). Skewness and asymmetry: Measures and orderings, *Annals of Statistics*, **14**, 994–1011.

Miscellaneous Collections (1934). *World Weather Records, 1921–1930*, **90**, Washington, DC, 146.

Miscellaneous Collections (1947). *World Weather Records, 1931–1940*, **105**, Washington, DC, 145.

Mooley, D. A. (1973). Gamma distribution probability model for Asian summer monsoon monthly rainfall, *Monthly Weather Review*, **101**, 160–176.

Mudholkar, G. S. and Hutson, A. (1998). LQ-moments: Analogs of L-moments, *Journal of Statistical Planning and Inference*, **71**, 191–208.

Mudholkar, G. S., Marchetti, C. E. and Lin, C. T. (1998). Independence characterizations and testing normality against restricted skewness-kurtosis alternatives, *Journal of Statistical Planning and Inference* (to appear).

Mudholkar, G. S., Natarajan, R. and Chaubey, Y. P. (1998). Independence characterization and inverse Gaussian goodness-of-fit composite hypothesis, *Technical Report*, University of Rochester, Rochester, NY. (abstract in *IMS Bulletin*, (1997) **26(4)**, 400).

Mudholkar, G. S. and Natarajan, R. (1998). The inverse Gaussian models: Analogs of symmetry, skewness and kurtosis, *Technical Report*, University of Rochester, NY.

Mudholkar, G. S., Srivastava, D. K. and Freimer, M. (1995). The exponentiated Weibull family: A reanalysis of the bus-motor-failure data, *Technometrics*, **37**, 436–445.

Ord, J. K. (1967). On a system of discrete distributions, *Biometrika*, **54**, 649–656.

Ord, J. K. (1972). *Families of Frequency Distributions*, Hafner, New York.

Sillitto, G. P. (1951). Interrelations bewteen certain linear systematic statistics of samples from continuous population, *Biometrika*, **38**, 377–382.

Serfling, R. J. (1980). *Approximation Thorems of Mathematical Statistics*, John Wiley & Sons, New York.

Seshadri, V. (1993). *The Inverse Gaussian Distribution: A Case Study in Exponential Families*, Clarendon Press, Oxford.

Seshadri, V. (1998). *The Inverse Gaussian distribution: Statistical Theory and Applications*, Springer-Verlag, New York.

Smithsonian Institution, Miscellaneous Collections (1927). *World Weather Records*, **79**, Washington, DC, 323.

U.S. Department of Commerce, Weather Bureau (1959). *World Weather Records, 1941-1950*, Washington DC, 398.

U.S. Department of Commerce, Weather Bureau (1967). *World Weather Records, 1951-1960*, **4**, Asia, Washington DC, 345.

U.S. Water Resources Council, Washington, Hydrology Comittee (1977). *Guidelines for determining flood flow frequency*, United States, Water Resources Council, Washington DC.

CHAPTER 6

MODELING WITH A BIVARIATE GEOMETRIC DISTRIBUTION

SUNIL K. DHAR

New Jersey Institute of Technology, Newark, NJ

Abstract: The discrete analog of the bivariate distributions of Freund's models is described, interpreting its assumptions. This discrete bivariate geometric distribution has applications to survival analysis, reliability theory and count data. Related models derived by incorporating the environmental effects are also discussed. The application to sports data is demonstrated.

Keywords and phrases: Survival function, environmental effect

6.1 INTRODUCTION

A number of authors have arrived at different bivariate geometric distributions (BVG). Azlarov and Volodin (1982) considered the discrete analog of Marshall-Olkin's (1967) bivariate and trivariate exponential distributions. However, Basu and Dhar (1995) looked at this problem from a totally independent approach and have arrived at the same discrete analog of Marshall-Olkin's (1967) for the general multivariate exponential (MVE) with no restriction on their parameters, except that probability parameters are between 0 and 1. This BVG is different from the one described in Marshall-Olkin (1985).

Dhar (1998) arrived at another bivariate geometric distribution that was developed using ideas from Freund (1961) reliability models. Though this newly obtained geometric model is named after the author Freund (1961), not all the distribution properties follow those of the Freund (1961) model.

101

In this paper, the assumptions leading to the bivariate geometric model are interpreted. New bivariate models are derived from this bivariate geometric distribution taking into account the environmental effects. Further, the method of Dhar (1998) to compute the moment estimator under the practical data set has been expressed clearly. The applicability of this model is demonstrated using a real data set in Section 6.4.

6.2 INTERPRETATION OF BVG MODEL ASSUMPTIONS

To clarify the assumptions the basic derivation from Dhar (1998) is repeated here. Let X and Y be the discrete lifetime distributions of components 1 and 2, respectively. The joint density

$$
\begin{aligned}
P(X = x, Y = y) \;=\; & P(Y = y | X = x < Y)P(X = x < Y) \\
& + P(X = x | Y = y < X)P(Y = y < X) \\
& + P(X = x, Y = y | X = Y)P(X = Y),
\end{aligned}
$$

$$(6.2.1)$$

$x, y = 1, 2, \ldots$. Note that if component represented by X fails at x, before Y fails, then the failed component is immediately replaced. Hence, the updated two component system will follow the distribution X*, Y*. The $(X^* = x)$ and $(x < Y^*)$ are then treated to be independent.

$$
\begin{aligned}
P(X = x < Y) \;=\; & P(X^* = x < Y^*) \\
=\; & P(X^* = x)P(x < Y^*) \\
=\; & p_2^x p_1^{x-1} q_1,
\end{aligned}
$$

$$(6.2.2)$$

$\left[P(Y = y < X) = p_1^y p_2^{y-1} q_2 \right]$, where $0 < p_i < 1$, $p_i + q_i = 1$, $i = 1, 2$ and p_i is the probability of survival of the replaced component i, for a unit time, when component i failed before its counterpart, $i = 1, 2$. Take the conditional survival distributions to be truncated geometric

$$
P(X = x | Y = y < X) = q_3 p_3^{x-y-1}, \qquad 1 \le y < x, \tag{6.2.3}
$$

$\left[P(Y = y | X = x < Y) = q_4 p_4^{y-x-1} 1 \le x < y \right]$, $\qquad x, y = 1, 2, \ldots$,

$0 < p_3, p_4 < 1$, $p_i + q_i = 1$, $i = 3, 4$. Here, p_j, $j = 3, 4$ represents the probability of survival of component $j - 2$ for a unit time, given that it has already survived for some time, at the end of which the adjacent component failed. Finally, visualizing the two components as a single interdependent unit in which both the components fail simultaneously, the probability of the failure occurring at time $X = Y$ is

$$P(X = x, Y = y | X = Y) = q_{12}p_{12}^{x-1}, \tag{6.2.4}$$

where $x = 1, 2, \ldots, 0 < p_{12} < 1$, $p_{12} + q_{12} = 1$. Here p_{12} represents the probability of simultaneous survival, for a unit time, of the two components treated as one component. Then, from (6.2.2) $P(X = Y) = 1 - P(X \neq Y) = 1 - \{[p_1 q_2 + p_2 q_1][1 - p_1 p_2]^{-1}\} = q_1 q_2 [1 - p_1 p_2]^{-1}$. This, along with (6.2.2) to (6.2.4) substituted in (6.2.1), gives

$$
\begin{aligned}
P(X = x, Y = y) &= \frac{q_2 q_3}{p_2 p_3} \left[\frac{p_1 p_2}{p_3} \right]^y p_3^x, && \text{if } y < x, \ x, y = 1, 2, \ldots, \\
&= \frac{q_1 q_4}{p_1 p_4} \left[\frac{p_1 p_2}{p_4} \right]^x p_4^y, && \text{if } x < y, \ x, y = 1, 2, \ldots, \\
&= \frac{q_1 q_2 q_{12}}{1 - p_1 p_2} p_{12}^{x-1}, && \text{if } x = y = 1, 2, \ldots.
\end{aligned}
$$

$$\tag{6.2.5}$$

In order for the above function to be a density, we need $p_1 p_2 < p_3$ and $p_1 p_2 < p_4$. These strict inequalities tell us that the joint survival of the replaced components 1 and 2, treated to be independent of each other, is less than the probability of survival of the component which is known to have out lived the other. In Dhar (1998) it is seen that the survival function corresponding to (6.2.5) satisfies the loss of memory property iff $p_{12} = p_1 p_2$. This condition suggests that the probability of simultaneous survival of the two components is equal to the probability of the joint survival of the replaced components 1 and 2, when they fail before their complimentary components and are treated to be independent of each other. The bivariate geometric model with the additional assumption $p_{12} = p_1 p_2$, in (6.2.5) will be referred to as BVG(p_1, p_2, p_3, p_4). Again from Dhar (1988, Lemma 3.2), the BVG(p_1, p_2, p_3, p_4) model has its marginals as a mixture of two geometric distributions, provided $p_3 > p_1$ and $p_4 > p_2$. These two conditions interpret as follows. The probability of survival of the component which is known to have out lived the other is greater than the probability of survival of that replaced component which failed before the other component. The two survival probabilities so compared are referring to the same component slots but under different circumstances.

6.3 THE MODEL UNDER THE ENVIRONMENTAL EFFECT

Consider η to be the random variable that counts the number of discrete time steps needed for 'a' number of events to occur. Then a density of η is given by $\binom{\eta-1}{a-1}b^{\eta-a}(1-b)^a$, where $0 < b < 1$ and $1 \leq a \leq \eta$ (the Negative Binomial [NB] distribution). The longer the waiting or elapsed time η in the environment, smaller is the probability of survival p^η of the component, where p is the initial survival probability. Multiplying the conditional density, given η, of the r.v. $BVG(p_1^\eta, p_2^\eta, p_3^\eta, \ p_4^\eta)$ by the density of η and then summing over all possible integer values of $\eta \geq a$ yields a new bivariate geometric distribution BVG-NB(p_1, p_2, p_3, p_4, a, b). Let

$$\Delta_i f(x_1, x_2, ..., x_i, ..., x_n)$$
$$= \ f(x_1, x_2, ..., x_i - 1, ..., x_n) - f(x_1, x_2, ..., x_i, ..., x_n),$$

i.e., the backward shift operator acting on the i-th variable. Further, let $x \wedge y$ and $x \vee y$ denote the minimum and maximum of the two real numbers x and y. Then, the joint density of BVG-NB(p_1, p_2, p_3, p_4, a, b) is given by

$$(1-b)^{-a}\{p_1 p_2\}^{-a(x \wedge y)} P(X = x, Y = y)$$
$$= \ \Delta_1 \left\{ p_3^{ax}[p_2 p_3^y - bp_1^y p_2^y p_3^x]^{-a} - p_3^{ax}[p_3^y - bp_1^y p_2^y p_3^x]^{-a} \right\},$$
$$\text{if } y < x, \ x, y = 1, 2, \ldots,$$
$$= \ \Delta_2 \left\{ p_4^{ay}[p_1 p_4^x - bp_1^x p_2^x p_4^y]^{-a} - p_4^{ay}[p_4^x - bp_1^x p_2^x p_4^y]^{-a} \right\},$$
$$\text{if } x < y, \ x, y = 1, 2, \ldots,$$
$$= \ [p_1 p_2 - bp_1^x p_2^x]^{-a} - [p_2 - bp_1^x p_2^x]^{-a} - [p_1 - bp_1^x p_2^x]^{-a} + [1 - bp_1^x p_2^x]^{-a},$$
$$\text{if } x = y = 1, 2, \ldots. \tag{6.3.6}$$

Here, $0 < p_1, \ p_2 < 1$, $p_1 p_2 < p_3$ and $p_1 p_2 < p_4$. The corresponding joint survival function computed using Dhar (1998, Equation 3.6) is given by

$$(1-b)^{-a} P(X > x, Y > y)$$
$$= \ \sum_{i=1}^{x-y} p_3^{a(x-y)}(p_1 p_2)^{a(y+i)} \left([p_2 p_3^i - bp_3^{x-y}(p_1 p_2)^{y+i}]^{-a} \right.$$
$$\left. - [p_3^i - bp_3^{x-y}(p_1 p_2)^{y+i}]^{-a} \right)$$
$$\times I[y < x] + \sum_{i=1}^{y-x} p_4^{a(y-x)}(p_1 p_2)^{a(x+i)} \left([p_1 p_4^i - bp_4^{y-x}(p_1 p_2)^{x+i}]^{-a} \right.$$
$$\left. - [p_4^i - bp_4^{y-x}(p_1 p_2)^{x+i}]^{-a} \right)$$
$$\times I[x < y] + [p_1 p_2]^{a(x \vee y)}[1 - b(p_1 p_2)^{x \vee y}]^{-a}, \tag{6.3.7}$$

where $x, y = 1, 2, \ldots$.

Lemma 6.3.1 *The marginal survival functions of (i) BVG-NB are*

$$(1 - b)^{-a} P(Y > y)$$

$$= \sum_{i=1}^{y} p_4^{ay} (p_1 p_2)^{ai} \left([p_1 p_4^i - b p_4^y (p_1 p_2)^i]^{-a} \right.$$

$$- [p_4^i - b p_4^y (p_1 p_2)^i]^{-a} \right) I[y > 0]$$

$$- [p_1 p_2]^{ay} [1 - b(p_1 p_2)^y]^{-a}, \quad y = 1, 2, \ldots$$

$$(1 - b)^{-a} P(X > x)$$

$$= \sum_{i=1}^{x} p_3^{ax} (p_1 p_2)^{ai} \left([p_2 p_3^i - b p_3^x (p_1 p_2)^i]^{-a} \right.$$

$$- [p_3^i - b p_3^x (p_1 p_2)^i]^{-a} \right) I[x > 0]$$

$$- [p_1 p_2]^{ax} [1 - b(p_1 p_2)^x]^{-a}, \quad x = 1, 2, \ldots.$$

PROOF. Obvious from (6.3.7). □

Lemma 6.3.2 *For BVG-NB, the distribution of $\min(X_1, X_2)$ belongs to the family of distributions with survival function $[(1 - b)p^x]^a [1 - bp^x]^{-a}, 1 \leq a, x = 0, 1, 2, \ldots$ the discrete analog of the Pareto type 2 distribution.*

PROOF. Note that

$$P[\min(X_1, X_2) > u] = P[X_1 > u, X_2 > u]$$

$$= [(1 - b)(p_1 p_2)^u]^a [1 - b(p_1 p_2)^u]^{-a},$$

when (X_1, X_2) has the BVG-NB distribution, $u = 0, 1, 2, \ldots$. This in turn follows by letting $u = x = y$ in (6.3.7). □

6.4 DATA ANALYSIS WITH BVG MODEL

Dhar (1998) demonstrated the applicability of the bivariate geometric model, using simulation and a practical data set. In this section, 1995 IX World Cup diving championship data is introduced to demonstrate the inference procedure and the practical applicability of the model. The data consists of scores given by seven judges from seven different countries recorded in a video that starts at the end of the fourth round, which is a random start, taken from NBC sports TV. The score given by each judge is a discrete

random variable taking integer values and also the midpoints of consecutive integers between zero and 10. After dropping the highest and the
lowest of the seven raw scores, the remaining 5 scores are averaged, mul-
tiplied by 3 times the degree of difficulty of the dive to give a score for
the dive. In this data set, we compare the scores given by two groups of
the judges. One group consists of Asian and Caucasus countries namely
Japan and Tajikistan, with the maximum of their scores as X. The other
group consists of western countries, United Kingdom, Australia, Canada,
France and Iceland, with their maximum score as Y. The scores given by
these judges to divers from the Asian and Caucasus group, which include
countries like China, Ukraine, Belarus, Russia will be looked at. We will
see the MLE estimate of $P(X < Y)$ and compare it with the MLE estimate
of $P(Y < X)$, to determine which maximum score is higher, with large
probability. The same procedure will be repeated for all divers including
the ones from western countries like Germany and USA, to see if there is
any change in the probabilistic inequality of the maximum scores of the two
sets of judges.

Given below in Table 6.1 is $2 \times (x, y)$ in the sequence in which Judges
scores were relayed. The 2 here is to convert the data into integer valued
random variable. The score corresponding to the dive of Michael Murphy
of Australia (item number 3) was not displayed by NBC sports.

To estimate the parameters of the bivariate geometric model the above
data is assumed to be a random sample. If we had the entire data, rounds
one to six, of the diving competition one could remove the earlier scores of
the same diver. Also, assuming that the maximums will follow the bivariate geometric distribution. We will compute the MLE estimate based on
thirteen observations, item numbers 1, 5, 6, 8, 9, 10, 11, 14, 15, 17 to 20,
excluding for the time being, divers from USA and Germany. Dhar (1998)
has shown through simulation that MLE gives smaller bias than method of
moment estimators of the p's. The MLE's so computed are $\hat{p}_1 = 162/174$,
$\hat{p}_2 = 168/174$, $\hat{p}_3 = 0.5$, $\hat{p}_4 = 0.3$, $\hat{P}(X > Y) = 27/850 = 0.317647058$
and $\hat{P}(X < Y) = 56/85 = 0.658823529$. This suggests that the maximum
score of the judges from Japan and Tajikistan is probabilistically lower than
the maximum score of the judges from the western countries. Again, this
computation is carried through, all the data points in Table 6.1 to give
$\hat{p}_1 = 245/262$, $\hat{p}_2 = 252/259$, $\hat{p}_3 = 1/6$, $\hat{p}_4 = 1/4$, $\hat{P}(X > Y) = 245/874 =$
0.280320366 and $\hat{P}(X < Y) = 612/874 = 0.700228833$, which implies that
the maximum score given the judges from the western countries continue to
probabilistically dominate those of the judges from Japan and Tajikistan.
There is no indication in this study of any partiality towards one region
over the other by these groups of judges. These probability estimates in

TABLE 6.1 This data is taken from a video recording during the summer of 1995 relayed by NBC sports TV, IX World Cup diving competition, Atlanta, Georgia. The data starts at the last dive of the fourth round of the diving competition

Item #	Diver	X: max score, Asian & Caucasus	Y: max score, West
1	Sun Shuwei, China	19	19
2	David Pichler, USA	15	15
4	Jan Hempel, Germany	13	14
5	Roman Volodkuv, Ukrain	11	12
6	Sergei Kudrevich, Belarus	14	14
7	Patrick Jeffrey, USA	15	14
8	Valdimir Timoshinin, Russia	13	16
9	Dimitry Sautin, Russia	7	5
10	Xiao Hailiang, China	13	13
11	Sun Shuwei, China	15	16
12	David Pichler, USA	15	15
13	Jan Hempel, Germany	17	18
14	Roman Volodkuv, Ukrain	16	16
15	Sergei Kudrevich, Belarus	12	13
16	Patrick Jeffrey, USA	14	14
17	Valdimir Timoshinin, Russia	12	13
18	Dimitry Sautin, Russia	17	18
19	Xiao Hailiang, China	9	10
20	Sun Shuwei, China	18	18

terms of relative magnitudes are consistent with their respective empirical estimates.

Consider the data set in Dhar (1998, Table 3) constructed by projecting consumers preference, from 1, the highest, to 10, the lowest. For the sake of clarity this table has been repeated here as Table 6.2 given below. The table contains scores given by 15 customers to the two most popular competing soft drinks, e.g., Coke (X) and Pepsi (Y).

TABLE 6.2 Projected consumers preference ranks, from 1, the highest preference, to 10, the lowest

X	Y	X	Y	X	Y
1	10	4	5	1	9
10	1	5	5	5	1
3	7	1	3	1	10
2	4	2	1	2	6
6	1	2	2	2	3

The method of moment estimators as described in Dhar (1998), using EX, EY, EX^2 and EY^2 in terms of the p's, is repeated here for the sake of clarification. These are four equations in p_1–p_4, because p_{12} is taken to be equal to $p_1 p_2$. This in turn gives a polynomial equation in $m = p_1 p_2$.

$$m\left[[(\bar{x} + \bar{x}_2)/2](1 - m) - \bar{x}\right]\left[[(\bar{y} + \bar{y}_2)/2](1 - m) - \bar{y}\right]$$
$$= \left[m\{[(\bar{x} + \bar{x}_2)/2](1 - m) - \bar{x}\} + [1 - \bar{x}(1 - m)]^2\right]$$
$$\times \left[m\{[(\bar{y} + \bar{y}_2)/2](1 - m) - \bar{y}\} + [1 - \bar{y}(1 - m)]^2\right].$$

Using MATHEMATICA one notices that this polynomial in $1 - m$ yields four real solutions for the data in Table 6.2, of which only one is extraneous, i.e., outside the range of $[0, 1]$. The largest solution among the remaining three gives the most meaningful estimate of $1 - p_1 p_2$ as $1 - m$.

The estimates of p_3 and p_4 can now be obtained from

$$\hat{p}_3 = 1 - \left\langle \left[\bar{x} - (1 - m)^{-1}\right]\left[[(\bar{x} + \bar{x}_2)/2] - \bar{x}(1 - m)^{-1}\right]^{-1} \right\rangle$$

and

$$\hat{p}_4 = 1 - \left\langle \left[\bar{y} - (1 - m)^{-1}\right]\left[[(\bar{y} + \bar{y}_2)/2] - \bar{y}(1 - m)^{-1}\right]^{-1} \right\rangle.$$

Here again, as described above, $\bar{y}_2$ is the second sample moment of Y values. Using the original moment equations, corresponding to $\bar{x}$ and $\bar{y}$, gives $\hat{p}_1 = (1-m)[(1-\hat{p}_3)\bar{x}-1]+\hat{p}_3$ and $\hat{p}_2 = (1-m)[(1-\hat{p}_4)\bar{y}-1]+\hat{p}_4$. Since $p_1 p_2 = m$, one could use this equation to eliminate extraneous solutions of p_1 to p_4, corresponding to various values of m.

The estimates corresponding to this method of moment estimation, when applied to Table 6.2 data, yield $\hat{p}_1 = 0.702492288$, $\hat{p}_2 = 0.866047395$, $\hat{p}_3 = 0.585531905$, $\hat{p}_4 = 0.667665863$, $\hat{P}(X > Y) = 0.240292793$ and $\hat{P}(X < Y) = 0.657942448$, i.e., brand X is more likely to be preferred over brand Y. Here, the MLE's are $\hat{p}_1 = 17/28 = 0.607142857$, $\hat{p}_2 = 22/28 = 0.785714285$, $\hat{p}_3 = 15/19 = 0.789473684$, $\hat{p}_4 = 31/40 = .775$, $\hat{P}(X > Y) = 0.2487860488$ and $\hat{P}(X < Y) = 0.590243902$, i.e., brand X is more likely to be preferred to brand Y. The last two conclusions are consistent with the empirical estimates $\hat{P}(X > Y) = 4/15 = .2\dot{6}$ and $\hat{P}(X < Y) = 9/15 = 0.6$ and are, therefore, the right conclusions.

REFERENCES

Azlarov, T. A. and Volodin, N. A. (1982). On the discrete analog of Marshall-Olkin's distribution, Stability Problems for Stochastic Models, *Proceedings of the 6th International Seminar* held in Moscow, USSR, April 1982 - *Lecture Notes in Mathematics*, **982**, 17–23.

Basu, A. P. and Dhar, S. K. (1995). Bivariate geometric distribution, *Journal of Applied Statistical Science*, **2**, 12.

Dhar, S. K. (1998). Data analysis with discrete analog of Freund's model, *Journal Applied Statistical Science*, **7**, 169–183.

Freund, J. E. (1961). A bivariate extension of the exponential distribution, *Journal of the American Statistical Association*, **56**, 971–977.

Marshall, A. W. and Olkin, I. (1967). A multivariate exponential distribution, *Journal of the American Statistical Association*, **62**, 30–44.

Marshall, A. W. and Olkin, I. (1985). A family of bivariate distributions generated by the bivariate Bernoulli distribution, *Journal of the American Statistical Association*, **80**, 332–338.

Part III
Estimation and Testing

CHAPTER 7

SMALL AREA ESTIMATION: UPDATES WITH APPRAISAL

J. N. K. RAO

Carleton University, Ottawa, Ontario, Canada

Abstract: Small area estimation has received a lot of attention in recent years due to growing demand for reliable small area estimators. Traditional area-specific direct estimators do not provide adequate precision because sample sizes in small areas are seldom large enough. This makes it necessary to employ indirect estimators that borrow strength from related areas; in particular, model–based indirect estimators. Ghosh and Rao (1994) provided a comprehensive review and appraisal of methods for small area estimation, covering the literature to 1992–3. This paper provides updates to Ghosh and Rao (1994) by covering the literature over the past five years or so on model–based estimation. In particular, we cover several small area models and empirical best linear unbiased prediction (EBLUP), empirical Bayes (EB) and hierarchical Bayes (HB) methods applied to these models. We also present several recent applications of small area estimation.

Keywords and phrases: Small area estimation, empirical best linear unbiased prediction, empirical Bayes, hierarchical Bayes.

7.1 INTRODUCTION

A geographical area or more generally any subpopulation (domain) is regarded as a "small area" if the number of domain–specific sample observations is small. Typically, the domain sample size tends to increase with the size of the domain, but this is not always true. For example, in the U.S. Third National Health and Nutrition Examination Survey (NHANESIII)

113

states with large Hispanic and black populations (e.g., California, Texas) were oversampled at the expense of very small samples or even no samples in other states (e.g., mid–western states). Yet reliable estimates are desired for all the states and sub-areas (e.g., counties) within states. Demand for reliable small area statistics from both public and private sectors has grown rapidly in recent years.

"Direct" estimators, based only on the domain–specific sample data, are typically used to estimate domain parameters. But sample sizes in small areas are rarely large enough for direct estimators to provide acceptable precision. This makes it necessary to "borrow strength" from related areas to find "indirect" estimators that increase the effective sample size and thus increase the precision. Such indirect estimators are based on either implicit or explicit models that provide a link to related small areas through supplementary data such as recent census counts and current administrative records. Indirect estimators based on implicit models include synthetic and composite estimators, while those based on explicit models incorporating area–specific effects include empirical Bayes (EB), empirical best linear unbiased prediction (EBLUP) and hierarchical Bayes (HB) estimators.

Ghosh and Rao (1994) presented a comprehensive overview and appraisal of methods for small area estimation, covering the literature to 1992–3. We refer the reader to Schaible (1996) for an excellent account of the use of indirect estimators in U.S. Federal Programs.

Ghosh and Rao (1994) provided a list of symposia and workshops on small area estimation that have been organized in recent years. We update that list by the following: (i) Conference on Small Area Estimation, U.S. Bureau of the Census, Washington, D.C., March 26-27, 1998; (ii) International Satellite Conference on Small Area Estimation, Riga, Latvia, August 20-21, 1999. Short courses have also been organized: (i) "Small Area Estimation" by J.N.K. Rao, W.A. Fuller, G. Kalton and W.L. Schaible, organized by the Joint Program in Survey Methodology and the Washington Statistical Society, Washington, D.C., May 22–23, 1995; (ii) "Introduction to Small Area Estimation" by J.N.K. Rao, organized by the International Association of Survey Statisticians, Riga, Latvia, August 19, 1999. In addition, numerous invited and contributed sessions on small area estimation have been organized at recent professional statistical meetings, including the American Statistical Association Annual Meetings, the International Statistical Institute bi-annual sessions and the International Indian Statistical Association Conference, Hamilton, Canada, 1998.

Singh, Gambino and Mantel (1994) discussed survey design issues that have an impact on small area statistics. In particular, they presented an excellent illustration of compromise sample size allocations to satisfy reliability requirements at the provincial level as well as sub provincial level. For the Canadian Labour Force Survey with a monthly sample of 59,000 households, optimizing at the provincial level yields a coefficient of variation (CV) for "unemployed" as high as 17.7% for some Unemployment Insurance (UI) regions. On the other hand, a two–step allocation with 42,000 households allocated at the first step to get reliable provincial estimates and the remaining 17,000 households allocated in the second step to produce best possible UI region estimates reduces the worst case of 17.7% CV for UI regions to 9.4% at the expense of a small increase in CV at the provincial and national levels: CV for Ontario increases from 2.8% to 3.4% and for Canada from 1.36% to 1.51%. Preventive measures, such as compromise sample allocations, should be taken at the design stage, whenever possible, to ensure precision for domains like the UI region. But even after taking such measures sample sizes may not be large enough for direct estimates to provide adequate precision for all small areas of interest. As noted before, sometimes the survey is deliberately designed to oversample specific areas (domains) at the expense of small samples or even no samples in other areas of interest.

This paper provides updates to Ghosh and Rao (1994) by covering recent work on model–based small area estimation; in particular, on empirical best linear unbiased prediction (EBLUP), empirical Bayes (EB) and hierarchical Bayes (HB) methods and their applications.

7.2   SMALL AREA MODELS

It is now generally accepted that when indirect estimators are needed they should be based on explicit models that relate the small areas of interest through supplementary data such as last census data and current administrative data. An advantage of the model approach is that it permits validation of models from the sample data. Interesting work on traditional indirect estimates (synthetic, sample-size dependent etc.), however, is also reported in the recent literature [see e.g., Falorsi, Falorsi and Russo (1994); Chaudhuri and Adhikary (1995); Schaible (1996); Marker (1999)].

Small area models may be broadly classified into two types: area level and unit level.

7.2.1   Area Level Models

Area–specific auxiliary data, x_i, are assumed to be available for the sampled areas i $(= 1, \ldots, m)$ as well as the nonsampled areas. A basic area level model assumes that the population small area total Y_i or some suitable function $\theta_i = g(Y_i)$, such as $\theta_i = \log(Y_i)$, is related to $\mathbf{x}_i$ through a linear model with random area effects v_i:

$$\theta_i = \mathbf{x}_i'\boldsymbol{\beta} + v_i, \quad i = 1,\ldots,m \qquad (7.2.1)$$

where $\boldsymbol{\beta}$ is the p–vector of regression parameters and the v_i's are uncorrelated with mean zero and variance σ_v^2. Normality of the v_i is also often assumed. The model (7.2.1) also holds for the non sampled areas. It is also possible to partition the areas into groups and assume separate models of the form (7.2.1) across groups.

We assume that direct estimators $\hat{Y}_i$ of Y_i are available whenever the area sample size $n_i > 1$. It is also customary to assume that

$$\hat{\theta}_i = \theta_i + e_i \qquad (7.2.2)$$

where $\hat{\theta}_i = g(\hat{Y}_i)$ and the sampling errors e_i are independent $N(0, \psi_i)$ with known ψ_i. Combining this sampling model with the "linking" model (7.2.1), we get the well–known area level linear mixed model of Fay and Herriot (1979):

$$\hat{\theta}_i = \mathbf{x}_i'\boldsymbol{\beta} + v_i + e_i. \qquad (7.2.3)$$

Note that (7.2.3) involves both design–based random variables e_i and model-based random variables v_i. In practice, sampling variances ψ_i are seldom known, but smoothing of estimated variances $\hat{\psi}_i$ is often done to get stable estimates ψ_i^* which are then treated as the true ψ_i. Other methods of handling unknown ψ_i are mentioned in Section 7.4. An advantage of the area–level model (7.2.3) is that the survey weights are accounted for through the direct estimators $\hat{\theta}_i$.

The assumption $E(e_i \mid \theta_i) = 0$ in the sampling model ref2.2) may not be valid if the sample size n_i is small and θ_i is a nonlinear function of Y_i, even if the direct estimator $\hat{Y}_i$ is design–unbiased, i.e. $E(\hat{Y}_i \mid Y_i) = Y_i$. A more realistic sampling model is given by

$$\hat{Y}_i = Y_i + e_i^* \qquad (7.2.4)$$

with $E(e_i^* \mid Y_i) = 0$, i.e., $\hat{Y}_i$ is design–unbiased for the total Y_i. In this case, however, we cannot combine (7.2.4) with the linking model to produce a linear mixed model. As a result, standard results in linear model theory do not apply, unlike in the case of (7.2.3). Alternative methods to handle this case are needed (see Section 7.3).

The basic area level model has been extended to handle correlated sampling errors, spatial dependence of random small area effects, vectors of parameters $\boldsymbol{\theta}_i$ (multivariate case), time series and cross–sectional data and others (see Ghosh and Rao, 1994). We discuss some of the recent models for combining cross–sectional and time series data. Suppose θ_{it} denotes a parameter of interest for small area i at time t and $\hat{\theta}_{it}$ is a direct estimator of θ_{it}. Ghosh et $al.$ (1996) assumed the sampling model $\hat{\theta}_{it} \mid \theta_{it} \overset{ind}{\sim} N(\theta_{it}, \psi_{it})$ with known sampling variances ψ_{it}, and the linking model

$$\theta_{it} \mid \mathbf{u}_t \sim N\big(\mathbf{x}'_{it}\boldsymbol{\beta} + \mathbf{z}'_{it}\mathbf{u}_t, \sigma_t^2\big) \tag{7.2.5}$$

and

$$\mathbf{u}_t \mid \mathbf{u}_{t-1} \sim N\big(\mathbf{u}_{t-1}, \mathbf{W}\big) \tag{7.2.6}$$

with known auxiliary variables $\mathbf{x}_{it}$ and $\mathbf{z}_{it}$; they have actually studied the multivariate case $\boldsymbol{\theta}_{it}$. Note that (7.2.6) is the well–known random walk model. The above model has the following limitations: (i) Independence of $\hat{\theta}_{it}$'s over t for each i may not be realistic because estimates are typically correlated over time. (ii) The linking model (7.2.5) does not include area–specific random effects. As a result, it can lead to "over shrinkage". Rao and Yu (1992, 1994) proposed more realistic sampling and linking models. They assumed the sampling model

$$\hat{\boldsymbol{\theta}}_i \mid \boldsymbol{\theta}_i \overset{ind}{\sim} N(\boldsymbol{\theta}_i, \boldsymbol{\Psi}_i) \tag{7.2.7}$$

with known sampling covariance matrix $\boldsymbol{\Psi}_i$, and the linking model

$$\theta_{it} = \mathbf{x}_{it}^T\boldsymbol{\beta} + v_i + u_{it} \tag{7.2.8}$$

with $v_i \overset{iid}{\sim} N(0, \sigma_v^2)$ and independent of u_{it}'s which are assumed to follow an AR(1) model:

$$u_{it} = \rho u_{i,t-1} + \varepsilon_{it}, \quad |\rho| < 1 \tag{7.2.9}$$

with $\varepsilon_{it} \overset{iid}{\sim} N(0, \sigma^2)$, where $\hat{\boldsymbol{\theta}}_i = (\hat{\theta}_{i1}, \ldots, \hat{\theta}_{iT})'$ and $\boldsymbol{\theta}_i = (\theta_{i1}, \ldots, \theta_{iT})'$. Models of the form (7.2.7)–(7.2.9) have been extensively studied in the econometric literature, ignoring sampling errors, i.e., treating $\hat{\theta}_{it}$ as θ_{it}. The above sampling model permits correlations among sampling errors over time and the linking model (7.2.9) includes both area–specific random effects v_i and area by time specific random effects u_{it}. Datta, Lahiri and Lu (1994), following Rao and Yu (1992), used the same sampling model (7.2.7) but assumed the following linking model:

$$\theta_{it} \mid v_i, \mathbf{u}_t \sim N(\mathbf{x}_{it}^T\boldsymbol{\beta}_i + v_i + \mathbf{z}_{it}^T\mathbf{u}_t, \sigma_i^2) \tag{7.2.10}$$

where $\boldsymbol{\beta}_i$'s and σ_i^2's are random and $\mathbf{u}_t$ follows the random walk model (7.2.6). This model allows area–specific random effects v_i and random slopes β_i, but does not contain area by time specific random effects u_{it}. Datta, Lahiri and Maiti (1999) used the Rao–Yu sampling and linking models (7.2.7) and (7.2.8) but replaced the AR(1) model (7.2.9) by a random walk model given by (7.2.9) with $\rho = 1$. Datta $et\ al.$ (1999) considered a similar model but added extra terms to $\mathbf{x}_{it}^T\beta + v_i$ to reflect seasonal variation in their application to estimating unemployment rates for the U.S. states. Singh $et\ al.$ (1994) also used time series/cross-sectional models, but assumed that the sample errors are uncorrelated over time.

Area level models have also been used in the context of disease mapping or estimating regional mortality and disease rates, as noted by Ghosh and Rao (1994). A simple model assumes that the observed small area disease counts $y_i \mid \theta_i \overset{ind}{\sim}$ Poisson $P(n_i\theta_i)$ and $\theta_i \overset{iid}{\sim}$ gamma $G(a,b)$, where θ_i is the true incidence rate and n_i is the number exposed in area i. Maiti (1998) used $\beta_i = \log\theta_i \overset{iid}{\sim} N(\mu, \sigma^2)$ instead of $\theta_i \overset{iid}{\sim} G(a,b)$. He also considered a spatial dependence model for β_i's, using conditional autoregression (CAR) that relates each β_i to a set of neighbourhood areas of area i; see also Ghosh $et\ al.$ (1997). Lahiri and Maiti (1996) modelled age–group specific area disease counts y_{ij}, using Clayton and Kaldor's (1987) approach. They assumed that $y_{i.} = \Sigma_j y_{ij} \mid \theta_i \overset{ind}{\sim} P(e_i\theta_i)$ and $\theta_i \overset{iid}{\sim} G(a,b)$, where $e_i = \Sigma_j \psi_j n_{ij}$ is the expected number of deaths in area i, ψ_j is the j-th group effect assumed to be known and n_{ij} is the number exposed in the j-th age group and area i. Nandram $et\ al.$ (1998) assumed that $y_{ij} \mid \theta_{ij} \overset{ind}{\sim} P(n_{ij}\theta_{ij})$ and $\log\theta_{ij} = \mathbf{x}_j'\beta + v_i$ with $v_i \overset{ind}{\sim} N(0, \sigma^2)$, where θ_{ij} is the area/age-specific mortality rate and $\mathbf{x}_j$ is a vector of covariates for age group j. They also considered random slopes β_i in the linking model.

7.2.2   Unit Level Models

A basis unit level population model assumes that the unit y–values y_{ij}, associated with the units j in the areas i, are related to auxiliary variables $\mathbf{x}_{ij}$ through a nested error regression model

$$y_{ij} = \mathbf{x}_{ij}'\beta + v_i + e_{ij}, \quad j = 1, \dots, N_i;\ i = 1, \dots, m \qquad (7.2.11)$$

where $v_i \overset{iid}{\sim} N(0, \sigma_v^2)$ are independent of $e_{ij} \overset{iid}{\sim} N(0, \sigma_e^2)$ and N_i is the number of population units in the i-th area. The parameters of interest are the totals Y_i or the means $\overline{Y}_i$.

The model (7.2.11) is appropriate for continuous variables y. To handle count or categorical (e.g., binary) y–variables, generalized linear mixed models with random small area effects, v_i, are often used. Ghosh $et\ al.$

(1998) assumed models of the form: (i) Given θ_{ij}'s, the y_{ij}'s are independent and belong to the exponential family with canonical parameter θ_{ij}; (ii) Linking model $g(\theta_{ij}) = \mathbf{x}'_{ij}\boldsymbol{\beta} + v_i$ where $v_i \overset{iid}{\sim} N(0, \sigma_v^2)$ and $g(\cdot)$ is a strictly increasing function. The linear mixed model (7.2.11) is a special case of this class with $g(a) = a$. The logistic function $g(a) = \log[a/(1-a)]$ is often used for binary y [see e.g., Farrell, MacGibbson and Tomberlin (1997)] although probit functions can also be used and offer certain advantages for hierarchical Bayes (HB) inference [Das, Rao and You (1999)].

The sample data $\{y_{ij}, \mathbf{x}_{ij}, \ j = 1, \ldots, n_i; \ i = 1, \ldots, m\}$ is assumed to obey the population model. This implies that the sample design is ignorable or selection bias is absent which, for example, is satisfied for simple random sampling within areas. For more general designs, the sample indicator variable, a_{ij}, should be unrelated to y_{ij}, condition on x_{ij}. Model–based estimators for unit level models do not depend on the survey weights, $\tilde{w}_{ij}$, so that design–consistency as n_i increases is forsaken except when the design is self-weighting, i.e., $\tilde{w}_{ij} = \tilde{w}$, as in the case of simple random sampling. The area level model (7.2.3) is free of these limitations but assumes that the sample variances ψ_i are known; if ψ_i's are assumed unknown the model becomes nonidentifiable or nearly nonidentifiable leading to highly unstable estimates of the parameters. The unit level model is free of the latter difficulty and survey weights can also be incorporated using model–assisted estimators; see Section 7.3.14.

Various extensions of the basic area level models have been studied over the past five years or so. Stukel and Rao (1999) studied two–fold nested error regression models which are appropriate for two–stage sampling within small areas. Following Kleffe and Rao (1992) Arora and Lahiri (1997) studied unit level models of the form (7.2.11) with random error variances σ_i^2 such that $\sigma_i^{-2} \overset{iid}{\sim} G(a, b)$; Kleffe and Rao (1992) assumed the existence of only mean and variance of σ_i^2, without specifying a parametric distribution on σ_i^2. Datta et al. (1999) extended the unit level model (7.2.11) to the multivariate case $\mathbf{y}_{ij}$, following Fuller and Harter (1987). This extension leads to a multivariate nested error regression model. Moura and Holt (1999) generalized (7.2.11) to allow some or all of the regression coefficients to be random and to depend on area level auxiliary variables, thus effectively integrating the use of unit level and area level covariates into a single model. You and Rao (1999a) also studied similar two–level models.

Malec, Davis and Cao (1996) and Malec et al. (1997) studied the binary case, using logistic linear mixed models with random slopes to link the small areas. Raghunathan (1993) specified only the first two moments of y_{ij}'s conditional on small area means θ_i's and the first moment of θ_i as

$\tau_i = h(\mathbf{z}_i'\boldsymbol{\beta})$ for known inverse "link" function $h(\cdot)$ and the second moment of θ_i is allowed to depend on τ_i.

Many of the small area linear mixed models studied in the literature are special cases of the following general linear mixed model with a block diagonal covariance structure, sometimes called longitudinal mixed linear models [Prasad and Rao (1990); Datta and Lahiri (1997)]:

$$\mathbf{y}_i^* = \mathbf{X}_i\boldsymbol{\beta} + \mathbf{Z}_i\mathbf{v}_i + \mathbf{e}_i, \quad i = 1, \ldots, m \qquad (7.2.12)$$

where $\mathbf{v}_i \overset{ind}{\sim} (\mathbf{0}, \mathbf{G}_i(\boldsymbol{\tau}))$ and independent of $\mathbf{e}_i \overset{ind}{\sim} (\mathbf{0}, \mathbf{R}_i(\boldsymbol{\tau}))$. For example, the basic area level (7.2.3) is of the form (7.2.12) with $\mathbf{y}_i^* = \hat{\theta}_i$, $\mathbf{Z}_i = 1$, $\mathbf{G}_i(\boldsymbol{\tau}) = \sigma_v^2$ and $\mathbf{R}_i(\boldsymbol{\tau}) = \psi_i$. Das and Rao (1999) studied general mixed ANOVA models of the form

$$\mathbf{y}^* = \mathbf{X}\boldsymbol{\beta} + \mathbf{Z}_1\mathbf{v}_1 + \cdots + \mathbf{Z}_q\mathbf{v}_q + \mathbf{e}_i \qquad (7.2.13)$$

where $\mathbf{Z}_i$ consists of only 0's and 1's such that there is exactly one 1 in each row and at least one 1 in each column, $\mathbf{v}_i \overset{ind}{\sim} (\mathbf{0}, \sigma_i^2\mathbf{I})$ and independent of $\mathbf{e} \sim (\mathbf{0}, \sigma^2\mathbf{I})$. This model relaxes the assumption a block diagonal covariance structure.

Ghosh and Rao (1994) reviewed some work on model diagnostics for models involving random effects. Jiang, Lahiri and Wu (1998) developed a chi–squared test for checking the normality of the random effects v_i and the errors e_{ij} in the basic unit level sample model $y_{ij} = \mathbf{x}_{ij}'\boldsymbol{\beta} + v_i + e_{ij}$, $j = 1, \ldots, n_i; i = 1, \ldots, m$.

7.3 MODEL-BASED INFERENCE

EBLUP, EB and HB methods have played a prominent role for model-based small area estimation. EBLUP is applicable for linear mixed models whereas EB and HB are more generally valid. EBLUP point estimators do not require distributional assumptions, but normality of random effects is often assumed for estimating the mean squared error (MSE) of the estimators. Also, EBLUP and EB estimators are identical under normality and nearly equal to the HB estimator, but measures of variability of the estimators may be different. To illustrate the methods, we consider the basic area level model (7.2.3), which is extensively used in practice, and then discuss recent methodological developments.

7.3.1 EBLUP Method

Appealing to general results for linear mixed models, the BLUP estimator of θ_i under (7.2.3) is given by

$$\tilde{\theta}_i(\sigma_v^2) = \gamma_i \hat{\theta}_i + (1 - \gamma_i) \mathbf{x}_i' \tilde{\boldsymbol{\beta}}(\sigma_v^2) \qquad (7.3.14)$$

where $\gamma_i = \sigma_v^2 / (\sigma_v^2 + \psi_i)$ and $\tilde{\boldsymbol{\beta}}(\sigma_v^2)$ is the weighted least squares (WLS) estimator of $\boldsymbol{\beta}$ with weights $(\sigma_v^2 + \psi_i)^{-1}$. It follows from (7.3.14) that the BLUP estimator is a weighted combination of the direct estimator $\hat{\theta}_i$ and the regression synthetic estimator $\mathbf{x}_i' \tilde{\boldsymbol{\beta}}(\sigma_v^2)$. The result (7.3.14) does not require the normality of v_i and e_i. Since σ_v^2 is unknown, we replace it by a suitable estimator $\hat{\sigma}_v^2$ to obtain a two-step or EBLUP estimator $\tilde{\theta}_i = \tilde{\theta}_i(\hat{\sigma}_v^2)$. The estimator of total Y_i is taken as $g^{-1}(\tilde{\theta}_i) = h(\tilde{\theta}_i)$. One could use either the method of fitting constants (not requiring normality) or the restricted maximum likelihood (REML) method under normality to estimate σ_v^2. Jiang (1996) showed that REML estimators of variance components in linear mixed models remain consistent under deviations from normality. Therefore, $\tilde{\theta}_i$ with REML estimator of σ_v^2 is also asymptotically valid under nonnormality.

As noted in Section 7.2.1, EBLUP estimation is not applicable if the sampling model (7.2.2) is changed to the more realistic model (7.2.4).

A measure of variability associated with EBLUP estimator is given by its MSE, but no closed form for MSE exists except in some special cases. As a result, considerable attention has been given in recent years to obtain accurate approximations to the MSE of EBLUP estimators. An accurate approximation to $\mathrm{MSE}(\tilde{\theta}_i) = E(\tilde{\theta}_i - \theta_i)^2$, for large m, under normality is given by

$$\mathrm{MSE}(\tilde{\theta}_i) \approx g_{1i}(\sigma_v^2) + g_{2i}(\sigma_v^2) + g_{3i}(\sigma_v^2) \qquad (7.3.15)$$

where

$$g_{1i}(\sigma_v^2) = \gamma_i \psi_i, \qquad (7.3.16)$$

$$g_{2i}(\sigma_v^2) = (1 - \gamma_i)^2 \mathbf{x}_i' \left[\sum_i \mathbf{x}_i \mathbf{x}_i' / (\sigma_v^2 + \psi_i) \right]^{-1} \mathbf{x}_i, \qquad (7.3.17)$$

$$g_{3i}(\sigma_v^2) = [\psi_i^2 / (\sigma_v^2 + \psi_i)^4] E(\hat{\theta}_i - x_i'\beta)^2 \overline{V}(\hat{\sigma}_v^2), \qquad (7.3.18)$$

$$= [\psi_i^2 / (\sigma_v^2 + \psi_i)^2] \overline{V}(\hat{\sigma}_v^2) \qquad (7.3.19)$$

and $\overline{V}(\hat{\sigma}_v^2)$ is the asymptotic variance of $\hat{\sigma}_v^2$ [Prasad and Rao (1990)]. The leading term $g_{1i}(\sigma_v^2) = \gamma_i \psi_i$ is of order $O(1)$ whereas $g_{2i}(\sigma_v^2)$, due to estimating $\boldsymbol{\beta}$, and $g_{3i}(\sigma_v^2)$, due to estimating σ_v^2, are both of order $O(m^{-1})$, for large m. Note that the leading term shows that $\text{MSE}(\tilde{\theta}_i)$ can be substantially smaller than $\text{MSE}(\hat{\theta}_i)$ under the model (7.2.3) when γ_i is small or the model variance σ_v^2 is small relative to the sampling variance ψ_i. The success of small area estimation, therefore, largely depends on getting good auxiliary information $\{\mathbf{x}_i\}$ that leads to a small model variance relative of ψ_i. Of course, one should also make a thorough validation of the assumed model.

An estimator of $\text{MSE}(\tilde{\theta}_i)$, correct to the same order of approximation as (7.3.15), is given by

$$\text{mse}(\tilde{\theta}_i) \approx g_{1i}(\hat{\sigma}_v^2) + g_{2i}(\hat{\sigma}_v^2) + 2g_{3i}(\hat{\sigma}_v^2), \qquad (7.3.20)$$

ie., the bias of (7.3.20) is of lower order than m^{-1} for large m. The approximation (7.3.20) is valid for both the method of fitting constants estimator and the REML estimator, but not for the ML estimator of σ_v^2 [Datta and Lahiri (1997); Prasad and Rao (1990)]. Using the fitting of constants estimator, Lahiri and Rao (1995) showed that (7.3.20) is robust to nonnormality of the small area effects v_i in the sense that approximate unbiasedness remains valid. Note that the normality of sampling errors e_i is still assumed but it is less restrictive due to the central limit theorem's effect on the direct estimators $\hat{\theta}_i$.

A criticism of the MSE estimator (7.3.20) is that it is not area–specific in the sense that it does not depend on $\hat{\theta}_i$ although $\mathbf{x}_i$ is involved through (7.3.17). But it is easy to find other choices using the form 7.3.18) for $g_{3i}(\sigma_v^2)$. For example, we can use

$$\begin{aligned} \text{mse}_1(\hat{\theta}_i) &= g_{1i}(\sigma_v^2) + g_{2i}(\sigma_v^2) + g_{3i}(\sigma_v^2) \\ &\quad + \left[\psi_i^2/(\hat{\sigma}_v^2 + \psi_i)^4\right](\hat{\theta}_i - \mathbf{x}_i'\hat{\boldsymbol{\beta}})^2 h_i(\hat{\sigma}_v^2), \qquad (7.3.21) \end{aligned}$$

where $\hat{\boldsymbol{\beta}} = \tilde{\boldsymbol{\beta}}(\hat{\sigma}_v^2)$ and $h_i(\sigma_v^2) = \overline{V}(\hat{\sigma}_v^2) = 2m^{-2}\Sigma_i(\sigma_v^2 + \psi_i)^2$ for the fitting of constants estimator $\hat{\sigma}_v^2$ [Rao (1998)]. The last term of (7.3.21) is less stable than $g_{3i}(\hat{\sigma}_v^2)$ but it is of lower order than the leading term $g_{1i}(\hat{\sigma}_v^2)$.

Stukel and Rao (1999) obtained EBLUP estimators and associated approximately unbiased (or second–order correct) MSE estimators under two–way nested error regression models. Moura and Holt (1999) obtained similar results for the two–level models. Simulation results of Stukel and Rao (1999) suggest that the behaviour of relative bias of MSE estimators is more complex than in the one–way case. Datta, Day and Basawa (1999) studied the multivariate nested error regression model and developed EBLUP estimators and associated second–order correct estimators of MSE, using REML or ML estimators. In the case of REML estimators, a formulae of the form (7.3.20) holds but for ML estimators an extra term of order $O(m^{-1})$ should be subtracted. Datta and Lahiri (1997) obtained similar results for the general linear mixed model with a block diagonal covariance structure, (7.2.12). Das and Rao (1999) extended these results to the general mixed ANOVA model (7.2.13) in which case the asymptotic set–up is more complex. Datta, Lahiri and Maiti (1999) and You (1999) obtained EBLUP estimators and associated second–order correct estimators of MSE for the time series/cross–sectional linking model (7.2.8) with a random walk model on u_{it}'s. Datta *et al.* used ML and REML estimators of model parameters while You employed the method of moments estimators.

As noted in Section 7.2, model–based estimators for unit level models do not depend on the survey weights. Prasad and Rao (1999) obtained model–assisted estimators for the nested error regression model that depend on survey weights $\tilde{w}_{ij}$ and remain design–consistent as the sample size, n_i, increases. The unit level sample model is first reduced to

$$\overline{y}_{iw} = \overline{\mathbf{x}}'_{iw}\boldsymbol{\beta} + v_i + \overline{e}_{iw}, \qquad (7.3.22)$$

where $\overline{y}_{iw} = \Sigma_j w_{ij} y_{ij}$ with $w_{ij} = \tilde{w}_{ij}/\Sigma_j \tilde{w}_{ij}$ and similar expressions for $\overline{\mathbf{x}}_{iw}$ and $\overline{e}_{iw}$. A pseudo–BLUP estimator of $\theta_i = \overline{\mathbf{X}}'_i\boldsymbol{\beta} + v_i$, for fixed σ_v^2 and σ_e^2, say $\hat{\theta}_{iw}(\sigma_v^2, \sigma_e^2)$ is then obtained from the reduced model (7.3.22), noting that $\overline{e}_{iw} \overset{ind}{\sim} N(0, \sigma_e^2 \Sigma_j w_{ij}^2)$, where $\overline{\mathbf{X}}_i$ is the vector of known population means and $\overline{Y}_i \approx \theta_i$ for large N_i (This estimator is called pseudo–BLUP because it is different from the BLUP estimator under the full unit–level sampling model). The unknown parameters σ_v^2 and σ_e^2 are then replaced by model–consistent estimators $\hat{\sigma}_v^2$ and $\hat{\sigma}_e^2$ under the full model to obtain the pseudo–EBLUP estimator $\hat{\theta}_{iw} = \hat{\theta}_{iw}(\hat{\sigma}_v^2, \hat{\sigma}_e^2)$. This estimator is model–assisted and it is approximately design and model unbiased even if the sample design is nonignorable. Prasad and Rao (1999) also obtained a second–order correct estimator of $\mathrm{MSE}(\hat{\theta}_{iw})$.

Rivest and Belmonte (1999) obtained an unbiased estimator of the conditional MSE of the EBLUP estimator $\tilde{\theta}_i = \tilde{\theta}_i(\hat{\sigma}_v^2)$ for the basic area level model, assuming only the sampling model, i.e., conditionally given θ_i's. Hwang and Rao (1987) obtained similar results and showed empirically that the model–based estimator of MSE, (7.3.20), is much more stable than the unbiased estimator and that it tracks the conditional MSE quite well even under moderate violations of the assumed linking model (7.2.1). Only in extreme cases, such as large outliers θ_i, the model–based estimator might perform poorly compared to the unbiased estimator.

7.3.2   EB Method

In the EB approach to the basic area level model, given by (7.2.1) and (7.2.2), the conditional distribution of θ_i given $\hat{\theta}_i$ and model parameters $\boldsymbol{\beta}$ and σ_v^2, denoted $f(\theta_i \mid \hat{\theta}_i, \boldsymbol{\beta}, \sigma_v^2)$, is first obtained. The model parameters are estimated from the marginal distribution of $\hat{\theta}_i$'s, and inferences are then based on the estimated conditional (or posterior) distribution of θ_i, $f(\theta_i \mid \hat{\theta}_i, \hat{\boldsymbol{\beta}}, \hat{\sigma}_v^2)$. In particular, the mean of the estimated posterior distribution is the EB estimator $\tilde{\theta}_i^{\mathrm{EB}}$. Under normality, $\tilde{\theta}_i^{\mathrm{EB}}$ is identical to the EBLUP estimator $\tilde{\theta}_i$, but the EB approach is generally for any joint distribution. It should be noted that the EB approach is essentially frequentist because it uses only the sampling model and the linking model which can be validated from the data; no priors on the model parameters are involved unlike in the HB approach.

As a measure of variability of $\tilde{\theta}_i^{\mathrm{EB}}$, the variance of the estimated posterior is used. Under normality, it is given by $g_{1i}(\hat{\sigma}_v^2) = \hat{\gamma}_i \psi_i$ which leads to severe underestimation of true variability as measured by MSE. Laird and Louis (1987) proposed a parametric bootstrap method to account for the variability in $\hat{\boldsymbol{\beta}}$ and $\hat{\sigma}_v^2$, but Butar and Lahiri (1997) showed that it is not second–order correct, ie. its bias involves terms of order m^{-1}, unlike the bias of (7.3.20) or (7.3.21). By correcting this bias, they obtained an estimator which is identical to the area–specific MSE estimator (7.3.21). Therefore, corrected EB and EBLUP lead to the same result under normality.

Farrell, MacGibbon and Tomberlin (1997) studied EB estimation for binary y, assuming the sampling model $y_{ij} \mid \theta_{ij} \overset{ind}{\sim} \mathrm{Bernoulli}(\theta_{ij})$ and the linking logistic model $\log\{\theta_{ij}/(1 - \theta_{ij})\} = \mathbf{x}'_{ij}\boldsymbol{\beta} + v_i$ with $v_i \overset{iid}{\sim} N(0, \sigma_v^2)$. The conditional distribution of θ_{ij}'s is approximated by a multivariate normal to get an EB estimator of local area proportion $\overline{Y}_i$. They employed the bootstrap method of Laird and Louis (1987) to get a bootstrap–adjusted estimate of variability associated with the EB estimator. But results of Butar and Lahiri (1997) for the linear case suggest that the bootstrap method may not be second–order correct in the nonlinear case as well. Jiang and Lahiri (1998) also studied EB estimation for the above model and obtained the EB estimator exactly through one–dimensional numerical integration. They called the EB estimator an empirical best predictor (EBP) which may be more appropriate because no priors on model parameters are involved. Employing method of moment estimators of model parameters $\boldsymbol{\beta}$ and σ_v^2, they also obtained an approximation to MSE of the EB estimator correct to terms of order m^{-1}. Jiang and Lahiri (1998) proposed a jackknife method of estimating MSE that is applicable to general longitudinal linear and generalized linear mixed models. This method leads to second–order correct MSE estimators and looks promising. But one needs to recompute the REML estimates of model parameters by deleting each area in turn. The computations can be significantly reduced by using a single step of the Newton–Raphson algorithm with the estimates from the full sample as starting values. Properties of this simplification remain to be studied. Booth and Hobert (1988) argued that the conditional MSE of the EBP given the i-th area data is more relevant as a measure of variability than the unconditional MSE because it is area–specific. Fuller (1989) earlier proposed a similar criterion in the context of linear mixed models. But the MSE estimator (7.3.21) shows that it is possible to obtain area–specific estimators of the unconditional MSE, at least in the linear model case. Also, it is not clear how one should proceed with the conditioning when two or more small area estimators need to be aggregated to obtain an estimator for a larger area. How would one define the conditional MSE of the larger area estimator?

Arora *et al.* (1997) studied the nested error regression model with random error variances σ_i^2 and assumed $\sigma_i^{-2} \overset{iid}{\sim} G(a, b)$. They obtained the EB estimator of small area mean $\overline{Y}_i$ and applied the Laird–Louis bootstrap to estimate its MSE, taking account of the variability due to estimation of model parameters.

7.3.3   HB Method

The HB approach has been extensively used for small area estimation. It is straightforward, inferences are exact and it can handle complex problems using recently developed Monte Carlo Markov Chain (MCMC) methods, such as the Gibbs sampler. A prior distribution on the model parameters (also called hyper parameters) is specified and the posterior distribution of the small area totals Y_i or $g(Y_i) = \theta_i$ is then obtained. Inferences are based on the posterior distribution; in particular, Y_i or θ_i is estimated by its posterior mean and its precision is measured by its posterior variance.

For the basic area level model, (7.2.1) and (7.2.2), with normality of v_i and e_i, the posterior mean $E(\theta_i \mid \hat{\boldsymbol{\theta}})$ and posterior variance $V(\theta_i \mid \hat{\boldsymbol{\theta}})$ are obtained in two stages, where $\hat{\boldsymbol{\theta}} = (\hat{\theta}_1, \ldots, \hat{\theta}_m)'$. In the first stage, we obtain $E(\theta_i \mid \hat{\boldsymbol{\theta}}, \sigma_v^2)$ and $V(\theta_i \mid \hat{\boldsymbol{\theta}}, \sigma_v^2)$ for fixed σ_v^2, assuming an improper prior, $f(\boldsymbol{\beta}) \propto$ const., on $\boldsymbol{\beta}$ to reflect absence of prior information on $\boldsymbol{\beta}$. The conditional posterior mean, given σ_v^2, is identical to the BLUP estimator $\tilde{\theta}_i(\sigma_v^2)$ and the conditional posterior variance is equal to $g_{1i}(\sigma_v^2) + g_{2i}(\sigma_v^2)$. At the second stage, we take account of the uncertainty about σ_v^2 by first calculating its posterior distribution $f(\sigma_v^2 \mid \hat{\boldsymbol{\theta}})$, assuming a prior distribution on σ_v^2 and prior independence of $\boldsymbol{\beta}$ and σ_v^2. The posterior mean and variance are then obtained as

$$\tilde{\theta}_i^{\text{HB}} = E(\theta_i \mid \hat{\boldsymbol{\theta}}) = E_{\sigma_v^2 \mid \hat{\boldsymbol{\theta}}}[\tilde{\theta}_i(\sigma_v^2)] \tag{7.3.23}$$

$$V(\theta_i(\hat{\boldsymbol{\theta}}) = E_{\sigma_v^2 \mid \hat{\boldsymbol{\theta}}}[g_{1i}(\sigma_v^2) + g_{2i}(\sigma_v^2)] + V_{\sigma_v^2 \mid \hat{\boldsymbol{\theta}}}[\tilde{\theta}_i(\sigma_v^2)] \tag{7.3.24}$$

where $E_{\sigma_v^2 \mid \hat{\boldsymbol{\theta}}}$ and $V_{\sigma_v^2 \mid \hat{\boldsymbol{\theta}}}$ denote the expectation and variance with respect to $f(\sigma_v^2 \mid \hat{\boldsymbol{\theta}})$. No closed form expressions for (7.3.23) and (7.3.24) exist, but in this simple case they can be evaluated numerically using only one–dimensional integration. For complex models, high–dimensional integration is often involved and it is necessary to use MCMC–type methods to overcome the computational difficulties.

It follows from (7.3.23) that $\tilde{\theta}_i^{\text{HB}} \approx \tilde{\theta}_i(\hat{\sigma}_v^2)$ but (7.3.24) shows that ignoring uncertainty about σ_v^2 and using $g_{1i}(\hat{\sigma}_v^2) + g_{2i}(\hat{\sigma}_v^2)$ as a measure of variability can lead to significant underestimation.

If the assumed prior $f(\sigma_v^2)$ is proper and informative, the HB approach encounters no difficulties. On the other hand, an improper prior $f(\sigma_v^2)$ could lead to an improper posterior [Hobert and Casella (1996)]. In the latter case, we cannot avoid the difficulty by choosing a diffuse proper prior on σ_v^2 because we will be simply approximating an improper posterior by a proper posterior.

To illustrate the use of Gibbs sampling, we again consider the basic area level model under normality. To implement Gibbs sampling assuming the prior $f(\tau_v = \sigma_v^{-2})$ is a gamma(a, b), $a > 0$, $b > 0$, we need the following Gibbs–conditional distributions:

(i) $\boldsymbol{\beta} \mid \boldsymbol{\theta}, \sigma_v^2, \hat{\boldsymbol{\theta}} \sim N_p[(\mathbf{X}'\mathbf{X})^{-1}\mathbf{X}'\boldsymbol{\theta}, \sigma_v^2(\mathbf{X}'\mathbf{X})^{-1}]$ \hfill (7.3.25)

(ii) $\theta_i \mid \boldsymbol{\beta}, \sigma_v^2, \hat{\boldsymbol{\theta}} \stackrel{ind}{\sim} N(\tilde{\theta}_i(\boldsymbol{\beta}, \sigma_v^2) = \gamma_i\theta_i + (1 - \gamma_i)\mathbf{x}_i'\boldsymbol{\beta}, \gamma_i\psi_i)$, \hfill (7.3.26)

$$i = 1, \ldots, m$$

(iii) $\sigma_v^{-2} \mid \boldsymbol{\beta}, \boldsymbol{\theta}, \hat{\boldsymbol{\theta}} \sim G\left[\dfrac{m}{2} + a, \dfrac{1}{2}\Sigma(\theta_i - \mathbf{x}_i'\boldsymbol{\beta})^2 + b\right]$, \hfill (7.3.27)

where $\mathbf{X}$ is the $m \times p$ matrix with $\mathbf{x}_i'$ as the i-th row and $\boldsymbol{\theta} = (\theta_1, \ldots, \theta_m)'$. The Gibbs algorithm is as follows: (a) Using starting values $\theta_i^{(0)}$ and $\sigma_v^{2(0)}$, draw $\boldsymbol{\beta}^{(1)}$ from (7.3.25). (b) Draw $\theta_i^{(1)}$, $i = 1, \ldots, m$ from (7.3.26) using $\boldsymbol{\beta}^{(1)}$ and $\sigma_v^{2(0)}$. (c) Draw $\sigma_v^{2(1)}$ from (7.3.27) using $\theta_i^{(1)}$, $i = 1, \ldots, m$ and $\boldsymbol{\beta}^{(1)}$. Steps (a)–(c) complete one cycle. Perform a large number of cycles, say t, called "burn-in period", until convergence and then treat

$(\boldsymbol{\beta}^{(t+j)}, \sigma_v^{2(t+j)}, \theta_i^{(t+j)}, \ j = 1, \ldots, J)$ as J samples from the joint posterior of $\boldsymbol{\beta}$, σ_v^2 and θ_i, $i = 1, \ldots, m$. Other methods use multiple parallel runs instead of a single long run as above. Parallel runs can be wasteful because initial "burn-in" periods are discarded from each run. But single long run may leave a significant portion of the space generated by the joint posterior unexplored.

The posterior mean and posterior variance are estimated as

$$\tilde{\theta}_i^{\text{HB}} \approx \frac{1}{J} \sum_j \tilde{\theta}_i \left[\sigma_v^{2(t+j)} \right] = \frac{1}{J} \sum_j \tilde{\theta}_i(j) = \tilde{\theta}_i(\cdot) \tag{7.3.28}$$

and

$$
\begin{aligned}
V(\theta_i \mid \hat{\boldsymbol{\theta}}) \quad &\approx \quad \frac{1}{J} \sum_j \left[g_{1i}(\sigma_v^{2(t+j)}) + g_{2i}(\sigma_v^{2(t+j)}) \right] \\
&+ \frac{1}{J} \sum_j \left[\tilde{\theta}_i(j) - \tilde{\theta}_i(\cdot) \right]^2.
\end{aligned}
\tag{7.3.29}
$$

The estimator $\tilde{\theta}_i(\cdot)$ has smaller simulation error than the estimator $J^{-1} \sum_j \theta_i^{(t+j)}$ because of the Rao–Blackwell property. It is therefore advisable to do analytical calculations first before applying Gibbs sampling.

For the basic area level model, all the conditional distributions, (7.3.25)–(7.3.27), are in a closed form and, therefore, samples can be generated directly. But for more complex models, some of the conditionals may not have closed form in which case alternative algorithms, such as Metropolis–Hastings within Gibbs, are needed to draw samples from the joint posterior distribution. We refer the reader to Brooks (1998) for an excellent review of MCMC methods. Software, called BUGS and CODA, are readily available for implementing MCMC and convergence diagnostics, but caution should be exercised in using MCMC methods. For example, Hobert and Casella (1996) demonstrated that the Gibbs sampler could lead to seemingly reasonable inferences about a nonexistent posterior distribution. This happens when the posterior is improper and yet all the Gibbs–conditional distributions are proper. Another difficulty with MCMC is that the convergence diagnostics tools can fail to detect the sorts of convergence failure that they were designed to identify [Cowles and Carlin (1996)]. Further difficulties include the choices of t for the burn–in period, number of simulated samples, J, and the starting values.

The HB methodology for the small area models discussed in Section 7.2 and other models has been developed in recent years and a variety of applications to real data have been reported. Some of these applications are given in Section 7.4.

To take account of survey weights, You and Rao (1999b) developed a pseudo-HB methodology which leads to estimators similar to the pseudo-EBLUP estimators of Prasad and Rao (1999).

Singh *et al.* (1998) made a comparison of frequentist and Bayesian measures of error, using analytical and empirical methods.

Datta and Lahiri (1995) considered robust HB estimation using a class of scale mixtures of normal distributions on the random effects v_i with basic area level model. This class includes t, Laplace and logistic distributions; Cauchy distribution for outlier areas was adopted.

You (1999) considered the more realistic sampling model (7.2.4) on $\hat{Y}_i$ with sampling errors e_i^* and the linking model (7.2.1). Assuming $V(e_i^* \mid Y_i) = \psi_i^2 Y_i^2$ and $\hat{\theta}_i = \log(\hat{Y}_i)$, he used HB methods to demonstrate that for small sample sizes the posterior inferences under the sample model (7.2.4) can be significantly different from those under the sampling model on $\hat{\theta}_i$.

7.4 SOME RECENT APPLICATIONS

In this section we present some recent applications of EBLUP, EB and HB approaches to small area estimation.

7.4.1 Area–level Models

Basic models

(1) Dick (1995) used the basic area level model (7.2.3) to estimate net under coverage rates in the 1991 Canadian Census. The goal is to estimate 96 adjustment factors $\theta_i = T_i/C_i$, corresponding to 2(sex) × 4(age) × 12(province) combinations, where T_i is the true (unknown) count and C_i is the census count in the i-th area (domain); the net undercoverage rate in the i-th area is given by $U_i = 1 - \theta_i^{-1}$. Direct estimates $\hat{\theta}_i$ were obtained from a post enumeration survey, and sampling variances ψ_i were derived through smoothing of estimated variances, assuming ψ_i is proportional to some power of C_i. Explanatory variables, $\mathbf{x}$, were selected from a set of 42 variables by backward stepwise regression. EBLUP (EB) estimates of θ_i were used and their MSE estimated using (7.3.20) with REML estimate of σ_v^2. The EB adjustment factors $\tilde{\theta}_i^{\text{EB}}$ were converted to estimates of missing persons, $M_i = T_i - C_i$, and these estimates were raked to ensure consistency with direct estimates of marginal totals. The raked EB estimates, $\tilde{\theta}_i^R$, were used as the final estimates of M_i's. MSE estimate of $\tilde{\theta}_i^R$ was obtained as $[\text{mse}(\tilde{\theta}_i^{\text{EB}})]\, (\tilde{\theta}_i^R/\tilde{\theta}_i^{\text{EB}})^2$. This somewhat ad hoc method ensures that the

coefficient of variation (CV) of $\tilde{\theta}_i^{EB}$ is retained by $\tilde{\theta}_i^R$, but properties of this method remains to be investigated.

(2) The basic area level model (7.2.3) with $\theta_i = \log Y_i$ has been recently used to produce model–based county estimates of poor school–age children in U.S.A. [National Research Council (1998)]. Using these estimates, the US Department of Education allocates over \$7 billion of federal funds annually to counties. The difficulty with unknown ψ_i was handled by using a model of the form (7.2.3) for the census year 1990, for which reliable estimates $\hat{\psi}_{ic}$ of sampling variances, ψ_{ic}, are available and assuming the census small area effects v_{ic} follow the same distribution as v_i, i.e., $N(0, \sigma_v^2)$. Under the latter assumption, an estimate of σ_v^2 was obtained from the census data assuming $\hat{\psi}_{ic} = \psi_{ic}$ and used in the current model (7.2.3), assuming $\psi_i = \sigma_e^2/n_i$, to get an estimate of σ_e^2. The resulting estimate, $\tilde{\psi}_i = \tilde{\sigma}_e^2/n_i$, was treated as the true ψ_i in developing EBLUP estimates, $\tilde{\theta}_i$, of θ_i. The small area (county) totals Y_i (number of school–age children in poverty) can then be estimated as $\tilde{Y}_i = \exp(\tilde{\theta}_i)$, but a more refined method based on the mean of lognormal distribution was used: $\tilde{\tilde{Y}}_i = \exp\{\tilde{\theta}_i + \frac{1}{2}\mathrm{mse}(\tilde{\theta}_i)\}$, ignoring the g_{3i}–term in (7.3.20) which was found to be small. The MSE of $\tilde{\tilde{Y}}_i$ was estimated using the approximation $\mathrm{MSE}(\tilde{\theta}_i) \approx \mathrm{CV}^2(\tilde{\tilde{Y}}_i)$. The estimates $\tilde{\tilde{Y}}_i$ were raked to agree with model–based state estimates obtained from a state model. The reader is referred to National Research Council (1998) for details on x–variables used in the county model and evaluation of the models. Several criteria were used for evaluating the models and the estimates, including regression diagnostics and comparisons to the 1990 Census counts.

(3) Other applications of the basic area level model include the following: (i) Estimation of unemployment rates at census tract level [Chand and Alexander (1995)]; (ii) Estimation of counts in employment categories and household income categories at the Congressional District level [Griffiths (1996)]; (iii) Estimation at the provincial level in the Italian Labour Force Survey [Falorsi, Falorsi and Russo (1995)].

Multivariate models

Datta *et al.* (1996) used multivariate area level (Fay–Herriot) models to develop HB estimators of median income of four–person families for U.S. states. Here $\boldsymbol{\theta}_i = (\theta_{i1}, \theta_{i2}, \theta_{i3})'$ with θ_{i1}, θ_{i2} and θ_{i3} denoting the true median incomes of four–, three– and five–person families in state i. Adjusted census income and base–year median census median income for the three groups were used as explanatory variables. Diffuse priors on model parameters were used along with Gibbs sampling. The resulting HB estimators,

HB^3, were compared to the direct Current Population Survey (CPS) estimators and univariate and bivariate model based HB estimators, HB^1 and HB^2, treating the 1979 estimates, available from the 1980 census data, as the true values. In terms of absolute relative bias averaged over the states, the three HB estimators performed similarly, but outperformed the direct CPS estimates. In this application, the univariate estimator HB^1 worked well and it is not necessary to use more complicated estimators based on multivariate models. Estimates of θ_{i1} are used for administering an energy assistance program to low–income families.

Time series models

(1) Ghosh *et al.* (1996) developed HB estimators under the time series linking model given by (7.2.5) and (7.2.6) and applied them to estimate median income of four person families using direct estimates $\hat{\theta}_{it}$, $i = 1, \ldots, 51$; $t = 1, \ldots, 10$ for the 51 states over a ten year period.

(2) Datta *et al.* (1994) used the time series model (7.2.10) with $\mathbf{u}_t$ following (7.2.6) and developed HB estimators. They also used methods for validating the model, based on cross–validation. They applied the methods to estimate monthly unemployment rates for U.S. states. HB estimators performed significantly better than the CPS estimates, as measured by the CPS and HB standard errors. We refer the reader to Datta *et al.* (1994) for details on the x–variables used. Datta *et al.* (1999) used the linking model (7.2.8) with a random walk model on the u_{it}'s, but added extra terms to (7.2.8) to reflect seasonal variation in unemployment rates.

(3) Datta *et al.* (1999) developed EB estimators to estimate median income of four–person families by U.S. states using time series and cross–sectional data. They employed the linking model (7.2.8) with a random walk model on u_{it}'s. Using the 1979 estimates available from the 1980 Census data as the true values, they compared the EB (EBLUP) estimates with the HB estimates of Ghosh *et al.* (1996) and the CPS direct estimates. In terms of absolute relative bias averaged over states, EB performed better than HB and both EB and HB performed much better than the CPS direct estimate. In terms of coefficient of variation, EB again performed better than HB and CPS; second–order correct estimate of MSE of EB was used.

Disease mapping models

Maiti (1998) used the model $y_i \mid \theta_i \overset{ind}{\sim} P(n_i\theta_i)$ and $\beta_i = \log \theta_i \overset{iid}{\sim} N(\mu, \sigma^2)$ with diffuse prior on μ and a gamma prior on σ^{-2}. He obtained HB estimators of θ_i and the posterior variance of θ_i, and applied the results to the well-known lip cancer data from Scottish Counties (small areas); see Clayton and Kaldor (1987) for details. He also studied HB estimation under the spatial dependence model for β_i's mentioned in Section 7.2.1. Estimates of θ_i's are very similar for both the models but standard errors for the spatial model are smaller than those under the first model. Lahiri and Maiti (1996) obtained EB estimators and second order correct estimators of MSE under the Clayton–Kaldor model mentioned in Section 7.2.1, and illustrated the method on the Clayton–Kaldor data set. Nandram et al. (1998) used the age–group specific models, mentioned in Section 7.2.1, to obtain HB estimators and also developed Bayesian methods to compare alternative models, using three different measures of fit. They applied the results to estimate age specific and age adjusted mortality rates for Health Service Areas (sets of counties based on where residents seek routine hospital care) for the disease category "all cancers for white males".

7.4.2   Unit Level Models

We now briefly describe some recent applications of unit level models.

Basic nested error regression models

Rao and Choudhry (1995) provided an overview of small area estimation in the context of business surveys. They also studied the performance of EBLUP estimator of small area total relative to traditional estimators through simulation using real and synthetic populations.

Multivariate nested error regression models

Datta, Day and Basawa (1999) obtained EBLUP (EB) estimators and second order correct estimators of MSE, as noted in Section 7.3.1, for the multivariate nested error regression models. They conducted a simulation study using the sample sizes and auxiliary variable values given by Battese, Harter and Fuller (1988). Further, they estimated model parameters for their multivariate model using Battese et al. data on crop areas under corn and soybeans for $m = 12$ counties in North–Central Iowa. Treating the estimated parameters as true values, they generated simulated samples and showed that the multivariate approach can achieve substantial improvement over the univariate approach.

Random error variance models

Arora and Lahiri (1997) obtained a reduced model from the unit level random error variance model by incorporating survey weights. They performed HB analysis on the reduced model with $\sigma_i^{-2} \overset{iid}{\sim} G(a, b)$, and applied the results to estimate the average weekly consumer expenditures of various items, goods and services for $m = 43$ publication areas (small areas) in U.S.A.

Two level models

Moura and Holt (1999) applied EBLUP estimators to data from a sample of 951 retail stores in Southern Brazil classified into 73 small areas. They compared the average second order correct MSE of the estimators to the average MSE value for the nested error regression model to demonstrate improvement in efficiency. You and Rao (1999a) applied HB methods to the Brazilian data. They studied three different two level models: (1) equal error variances; (2) unequal error variances; (3) random error variances. Bayesian diagnostics revealed that model (2) fits the data better than models (1) and (3).

Logistic linear mixed models

Malec et al. (1997) used logistic linear mixed models and the HB approach to estimate proportions for demographic groups within U.S. states. Data from the National Health Interview Survey were used for this purpose. Cross–validation methods were used to evaluate the model fit. For one of the binary variables observed for respondents to the 1990 census long form, they compared the estimates from alternative methods and models with the very accurate census estimates of true values. For logistic linear mixed models, not all the conditional distributions for Gibbs sampling have closed form unlike those obtained for the probit linear mixed model derived from a latent variable approach [Das et al. (1999)].

Malec, Davis and Cao (1996) studied logistic linear mixed models to estimate overweight prevalence for subgroups (small areas) using National Health and Nutrition Examination Survey (NHANES III) data. Again, HB methods were used but survey weights were incorporated using a pseudo–likelihood. Folsom, Shah and Vaish (1999) studied general logistic mixed linear models in the context of estimating substance abuse in U.S. states from the 1994-6 National Household Surveys on Drug Abuse. They developed survey–weighted pseudo HB estimators and associated posterior variance, using MCMC methods.

Ghosh *et al.* (1998) applied the HB approach to generalized linear mixed models and used the results on two real data sets. The first data set, based on a 1991 sample of all persons in 15 geographical regions of Canada consists of responses classified into four categories to the question "Have you experienced any negative impact of exposure to health hazards in the work place?" Objective here is to estimate the proportion of workers in each of the four response categories for every one of 60 groups cross–classified by 16 geographical regions and 4 demographic (age × sex) groups. The second data set relates to cancer mortality rates for the 115 counties in Missouri during 1972–81.

Acknowledgment This work was supported by a research grant from the Natural Sciences and Engineering Research Council of Canada.

REFERENCES

Arora, V. and Lahiri, P. (1997). On the superiority of the Bayesian method over the BLUP in small area estimation problems, *Statistica Sinica*, **7**, 1053–1063.

Arora, V. and Lahiri, P. and Mukherjee, K. (1997). Empirical Bayes estimation of finite population means from complex surveys, *Journal of the American Statistical Association*, **92**, 1555–1562.

Battese, G. E., Harter, R. M. and Fuller, W. A. (1988). An error component model for prediction of county crop areas using survey and satellite data, *Journal of the American Statistical Association*, **83**, 28–36.

Booth, J. G. and Hobert, J. P. (1998). Standard errors of predictors in generalized linear mixed models, *Journal of the American Statistical Association*, **93**, 362–372.

Brooks, S. P. (1998). Markov chain Monte Carlo method and its application. *The Statistician*, **47**, 69–100.

Butar, P. B. and Lahiri, P. (1997). On the measures of uncertainty of empirical Bayes small area estimators, *Technical Report*, Department of Mathematics and Statistics, University of Nebraska–Lincoln.

Chand, N. and Alexander, C. H. (1995). Indirect estimation of rates and proportions for small areas with continuous measurement, *Proceedings Sec. Survey Res. Meth.*, American Statistical Association, pp. 549–54.

Chaudhuri, A. and Adhikary, A. K. (1995). On generalized regression estimators of small domain totals – an evaluation study, *Pakistan Journal of Statistics*, **11**, 173–189.

Clayton, D. and Kaldor, J. (1987). Empirical Bayes estimates of age–standardized relative risks for use in disease mapping, *Biometrics*, **43**, 671–681.

Cowles, M. K. and Carlin, B. P. (1996). Markov chain Monte Carlo convergence diagnostics: a comparative review, *Journal of the American Statistical Association*, **91**, 883–904.

Das, K. and Rao, J. N. K. (1999). Second order approximations for standard errors of empirical BLUP estimators in general mixed ANOVA models, Paper under preparation.

Das, K. and Rao, J. N. K. and You, Y. (1999). Small area estimation for binary variables using probit linear mixed models. Paper under preparation.

Datta, G. S., Lahiri, P. and Lu, K. L. (1994). Hierarchical Bayes time series modeling in small area estimation with applications. *Technical Report*, Department of Mathematics and Statistics, University of Nebraska–Lincoln.

Datta, G. S. and Lahiri, P. (1995). Robust hierarchical Bayes estimation of small area characteristics in the presence of covariates and outliers, *Journal of Multivariate Analysis*, **54**, 310–328.

Datta, G. S., Ghosh, M., Nangia, N. and Natarajan, K. (1996). Estimation of median income of four–person families: a Bayesian approach, In *Bayesian Analysis in Statistics and Econometrics* (Eds., D. A. Berry, K. M. Chaloner, J. K. Geweke), pp. 129–140, John Wiley & Sons, New York.

Datta, G. S. and Lahiri, P. (1997). A unified measure of uncertainty of estimated best linear unbiased predictor in small–area estimation problems, *Technical Report*, Department of Statistics, University of Georgia–Athens.

Datta, G. S., Lahiri, P. and Maiti, T. (1999). Empirical Bayes estimation of median income of four person families by states using time series and cross–sectional data, *Technical Report*, Department of Statistics, University of Georgia–Athens.

Datta, G. S., Lahiri, P., Maiti, T. and Lu, K. L. (1999). Hierarchical Bayes estimation of unemployment rates for the U.S. states, *Journal of the American Statistical Association*, **94**.

Datta, G. S., Day, B. and Basawa, I. (1999). Empirical best linear unbiased and empirical Bayes prediction in multivariate small area estimation, *Journal of Statistical Planning and Inference*, **75**, 269–279.

Dick, P. (1995). Modelling net undercoverage in the 1991 Canadian Census, *Survey Methodology*, **21**, 45–54.

Falorsi, P. P., Falorsi, S. and Russo, A. (1994). Empirical comparison of small area estimation methods for the Italian labour force survey, *Survey Methodology*, **20**, 171–176.

Falorsi, P. D., Falorsi, S. and Russo, A. (1995). Small area estimation at provincial level in the Italian Labour Force Survey, *Proceedings of the Annual Research Conference, U.S. Census Bureau*, pp. 617–635.

Farrell, P. J., MacGibbon, B. and Tomberlin, T. J. (1997). Empirical Bayes estimates of small area proportions in multistage designs, *Statistica Sinica*, **7**, 1065–1083.

Fay, R. E. and Herriot, R. A. (1979). Estimates of income for small places: an application of James–Stein procedures to census data, *Journal of the American Statistical Association*, **74**, 269–277.

Folsom, R., Shah, B. and Vaish, A. (1999). Substance Abuse in States: Model Based Estimates from the 1994–96 National Household Surveys on Drug Abuse: Methodology Report, Research Triangle Institute.

Fuller, W. A. and Harter, R. M. (1987). The multivariate components of variance model for small area estimation, In *Small Area Statistics* (Eds., R. Platek, J. N. K. Rao, C. E. Särndal and M. P. Singh), pp. 103–123, John Wiley & Sons, New York.

Fuller, W. A. (1989). Prediction of true values for the measurement error model. Paper presented at the *Conference on Statistical Analysis of Measurement Error Models*, Humboldt State University, Humboldt, Germany.

Ghosh, M. and Rao, J. N. K. (1994). Small area estimation: an appraisal, *Statistical Science*, **9**, 55–93.

Ghosh, M., Natarajan, K., Kim, D. and Waller, L. A. (1997). Hierarchical Bayes GLM's for the analysis of spatial data: an applciation to disease mapping, *Technical Report*, University of Florida–Gainsville.

Ghosh, M., Nangia, N. and Kim, D. H. (1996). Estimation of median income of four–person families: a Bayesian approach, *Journal of the American Statistical Association*, **91**, 1423–1431.

Ghosh, M., Natarajan, K., Stroud, T. W. F. and Carlin, B. P. (1998). Generalized linear models for small area estimation, *Journal of the American Statistical Association*, **93**, 273–282.

Hobert, J. P. and Casella, G. (1996). The effect of improper priors on Gibbs sampling in hierarchical linear mixed models, *Journal of the American Statistical Association*, **91**, 1461–1479.

Griffiths, R. (1996). Current Population Survey small area estimation for congressional districts, *Proceedings of the Section on Survey Research Methods, American Statistical Association*, pp. 314–319.

Jiang, J. (1996). REML estimation: asymptotic behaviour and related topics, *Annals of Statistics*, **24**, 255–286.

Jiang, J. and Lahiri, P. (1998). Empirical best prediction for small area inference with binary data, *Technical Report*, Department of Mathematics and Statistics, University of Nebraska–Lincoln.

Jiang, J., Lahiri, P. and Wu, C. (1998). On Pearson–χ^2 testing with unobservable frequencies and mixed model diagnostics. *Technical Report*, Department of Statistics, Case Western Reserve University, Cleveland.

Jiang, P., Lahiri, P. and Wan, S. (1999). Jackknifing the mean squared error of empirical best predictor. *Technical Report*, Department of Statistics, Case Western Reserve University, Cleveland.

Kleffe, J. and Rao, J. N. K. (1992). Estimation of mean square error of empirical best linear unbiased predictors under random error variance linear model, *Journal of Multivariate Analysis*, **43**, 1–15.

Lahiri, P.A. and Rao, J. N. K. (1995). Robust estimation of mean squared error of small area estimators, *Journal of the American Statistical Association*, **82**, 758–766.

Lahiri, P.A. and Maiti, T. (1996). Empirical Bayes estimation of mortality from diseases for small area, *Technical Report*, Department of Mathematics and Statistics, University of Nebraska–Lincoln.

Laird, N.M. and Louis, T.A. (1987). Empirical Bayes confidence intervals based on bootstrap samples, *Journal of the American Statistical Association*, **82**, 739–750.

Maiti, T. (1998). Hierarchical Bayes estimation of mortality rates for disease mapping, *Journal of Statistical Planning and Inference*, **69**, 339–348.

Malec, D., Davis, W. and Cao, X. (1996). Small area estimates of overweight prevalence using the third National Health and Nutrition Examination Survey (NHANESIII), *Proceedings of the Section on Survey Research Methods, American Statistical Association*, pp. 326–331.

Malec, D., Sedransk, J., Moriarity, C. L. and LeClere, F. (1997). Small area inference for binary variables in the National Health Interview Survey, *Journal of the American Statistical Association*, **92**, 815–826.

Marker, D. A. (1999). Organization of small area estimators using generalized linear regression framework, *Journal of Official Statistics*, **15**, 1–24.

Moura, F. and Holt, D. (1999). Small area estimation using survey data multi level models, *Survey Methodology*, **25**, 73–80.

Nandram, B., Sedransk, J. and Rickle, L. (1998). Regression analysis of mortality rates for U.S. Health Service Areas, *Technical Report*, Worcester Polytechnic Institute.

National Research Council (1998). *Small Area Estimation of School–Age Children in Poverty*, Interim Report 2, National Research Council, Washington, D.C.

Prasad, N. G. N. and Rao, J. N. K. (1990). The estimation of mean squared errors of small–area estimators, *Journal of the American Statistical Association*, **85**, 163–171.

Prasad, N. G. N. and Rao, J. N. K. (1999). On robust estimation using a simple random effects model, *Survey Methodology*, **25**, 67-72.

Raghunathan, T. E. (1993). A quasi–empirical bayes method for small area estimation, *Journal of the American Statistical Association*, **88**, 1444–1448.

Rao, J. N. K. and Yu, M. (1992). Small area estimation by combining time series and cross–sectional data, *Proceedings of the Section on Survey Research Methods, American Statistical Association*, pp. 1–9.

Rao, J. N. K. and Yu, M. (1994). Small area estimation by combining time series and cross–sectional data, *The Canadian Journal of Statistics*, **22**, 511–528.

Rao, J. N. K. and Choudhry, G. H. (1995). Small area estimation: overview and empirical study, In *Business Survey Methods* (Eds., B. G. Cox *et al.*), pp. 527–542, John Wiley & Sons, New York.

Rao, J. N. K. (1998). EB and EBLUP in small area estimation. *Technical Report*, Laboratory for Research in Statistics and Probability, Carleton University, Ottawa, Ontario.

Rivest, L. P. and Belmonte, E. (1999). The conditional mean squared errors of small area estimators in survey sampling. *Technical Report*, Laval University, Laval, Quebec.

Schaible, W. L. (1996) (Ed.). *Indirect Estimators in U.S. Federal Programs*. Lecture Notes in Statistics No. 108, Springer-Verlag, New York.

Singh, A. C., Mantel, H. J. and Thomas, B. W. (1994). Time series EBLUPs for small areas using survey data, *Survey Methodology*, **20**, 33–43.

Singh, A. C., Stukel, D. and Pfeffermann, D. (1998). Bayesian versus frequentist measures of errors in small area estimation, *Journal of the Royal Statistical Society, Series B*, 60, 377-396.

Singh, M. P., Gambino, J. and Mantel, H. J. (1994). Issues and strategies for small area data, *Survey Methodology*, **20**, 3-22.

Stukel, D. M. and Rao, J. N. K. (1999). On small–area estimation under two–fold nested error regression models, *Journal of Statistical Planning and Inference*, **78**, 131–147.

You, Y. (1999). Hierarchical Bayes and Related Methods for Model–Based Small Area Estimation, *Ph.D. Thesis*, School of Mathematics and Statistics, Carleton University, Ottawa, Ontario.

You, Y. and Rao, J. N. K. (1999a). Hierarchical Bayes estimation of small area means using multi–level models. Invited paper, *Conference*

Proceedings, IASS Satellite Conference on Small Area Estimation, 171–185, Riga, Latvia.

You, Y. and Rao, J. N. K. (1999b). Pseudo hierarchical Bayes small area estimation using sampling weights, *Proceedings of the Survey Methods Section,* SSC Annual Meetings, Regina, Saskatchewan, June, 1999.

CHAPTER 8

UNIMODALITY IN CIRCULAR DATA: A BAYES TEST

SANJIB BASU

Northern Illinois University, DeKalb, IL

S. RAO JAMMALAMADAKA

University of California, Santa Barbara, CA

Abstract: Circular data which represent directions in two dimensions, may be measured as angles. Unimodality, which is often assumed, is a crucial issue since modeling and further inference depend on it. Just as on the real line, descriptive as well as inference tools are different for unimodal data as opposed to multi-modal data. We propose a Bayesian test for unimodality of circular data using mixtures of von-Mises distribution as the alternative. The proposed test is performed and evaluated using Markov Chain Monte-Carlo methodology.

Keywords and phrases: Directional data, von Mises distribution, mixture distribution, Bayes approach

8.1 INTRODUCTION

Suppose we have a set of independent and identically distributed measurements on 2-dimensional directions, say α_1, α_2, ..., α_n. These measurements, called angular or circular data, can be represented as points on the circumference of a circle with unit radius. They may be wind directions, the vanishing angles at the horizon for a group of birds or the times of arrival at a hospital emergency room where the 24 hour cycle is represented as a circle. Such data may have one or more peaks or show no preferred direction at all, i.e., correspond to an isotropic or uniform distribution. Most circular statistical inference about preferred directions or modes starts after eliminating the last possibility namely that the data has no preferred direction i.e., that it is not uniformly distributed. The next step is to ask if there is just a single mode or if the data is multimodal, which is the subject of this paper.

As an example, consider a meteorologist studying wind directions. Based on past data, (s)he might be interested in knowing if the wind direction is predominantly in one direction or whether it is indeed different say at different times of the day or week. Similarly in calculating the directional spectrum of ocean waves, it is crucial to know whether we are dealing a unimodal or multimodal spectrum.

Circular data involves observations θ which are angles, i.e, $0 \leq \theta < 2\pi$. Such data are inherently periodic, i.e., $\theta \equiv (\theta + 2\pi k)$ for any integer k. This inherent periodicity sets apart circular statistical analysis, from the more common "linear" statistical analysis where one uses methods and models based on the mean, variance, etc. Such models and methods are not appropriate in circular statistics.

In standard(linear) statistics, a univariate density f is unimodal or has a single mode if f is non-decreasing up to a point M and non-increasing thereafter. In circular statistics, however, due to the circular nature and lack of well-defined left and right endpoints (such as $-\infty$ and ∞ in real line), the definition of unimodality also requires an antimode A. We will say that a circular probability density $p(\theta)$ is unimodal with mode at M if there exists an antimode A such that $p(\theta)$ is non-decreasing for $A \leq \theta \leq M$ and is non-increasing for $M \leq \theta \leq A$.

Knowledge of the number of modes of $p(\theta)$ is clearly of importance in circular statistics. For instance, a common example of circular data involves the vanishing directions of pigeons when they are released some distance away from their "home". The underlying scientific question relates to how these birds orient themselves. Are they flying towards their "home-direction"? Unimodality of the density $p(\theta)$ here would imply that pigeons have a preferred vanishing direction and is a hypothesis of considerable scientific interest.

As another example, several stations measure the mean wave direction every hour which corresponds to the dominant energy of the period. The wave directions depend on weather conditions, ocean currents and many other natural factors. The daily variation of the wave directions is an example of circular data on a 24-hour cycle. The hypothesis of unimodality here would imply that there is an overall preferred direction around which the daily variations of the wave directions are distributed.

In linear statistics, the problem of estimating the number of modes and/or statistical tests for discovering the presence of more than one mode are considered by many authors. The earliest approach involve modeling multimodality through mixtures of distributions, see Wolfe (1970). Later works include several different approaches, density estimation and bump hunting [Good and Gaskins (1981), Silverman (1981)], distance of empirical distribution from the closest unimodal distribution [Hartigan and Hartigan (1985)] and the approach of excess mass functional [Müller and Swatzki (1991)]. Recently, Basu (1995) proposed a Bayesian test for unimodality using the Khintchine representation which states that every unimodal distribution on the real line can be represented as a mixture of uniform distributions.

We address a similar problem here but in the context of circular data. Let $\theta_1, \ldots, \theta_n$ be i.i.d. observations from the circular density $p(\theta)$. We want to test $H_0 : p(\theta)$ is a unimodal against the alternative that it is not. We propose a Bayesian test which incorporates observed data and prior information. In this test, we restrict ourselves to the class of models whose density $p(\theta)$ can be represented parametrically as a mixture of two von-Mises distributions. After observing the data $\theta_1, \ldots \theta_n$, the joint prior distribution of the mixing proportion and the location and scale parameters of the two components are updated to their joint posterior distribution. The posterior probability of $p(\theta)$ being unimodal is then compared to the prior probability of unimodality to make a decision between H_0 and H_1. These probabilities are computed by Markov Chain Monte Carlo sampling. In fact, one of the major strength of the proposed method is the simplicity of the computations involved; they are mostly direct simulations from popular densities which can be routinely implemented.

8.2   EXISTING LITERATURE

Many excellent books discuss statistical analysis of circular data, including Mardia (1972), Batschelet (1981) and Fisher (1993). We refer the reader to these books and the references therein. However, there does not seem to be any work on tests for unimodality of circular data.

The circular data literature, related to this article can be broadly divided into three groups. The first group involves tests for randomness against a unimodal alternative. Here the null hypothesis is isotropy, modeled by the uniform distribution on the circle. A common test for this against the von-Mises distribution, is known as the Rayleigh test. This test based on the length of the sample resultant, is known to be uniformly most powerful invariant test; see, for instance Mardia (1972, pp. 180–182). For other tests which are more nonparametric, see Section 3.1 of Rao (Jammalamadaka) (1984).

Another set of references which could be related to the question we are studying comes from the density estimation point of view. Semiparametric and nonparametric density estimation for circular data are studied by many, for example, see Bai *et al.* (1988). From a density estimate one can determine the number of modes. However, tests of hypotheses are harder to come by since this involves the much harder problem of density estimation.

Other related work is on mixture distributions and estimating the number of mixing components, etc; see Mardia (1972), Bartel (1984). Spurr and Koutbeiy (1991) proposed a stepwise procedure for testing for the number of components in a von-Mises mixture, by first testing for one component against more than one, then two components against more than two and so on. This is a Bootstrap based test and can be computationally intensive. We point out here that a two or more component mixture can still be unimodal and hence the problem we are addressing is clearly distinct from these articles.

We also mention here that Mardia and Spurr (1973) developed a multi-sample test for data drawn from a L-modal population which they model by a mixture of a scaled von-Mises distribution on $[0, 2\pi/L)$, another scaled von-Mises distribution on $[2\pi/L, 4\pi/L)$ and so on. Our approach is also quite distinct from this work.

8.3 MIXTURE OF TWO VON-MISES DISTRIBUTIONS

We are given n i.i.d. circular observations $\theta_1, \ldots, \theta_n$ $(0 \leq \theta < 2\pi)$ from an unknown circular density $p(\theta)$ and want to test $H_0 : p(\theta)$ is unimodal against $H_1 : p(\theta)$ is not unimodal. We model $p(\theta)$ parametrically as a mixture of two von-Mises distribution,

$$p(\theta) = \pi \, vm(\theta|\mu_1, \kappa_1) + (1 - \pi) \, vm(\theta|\mu_2, \kappa_2) \qquad (8.3.1)$$

where $vm(\theta|\mu, \kappa) = \exp(\kappa \cos(\theta - \mu))/\{2\pi I_0(\kappa)\}$, $0 \leq \theta < 2\pi$ denotes the density of a von-Mises distribution. Here $I_0(\kappa)$ is the modified Bessel function of order 0.

The von-Mises distribution $vm(\theta|\mu, \kappa)$ plays a central role in circular statistics, quite similar to that of the normal distribution in linear statistics. The parameter μ $(0 \leq \mu < 2\pi)$ is called the mean direction. The von-Mises distribution is symmetric and unimodal about μ. The parameter $\kappa > 0$ is the concentration parameter (similar to a precision parameter) and measures the concentration of mass around μ. Note that while the mean direction parameter μ is angular (i.e., $0 \leq \mu < 2\pi$), the concentration κ is a positive real parameter, a fact useful our posterior simulations. As $\kappa \to 0$, the von-Mises distribution converges to the uniform distribution on the circle whereas as $\kappa \to \infty$, the von-Mises distribution converges to a degenerate distribution at μ. The popularity of von-Mises distribution in circular statistics stems from the fact that closed form results are often available for the sampling distributions of statistics from this model which are almost impossible for most other circular distributions. We refer the reader to books by Mardia (1972), Fisher (1993) for further properties of this distribution.

There are several advantages to modeling $p(\theta)$ parametrically as von-Mises mixture. A 2-component von-Mises mixture allows a wide variety shapes (based on various choices of the parameters $\mu_1, \mu_2, \kappa_1, \kappa_2$ and π) which includes symmetric and asymmetric, as well as both unimodal and bimodal densities. In Figure 8.1 we show three such mixtures to illustrate the different shapes and modality choices that are possible. Secondly, the conjugate prior for the mean direction of a von-Mises distribution is known and is another von-Mises distribution. This structure provides a flexible and at the same time, a mathematically convenient prior structure. Thirdly, if $p(\theta)$ is a 2-component von-Mises mixture, then a complete mathematical characterization is available about when $p(\theta)$ is unimodal and when it is not. This characterization, due to Mardia and Sutton (1975), is described next.

Let $p(\theta) = \pi\, vm(\theta|\mu_1, \kappa_1) + (1-\pi)\, vm(\theta|\mu_2, \kappa_2)$. By appropriate choice of the zero direction, one can assume that $\mu_1 = 0$ and $0 \leq \mu_2 \leq \pi$. The characterization is stated in this parametrization.

Case 1. This is a boundary case when $\mu_2 = \pi$, i.e., the two means are at the opposite ends of the circle. Then the density $p(\theta)$ is bimodal if and only if the mixing proportion π satisfies $p_1 \leq \pi \leq p_2$ where $p_1 = \{1 + \kappa^* \exp(\kappa_1 + \kappa_2)\}^{-1}$, $p_2 = \{1 + \kappa^* \exp(-\kappa_1 - \kappa_2)\}^{-1}$, and $\kappa^* = \{\kappa_1 I_0(\kappa_2)\}/\{\kappa_2 I_0(\kappa_1)\}$. In this case, the two modes are at π and 2π.

Case 2. This is the important case when $0 < \mu_2 < \pi$. Then the density $p(\theta)$ is bimodal if and only if $p'_1 \leq \pi \leq p'_2$ and $\sin\mu_2 \leq h(\widehat{\theta})$ Here $p'_j = \{1 - \kappa^*/u(\theta_i)\}^{-1}$, $j = 1, 2$; κ^* is as defined above, $u(\theta) = \{\sin(\theta - \mu_2)/\sin\theta\} \exp(\kappa_2 \cos(\theta - \mu_2) - \kappa_1 \cos\theta)$ and $0 < \theta_1 < \theta_2 < \mu_2$ are the two solutions of the equation $h(\theta) = \sin\mu_2$. Finally, $h(\theta) = \sin\theta \sin(\theta - \mu_2)\{\kappa_2 \sin(\theta_m u_2) - \kappa_1 \sin\theta\}$ and $\widehat{\theta}$ maximizes $h(\theta)$ within $0 < \theta < \mu_2$ which further can be obtained as the real root of a cubic equation.

Mardia and Sutton (1975) also provide information on the location of the mode(s) and antimode(s) of the mixture density $p(\theta)$ which are omitted here as they are not of primary importance in our context.

8.4   PRIOR SPECIFICATION

We next specify the prior models for the parameters of the mixture density $p(\theta)$. Note that this density has five parameters, the mixing proportion π, the two mean directions μ_1 and μ_2 and the two concentrations κ_1 and κ_2. For the mixing proportion π, we assume a Uniform[0, 1] prior which reflects our prior uncertainty about its value.

We next describe the prior distributions for the mean directions μ_i, $i = 1, 2$. Note that the von-Mises distribution can alternatively be written as $vm(\theta|\mu, \kappa) = \{2\pi I_0(\kappa)\}^{-1} \exp(\kappa \eta^T \underset{\sim}{l})$ where $\eta^T = (\eta_1, \eta_2) = (\cos\mu, \sin\mu)$ and $\underset{\sim}{l}^T = (l_1, l_2) = (\cos\theta, \sin\theta)$. This alternative representation shows the exponential family structure of the von-Mises distribution. Using this structure, Mardia and El-Atoum (1976) showed that a conjugate prior for μ in the $vm(\theta|\mu, \kappa)$ sampling density is another von-Mises distribution, say $\pi(\mu) = vm(\nu, \tau)$.

We use this convenient conjugate structure in our formulation and assume that the two mean directions, μ_j, $j = 1, 2$, have two independent von-Mises prior, $vm(\nu_j, \tau_j)$, $j = 1, 2$. Within the von-Mises parametric structure, the choice of the hyperparameters ν_j, τ_j actually provides considerable flexibility in modeling different prior opinion. Note that as τ_j tends to zero, the $vm(\nu_j, \tau_j)$ prior tends to the uniform distribution on the circle. Thus, one can specify small values for the hyperparameter τ_j to reflect prior ignorance about μ_j.

Finally we specify the priors for the two concentration parameters κ_1 and κ_2. The concentration parameter κ of a von-Mises distribution does not have a conjugate prior (due to the presence of the modified Bessel function $I_0(\kappa)$ term). However, as we noted before, κ is not restricted to a circular domain and can take any positive real value. A popular prior choice for precision parameter is a Gamma prior. We assume that the two concentration parameters κ_j, $j = 1, 2$ have two independent Gamma(α_j, β_j), $j = 1, 2$ priors. In fact, many other prior choices for κ_j are possible. Basu and Jammalamadaka (1999) describe a broad class of prior choices for κ_j and describe how the unimodality test can be carried out for any prior selection from this broad class.

8.5   PRIOR AND POSTERIOR PROBABILITY OF UNIMODALITY

In the following, we first repeat the complete model structure.

- We observe circular observations $\theta_1, \ldots, \theta_n$ i.i.d. from the the density $p(\theta) = \pi\, vm(\theta|\mu_1, \kappa_1) + (1 - \pi)\, vm(\theta|\mu_2, \kappa_2)$. This results in the likelihood

$$L(\pi, \mu_1, \mu_2, \kappa_1, \kappa_2)$$
$$= \prod_{i=1}^{n} p(\theta_i) = \prod_{i=1}^{n} \{\pi\, vm(\theta_i|\mu_1, \kappa_1) + (1 - \pi)\, vm(\theta_i|\mu_2, \kappa_2)\}$$

$$(8.5.2)$$

- The mixing proportion π has a Uniform$[0, 1]$ prior distribution.

- The prior for μ_j is $p(\mu_j) = vm(\nu_j, \tau_j)$, $j = 1, 2$.

- The prior for κ_j is $p(\kappa_j) = \text{Gamma}(\alpha_j, \beta_j)$, $j = 1.2$.

Our proposed Bayesian test of unimodality is performed by comparing the prior probability from this model with the posterior probability of unimodality. The computation for these probabilities are described next.

The prior probability of unimodality is the integral of the joint prior density of $(\pi, \mu_1, \mu_2, \kappa_1, \kappa_2)$ over the region of the joint parameter space (as described in the Mardia and Sutton (1975) result of Section 8.3) on which the mixture density $p(\theta)$ is unimodal. While the joint prior density $p(\pi, \mu_1, \mu_2, \kappa_1, \kappa_2)$ can be easily written down, the form of the unimodality region in the five-dimensional space of $(\pi, \mu_1, \mu_2, \kappa_1, \kappa_2)$ described in the Mardia and Sutton (1975) result is highly complicated and hence the resulting integral is analytically intractable. We instead obtain a Monte Carlo estimate of the prior probability of unimodality as follows. (i) Let $\underset{\sim}{\phi} = (\pi, \mu_1, \mu_2, \kappa_1, \kappa_2)$. We generate i.i.d. samples $\{\underset{\sim}{\phi}^{(t)} : t = 1, \ldots, T_1\}$ from the joint prior distribution of $\underset{\sim}{\phi}$. (ii) For each generated $\underset{\sim}{\phi}^{(t)}$, we examine if the resulting mixture density $p(\theta)$ is unimodal by checking the Mardia-Sutton condition. (iii) Finally, we obtain a simulation-consistent estimator of the prior probability of unimodality as {Number of generated $\underset{\sim}{\phi}^{(t)}$ for which the resulting $p(\theta)$ is unimodal}$/T_1$. Due to the independence structure, simulation from the joint prior can be done componentwise, which only involves random variate generation from some common densities. Further, checking the Mardia-Sutton condition for a given value of $\underset{\sim}{\phi}$ is also relatively straightforward.

The other probability required for the assessment of modality is the joint posterior probability of the parameter region on which the mixture density $p(\theta)$ is unimodal. We plan to estimate this probability also as a Monte Carlo average, i.e., once we have samples $\{\phi^{(t)} : t = 1, \ldots, T_2\}$ from the joint posterior distribution, we can simply estimate the posterior probability of unimodality by following the method outlined in the previous paragraph.

The joint posterior distribution $p(\pi, \mu_1, \mu_2, \kappa_1, \kappa_2 \mid \text{data})$ is, however, analytically intractable and hence direct generation $\phi = (\pi, \mu_1, \mu_2, \kappa_1, \kappa_2)$ is very hard. We, instead, take recourse to Gibbs Sampling. We refer the reader to Gelfand and Smith (1990), Casella and George (1992) and the collection of papers by Gilks *et al.* (1995) for the theory, implementation and convergence issues of the Gibbs sampler. The main idea of the Gibbs sampler is to simulate alternately and iteratively for the conditional posterior distributions of each unobservable given the data and other observables.

The details of the Gibbs sampler for our model including the form of the full conditional distributions and how to simulate from this conditional distributions is described in Basu and Jammalamadaka (1999). We note here that latent variables $I_1, \ldots, I_n$ are introduced in the implementation of the Gibbs sampler. I_i is an indicator variable denoting the component from which θ_i is coming, i.e., $\theta_i | I_i = 1 \sim vm(\mu_1, \kappa_1)$, $\theta_i | I_i = 2 \sim vm(\mu_2, \kappa_2)$, $i = 1, \ldots, n$ and $I_1, \ldots, I_n$ are i.i.d. with $P(I_i = 1) = \pi$, $P(I_i = 2) = 1 - \pi$ a priori. For further details of the Gibbs sampler, the reader is referred to Basu and Jammalamadaka (1999).

8.6   THE BAYES FACTOR

Standard Bayesian solution to a hypothesis testing problem involves formulating parametric models for null (H_0) and alternative (H_1) hypotheses and subsequently choosing one over the other in the light of the data and prior opinion. Perhaps the most widely used selection criterion used in this context is the 'Bayes factor of H_1 against H_0' formally defined as the ratio $B_{10} = \dfrac{\text{posterior odds}}{\text{prior odds}} = \dfrac{P(H_1|\text{data})P(H_0)}{P(H_0|\text{data})P(H_1)}$ The Bayes factor is used as a summary of evidence of H_1 against H_0 provided by the data. Thus, operationally, Bayes factor has the same role as that of a P-value in classical hypothesis testing scenario (see the review article by Kass and Raftery (1995) for an illuminating discussion comparing Bayes factor and P-values). From another perspective, the Bayes factor is similar to the likelihood ratio statistics as the former is the ratio of the marginal likelihood under H_1 against that over H_0. The following Table 8.1 provided by Kass and

Raftery (1995) gives a rounded scale for interpretation of B_{10} and $\log B_{10}$.

TABLE 8.1 Evidence in support of alternative model from Bayes factor

B_{10}	$\log B_{10}$	evidence for M_1
< 1	< 0	negative (supports M_0
1 to 3	0 to 1	barely worth mentioning
3 to 12	1 to 2.5	positive
12 to 150	2.5 to 5	strong
> 150	> 5	very strong

For our test of $H_0 : p(\theta)$ is unimodal against $H_1 : p(\theta)$ is non-unimodal, we report the estimated Bayes factor B_{10}. This is easily obtained once we estimate $P(H_0) =$ prior probability of unimodality and $P(H_0 |\text{data}) =$ posterior probability of unimodality by the Monte Carlo methods mentioned in section 8.5.

8.7 APPLICATION

We consider data collected by Schmidt-Koenig (1963) in an experiment to determine how do birds determine directions and orient themselves. In this experiment, 15 homing pigeons were released about 16.25 kilometers northwest from their loft. The measurements listed in Table 8.2 are their vanishing directions measured in degrees. The direction of the loft is 149°.

TABLE 8.2 Vanishing direction of 15 homing pigeons. The loft direction is 149°

| 85 | 135 | 135 | 140 | 145 | 150 | 150 | 150 | 160 | 285 | 200 | 210 | 220 | 225 | 270 |

These data have been analyzed by several authors, including Mardia (1972) and Fisher (1993). As can be seen in Table 8.2, most of the observations are concentrated around south (180°) with two observations in the east and west direction. Fisher (1993) reports that a goodness-of-fit test reveals there is some evidence a von-Mises distribution may not be a totally adequate description of the data.

We apply our proposed Bayesian modality test to these circular data. The following priors are used: (*i*) $\pi \sim \text{Uniform}(0,1)$, (*ii*) $\mu_1 \sim vm(0°, 0.25)$ and $\mu_2 \sim vm(180°, 0.25)$, and (*iii*) $\kappa_j \sim \text{Gamma}(1,5)$, $j = 1, 2$ where Gamma(α, β) has density proportional to $x^{\alpha-1} \exp(-x/\beta)$. These are moderately flat priors and the mean directions for μ_1 and μ_2 are chosen in opposite directions.

The prior probability of unimodality is estimated based on 100,000 samples generated from the above prior model and then following the method outlined in Section 8.5. We estimate the prior probability of unimodality to be 0.48867.

The posterior probability is estimated from 50,000 MCMC samples generated from the posterior after an initial burn-in of 10,000 and then once again following the method of section 8.5. Figure 8.2 shows the kernel density estimates of the posterior density for the two mean directions μ_1, μ_2, the two concentration parameters κ_1, κ_2 and the mixing proportion π. These density estimates are obtained using the CODA software [Best et al. (1996)]. The posterior summary estimates (posterior mean, standard deviation and percentiles) of these parameters are shown in Table 8.3. We check convergence of the MCMC sampler using different convergence and stationarity checks available in CODA. The autocorrelation plots at different lags based on the simulated samples of $\kappa_1, \kappa_2, \mu_1, \mu_2$ and π are shown in Figure 8.3. High autocorrelations typically imply slow mixing and slow convergence. In Figure 8.3, the autocorrelations for κ_1 and κ_2 die out quickly. The autocorrelations for μ_1 and μ_2 do not die so quickly whereas π has significant autocorrelations till lag 20.

TABLE 8.3 Estimated posterior mean, standard deviation and percentiles of $\mu_1, \mu_2, \kappa_1, \kappa_2$ and π

	Posterior Mean	Posterior Std. Dev.	Percentiles (2.5%, 25%, 50%, 75%, 97.5%)
μ_1	185	56.4	(61.8,152,180,220,311)
μ_2	184	51.1	(81.2, 152, 179, 216, 296)
κ_1	3.18	3.24	(0.22, 1.39, 2.22, 3.75, 12.2)
κ_2	3.21	3.20	(0.28, 1.40, 2.24, 3.81, 12.1)
π	0.50	0.28	(0.03, 0.27, 0.50, 0.73, 0.98)

The posterior probability of unimodality from the generated MCMC samples is estimated to be 0.68382. Based on these prior and posterior probabilities of unimodality, the Bayes factor for non-unimodality against unimodality is estimated as $B_{10} = 0.44188$. Thus, the data do not provide almost any evidence against the null hypothesis of unimodality. This is also evident from the posterior density estimates in Figure 8.2. The posterior density estimates of μ_1, μ_2 and κ_1, κ_2 are almost identical, which probably indicate that the two components of the mixture density are close to identical or that the mixture density is just a single von-Mises distribution. If this is true, then the mixing proportion π becomes redundant which could explain the large spread in its posterior density estimate and its autocorrelations staying on for up to lag 20.

We further compute the predictive density of a new circular observation $p(\theta \mid \text{data})$ conditioned on the 15 vanishing directions already observed. This is obtained as

$$p(\theta \mid \text{data}) = \int p(\theta \mid \pi, \mu_1, \mu_2, \kappa_1, \kappa_2) \, d\pi(\pi, \mu_1, \mu_2, \kappa_1, \kappa_2 \mid \text{data})$$

where $\pi(\pi, \mu_1, \mu_2, \kappa_1, \kappa_2 \mid \text{data})$ is the posterior distribution of the parameters. This predictive density is estimated on a grid of θ values where the integral above is estimated by the Monte Carlo average of the generated MCMC samples. This estimated predictive density in Figure 8.4 exhibits a clear unimodal structure and provide further evidence to our test result.

8.8 SOME ISSUES

A. Identifiability: In mixture modeling, identifiability of parameters is typically of concern. To see how identifiability issues may arise in our two-component von-Mises mixture model, consider the likelihood function defined in (8.5.2). It is easy to see from (8.5.2) that $L(\pi, \mu_1, \mu_2, \kappa_1, \kappa_2) = L(1 - \pi, \mu_2, \mu_1, \kappa_2, \kappa_1)$, i.e., $(\pi, \mu_1, \mu_2, \kappa_1, \kappa_2)$ and $(1 - \pi, \mu_2, \mu_1, \kappa_2, \kappa_1)$ provide identical likelihood. In Bayesian analysis, non-identifiability is often avoided by bringing in separation of parameter values in the prior modeling. However, if (μ_1, μ_2) and (κ_1, κ_2) have exchangeable priors and if the prior for π is symmetric around $1/2$, then the prior and hence the posterior also fails to identify between $(\pi, \mu_1, \mu_2, \kappa_1, \kappa_2)$ and $(1 - \pi, \mu_2, \mu_1, \kappa_2, \kappa_1)$. While non-identifiability is not a formal problem in Bayesian inference, it may lead to very slow convergence of the MCMC sampler. The resulting inference could also be troublesome. for example, the posterior distribution of μ_1 may appear to be bimodal due to concentration of mass around the mean directions of both components.

One way to ensure identifiability is to put some prior constraints on the parameter space. For example, a common constraint put in two component mixture is $\mu_1 \leq \mu_2$. In Bayesian analysis, this constraint can be brought in very naturally by simply defining the prior support to be the constrained space. This constraint makes all the parameters identifiable. Bayesian analysis under this constraint can be carried out in a straightforward manner, however it does bring in complications within the MCMC sampler. The full conditional distributions of both μ_1 and μ_2 are now constrained by the other parameter. Robert (1996) discuss the issue of parameterizations and constraints in the context of normal mixture models and suggests the reparametrization $\mu = \mu_1$ and $\lambda = \mu_2 - \mu_1$ where λ is assumed to non-negative a priori. This reparametrization generally achieves more stability within the MCMC sampler and faster convergence.

B. Choice of two component mixture. We model the sampling distribution as a two component mixture of von-Mises distribution. This model allows substantial flexibility as the mixing proportion π, the two mean directions μ_1, μ_2 and the two concentration parameters κ_1, κ_2 are allowed to vary thus resulting in different shapes and scales of the mixture density. However, a two-component mixture can at most produce a bimodal density. Thus, if data generated from a tri-modal or multi-modal distribution is fed into our model, it is not obvious how our proposed test will behave. Secondly, the unimodal or non-unimodal densities that can be obtained within our model are only those which can be characterized as two-component mixtures of von-Mises distributions. We thus do not have extensive flexibility on the functional form of the density.

The problem of more than two modes can be addressed by considering a k-component mixture density model: $p(\theta) = \sum_{j=1}^{k} \pi_j \, vm(\theta|\mu_j, \kappa_j)$ where $\sum_{j=1}^{k} \pi_j = 1$. The analysis for such a model can be performed in an analogous manner with some minor modifications in the full conditional distributions of the parameters. However, the identifiability issues discussed above becomes more severe and convergence issues in the MCMC sampler becomes more critical. Another problem is how to determine the value of k. One can put a hierarchical structure to the problem by assuming a prior distribution supported on positive integer values for k. This, however, makes the problem very hard as it now becomes a variable dimension problem and one may need to use the reversible jump algorithm to move from one dimension to another within the MCMC sampler. Green and Richardson (1997) recently addressed this variable dimension problem in the context of normal mixtures.

Mixtures of more than two components allows somewhat more flexibility in the functional form of the mixture density. Further flexibility can be obtained by semiparametric modeling. For example, in real line, all univariate unimodal distributions can be characterized as mixtures of uniform distributions (known as the Khintchine representation). This mixture representation is often used in modeling univariate unimodal distribution, one then assumes a prior on the mixing distribution of the uniforms. A similar representation also exists for unimodal distributions on the circle, they can also be written as mixtures of uniform distributions [see Fang et al. (1989)]. We are currently developing a semiparametric test for unimodality of circular data based on this representation.

REFERENCES

Bai, Z. D., Rao, C. R. and Zhao, L. C. (1988). Kernel estimators of density function of directional data, *Journal of Multivariate Analysis*, **27**, 24–39.

Bartels, R. (1984). Estimation in a bidirectional mixture of von-Mises distributions, *Biometrics*, **40**, 777–784.

Basu, S. (1995). Bayesian tests for unimodality, *Technical Report*.

Basu, S. and Jammalamadaka, S. Rao (1999). Parametric and semiparametric Bayesian tests for unimodality in directional data, Manuscript under preparation.

Batschelet, E. (1981). *Circular Statistics in Biology*, Academic Press, London.

Best, N., Cowles, M. K. and Vines, K. (1996). *CODA: Convergence Diagnosis and Output Analysis Software for Gibbs Sampling Output*, Version 0.30.

Fang, K-T., Kotz, S. and Ng, K-W. (1989). *Symmetric Multivariate and Related Distributions*, Chapman and Hall, London.

Fisher, N. I. (1993). *Statistical Analysis of Circular Data*, Cambridge University Press, Cambridge.

Gilks, W. R., Richardson, S. and Spiegelhalter, D. J. (Eds) (1995). *Markov Chain Monte Carlo in Practice*, Chapman and Hall, New York.

Good, I. J. and Gaskins, R. A. (1980). Density estimation and bump-hunting by the penalized likelihood method exemplified by scattering and meteorite data, *Journal of the American Statistical Association*, **75**, 42–73.

Green, P. and Richardson, S. (1997). On Bayesian analysis of mixtures with an unknown number of components, *Journal of the Royal Statistical Society, Series B*, **59**, 731–792.

Hartigan, J. A. and Hartigan, P. M. (1985). The dip test for unimodality, *Annals of Statistics*, **13**, 70–84.

Kass, R. E. and Raftery, A. E. (1995). Bayes factors, *Journal of the American Statistical Association*, **90**, 773–795.

Mardia, K. V. (1972). *Statistical Analysis of Directional Data*, Academic Press, London.

Mardia, K. V. and El-Atoum, S. A. M. (1976). Bayesian inference for the von Mises-Fisher distribution, *Biometrika*, **63**, 203–205.

Mardia, K. V. and Spurr, B. D. (1973) Multisample tests for multimodal and axial circular populations, *Journal of the Royal Statistical Society, Series B*, **35**, 422–436.

Mardia, K. V. and Sutton, T. W. (1975). On the modes of mixture of two von-Mises distributions, *Biometrika*, **62**, 699–701.

Müller, D. W. and Sawitzki, G. (1991). Excess mass estimates and test for multimodality, *Journal of the American Statistical Association*, **86**, 738–746.

Rao, J. S. (1984). Nonparametric methods in directional data, In *Handbook of Statistics, Volume 4* (Eds., P. R. Krishnaiah and P. K. Sen), pp. 755–770, Elsevier Science Publishers, Amsterdam.

Robert, C. P. (1996). Mixtures of distributions: inference and estimation, In *Markov Chain Monte Carlo in practice* (Eds., W. Gilks, S. Richardson and D. Spiegelhalter), pp. 441–461, Chapman and Hall, London.

Silverman, B. W. (1981). Using kernel density estimates to investigate multimodality, *Journal of the Royal Statistical Society, Series B*, **43**, 97–99.

Spurr, B. D. and Koutbeiy, M. A. (1991). A comparison of various methods for estimating the parameters in mixtures of von-Mises distributions, *Communications in Statistics—Simulation and Computation*, **20**, 725–741.

Wolfe, J. H. (1970). Pattern clustering by multivariate mixture analysis, *Multivariate Behavioural Research*, **5**, 329–350.

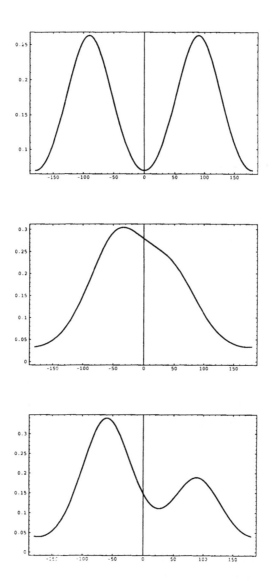

FIGURE 8.1 Three von-Mises mixtures: Top $= 0.5 \, \mathrm{vm}(\theta \,|\, -90°, 2) +$
$0.5 \, \mathrm{vm}(\theta \,|\, 90°, 2)$, Middle $= 0.6 \, \mathrm{vm}(\theta \,|\, -45°, 1.5) + 0.4 \, \mathrm{vm}(\theta \,|\, 45°, 1.5)$, Bottom
$= 0.65 \, \mathrm{vm}(\theta \,|\, -60°, 2) + 0.35 \, \mathrm{vm}(\theta \,|\, 90°, 2)$

SANJIB BASU and S. RAO JAMMALAMADAKA

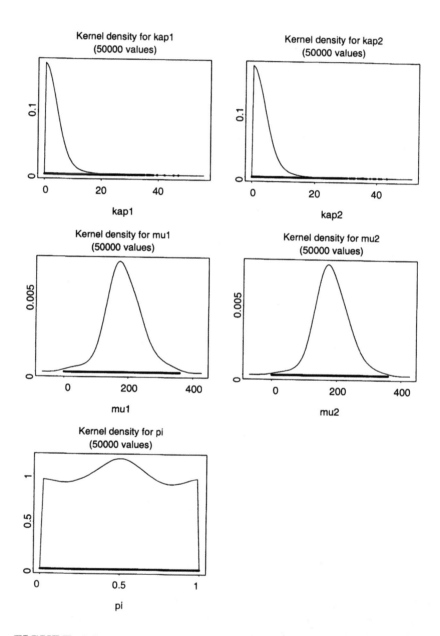

FIGURE 8.2 Kernel estimates of the posterior density of the parameters $\kappa_1, \kappa_2, \mu_1, \mu_2, \pi$

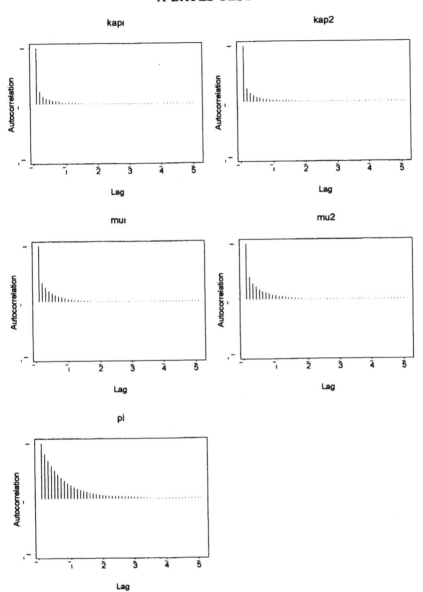

FIGURE 8.3 Autocorrelation plot at different lags for the five parameters: $\kappa_1, \kappa_2, \mu_1, \mu_2, \pi$

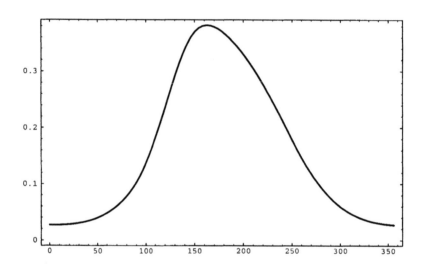

FIGURE 8.4 Predictive density of a new circular obsevation for the pigeon data

CHAPTER 9

MAXIMUM LIKELIHOOD ESTIMATION OF THE LAPLACE PARAMETERS BASED ON PROGRESSIVE TYPE-II CENSORED SAMPLES

RITA AGGARWALA

University of Calgary, Calgary, Alberta, Canada

N. BALAKRISHNAN

McMaster University, Hamilton, Ontario, Canada

Abstract: In this paper, we derive the maximum likelihood estimators of the location and scale parameters of a Laplace distribution based on progressive Type-II right censored samples. The results obtained here are a generalization of those given in Balakrishnan and Cutler (1995) for one-sided Type-II censoring.

Keywords and phrases: Progressive type-II censoring, order statistics, Laplace distribution, maximum likelihood estimators

9.1 INTRODUCTION

Maximum likelihood estimation for the Laplace distribution based on full samples has been discussed by a number of authors; see, for example, Johnson, Kotz and Balakrishnan (1995). Balakrishnan and Cutler (1995) have discussed maximum likelihood estimation for parameters of the Laplace distribution based on conventional Type-II censored samples. They consider both symmetric and one-sided (right) censoring.

The scheme of progressive Type-II right censoring arises naturally in life-testing experimentation, as it is often desirable to remove live items from experimentation at points other than the final termination point. In this scheme, we begin the test at time zero with n independent live items on test. Immediately following the first observed failure, a fixed number R_1 of surviving items are removed at random from the test. Immediately following the next observed failure, a fixed number R_2 of surviving items are removed at random from the test. This process continues until, immediately following the time of the mth observed failure, the remaining $R_m = n - R_1 - R_2 - \cdots R_{m-1} - m$ items are removed from the test. We will denote the m ordered observed failure times by $Y_{i:m:n}^{(R_1,R_2,\ldots,R_m)}$, $i = 1,\ldots,m$ and call them the progressive Type-II right censored order statistics of size m from a sample of size n with progressive censoring scheme $(R_1, R_2, \ldots, R_m)$. This type of censoring scheme may be desirable, for example, in destructive testing of mechanical components; see, for example, Montanari and Cacciari (1988). A number of other authors have studied problems of inference pertaining to progressive censoring, including Cohen (1963, 1975, 1991), Mann (1969, 1971), Thomas and Wilson (1972), and Viveros and Balakrishnan (1994). Balakrishnan and Sandhu (1995, 1996), and Aggarwala and Balakrishnan (1998) discuss some mathematical properties of these progressive Type-II censored order statistics arising from general continuous, exponential and uniform distributions. A thorough overview of the subject of progressive censoring is given in Balakrishnan and Aggarwala (2000).

It is well documented [see, for example, Lawless (1982) or any of the references mentioned above] that if the failure times of the n items originally on test with progressive censoring scheme $(R_1, R_2, \ldots, R_m)$ are from a continuous population with cumulative distribution function $F(x)$ and probability density function $f(x)$, then the joint probability density function of $Y_{i:m:n}^{(R_1,R_2,\ldots,R_m)}$, $i = 1,\ldots,m$ is given by

$$f_{1,2,\ldots,m:m:n}(y_1, y_2, \ldots, y_m) = c \prod_{i=1}^{m} f(y_i)\,[1 - F(y_i)]^{R_i},$$
$$-\infty < y_1 < \cdots < y_m < \infty, \qquad (9.1.1)$$

where

$$c = n(n - R_1 - 1)(n - R_1 - R_2 - 2) \cdots$$
$$\cdots (n - R_1 - R_2 - \cdots - R_{m-1} - m + 1).$$

For convenience in notation, since in this discussion, it will not be unclear as to what the progressive Type-II right censoring scheme is, we will denote the m progressive Type-II right censored order statistics by $Y_{i:m:n}$, $i = 1, \ldots, m$.

In this paper, we assume that the underlying failure times follow a two-parameter double exponential, or Laplace, distribution, with probability density function given by

$$f(y) = \frac{1}{2\theta_2} e^{-|y - \theta_1|/\theta_2}, \quad -\infty < y < \infty, \ -\infty < \theta_1 < \infty, \ \theta_2 > 0, \quad (9.1.2)$$

and cumulative distribution function given by

$$F(x) = \frac{1}{2} e^{(x-\theta_1)/\theta_2}, \ x \leq \theta_1,$$
$$= 1 - \frac{1}{2} e^{-(x-\theta_1)/\theta_2}, \ x > \theta_1. \quad (9.1.3)$$

We derive the maximum likelihood estimators of the location and scale parameters of the above Laplace distribution based on progressive Type-II right censored samples. The results obtained are generalizations of those given in Balakrishnan and Cutler (1995), where it is shown that for conventional Type-II right censored samples, where only the first m failure times are observed, the maximum likelihood estimator of θ_1 is simply the usual sample median based on the full sample, provided $m \geq \frac{n}{2}$. For $m < \frac{n}{2}$, the MLE of θ_1 turns out to be a linear function of the observed order statistics. In both cases, they show that the MLE of θ_2 is a linear function of the observed order statistics. The results presented in this paper for maximum likelihood estimation based on progressive Type-II right censored samples from the Laplace distribution simplify to those presented by Balakrishnan and Cutler (1995) for the special case when $R_1 = R_2 = \cdots = R_{m-1} = 0$ and $R_m = n - m$, in which case we are left with a conventional progressive Type-II right censored sample.

9.2   EXAMINING THE LIKELIHOOD FUNCTION

Consider a progressive Type-II right censored sample of size m with censoring scheme $(R_1, R_2, \ldots, R_m)$ from a random sample of size n from the

Laplace distribution with probability density function given in (9.1.2). The likelihood function, L, for a progressive Type-II right censored sample $(Y_{i:m:n},\ i = 1, 2, \ldots, m)$ is given by (ignoring the constant c)

$$L(\theta_1, \theta_2) = \prod_{i=1}^{m} f(Y_{i:m:n}) [1 - F(Y_{i:m:n})]^{R_i}, \qquad (9.2.4)$$

where $f(\cdot)$ and $F(\cdot)$ are as given in (9.1.2) and (9.1.3), respectively.

We will first maximize with respect to θ_1.

Notice that, for values of $\theta_1 \le Y_{1:m:n}$, the likelihood function reduces to

$$L_0(\theta_1, \theta_2) = \frac{1}{2^n \theta_2^m} e^{-\sum_{i=1}^{m}(R_i+1)(Y_{i:m:n}-\theta_1)/\theta_2}, \qquad (9.2.5)$$

which is a monotonically increasing function of θ_1. Next, we consider values of $\theta_1 > Y_{m:m:n}$. For these values, the likelihood function reduces to

$$L_m(\theta_1, \theta_2) = \frac{1}{(2\theta_2)^m} e^{\sum_{i=1}^{m}(Y_{i:m:n}-\theta_1)/\theta_2} \prod_{i=1}^{m} \left[1 - \frac{1}{2}e^{(Y_{i:m:n}-\theta_1)/\theta_2}\right]^{R_i}. \qquad (9.2.6)$$

Upon taking the logarithm of L_m and differentiating with respect to θ_1, we obtain

$$\frac{\partial \ln L_m}{\partial \theta_1} = -\frac{m}{\theta_2} + \sum_{i=1}^{m} \frac{R_i\, e^{(Y_{i:m:n}-\theta_1)/\theta_2}}{2\theta_2 \left[1 - \frac{1}{2}e^{(Y_{i:m:n}-\theta_1)/\theta_2}\right]} \qquad (9.2.7)$$

Now, if $R_1 = R_2 = \cdots = R_m = 0$, then $m = n$, so that the right hand side of (9.2.7) is simply $-\frac{n}{\theta_2}$ which is strictly less than 0. If some $R_i > 0$, $i = 1, 2, \ldots, m$, then

$$\frac{\partial \ln L_m}{\partial \theta_2} < -\frac{m}{\theta_2} + \frac{\sum_{i=1}^{m} R_i}{\theta_2} \le 0 \text{ if } \sum_{i=1}^{m} R_i \le m, \text{ ie. } n - m \le m, \text{ ie. } m \ge \frac{n}{2},$$

so that L_m is monotonically decreasing for these values of m. Thus, if the observed number of failures $m \ge \frac{n}{2}$, the maximum likelihood estimator of θ_1 lies in the interval $[Y_{1:m:n},\ Y_{m:m:n}]$.

Consider now the values of θ_1 such that $Y_{j:m:n} < \theta_1 \le Y_{j+1:m:n}$ for $j \in \{1, 2, \ldots,\ m-1\}$. In this case, the likelihood function reduces to

$$\begin{aligned}
L_j(\theta_1, \theta_2) &= \frac{1}{2^m \theta_2^m} e^{\sum_{i=1}^{j}(Y_{i:m:n}-\theta_1)/\theta_2} \prod_{i=1}^{j} \left[1 - \frac{1}{2}e^{(Y_{i:m:n}-\theta_1)/\theta_2}\right]^{R_i} \\
&\quad \times e^{-\sum_{i=j+1}^{m}(Y_{i:m:n}-\theta_1)/\theta_2} \prod_{i=j+1}^{m} \left[\frac{1}{2}e^{-(Y_{i:m:n}-\theta_1)/\theta_2}\right]^{R_i}.
\end{aligned}$$

$$(9.2.8)$$

Notice that the likelihood function is a continuous function in θ_1 $(-\infty < \theta_1 < \infty)$, so that

$$L_j\left(Y_{j+1:m:n}, \theta_2\right) = L_{j+1}\left(Y_{j+1:m:n}, \theta_2\right), \quad j = 0, 1, \ldots, m-1.$$

Upon taking the logarithm of L_j and differentiating with respect to θ_1, we obtain

$$\frac{\partial \ln L_j}{\partial \theta_1} = \frac{1}{\theta_2} \left[\sum_{i=1}^{j} \frac{R_i \, e^{(Y_{i:m:n}-\theta_1)/\theta_2}}{2 - e^{(Y_{i:m:n}-\theta_1)/\theta_2}} + n - \sum_{i=1}^{j} R_i - 2j \right]. \qquad (9.2.9)$$

Now, if $R_i = 0$, $i = 1, 2, \ldots, j$, then the right hand side of (9.2.9) becomes simply $\frac{n-2j}{\theta_2}$, which is strictly negative, provided $j > \frac{n}{2}$. If some $R_i > 0$, $i = 1, 2, \ldots, j$, then the right hand side of (9.2.9) is strictly less than $\frac{n-2j}{\theta_2} \le 0$ if $j \ge \frac{n}{2}$. Thus, in general,

$$\frac{\partial L_j}{\partial \theta_1} < 0 \text{ provided } j > \frac{n}{2}$$

so that L_j is monotonically decreasing for these values of j. Thus, if the observed number of failures, $m > \frac{n}{2}$, the maximum likelihood estimator of θ_1 lies in the interval $\left[Y_{1:m:n}, Y_{[n/2]+1:m:n}\right]$.

Upon further inspection of (9.2.9), it is evident that if $n - R_1 - R_2 - \cdots - R_j - 2j > 0$, then the right hand side is strictly positive. This can only be possible if $j < \frac{n}{2}$, in which case $n - R_1 - R_2 - \cdots - R_j - 2j > \frac{n}{2} - j - R_1 - R_2 - \cdots - R_j \ge 0$ if $R_1 + R_2 + \cdots + R_j + j \le \frac{n}{2}$. Thus,

$$\frac{\partial L_j}{\partial \theta_1} > 0 \text{ provided } j < \frac{n}{2} \text{ and } R_1 + R_2 + \cdots + R_j + j \le \frac{n}{2},$$

so that L_j is monotonically increasing for these values of j.

9.3 ALGORITHM TO FIND MLE'S

At this point, we can formulate the following algorithm to narrow our search for, and obtain the maximum likelihood estimates of the location and scale parameters from a Laplace distribution when a progressive Type-II right censored sample of size m from a sample of size n is observed, with censoring scheme $(R_1, R_2, \ldots, R_m)$. We will denote the progressive Type-II right censored order statistics from this sample by $Y_{i:m:n}$, $i = 1, 2, \ldots, m$, and the corresponding observed values of the order statistics by y_i, $i = 1, 2, \ldots, m$.

1. Find the largest number $k < \frac{n}{2}$ such that $R_1 + R_2 + \cdots + R_k + k \le \frac{n}{2}$, assuming $R_0 = 0$.

2. (a) If $m < \frac{n}{2}$, the maximum likelihood estimates are those corresponding to

$$\max \left\{ \begin{array}{l} \max_{y_{k+1} \le \theta_1 \le y_{k+2}, \; \theta_2 > 0} L_{k+1}, \\ \max_{y_{k+2} < \theta_1 \le y_{k+3}, \; \theta_2 > 0} L_{k+2}, \ldots, \\ \max_{y_{m-1} < \theta_1 \le y_m, \; \theta_2 > 0} L_{m-1} \\ \max_{\theta_1 > y_m, \; \theta_2 > 0} L_m \end{array} \right\}.$$

(b) If $m > \frac{n}{2}$, the maximum likelihood estimates are those corresponding to

$$\max \left\{ \begin{array}{l} \max_{y_{k+1} \le \theta_1 \le y_{k+2}, \; \theta_2 > 0} L_{k+1}, \\ \max_{y_{k+2} < \theta_1 \le y_{k+3}, \; \theta_2 > 0} L_{k+2}, \ldots, \\ \max_{y_{n/2} < \theta_1 \le y_{(n/2)+1}, \; \theta_2 > 0} L_{n/2} \end{array} \right\}$$

if n is even, and

$$\max \left\{ \begin{array}{l} \max_{y_{k+1} \le \theta_1 \le y_{k+2}, \; \theta_2 > 0} L_{k+1}, \\ \max_{y_{k+2} < \theta_1 \le y_{k+3}, \; \theta_2 > 0} L_{k+2}, \ldots, \\ \max_{y_{(n-1)/2} < \theta_1 \le y_{(n+1)/2}, \; \theta_2 > 0} L_{(n-1)/2} \end{array} \right\}$$

if n is odd.

For the case n is odd, notice that we may have $k = \frac{n-1}{2}$. This simply means that the likelihood function is monotonically increasing for $\theta_1 < Y_{(n+1)/2:m:n}$ and the likelihood function is monotonically decreasing for $\theta_1 > Y_{(n+1)/2:m:n}$. Thus, the maximum likelihood estimator of θ_1 is $Y_{(n+1)/2:m:n}$, which we can use to solve for the maximum likelihood estimator of θ_2. The resulting likelihood function to be maximized with respect to θ_2 is proportional to

$$\frac{1}{\theta_2^m} e^{\sum_{i=1}^{(n-1)/2} \left(Y_{i:m:n} - Y_{(n+1)/2:m:n} \right)/\theta_2}$$

$$\times \prod_{i=1}^{(n-1)/2} \left\{ 1 - \frac{1}{2} e^{[Y_{i:m:n} - Y_{(n+1)/2}]/\theta_2} \right\}^{R_i}$$

$$\times e^{-\sum_{i=(n+3)/2}^{m} (R_i+1)[Y_{i:m:n} - Y_{(n+1)/2}]/\theta_2}.$$

(This corresponds to maximizing either $L_{(n-1)/2}$ or $L_{(n+1)/2}$, due to the continuity of the likelihood function discussed earlier.)

(c) If $m = \frac{n}{2}$, the maximum likelihood estimates are those corresponding to

$$\max\left\{\begin{array}{c}\max\limits_{y_{k+1}\leq\theta_1\leq y_{k+2},\ \theta_2>0} L_{k+1},\quad \max\limits_{y_{k+2}<\theta_1\leq y_{k+3},\ \theta_2>0} L_{k+2},\cdots, \\ \max\limits_{y_{(n/2)-1}<\theta_1\leq y_{(n/2)},\ \theta_2>0} L_{(n/2)-1} \end{array}\right\}$$

Notice that here, it is possible to obtain $k = \frac{n}{2} - 1 = m - 1$. This means that for $\theta_1 < Y_{m:m:n}$, the likelihood function is monotonically increasing and for $\theta_1 > Y_{m:m:n}$, the likelihood function is monotonically decreasing. Therefore, the maximum likelihood estimator of θ_1 is $Y_{m:m:n}$. This can be used to solve for the maximum likelihood estimator of θ_2. The resulting likelihood function to be maximized with respect to θ_2 is proportional to

$$\frac{1}{\theta_2^m} e^{\sum_{i=1}^{m-1}(Y_{i:m:n}-Y_{m:m:n})/\theta_2} \prod_{i=1}^{m-1}\left[1 - \frac{1}{2}e^{(Y_{i:m:n}-Y_{m:m:n})/\theta_2}\right]^{R_i}$$

Remark 9.3.1 For the special case of conventional Type-II right censoring, where $R_1 = R_2 = \cdots = R_{m-1} = 0$ and $R_m = n - m$, this algorithm reduces to that given in Balakrishnan and Cutler (1995): for $m < \frac{n}{2}$, $k = m - 1$, and we just maximize L_m. For $m > \frac{n}{2}$, for n odd, $k = \frac{n-1}{2}$, and the maximum likelihood estimator for θ_1 is $Y_{(n+1)/2:m:n}$. For n even, $k = \frac{n}{2} - 1$, and we must maximize $L_{n/2}$. From (9.2.9), $\frac{\partial L_{n/2}}{\partial \theta_1}$ is obviously zero, so the maximum likelihood estimator of θ_1 is any value in $\left[Y_{n/2:m:n}, Y_{(n/2)+1:m:n}\right]$. Finally, for $m = \frac{n}{2}$, $k = \frac{n}{2} - 1$, so that the maximum likelihood estimator of θ_1 is $Y_{n/2:m:n}$. These estimates of θ_1 may then be used to obtain maximum likelihood estimates of θ_2.

9.4   NUMERICAL EXAMPLE

Using the simulational algorithm given in Balakrishnan and Sandhu (1995), a progressive Type-II right censored sample of size $m = 10$ from a sample of size $n = 20$ from the Laplace distribution with $\theta_1 = 25$ and $\theta_2 = 5$ was simulated, with censoring scheme $\mathbf{R} = (2,0,0,2,0,0,0,2,0,4)$. The simulated progressive Type-II right censored sample is as follows:

19.21167876, 21.97364262, 23.41776818, 23.66253070, 23.80222832,

24.23017797, 25.62072188, 25.86990938, 26.47997028, 27.55344134.

From part (1) of the algorithm presented above, we find $k = 6$. Thus, from part (2c), we must find θ_1 and θ_2 which correspond to

$$\max\left\{\max\limits_{y_7\leq\theta_1\leq y_8,\ \theta_2>0} L_7,\quad \max\limits_{y_8<\theta_1\leq y_9,\ \theta_2>0} L_8,\quad \max\limits_{y_9<\theta_1\leq y_{10},\ \theta_2>0} L_9\right\}.$$

Using Maple V Release 3, the maximum value of the likelihood function is obtained when we maximize $L_8(\theta_1, \theta_2)$ over the region specified above. The corresponding maximum likelihood estimates are $\widehat{\theta}_1 = 26.31069$ and $\widehat{\theta}_2 = 2.67091$. It has been shown by the authors [see Aggarwala and Balakrishnan (1999)] that the best linear unbiased estimates and their standard errors for the two parameters in this case are

$$\theta_1^* = 26.26607, \ SE(\theta_1^*) = 0.72333, \qquad \theta_2^* = 2.64071, \ SE(\theta_2^*) = 0.28972.$$

These values agree well with the MLE's which we have just obtained.

Remark 9.4.1 It should be noted here that to obtain standard errors of the MLE's, a simulational study needs be conducted. Furthermore, since the class of distributions under study does not possess "regularity" properties, due to its lack of differentiability, it may not be appropriate to approximate the asymptotic variance-covariance matrix of the MLE's using the method of inverting the matrix of second derivatives.

Acknowledgements The authors wish to thank the Natural Sciences and Engineering Research Council of Canada for funding this research.

REFERENCES

Aggarwala, R. and Balakrishnan, N. (1998). Some properties of progressive censored order statistics from arbitrary and uniform distributions, with applications to inference and simulation, *Journal of Statistical Planning and Inference*, **70**, 35–49.

Balakrishnan, N. and Aggarwala, R. (2000). *Progressive Censoring: Theory, Methods and Applications*, Birkhäuser, Boston.

Balakrishnan, N. and Sandhu, R. A. (1995). A simple Simulational algorithm for generating progressive type-II censored samples, *The American Statistician*, **49**, 229–230.

Balakrishnan, N. and Sandhu, R. A. (1996). Best linear unbiased and maximum likelihood estimation for exponential distributions under general progressive Type-II censored samples, *Sankhyā, Series B*, **58**, 1–9.

Cohen, A. C. (1963). Progressively censored samples in life testing, *Technometrics*, **5**, 327–329.

Cohen, A. C. (1975). Multi-censored sampling in the three-parameter Weibull distribution, *Technometrics*, **17**, 347–351.

Cohen, A. C. (1991). *Truncated and Censored Samples: Theory and Applications*, Marcel Dekker, New York.

David, H. A. (1981). *Order Statistics*, Second edition, John Wiley & Sons, New York.

Johnson, N. L., Kotz, S. and Balakrishnan, N. (1995). *Continuous Univariate Distributions, Vol. 2*, John Wiley & Sons, New York.

Lawless, J. F. (1982). *Statistical Models and Methods for Lifetime Data*, John Wiley & Sons, New York.

Mann, N. R. (1969). Exact three-order-statistic confidence bounds on reliable life for a Weibull model with progressive censoring, *Journal of the American Statistical Association*, **64**, 306–315.

Mann, N. R. (1971). Best linear invariant estimation for Weibull parameters under progressive censoring, *Technometrics*, **13**, 521–533.

Montanari, G. C. and Cacciari, M. (1988). Progressively-censored aging tests on XLPE-insulated cable models, *IEEE Transactions on Electrical Insulation*, **23**, 365–372.

Thomas, D. R. and Wilson, W. M. (1972). Linear order statistic estimation for the two-parameter Weibull and Extreme-value distributions from Type-II progressively censored samples, *Technometrics*, **14**, 679–691.

Viveros, R. and Balakrishnan, N. (1994). Interval estimation of parameters of life characteristics from progressively censored data, *Technometrics*, **36**, 84–91.

CHAPTER 10

ESTIMATION OF PARAMETERS OF THE LAPLACE DISTRIBUTION USING RANKED SET SAMPLING PROCEDURES

DINESH S. BHOJ

Rutgers University, Camden, NJ

Abstract: The estimators of the parameters of Laplace distribution are obtained by using (i) ranked set sampling (RSS) proposed by McIntyre (1952), (ii) modified ranked set sampling (MRSS), and (iii) new ranked set sampling (NRSS) proposed by Bhoj (1997c). The coefficients to compute the estimators by using these procedures are reported. These estimators are compared with the ordered least squares estimators given by Govindarajulu (1966), and among themselves. It is demonstrated that the relative precisions of the estimators based on NRSS are higher than those based on the least squares method and the other two ranked set sampling procedures.

Keywords and phrases: Laplace distribution, least squares estimators, minimum variance estimators, modified ranked set sample, new ranked set sample, ranked set sample, relative precision, unbiased estimators

10.1 INTRODUCTION

Ranked set sampling (RSS) is a method of sampling that is advantageous when quantification of all sampling units is costly but where small sets of units can be ranked by means of visual inspection or other methods not

169

requiring actual measurements. In order to draw a ranked set sample of size n, n sets each of size n are drawn from a population. The n units from a single set are ranked on the basis of the magnitude of the variable under investigation without actually quantifying them. This ranking procedure is applied to all n sets. The n^2 ordered observations in the n sets can be displayed as:

$$
\begin{array}{cccc}
x_{(11)} & x_{(12)} & \cdots & x_{(1n)} \\
x_{(21)} & x_{(22)} & \cdots & x_{(2n)} \\
\vdots & \vdots & \cdots & \\
x_{(n1)} & x_{(n2)} & \cdots & x_{(nn)}
\end{array}
$$

The unit with the lowest rank is quantified from the first set, the unit with the second lowest rank is quantified from the second set, and this procedure is continued until the unit with the highest rank is quantified from the nth set. Thus only n observations $x_{(11)}, x_{(22)}, \ldots, x_{(nn)}$ are measured accurately and they constitute the RSS. Note that $x_{(ii)}$ is the ith ordered observation in the ith sample, and $x_{(11)}, x_{(22)}, \ldots, x_{(nn)}$ are independently distributed. The method of RSS was introduced by McIntyre (1952) to estimate mean pasture yields with greater efficiency than simple random sampling (SRS). McIntyre's goal was to maintain unbiasedness of SRS while effectively incorporating into the estimates the information given by ranking. It appears that RSS was not used by other investigators for over a decade. Then Halls and Dell (1966) first used this method in estimating forage yields in a pine hardwood forest. They found empirically that RSS was more efficient than SRS to estimate the population mean. But the required mathematical foundation for RSS was provided by Takahasi and Wakimoto (1968) and Dell and Clutter (1972). Dell and Clutter (1972) also considered theoretically the performance of RSS when there are errors in ranking. They showed that the RSS estimator for population mean is unbiased and is at least as effective as the SRS estimator with the same number of quantifications even when there are ranking errors. The relative precision of the two methods is equal to unity only if the ranking is no better than random. David and Levine (1972) considered the case where the ranking is done on the basis of a covariate instead of judgment. Under certain assumptions, they obtained a formula expressing relative precision in terms of the squared correlation coefficient between the covariate and the variate of interest. Stokes (1977) explored this model further. Stokes (1980) proposed an estimator for the population variance based on RSS. She showed that the estimator is asymptotically unbiased even in the presence of errors in ranking. Stokes and Sager (1988) developed an RSS estimator for a cumulative distribution function. They showed that an empirical distribution function based on RSS is an unbiased estimator for a distribution function and the estimator has smaller variance than the one based on SRS. Then they used the RSS estimator to construct confidence bands for the distribution function by using the Kolmogorov-Smirnov statistic. Bohn and Wolfe (1992) used the ranked set empirical distribution to derive an RSS version of the Mann-Whitney statistic and obtained some of its distributional properties. They compared the asymptotic relative efficiency of the RSS Mann- Whitney test with its corresponding SRS counterpart. They concluded that RSS approach was preferable. Patil, Sinha and Taillie (1993a) used the RSS method when sampling is from a finite population. They obtained explicit expressions for the variance and relative precision of RSS estimator for several set sizes when the population follows a linear or quadratic trend. These authors (1993b) studied the relative precision of RSS estimator with the regression estimator when the ranking is done on the basis of an auxiliary variable. Bhoj (1997a) obtained the estimators of parameters of the extreme value distribution using RSS. Bhoj and Ahsanullah (1996) derived the minimum variance linear unbiased estimators for the parameters of the generalized geometric distribution using RSS. Recently Bhoj (1997b, 1997c) proposed modified ranked set sampling (MRSS) and new ranked set sampling (NRSS) procedures, respectively for estimating the parameters. In this paper we use RSS, MRSS and NRSS to derive the estimators of mean and standard deviation of the Laplace distribution. These estimators are then compared with the other competing estimators and among themselves.

10.2 ESTIMATION OF PARAMETERS BASED ON THREE PROCEDURES

10.2.1 Ranked Set Sampling

In this section we use the ranked set sample of n observations $x_{(11)}, x_{(22)}, \cdots, x_{(nn)}$ to estimate the location and scale parameters of the distribution. Let μ and σ denote the location and scale parameters of the distribution. We define

$$y_{(ii)} = (x_{(ii)} - \mu)/\sigma, \ E(y_{(ii)}) = \alpha_i \text{ and } Var(y_{(ii)}) = v_{(ii)}.$$

In terms of original x(ii)'s we have

$$E(x_{(ii)}) = \mu + \sigma\alpha_i \text{ and } Var(x_{(ii)}) = v_{(ii)}\sigma^2.$$

Let $\mathbf{X}' = (x_{(11)}, x_{(22)}, \ldots, x_{(nn)})$,

$$\mathbf{1}' = (1, 1, ..., 1),$$

$$\alpha' = (\alpha_1, \alpha_2, \ldots, \alpha_n),$$

and Variance-Covariance matrix of $\mathbf{X} = V\sigma^2$, where V is an $n \times n$ diagonal matrix with v_{ii} as (i,i)th element. Then we can write

$$E(\mathbf{X}) = \mu 1 + \sigma\alpha = A\theta,$$

where $A' = \begin{pmatrix} 1 & 1 & \cdots & 1 \\ \alpha_1 & \alpha_2 & \cdots & \alpha_n \end{pmatrix}$ and $\theta' = (\mu, \sigma)$.

The minimum variance linear unbiased estimators (MVLUE) of θ is obtained by the least squares theorem of Gauss and Markoff. If $\tilde{\theta}$ denotes the MVLUE of θ, then $\tilde{\theta} = (A'V^{-1}A)^{-1}A'V^{-1}X$. After some simplifications, we can write

$$\tilde{\mu} = \sum_{i=1}^{n} w_{1i} x_{(ii)} \tag{10.2.1}$$

$$\tilde{\sigma} = \sum_{i=1}^{n} w_{2i} x_{(ii)}, \tag{10.2.2}$$

where

$$w_{1i} = (T_1 - \alpha_i T_3)/(Dv_{ii}),$$
$$w_{2i} = (\alpha_i T_2 - T_3)/(Dv_{ii}),$$

$$T_1 = \sum_{i=1}^{n} (\alpha_i^2/v_{ii}), \qquad T_2 = \sum_{i=1}^{n} (1/v_{ii}), \tag{10.2.3}$$

$$T_3 = \sum_{i=1}^{n} (\alpha_1/v_{ii}) \quad \text{and} \quad D = T_1 T_2 - T_3^2. \tag{10.2.4}$$

The variances and covariance of these estimators are given by

$$Var(\tilde{\mu}) = \frac{\sigma^2 T_1}{D}, \qquad Var(\tilde{\sigma}) = \frac{\sigma^2 T_2}{D} \quad \text{and}$$

$$Cov(\tilde{\mu}, \tilde{\sigma}) = -\frac{\sigma^2 T_3}{D}. \tag{10.2.5}$$

10.2.2 Modified Ranked Set Sampling

We assume that n $(n = 2m)$ is even so that direct comparison with the estimators based on new ranked set sampling procedure can be made. In the modified ranked set sampling we select only two appropriate order statistics. For notational convenience, we assume that the jth order statistic is selected from the first m samples and the kth order statistics is selected from the last m samples. The choices of jth and kth order statistics depend upon the distribution under investigation and the parameter(s) to be estimated.

Let μ^* and σ^* denote the estimators of μ and σ based on MRSS. These estimators can be derived by using the method described for the RSS procedure. These estimators and their variances and covariance are given by

$$\mu^* = \frac{\sum_{i=1}^{m}(c_j(x_{(ij)} + c_k x_{(m+ik)}))}{m} , \qquad (10.2.6)$$

$$\sigma^* = \frac{\sum_{i=1}^{m}(d_j(x_{(ij)} + d_k x_{(m+ik)}))}{m} , \qquad (10.2.7)$$

$$Var(\mu^*) = \frac{(\alpha_k^2 v_{jj} + \alpha_j^2 v_{kk})}{(\alpha_k - \alpha_j)^2} \frac{\sigma^2}{m} , \qquad (10.2.8)$$

$$Var(\sigma^*) = \frac{(v_{kk} + v_{jj})}{(\alpha_k - \alpha_j)^2} \frac{\sigma^2}{m} , \qquad (10.2.9)$$

$$Cov(\mu^*, \sigma^*) = -\frac{(\alpha_j v_{kk} + \alpha_k v_{jj})}{(\alpha_k - \alpha_j)^2} \frac{\sigma^2}{m} , \qquad (10.2.10)$$

where

$$c_j = \frac{\alpha_k}{\alpha_k - \alpha_j} , \quad c_k = \frac{-\alpha_j}{\alpha_k - \alpha_j} ,$$

$$d + k = \frac{1}{(\alpha_k - \alpha_j)} \quad \text{and} \quad d_j = -d_k. \qquad (10.2.11)$$

10.2.3   New Ranked Set Sampling

In new ranked set sampling (NRSS), we take m $(n = 2m)$ samples, each of size $2n$, from the population, and measure appropriate jth and kth order statistics from each sample. The n^2 ranked set sampling units can be displayed as:

$$
\begin{array}{cccc}
x_{(11)} & x_{(12)} & \cdots & x_{(1\ 2n)} \\
x_{(21)} & x_{(22)} & \cdots & x_{(2\ 2nn)} \\
\vdots & \vdots & \cdots & \\
x_{(n1)} & x_{(n2)} & \cdots & x_{(n\ 2n)}.
\end{array}
$$

The n measured observations $(x_{(ij)}, x_{(i,k)})$, $i = 1, 2, \ldots, m$ and $j < k$ constitute the NRSS sample. Now $x_{(ij)}$ and $x_{(ik)}$ are not independently distributed. However, $(x_{(ij)}, x_{(ik)})$ and $(x_{(i',j)}, x_{(i',k)})$ for $i \not\ll i'$ are independently distributed.

Let μ^{**} and σ^{**} denote the estimators of μ and σ based on NRSS. Then μ^{**} and σ^{**} are given by

$$\mu^{**} = \frac{\sum_{i=1}^{m}(c_j x_{(ij)} + c_k x_{(ik)})}{m}, \tag{10.2.12}$$

$$\sigma^{**} = \frac{\sum_{i=1}^{m}(d_j x_{(ij)} + d_k x_{(ik)})}{m}, \tag{10.2.13}$$

where the coefficients c_j, c_k, d_j and d_k are given in (10.2.11). The variances and covariance of these estimators are given by

$$Var(\mu^{**}) = \frac{(\sigma^2(\alpha_k^2 v_{jj} + \alpha_j^2 v_{kk} - 2\alpha_j \alpha_k v_{jk})}{m(\alpha_k - \alpha_j)^2}, \tag{10.2.14}$$

$$Var(\sigma^{**}) = \frac{\sigma^2(v_{jj} + v_{kk} - 2v_{jk})}{m(\alpha_k - \alpha_j)^2}, \tag{10.2.15}$$

$$Cov(\mu^{**}, \sigma^{**}) = \frac{-\sigma^2\{\alpha_k v_{jj} + \alpha_k v_{kk} - (\alpha_j + \alpha_k)v_{jk}\}}{m(\alpha_k - \alpha_j)^2}, \tag{10.2.16}$$

where
$$v_{jk} = Cov(y_{(ij)}, y_{(ik)}).$$

In Section 10.4, we compare the three sets of estimators of mu and sigma based on three ranked set sampling procedures, and one more set of estimators based on the ordered observations given by Govindarajulu (1966) for the Laplace distribution.

10.3 LAPLACE DISTRIBUTION

The random variable X has a Laplace (Double Exponential) distribution if it has a probability distribution function (pdf) of form

$$f(x) = \begin{cases} (2\sigma)^{-1}\exp[-|x - \mu|/\sigma], & -\infty < x < \infty, \ \sigma > 0 \\ 0, & \text{otherwise.} \end{cases}$$

This distribution is also known as the first law of Laplace. It is known that $E(x) = \mu$ and standard deviation, $\sigma' = \sqrt{2}\sigma$. Although we concentrate on the estimation of μ and σ, it is clear that the estimation of standard deviation, σ', needs minor adjustment by using the above relationship.

Govindarajulu (1966) derived the least squares estimates of and σ based on the ordered observations $x_{(1)} < x_{(2)} < \cdots < x_{(n)}$ by using Lloyd's (1952) method. Note that these observations are positively correlated. Govindarajulu defined

$$Z_{(i)} = (X_{(i)} - \mu)/\sigma, \qquad i = 1, 2, ..., n,$$

and showed that

$$\begin{aligned}
2^n \mu_i &= 2^n E(Z_{(i)}) \\
&= \sum_{m=n-i+1}^{n} \binom{n}{m} S_1(m, n-i+1) - \sum_{m=i}^{n} \binom{n}{m} S_1(m, i)
\end{aligned}$$

(10.3.17)

$$\begin{aligned}
2^n \mu_i^{(2)} &= 2^n E(Z_{(i)}^2) \\
&= \sum_{m=n-i+1}^{n} \binom{n}{m} \{S_2(m, n-i+1) + S_1^2(m, n-i+1)\} \\
&\quad + \sum_{m=i}^{n} \binom{n}{m} \{S_2(m, i) + S_1^2(m, i)\}
\end{aligned}$$

(10.3.18)

$$\begin{aligned}
2^n \mu_{i,j} &= 2^n E(Z_{(i)} Z_{(j)}) \\
&= \sum_{m=n-i+1}^{n} \binom{n}{m} \{S_2(m, n-i+1) \\
&\qquad + S_1(m, n-i+1) S_1(m, n-j+1)\} \\
&\quad - \sum_{m=i}^{j-1} \binom{n}{m} S_1(m, i) S_1(n-j+1, n-m) \\
&\quad + \sum_{m=j}^{n} \binom{n}{m} \{S_2(m, j) + S_1^2(m, i) S_1(m, j)\} \quad (10.3.19)
\end{aligned}$$

where $S_r(j, k) = \sum_{N=j}^{k} N^{-r}$, $r = 1, 2$ and $S_r(j_2, j_1)$ for $j_2 \geq j_1$ is interpreted as $S_r(j_1, j_2)$. Let

$$\sigma_{ij} = E\{(Z_{(i)} - \mu_i)(Z_{(j)} - \mu_j)\}, \qquad 1 \leq i \leq j \leq n,$$

and $\sum = (\sigma_{ij})$. Then the formulae for σ_{ii} and σ_{ij} can be obtained from (10.3.17), (10.3.18) and (10.3.19). Govindarajulu gave the values of μ_i

accurate to six decimal places for $i = 1 + [(n+1)/2], \ldots, n$, where $[\cdot]$ denotes the largest integer contained in $[\cdot]$. The values of μ_i for $i \leq [(n+1)/2]$ can be obtained from the relation:

$$\mu_i = -\mu_{n-i+1}, \qquad i \leq [(n+1)/2]$$

with $\mu_{[\frac{n}{2}]+1} = 0$ for odd n.

Let $\hat{\mu}$ and $\hat{\sigma}$ denote the least squares estimates of μ and σ based on order statistics. Govindarajulu showed that

$$Var(\hat{\mu}) = \frac{\sigma^2}{1'\Sigma^{-1}1}, \qquad Var(\hat{\sigma}) = \frac{\sigma^2}{\mu'\Sigma^{-1}\mu} \quad \text{and} \quad Cov(\hat{\mu}, \hat{\sigma}) = 0.$$

where $\mu' = (\mu_1, \mu_2, \ldots, \mu_n)$.

In terms of our notation from Section 10.2,

$$\alpha_1 = \mu_i \qquad\qquad (10.3.20)$$

$$v_{ii} = \mu_i^{(2)} - \mu_i^2, \quad i = 1, 2, \ldots, n. \qquad (10.3.21)$$

The computed values of α_i and v_{ii} are used in (10.2.3) and (10.2.4) to calculate w_{1i} and w_{2i}. To facilitate computations of the estimators $\tilde{\mu}$ and $\tilde{\sigma}$, the coefficients w_{1i} and w_{2i} are given in Table 10.1 for $2 \leq n \leq 15$. We note that $w_{1i} = w_{1\ n-i+1}$ and more weight in the center and less weight in the tails, and all weights are positive. In the case of w_{2i}, $w_{2i} = -w_{2\ n-i+1}$, and zero weight in the middle when n is odd.

10.4   COMPARISON OF ESTIMATORS

10.4.1   Joint Estimation of μ and σ

The variances of our estimators $\tilde{\mu}$ and $\tilde{\sigma}$ are computed from (10.2.5), and the covariance between them is zero. The variances of our estimators are compared with those based on ordered least squares estimators to assess the effectiveness of the ranked set sampling. The variances of $\hat{\mu}$ and $\hat{\sigma}$ are taken from Govindarajulu (1966). Table 10.2 gives the variances of both sets of estimators and the following two relative precisions:

$$RP_1 = Var(\hat{\mu})/Var(\tilde{\mu}) \qquad \text{and} \qquad RP_2 = Var(\hat{\sigma})/Var(\tilde{\sigma}).$$

We note that $\tilde{\mu}$ is uniformly better than $\hat{\mu}$ and $\tilde{\sigma}$ is better than $\hat{\sigma}$ for $n > 4$.

In the case of MRSS and NRSS procedures, we choose the values of j and k which minimize the generalized variance of the estimators, where generalized variance of μ^* and σ^* is given by

$$GVar(\mu^*, \sigma^*) = Var(\mu^*)Var(\sigma^*) - (Cov(\mu^*, \sigma^*))^2.$$

The generalized variance of μ^{**} and σ^{**} is similarly defined. In the case of MRSS, the values of (j, k) are (1,2), (1,3), (2,5), (2,6) and (2,7) for $n = 2, 4, \ldots, 10$, respectively. In the case of NRSS, the values of (j, k) are (1,3), (2,6), (2,8), (3,10) and (3,12) for $n = 2, 4, \ldots, 10$, respectively. We also note that we get the same minimum generalized variance for the two pairs of (j, k) and $(n' - k + 1, n' - j + 1)$, where $n' = n$ and $2n$ for MRSS and NRSS, respectively. Tables 10.3 and 10.4 give the coefficients c_j, d_k, variances and covariance of the estimators for MRSS and NRSS respectively. From Tables 10.2, 10.3 and 10.4 we note that the estimators for μ and σ based on RSS and MRSS are identical for $n = 2$. However, for $n > 2$, μ^* and σ^* are better than $\tilde{\mu}$ and $\tilde{\sigma}$. Furth $\hat{\sigma}$ for all n. We computed the following six relative efficiencies to assess the merit of NRSS compared to all other estimators:

$$RP_3 = \frac{Var(\hat{\mu})}{Var(\mu^{**})}, \quad RP_4 = \frac{Var(\tilde{\mu})}{Var(\mu^{**})}, \quad RP_5 = \frac{Var(\mu^*)}{Var(\mu^{**})},$$

$$RP_6 = \frac{Var(\hat{\sigma})}{Var(\sigma^{**})}, \quad RP_7 = \frac{Var(\tilde{\sigma})}{Var(\sigma^{**})}, \quad \text{and} \quad RP_6 = \frac{Var(\sigma^*)}{Var(\sigma^{**})}.$$

We note that there are substantial gains in relative precisions of μ^{**} and σ^{**} over the other estimators. The gains in relative precision of μ^{**} over $\hat{\mu}$ is much higher than that of σ^{**} over $\hat{\sigma}$. However, the gains in relative precision of σ^{**} over $\tilde{\sigma}$ and σ^* are larger than those of μ^{**} over $\hat{\mu}$ and $\tilde{\mu}$, respectively.

10.4.2 Estimation of μ

In the previous section, we considered the joint estimation of the two parameters based on n observations. Here we are interested in estimating μ only without assuming any knowledge on σ or its estimator. In the case of MRSS, $j = n/2$ and $k = (n/2) + 1$ minimizes the variance of μ^*. For NRSS procedure, we choose the jth and kth order statistics so that the variance of μ^{**} will be minimized. The optimal estimator of μ is based on quasi-mid-range $(x_{(j)} + x_{(2n-j+1)})/2$, where $j = 2, 4, 6, 7$ and 9 for $2n = 4, 8, 12, 16$ and 20, respectively. These are exactly the same as reported by Raghunandanan and Srinivasan (1971). Table 10.6 gives the variances μ^* and μ^{**} and the relative efficiency $RP_9 = Var(\mu^*)/Var(\mu^{**})$. We note the estimator μ^{**} is quite superior to μ^* for all n.

10.4.3 Estimation of σ

In this section, using MRSS and NRSS, we estimate sigma only without assuming any knowledge on μ or its estimator. We choose the appropriate values of j and k to minimize the variances of σ^* and σ^{**}. In the case of MRSS, the optimal values are $j = 1$ and $k = n$ for the range of values of n considered here. In the case of NRSS, the optimal estimator for sigma is based on jth quasi- range, $x_{(2n-j+1)} - x_{(j)}$, where $j = 1$, 1, 2, 2 and 3 for $n = 2$, 4, 6 and 10, respectively. The values of coefficients and variances of $\sigma*$ and σ^{**} are presented in Table 10.6. The relative precision, $RP_{10} = Var(\sigma^*)/Var(\sigma^{**})$, is also provided in Table 10.6 to compare σ^{**} with σ^*. It is clear that the estimators based on NRSS are superior to those based on MRSS. RP_9 decreases as n increases. However, RP_{10} decreases and then increases as n increases. The high values of the relative efficiencies for small n are important since the use of NRSS is recommended in practice for small n. Therefore, we recommend that the estimators based on NRSS should be used particularly for small n.

REFERENCES

Bhoj, D. S. (1997a). Estimation of parameters of the extreme value distribution using ranked set sampling, *Communications in Statistics—Theory and Methods*, **26**, 653–667.

Bhoj, D. S. (1997b). Estimation of parameters using modified ranked set sampling, In *Applied Statistical Science, II* (Ed., M. Ahsanullah), pp. 154–163, Nova Science Publishers, New York.

Bhoj, D. S. (1997c). New parametric ranked set sampling, *Journal of Applied Statistical Science*, **6**, 275–289.

Bhoj, D. S. and Ahsanullah, M. (1996). Estimation of parameters of the generalized geometric distribution using ranked set sampling, *Biometrics*, **52**, 685–694.

Bohn, L. L. and Wolfe, D. A. (1992). Nonparametric two-sample procedures for ranked-set samples data, *Journal of the American Statistical Association*, **87**, 552–561.

David, H. A. and Levine, D. N. (1972). Ranked set sampling in the presence of judgment error, *Biometrics*, **28**, 553–555.

Dell, T. R. and Clutter, J. L. (1972). Ranked-set sampling theory with order statistics background, *Biometrics*, **28**, 545–555.

Govindarajulu, Z. (1966). Best linear estimates under symmetric censoring of the parameters of a double exponential population, *Journal of the American Statistical Association*, **61**, 248–258.

Halls, L. K. and Dell, T. R. (1996). Trial of ranked set sampling for forage yields, *Forest Science*, **12**, 22–26.

Lloyd, E. H. (1952). Least-squares estimation of location and scale parameters using order statistics, *Biometrika*, **39**, 88–95.

McIntyre, G. A. (1952). A method of unbiased selective sampling, using ranked sets, *Australian Journal of Agricultural Research*, **3**, 385–390.

Patil, G. P., Sinha, A. K. and Taillie, C. (1993a). Ranked set sampling from a finite population in the presence of a trend on a site, *Journal of Applied Statistical Science*, 1, 51–65.

Patil, G. P., Sinha, A. K. and Taillie, C. (1993b). Relative precision of ranked set sampling: A comparison with the regression estimator, *Environmetrics*, **4**, 399–412.

Raghunandanan, K. and Srinivasan, R. (1971). Simplified estimation of parameters in a double exponential distribution, *Technometrics*, **13**, 689–691.

Stokes, S. L. (1977). Ranked set sampling with concomitant variables, *Communications in Statistics—Theory and Methods*, **6**, 1207–1211.

Stokes, S. L. and Sager, T. W. (1988). Characterization of a ranked-set sample with application to estimating distribution functions, *Journal of the American Statistical Association*, **83**, 374–381.

Takahasi, K. and Wakimoto, K. (1968). On unbiased estimates of the population mean based on the sample stratified by means of ordering, *Annals of the Institute of Statistical Mathematics*, **20**, 1–31.

TABLE 10.1 Coefficients for computing $\tilde{\mu}$ and $\tilde{\sigma}$

n	i	1	2	3	4	5	6	7	8
2	w_{1i}	0.50000							
	w_{2i}	−0.66667							
3	w_{1i}	0.23727	0.52547						
	w_{2i}	−0.44444	0.00000						
4	w_{1i}	0.13267	0.36733						
	w_{2i}	−0.30834	−0.21182						
5	w_{1i}	0.08306	024306	0.34776					
	w_{2i}	−0.22798	−0.24059	0.00000					
6	w_{1i}	0.05639	0.16586	0.27775					
	w_{2i}	−0.17653	−0.22172	−0.10829					
7	w_{1i}	004059	0.11829	0.21082	026059				
	w_{2i}	−0.14147	−0.19467	−0.14808	000000				
8	w_{1i}	003053	0.08788	0.15981	0.22178				
	w_{2i}	−0.11638	−0.16937	−0.15558	−0.06647				
9	w_{1i}	0.02375	006757	0.12317	0.18105	0.20892			
	w_{2i}	−0.09773	−0.14779	−0.15003	−0.09994	0.00000			
10	w_{1i}	0.01898	0.05344	0.09697	0.14621	0.18439			
	w_{2i}	−0.08343	−0.12981	−0.13995	−0.11338	−0.04517			
11	w_{1i}	0.01550	0.04326	0.07796	0.11859	0.15733	0.17469		
	w_{2i}	−0.07219	−0.11485	−0.12885	−0.11613	−0.07193	0.00000		
12	w_{1i}	0.01289	0.03571	0.06388	0.09724	0.13246	0.15783		
	w_{2i}	−0.06317	−0.10235	−0.11813	−0.11359	−0.08581	−0.03277		
13	w_{1i}	0.01089	0.02995	0.05322	0.08076	0.11135	0.13868	0.15032	
	w_{2i}	−0.05582	−0.09182	−0.10827	−0.10866	−0.09167	−0.05423	0.00000	
14	w_{1i}	0.00931	0.02547	0.04498	0.06795	0.09406	0.12020	0.13803	
	w_{2i}	−004973	−0.08288	−0.09938	−0.10279	−0.09284	−0.06704	−0.02490	
15	w_{1i}	0.00805	0.02192	0.03849	0.05786	0.08009	0.10373	0.12382	0.13207
	w_{2i}	−0.04463	−0.07522	−0.09145	−0.09671	−0.09140	−0.07387	−0.04235	0.00000

TABLE 10.2 Variances and relative precisions

n	$\dfrac{Var(\tilde{\mu})}{\sigma^2}$	$\dfrac{Var(\tilde{\mu})}{\sigma^2}$	$\dfrac{Var(\tilde{\sigma})}{\sigma^2}$	$\dfrac{Var(\tilde{\sigma})}{\sigma^2}$	RP_1	RP_2
2	1.0000	0.7188	0.7778	1.2778	1.391	0.609
3	0.5895	0.3357	0.4321	0.5590	1.756	0.773
4	0.4155	0.1913	0.2986	0.3209	2.172	0.931
5	0.3169	0.1221	0.2290	0.2110	2.595	1.085
6	0.2548	0.0842	0.1858	0.1502	3.024	1.237
7	0.2122	0.0614	0.1565	0.1127	3.456	1.389
8	0.1814	0.0466	0.1351	0.0878	3.889	1.538
9	0.1581	0.0366	0.1190	0.0705	4.322	1.689
10	0.1399	0.0294	0.1062	0.0578	4.756	1.836
11	0.1253	0.0242	0.0960	0.0483	5.187	1.986
12	0.1134	0.0202	0.0876	0.0410	5.630	2.135
13	0.1035	0.0171	0.0805	0.0353	6.052	2.283
14	0.0952	0.0147	0.0745	0.0306	6.488	2.432
15	0.0880	0.0127	0.0693	0.0269	6.917	2.579

TABLE 10.3 Coefficients, variances and covariance of estimators for MRSS

n	c_j	d_k	$\frac{Var(\mu^*)}{\sigma^2}$	$\frac{Var(\sigma^*)}{\sigma^2}$	$\frac{Cov(\mu^*,\sigma^*)}{\sigma^2}$
2	0.5000	0.6667	0.7188	1.2778	0.0000
4	0.1988	0.5783	0.1956	0.3282	0.0378
6	0.5000	0.6667	0.0847	0.1505	0.0000
8	0.3556	0.6491	0.0472	0.0867	0.0026
10	0.7580	0.6135	0.0296	0.0566	0.0024

TABLE 10.4 Coefficients, variances and covariance of estimators for NRSS

n	c_j	d_k	$\frac{Var(\mu^{**})}{\sigma^2}$	$\frac{Var(\sigma^{**})}{\sigma^2}$	$\frac{Cov(\mu^{**},\sigma^{**})}{\sigma^2}$
2	0.1900	0.5783	0.4788	0.4725	−0.0202
4	0.3356	0.6491	0.1221	0.1208	−0.0082
6	0.1843	0.5785	0.0487	0.0611	−0.0065
8	0.1621	0.7055	0.0244	0.0382	−0.0023
10	0.1133	0.6313	0.0149	0.0253	−0.0016

TABLE 10.5 Relative efficiencies of the estimators

n	RP_3	RP_4	RP_5	RP_6	RP_7	RP_8
2	2.089	1.051	1.501	1.646	2.704	2.704
4	3.404	1.567	1.603	2.471	2.665	2.715
6	5.232	1.730	1.738	3.042	2.459	2.464
8	7.428	1.910	1.931	3.540	2.302	2.271
10	9.392	1.975	1.985	4.195	1.284	1.235

TABLE 10.6 Relative efficiencies of the estimators based on MRSS and NRSS

n	$\frac{Var(\mu^*)}{\sigma^2}$	d_k	$\frac{Var(\sigma^*)}{\sigma^2}$	$\frac{Var(\mu^{**})}{\sigma^2}$	d_k	$\frac{Var(\sigma^{**})}{\sigma^2}$	RP_9	RP_{10}
2	0.7188	0.6667	1.2778	0.4201	0.3609	0.3152	1.711	4.054
4	0.1302	0.3609	0.1878	0.0937	0.2470	0.0860	1.390	2.183
6	0.0506	0.2847	0.0807	0.0394	0.3546	0.0407	1.284	1.985
8	0.0263	0.3935	0.0466	0.0213	0.2963	0.0230	1.234	2.027
10	0.106	0.2236	0.0310	0.0131	0.3560	0.0151	1.215	2.057

CHAPTER 11

SOME RESULTS ON ORDER STATISTICS ARISING IN MULTIPLE TESTING

SANAT K. SARKAR

Temple University, Philadelphia, PA

Abstract: Results on increasing sequences of critical values for order statistics, which are useful in multiple testing, are discussed. Some new results on probability distribution of ordered components are also presented.

Keywords and phrases: Order statistics, critical values, multiple test procedure, multivariate totally positive of order two, positively dependent, step-up test, step-down test

11.1 INTRODUCTION

A sudden upsurge of research has taken place in the area of multiple testing in the recent years that has resulted in some newer results and raised interesting questions related to probability distributions of ordered components of dependent random variables. A review of these and presentation of some additional new results are the main focus of this paper.

Given a family of null hypotheses $H_1, \ldots, H_n$, a multiple test procedure is designed to simultaneously test the hypotheses based on p-values associated with them. There are two types of multiple test procedure – single-step and stepwise. A single-step procedure tests a hypothesis without reference to one another in the family; whereas, in a stepwise procedure, the tests are conducted step by step in a certain order until a decision, either acceptance

183

or rejection, is reached for each hypothesis in the family. The two most commonly used stepwise procedures are step-down and step-up procedures. Let $P_{(1)} \leq \cdots \leq P_{(n)}$ denote the ordered versions of $P_1, \ldots, P_n$, the p-values corresponding to $H_1, \ldots, H_n$, respectively. Suppose that the hypotheses which correspond to these ordered p- values are $H_{(1)}, \ldots, H_{(n)}$, respectively. Then *a step-down procedure*, based on n constants $0 < a_1 \leq \cdots \leq a_n < 1$, proceeds with $P_{(1)}$, the most significant p-value. If $P_{(1)} > a_1$, testing stops and accepts all the hypotheses; otherwise, it rejects $H_{(1)}$ and goes to the next step. In general, if testing continues upto the ith step $(1 \leq i \leq n)$ and if $P_{(i)} > a_i$, testing stops by accepting all the remaining hypotheses $H_{(i)}, \ldots, H_{(n)}$; otherwise, rejects $H_{(i)}$ and goes to the $(i+1)$th step. A *step-up procedure*, based on n constants $0 < b_1 \leq \cdots \leq b_n < 1$, however, proceeds with $P_{(n)}$, the least significant p-value. If $P_{(n)} \leq b_n$, testing stops and rejects all the hypotheses; otherwise, it accepts $H_{(n)}$ and goes to the next step. In general, if testing continues upto the ith step $(1 \leq i \leq n)$ and if $P_{(n-i+1)} \leq b_{n-i+1}$, testing stops by rejecting all the remaining hypotheses $H_{(n-i+1)}, \ldots, H_{(1)}$; otherwise, accepts $H_{(n-i+1)}$ and goes to the $(i+1)$th step.

The idea of controlling the familywise error (FWE) rate, that is, the probability of rejecting any true null hypothesis, at a pre-specified level $\alpha \epsilon (0, 1)$ is a widely accepted concept in multiple testing [Hochberg and Tamhane (1993), Hsu (1996) and Westfall and Young (1993)]. The critical values in the above stepwise tests are chosen subject to such a requirement. We will assume that the p-values $P_1, \ldots, P_n$ correspond to right-tailed tests based on the test statistics $X_1, \ldots, X_n$, respectively, and that these statistics are all continuous having an exchangeable joint probability distribution under the null hypotheses. Then, in terms of $X_{(1):n} \leq \cdots \leq X_{(n):n}$, the ordered values of $X_1, \cdots, X_n$, the determination of the critical values in a step-down test becomes equivalent to finding constants $c_1 \leq \cdots \leq c_n$ satisfying

$$P\{X_{(k):k} \leq c_k\} \geq 1 - \alpha, \qquad \text{for } k = 1, \ldots, n; \qquad (11.1.1)$$

whereas, for a step-up test, the critical values correspond to constants $d_1 \leq \cdots \leq d_n$ satisfying the following set of inequalities:

$$P\{X_{(1):k} \leq d_1, \ldots, X_{(k):k} \leq d_k\} \geq 1 - \alpha, \quad \text{for } k = 1, \ldots, n. \quad (11.1.2)$$

For discussions on step-down and step-up tests, the readers are referred to Dunnett and Tamhane (1991, 1992a,b, 1993, 1995), Finner and Roter (1998), Hochberg (1988), Hochberg and Tamhane (1987), Rom (1990), Liu (1996, 1997a,b) and Tamhane, Liu and Dunnett (1998).

Do increasing sequences of critical values defined in (11.1.1)–(11.1.2) really exist? Although the answer is yes in case of (11.1.1), which can be easily seen by using the fact that $X_{(k):k}$ stochastically increases with k, it is however not so immediate for (11.1.2). The existence of increasing $d_1, \ldots, d_n$ satisfying (11.1.2) has been verified in some particular situations, and in that process some new results on probability distribution of ordered components of a random vector have been developed. These recent results will be reviewed in the following sections, in addition to presenting a few new ones.

11.2   THE MONOTONICITY OF d_i's

Towards finding a set of increasing d_i's satisfying (11.1.2), Hochberg (1988) first noted that with $d_i = F^{-1}(1 - \frac{\alpha}{i})$, $i = 1, \ldots, n$, where F^{-1} is the inverse of the common marginal cdf F, since

$$P\{X_{(1):k} \leq d_1, \ldots, X_{(k):k} \leq d_k\}$$
$$> P\{X_{(1):k} \leq d_{1,k}, \ldots, X_{(k):k} \leq d_{k,k}\}, \qquad (11.2.3)$$

where $F(d_{i,k}) = 1 - \frac{k-i+1}{k}\alpha$ for $i = 1, \ldots, k$, and the right-hand side of (11.2.3) is $1 - \alpha$ when X_i's are iid [see, for example, Karlin (1969), Sarkar and Chang (1997), and Simes (1986)], an increasing sequence of d_i's does exist in the iid case. The fact that these same d_i's also provide a solution to (11.1.2) in a more general situation where the X_i's are positively dependent was theoretically established in Sarkar (1998), and Sarkar and Chang (1997). That is, the following inequality:

$$P\{X_{(1):k} \leq d_{1,k}, \ldots, X_{(k):k} \leq d_{k,k}\} \geq 1 - \alpha, \qquad (11.2.4)$$

still holds even if X_i's are positively dependent. It was initially conjectured in Simes (1986) based on simulations.

In many multiple testing situations, when the null hypotheses are true, the underlying test statistics posses multivariate distributions that are positively dependent in the sense of satisfying the multivariate totally positive of order two (MTP$_2$) condition [due to Karlin and Rinott, (1980, 1981)]. An n-dimensional random vector $\boldsymbol{X} = (X_1, \ldots, X_n)'$ is said to be MTP$_2$, and TP$_2$ when $n = 2$, if its probability density, $f(\boldsymbol{x})$, satisfies the following condition:

$$f(\boldsymbol{x} \vee \boldsymbol{y})f(\boldsymbol{x} \wedge \boldsymbol{y}) \geq f(\boldsymbol{x})f(\boldsymbol{y}) \text{ for all } \boldsymbol{x}, \boldsymbol{y} \in \mathcal{R}^n,$$

where, with $\boldsymbol{x} = (x_1, \ldots, x_n)'$ and $\boldsymbol{y} = (y_1, \ldots, y_n)'$, $\boldsymbol{x} \vee \boldsymbol{y} = (\max(x_1, y_1), \ldots, \max(x_n, y_n))$ and $\boldsymbol{x} \wedge \boldsymbol{y} = (\min(x_1, y_1), \ldots, \min(x_n, y_n))$. This condition is satisfied by a large family of multivariate distributions, such as those with densities of the form

$$\int \prod_{i=1}^{n} f(x_i, z) g(z) dz, \qquad (11.2.5)$$

for some probability densities $f(x, z)$ and $g(z)$, where $f(x, z)$ is TP$_2$ in (x, z), listed in Sarkar and Chang (1997), in addition to the multivariate normal with zero means and nonnegative correlations, and the absolute-valued multivariate normal with zero means and some specific covariance structures.

The proof of (11.2.4) for general MTP$_2$ distributions and certain mixtures of MTP$_2$ distributions given in Sarkar (1998b) relies heavily on a new identity involving joint probability distribution of ordered components of a random vector $Y = (Y_1, \ldots Y_n)$, not necessarily MTP$_2$. This identity reduces to the one proved in Sarkar and Chang (1997) for iid random variables that was used to prove (11.2.4) for distributions of the type (11.2.5). These identities are presented in the next section along with other interesting results.

The problem of verifying the existence of increasing d_i's satisfying (11.1.2) becomes much more difficult for $n > 3$ if we insist on the equalities, rather than the inequalities, in (11.1.2); that is, if we want our step-up test to control the familywise error rate exactly at α. Note that, the monotonicity of the d_i's will follow using an induction argument if we can show that for

$$\psi_n(d) = P\{X_{(1):n} \le d_1, \ldots, X_{(n-1):n} \le d_{n-1}, X_{(n):n} \le d\}, \qquad (11.2.6)$$

which is a nondecreasing function of d, $\psi_n(d_{n-1}) \le 1 - \alpha \le \psi_n(\infty)$, assuming that there exist $d_1 \le \cdots \le d_{n-1}$ satisfying

$$P\{X_{(1):k} \le d_1, \ldots, X_{(k):k} \le d_k\} = 1 - \alpha \text{ for } k = 1, \ldots, n-1. \quad (11.2.7)$$

Since $X_{(i):n} \le X_{(i):n-1}$ for $i = 1, \ldots, n-1$,

$$\begin{aligned}
\psi_n(\infty) &= P\{X_{(1):n} \le d_1, \ldots, X_{(n-1):n} \le d_{n-1}\} \\
&\ge P\{X_{(1):n-1} \le d_1, \ldots, X_{(n-1):n-1} \le d_{n-1}\} \\
&= 1 - \alpha.
\end{aligned}$$

Hence, proving the other inequality $\psi_n(d_{n-1}) \le 1 - \alpha$ assuming (11.2.7) for $n > 3$ is the major problem here. Dalal and Mallows (1992) have given a proof of this in the special case of iid X_i's, but it remains to be a challenging open problem in the more general case of dependent X_i's.

Finner and Roters (1998) gave a counterexample which makes the point that one should not hope for increasing d_i's for any set of dependent X_i's. Sarkar (1998a) showed that the random variables considered in Finner and Roter's (1998) counterexample is not MTP$_2$, and proved that for equicorrelated trivariate normal with a nonnegative common correlation not exceeding a certain value (depending on α), which is MTP$_2$, the increasing property of the d_i's does hold. This raises the hope that the required monotonicity property might hold for positively dependent X_i's.

11.3   RESULTS ON ORDERED COMPONENTS OF A RANDOM VECTOR

We will state in this section a number of results involving joint probability distribution of the ordered components of a random vector $Y = (Y_1, \ldots, Y_n)$. First, we have some identities which played key roles in proving (11.2.4).

Theorem 11.3.1 *[Sarkar (1998b)] Let $Y_{(1)} \leq \cdots \leq Y_{(n)}$ be the ordered components of $Y = (Y_1, \ldots, Y_n)'$, and F_i be the marginal of Y_i. Then, (i) for any fixed $a_1 \leq \ldots, \leq a_n$,*

$$
\begin{aligned}
&P\{Y_{(1)} \leq a_1, \ldots, Y_{(n)} \leq a_n\} \\
&= \frac{1}{n}\sum_{i=1}^{n} F_i(a_1) + \sum_{i=1}^{n}\sum_{j=1}^{n-1} E\Big[\Big\{\frac{I(Y_i > a_{n-j})}{j+1} - \frac{I(Y_i > a_{n-j+1})}{j}\Big\} \\
&\qquad \times P\{Y_{(1)}^{(-i)} \leq a_1, \ldots, Y_{(n-j)}^{(-i)} \leq a_{n-j}|Y_i\}\Big],
\end{aligned}
\tag{11.3.8}
$$

and
(ii) for any fixed $b_1 \leq \ldots, \leq b_n$,

$$
\begin{aligned}
&P\{Y_{(1)} > b_1, \ldots, Y_{(n)} > b_n\} \\
&= 1 - \frac{1}{n}\sum_{i=1}^{n} F_i(b_n) + \sum_{i=1}^{n}\sum_{j=1}^{n-1} E\Big[\Big\{\frac{I(Y_i \leq b_{j+1})}{j+1} - \frac{I(Y_i \leq b_j)}{j}\Big\} \\
&\qquad \times P\{Y_{(j)}^{(-i)} > b_{j+1}, \ldots, Y_{(n-1)}^{(-i)} > b_n|Y_i\}\Big],
\end{aligned}
\tag{11.3.9}
$$

where, for each $i = 1, \ldots, n$, $Y_{(1)}^{(-i)} \leq \cdots \leq Y_{(n-1)}^{(-i)}$ denote the ordered components of the (n-1)-dimensional random vector $Y^{(-i)}$ obtained by ignoring Y_i from Y.

A number of interesting results follow from these identities.

Corollary 11.4 *If $(Y_1, \ldots, Y_n)$ is exchangeable with the common marginal F, then*

(i)

$$P\{Y_{(1)} \leq a_1, \ldots, Y_{(n)} \leq a_n\}$$

$$= F(a_1) + n \sum_{j=1}^{n-1} E\Big[\Big\{\frac{I(Y_n > a_{n-j})}{j+1} - \frac{I(Y_n > a_{n-j+1})}{j}\Big\}$$

$$\times P\{\mathbf{Y}_{(1)}^{(-n)} \leq a_1, \ldots, \mathbf{Y}_{(n-j)}^{(-n)} \leq a_{n-j}|Y_n\}\Big], \qquad (11.4.10)$$

and

(ii)

$$P\{Y_{(1)} > b_1, \ldots, Y_{(n)} > b_n\}$$

$$= 1 - F(b_n) + n \sum_{j=1}^{n-1} E\Big[\Big\{\frac{I(Y_n \leq b_{j+1})}{j+1} - \frac{I(Y_n \leq b_j)}{j}\Big\}$$

$$\times P\{\mathbf{Y}_{(j)}^{(-n)} > b_{j+1}, \ldots, \mathbf{Y}_{(n-1)}^{(-n)} > b_n|Y_n\}\Big]. \qquad (11.4.11)$$

Corollary 11.5 *[Sarkar and Chang (1997)] If $(Y_1, \ldots, Y_n)$ are iid, then*

(i)

$$P\{Y_{(1)} \leq a_1, \ldots, Y_{(n)} \leq a_n\}$$

$$= F(a_1) + n \sum_{j=1}^{n-1} \Big\{\frac{\bar{F}(a_{n-j})}{j+1} - \frac{\bar{F}(a_{n-j+1})}{j}\Big\}$$

$$\times P\{Y_{(1):n-1} \leq a_1, \ldots, Y_{(n-j):n-1} \leq a_{n-j}\},$$

$$(11.5.12)$$

and

(ii)

$$P\{Y_{(1)} > b_1, \ldots, Y_{(n)} > b_n\}$$

$$= \bar{F}(b_n) + n \sum_{j=1}^{n-1} \Big\{\frac{F(b_{j+1})}{j+1} - \frac{F(b_j)}{j}\Big\}$$

$$\times P\{Y_{(j):n-1} > b_{j+1}, \ldots, Y_{(n-1):n-1} > b_n\}, \quad (11.5.13)$$

where $\bar{F} = 1 - F$.

Corollary 11.6 *Let* $U_{(1):n} \leq \cdots \leq U_{(n):n}$ *be the order statistics based on* n *iid observations from* $U(0,1)$, *then* $\min_{1 \leq i \leq n}\{nU_{(i):n}/i\}$ *is also* $U(0,1)$

Alternative expressions for probability distributions of the extreme values can be obtained from the above identities. For instance,

$$P\{Y_{(n):n} \leq a\}$$

$$= \frac{1}{n}\sum_{i=1}^{n} F_i(a) - \sum_{i=1}^{n}\sum_{j=1}^{n-1} \frac{1}{j(j+1)}P\{Y_{(n-j)}^{(-i)} \leq a, Y_i > a\}$$

$$= 1 - \sum_{i=1}^{n} \bar{F}_i(a) + \sum_{i=1}^{n}\sum_{j=1}^{n-1} \frac{1}{j(j+1)}P\{\min(Y_{(n-j)}^{(-i)}, Y_i) > a\}.$$

$$(11.6.14)$$

The probability distribution of the other extreme can be similarly expressed in terms of marginal and the bivariate probabilities involving Y_i and $Y_{(j)}^{(-i)}$, $j \neq i$. It is to be noted that the term $1 - \sum_{i=1}^{n} \bar{F}_i(a)$ in (11.6.14) provides a lower bound to the probability $P\{Y_{(n):n} \leq a\}$ which one would get by applying the Bonferroni inequality [Miller (1981, p. 8)]. An improvement of this bound could be obtained by including the additional bivariate probabilities involving order statistics. This type of improvement is different from those known in the literature [see, for example, Efron (1997)].

We will now present some results relating to the question of finding increasing d_i's satisfying (11.1.2) with the equalities; that is, to the problem of proving the inequality $\psi_n(d_{n-1}) \leq 1 - \alpha$ as stated in the above section. First note that in the case of iid X_i's,

$$\psi_n(d_{n-1})$$

$$= nE[(F(d_{n-1}) - U_{(n-1):n-1})I(U_{(1):n-1} \leq F(d_1), \ldots, U_{(n-1):n-1}$$

$$\leq F(d_{n-1}))],$$

where U_i's are iid uniform on $(0,1)$. The required inequality in the iid case then follows by using the fact that $F(d_i) \longrightarrow 1$ for each $i = 1, \ldots, n-1$ as $F(d_1) \longrightarrow 1$ in the following theorem.

Theorem 11.6.1 *[Dalal and Mallows (1992)] Let* $U_{(1):n} \leq \cdots \leq U_{(n):n}$ *be the order statistics based on* n *iid observations from* $U(0,1)$, *and let* $0 < c_1 \leq \cdots \leq c_n < 1$ *be such that* $P\{U_{(1):k} \leq c_1, \ldots, U_{(k):k} \leq c_k\} = c_1$ *for all* $k = 1, \ldots, n$. *Then,* $c_1^{-1}E[(c_n - U_{(n):n})I(U_{(1):n} \leq c_1, \ldots, U_{(n):n} \leq c_n)]$ *is increasing in* c_1.

The following theorem gives an idea how close the constants in Theorem 11.6.1 are to each other.

Theorem 11.6.2 *The constants* $c_1 \leq \cdots \leq c_n$ *in Theorem 11.6.1 satisfy*

$$
c_k - c_{k-1} \;=\; \frac{k-1}{2}[\mathrm{Var}(U_{(k-2):k-2})
$$
$$
- \,\mathrm{Var}(U_{(k-2):k-2}|U_{(1):k-2} \leq c_1, \ldots, U_{(k-2):k-2} \leq c_{k-2})],
$$
$$
(11.6.15)
$$

for all $k = 3, \ldots, n-2$.

PROOF. It is enough to prove this for $k = n$.

$$
P\{U_{(1):n} \leq c_1, \ldots, U_{(n):n} \leq c_n\}
$$
$$
= \; nE\{(c_n - U_{(n-1):n-1})I(U_{(1):n-1} \leq c_1, \ldots, U_{(n-1):n-1} \leq c_{n-1})\}
$$
$$
= \; n(c_n - c_{n-1})P\{U_{(1):n-1} \leq c_1, \ldots, U_{(n-1):n-1} \leq c_{n-1}\}
$$
$$
+ \, nE\{(c_{n-1} - U_{(n-1):n-1})I(U_{(1):n-1} \leq c_1, \ldots, U_{(n-1):n-1} \leq c_{n-1})\}
$$
$$
= \; n(c_n - c_{n-1})P\{U_{(1):n-1} \leq c_1, \ldots, U_{(n-1):n-1} \leq c_{n-1}\}
$$
$$
+ \frac{n(n-1)}{2}E\{(c_{n-1} - U_{(n-2):n-2})^2 I(U_{(1):n-2}
$$
$$
\leq c_1, \ldots, U_{(n-2):n-2} \leq c_{n-2})\}
$$
$$
= \; n(c_n - c_{n-1})P\{U_{(1):n-1} \leq c_1, \ldots, U_{(n-1):n-1} \leq c_{n-1}\}
$$
$$
+ \frac{n(n-1)}{2}[\mathrm{Var}((c_{n-1} - U_{(n-2):n-2})|U_{(1):n-2}
$$
$$
\leq c_1, \ldots, U_{(n-2):n-2} \leq c_{n-2})
$$
$$
+ E^2\{(c_{n-1} - U_{(n-2):n-2})|U_{(1):n-2} \leq c_1, \ldots, U_{(n-2):n-2} \leq c_{n-2}\}]
$$
$$
\times P\{U_{(1):n-2} \leq c_1, \ldots, U_{(n-2):n-2} \leq c_{n-2}\}
$$
$$
= \; n(c_n - c_{n-1})P\{U_{(1):n-1} \leq c_1, \ldots, U_{(n-1):n-1} \leq c_{n-1}\}
$$
$$
+ \frac{n(n-1)}{2}[\mathrm{Var}((c_{n-1} - U_{(n-2):n-2})|U_{(1):n-2}
$$
$$
\leq c_1, \ldots, U_{(n-2):n-2} \leq c_{n-2})
$$
$$
+ \frac{1}{(n-1)^2}]P\{U_{(1):n-2} \leq c_1, \ldots, U_{(n-2):n-2} \leq c_{n-2}, \}.
$$

Now, since $P\{U_{(1):k} \leq c_1, \ldots, U_{(k):k} \leq c_k\}$ is the same for $k = n-2, n-1$ and n, we get

$$n(c_n - c_{n-1})$$

$$= \frac{n(n-1)}{2}[(\frac{2}{n(n-1)} - \frac{1}{(n-1)^2})$$

$$\quad - \mathrm{Var}((c_{n-1} - U_{(n-2):n-2})|U_{(1):n-2} \leq c_1, \ldots, U_{(n-2):n-2} \leq c_{n-2})]$$

$$= \frac{n(n-1)}{2}[\mathrm{Var}(U_{(n-2):n-2})$$

$$\quad - \mathrm{Var}(U_{(n-2):n-2}|U_{(1):n-2} \leq c_1, \ldots, U_{(n-2):n-2} \leq c_{n-2})].$$

This proves the result. □

REFERENCES

Dalal, S. R. and Mallows, C. L. (1992). Buying with exact confidence, *Annals of Applied Probability*, **2**, 752–765.

Dunnett, C. W. and Tamhane, A. C. (1991). Step-down multiple tests for comparing treatments with a control in unbalanced one-way layouts, *Statistics in Medicine*, **10**, 939–947.

Dunnett, C. W. and Tamhane, A. C. (1992a). A step-up multiple test procedure, *Journal of the American Statistical Association*, **87**, 162–170.

Dunnett, C. W. and Tamhane, A. C. (1992b). Comparisons between a new and active and placebo controls in an efficacy trial, *Statistics in Medicine*, **11**, 1057–1063.

Dunnett, C. W. and Tamhane, A. C. (1993). Power comparisons of some step-up multiple test procedures, *Statistics & Probability Letters*, **16**, 55–58.

Dunnett, C. W. and Tamhane, A. C. (1995). Step-up multiple testing of parameters with unequally correlated estimates, *Biometrics*, **51**, 217–227.

Efron, B. (1997). The length heuristic for simultaneous hypothesis tests, *Biometrika*, **84**, 143–157.

Finner, H. and Roters, M. (1998). Asymptotic comparisons of step-up and step-down multiple test procedures based on exchangeable test statistics, *Annals of Statistics*, **26**, 505–520.

Hochberg, Y. (1988). A sharper Bonferroni procedure for multiple tests of significance, *Biometrika*, **75**, 800–802.

Hochberg, Y. and Tamhane, A. C. (1987). *Multiple Comparison Procedures*, John Wiley & Sons, New York.

Hsu, J. C. (1996). *Multiple Comparisons: Theory and Methods*, Chapman and Hall, New York.

Karlin, S. (1969). *A First Course in Stochastic Processes*, Academic Press, New York.

Karlin, S., and Rinott, Y. (1980). Classes of orderings of measures and related correlation inequalities I: Multivariate totally positive distributions, *Journal of Multivariate Analysis*, **10**, 467–498.

Karlin, S. and Rinott, Y. (1981). Total positivity properties of absolute value multinormal variables with applications to confidence interval estimates and related probabilistic inequalities, *Annals of Statistics*, **9**, 1035–1049.

Liu, W. (1996). Multiple tests of a non-hierarchical finite family of hypotheses, *Journal of the Royal Statistical Society, Series B*, **58**, 455–461.

Liu, W. (1997a). Some results on step-up tests for comparing treatments with a control in unbalanced one-way layouts, *Biometrics*, **53**, 1508–1512.

Liu, W. (1997b). Stepwise tests when the statistics are independent, *Australian Journal of Statistics*, **39**, 169–177.

Miller, R. G., Jr. (1981). *Simultaneous Statistical Inference*, Springer-Verlag, New York.

Rom, D. M. (1990). A sequentially rejective test procedure based on a modified Bonferroni inequality, *Biometrika*, **77**, 663–665.

Sarkar, S. K. (1998a). A note on the monotonicity of the critical values of a step-up test, *Technical Report No. 98-2*, Temple University, Philadelphia.

Sarkar, S. K. (1998b). Some probability inequalities for ordered MTP_2 random variables: A proof of the Simes conjecture, *Annals of Statistics*, **26**, 494–504.

Sarkar, S. K. and Chang, C-K. (1997). The Simes method for multiple hypothesis testing with positively dependent test statistics, *Journal of the American Statistical Association*, **92**, 1601–1608.

Simes, R. J. (1986), An improved Bonferroni procedure for multiple tests of significance, *Biometrika*, **73**, 751–754.

Tamhane, A. C., Liu, W. and Dunnett, C. W. (1998). A generalized step-up-down multiple test procedure, *Canadian Journal of Statistics*, **26**, 353–363.

Westfall, P. H. and Young, S. S. (1993). *Resampling Based Multiple Testing*, John Wiley & Sons, New York.

Part IV
Robust Inference

CHAPTER 12

ROBUST ESTIMATION VIA GENERALIZED L-STATISTICS: THEORY, APPLICATIONS, AND PERSPECTIVES

ROBERT SERFLING

University of Texas at Dallas, Richardson, TX

Abstract: Generalized L-statistics, introduced in Serfling (1984) and including classical U-statistics and L-statistics, are linear functions based on the ordered evaluations of a kernel over subsets of the sample observations. In particular, generalized median statistics fall within this class and are found to fulfill an interesting and potent principle, that "smoothing" followed by "medianing" yields a very favorable combination of efficiency and robustness. Extensive asymptotic theory now available for generalized L-statistics is reviewed, including asymptotic normality, strong convergence, large deviation, sequential fixed-width confidence interval, jackknife, and bootstrap results, as well as Glivenko-Cantelli theory for associated empirical processes of U-statistic structure. Illustrative applications are treated, including nonparametric and robust location and spread estimation, nonparametric analysis of linear models, nonparametric regression, and robust parametric scale estimation for exponential distributions, equivalently tail index estimation for Pareto distributions.

Keywords and phrases: Generalized L-statistics, robust estimation

12.1   INTRODUCTION

The notion of generalized L-statistics (GL-statistics) unifies the simpler classes of L- and U-statistics while maintaining a nice level of mathematical tractability. In applications, the notion leads to formulation of highly competitive estimators in both nonparametric and robust parametric estimation contexts. Here we review the theory and applications of GL-statistics and illustrate through several examples an interesting and potent principle, that "smoothing" followed by "medianing" yields a very favorable combination of efficiency and robustness.

Initially we consider the setting of a sample of i.i.d. real-valued observations $X_1, \ldots, X_n$ having cdf F. Denote the ordered observations by $X_{n1} \leq \cdots \leq X_{nn}$. We ask

> *What common or unifying feature is shared by the sample mean, sample variance, sample median, 5% trimmed mean, Hodges-Lehmann location estimator (i.e., median of pairwise averages $(X_i + X_j)/2$), median of three-way averages $(X_i + X_j + X_k)/3$), Theil's nonparametric regression slope estimator (i.e., median of pairwise slopes $(Y_i - Y_j)/(X_i - X_j)$), and median of absolute differences $|X_i - X_j|$ $(i \neq j)$?*

Note that among these the sample mean, sample median, and 5% trimmed mean are *L-statistics*, i.e., linear functions of order statistics given by $\sum_{i=1}^{n} c_{ni} X_{ni}$ for some choice of constants c_{ni}. Also, the sample mean and sample variance are *U-statistics*: i.e., for particular choices of real-valued "kernel" $h(x_1, \ldots, x_m)$ defined on $\mathbf{R}^m$, they can be represented in the form $n_{(m)}^{-1} \sum h(X_{i_1}, \ldots, X_{i_m})$, where the sum is over all $n_{(m)} = n(n-1) \cdots (n-m+1)$ m-tuples $(i_1, \ldots, i_m)$ of *distinct* indices from $\{1, \ldots, m\}$. Finally, the Hodges-Lehmann location estimator can be represented as an *R-statistic*, i.e., a function of the ranks of the X_i's. [General background on L-, U-, and R-statistics may be found in Huber (1981) and Serfling (1980).] The remainder of the above statistics, however, are neither L- nor U- nor R-statistics, nor do they fall within any other traditional class of statistics.

12.1.1   A Unifying Structure

We can, however, draw together *all* of the above statistics into a *single coherent class*, as follows. Consider again a kernel $h(x_1, \ldots, x_m)$ defining a U-statistic, denote the ordered values of the summands $h(X_{i_1}, \ldots, X_{i_m})$ appearing in the associated U-statistic by

$$W_{n1} \leq \cdots \leq W_{n, n_{(m)}},$$

and with these associate the class of all *linear combinations of the ordered* W_{ni}'s, i.e., all statistics having the form

$$\sum_{i=1}^{n_{(m)}} c_{ni} W_{ni} \qquad (12.1.1)$$

for some choice of constants c_{ni}. We call statistics of form (12.1.1) *generalized L-statistics* (GL-statistics).

Note that each of the statistics considered in the above question may be expressed in this form for suitable choice of h and c_{ni}. Also, in particular, the entire class of *L-statistics* is obtained by taking kernel $h(x) = x$, and the entire class of *U-statistics* is obtained by taking $c_{ni} \equiv 1/n_{(m)}$. Moreover, interesting *new varieties* of statistic are included in this structure:

- *trimmed U-statistics* (i.e., eliminate the upper proportion α and lower proportion α W_{ni}'s and average the rest)

- *Winsorized U-statistics*

- *median of m-wise averages*, i.e., median$\{(X_{i_1} + \cdots X_{i_m})/m\}$ (which gives for $m = 1$ the usual sample median, for $m = 2$ a version of the Hodges-Lehmann location estimator, and for $m > 2$ new competitors to these estimators).

Various examples will be treated formally in Section 12.5.

The setting of GL-statistics may be extended in two ways. (i) The X_i's may be random elements of an *arbitrary space* as long as the kernel h is real-valued. (In the case $h(x) = x$, this reduces to requiring the X_i's to be real-valued.) (ii) In Section 12.2, after introducing a representation of GL-statistics in terms of *statistical functionals*, we widen this class of statistics by introducing a more general form of functional.

In order for the GL-statistic generalization to be useful in practice, the usual battery of theoretical results are needed, including asymptotic normality, strong convergence, Berry-Esséen rates, large deviation theory, sequential fixed-width confidence intervals, and jackknife and bootstrap results. These are obtained as follows. In Section 12.2 GL-statistics are formulated as *statistical functionals*, specifically as *L-functionals* evaluated at generalized empirical df's of *U-statistic structure*. This representation enables us in Section 12.3 to combine functional analysis for L-functionals with probabilistic analysis (specifically, *Glivenko-Cantelli theory*) for the generalized empirical df's, establishing a foundation for developing in Section 12.4 the above-mentioned theoretical results for GL-statistics. Also, in Section 12.4, some extensions to broader contexts are indicated. In Section 12.5 we examine a variety of illustrative applications in nonparametric estimation and robust parametric estimation.

12.2   BASIC FORMULATION OF GL-STATISTICS

Here we represent GL-statistics as statistical functionals. This enables a characterization of the parameter estimated by a GL-statistic as well as of the estimation error, thus providing a foundation for theoretical analysis by the method of differentiable statistical functions.

12.2.1   Representation of GL-Statistics as Statistical Functionals

Our representation of a GL-statistic as a *"differentiable statistical functional"* entails

- the use of *L-functionals* T, and

- the evaluation of such a $T(\cdot)$ at an empirical df of *U-statistic structure.*

We first review the nature of L-functionals $T(\cdot)$, then define the appropriate empirical df, and then put these together.

L-Statistics as statistical functionals

A functional $T(\cdot)$ defined on real-valued df's G and having the form

$$T(G) = \int_0^1 G^{-1}(t)J(t)dt \ + \ \sum_{j=1}^d a_j G^{-1}(p_j)$$

for some choice of function $J(\cdot)$ on $[0,1]$, integer $d \geq 0$, values $p_j \in [0,1]$ and constants a_j, is called an *L-functional*. It represents a *weighting of the quantiles* of G, combining a *continuous* weighting of all quantiles via J with a *discrete* weighting of selected quantiles. In connection with a sample of real-valued $X_1, \ldots, X_n$ having df F, evaluation of such a $T(\cdot)$ at the usual empirical cdf

$$\widehat{F}_n(x) = n^{-1}\sum_{i=1}^n \mathbf{1}\{X_i \leq x\}, \quad -\infty < x < \infty,$$

yields

$$T(\widehat{F}_n) = \sum_{i=1}^n \left[\int_{(i-1)/n}^{i/n} J(t)dt \right] \widehat{F}_n^{-1}(i/n) \ + \ \sum_{j=1}^d a_j \widehat{F}_n^{-1}(p_j),$$

which we recognize as an L-statistic because $\widehat{F}_n^{-1}(p) = X_{ni}$ for $(i-1)/n < p \leq i/n$. Thus a wide class of L-statistics is generated by evaluating various L-functionals at $\widehat{F}_n$.

Empirical CDF of U-statistic structure

Analogous to the above empirical df $\widehat{F}_n$ which jumps $1/n$ at the order statistics X_{ni}, we define an empirical df associated with the W_{ni}'s given above, namely the step function with jumps of size $1/n_{(m)}$:

$$\widehat{H}_n(y) = n_{(m)}^{-1} \sum \mathbf{1}\{h(X_{i_1}, \ldots, X_{i_m}) \leq y\}, \quad -\infty < y < \infty.$$

For each fixed y, $\widehat{H}_n(y)$ is a *U-statistic* as defined above. Thus, although this generalization of the usual empirical cdf has complex structure, it is of a familiar type. Note that $\widehat{H}_n$ estimates the df H_F of $h(X_1, \ldots, X_m)$:

$$E\widehat{H}_n(y) = H_F(y) = P(h(X_1, \ldots, X_m) \leq y), \quad -\infty < y < \infty.$$

For the kernel $h(x) = x$, H_F reduces to F and $\widehat{H}_n$ to $\widehat{F}_n$.

GL-statistics as statistical functionals

In the same way that L-functionals evaluated at F_n yield L-statistics, we generate GL-statistics by evaluating these *same* L-functionals at the *generalized* empirical df $\widehat{H}_n$, producing

$$T(\widehat{H}_n) = \sum_{i=1}^{n_{(m)}} \left[\int_{(i-1)/n_{(m)}}^{i/n_{(m)}} J(t)dt \right] \widehat{H}_n^{-1}(i/n_{(m)}) + \sum_{j=1}^{d} a_j \widehat{H}_n^{-1}(p_j).$$

$$(12.2.1)$$

A wide class of linear combinations of the W_{ni}'s is thus generated. Moreover, through this representation we easily characterize the *parameter* that is estimated by a GL-statistic. Quite simply, since $\widehat{H}_n$ estimates H_F, $T(\widehat{H}_n)$ estimates

$$T(H_F) = \int_0^1 H_F^{-1}(t)J(t)dt + \sum_{j=1}^{d} a_j H_F^{-1}(p_j).$$

In the following we shall treat GL-statistics in the form (12.2.1) as well as in an extended form now to be introduced.

12.2.2   A More General Form of Functional

Let us generalize the above L-functional to:

$$T(G) = \int_0^1 q \circ T_t(G) \ J^*(t) \ dt \ + \ \sum_{j=1}^D A_j \ q \circ T_{P_j}(G), \qquad (12.2.2)$$

where

- for each $t \in (0,1)$, $T_t(\cdot)$ denotes a particular L-functional as defined above (with $J(\cdot)$ replaced by a function $J_t(\cdot)$, d replaced by d_t, each a_j by a_{tj}, each p_j by p_{tj})

- $q : \mathbf{R} \mapsto \mathbf{R}$ is a Borel-measurable function.

With $q(x) = x$ and $T_t(G) = G^{-1}(t)$, each t, we recover the case of simple L-functionals. Below we shall see other useful cases of $q(\cdot)$ and $T_t(\cdot)$.

Two examples: Spread measures of Bickel and Lehmann

Evaluation of the functional (12.2.2) at either the classical empirical df $\widehat{F}_n$ or the more general empirical df $\widehat{H}_n$ brings further statistics of interest into our scope. As examples, we mention two spread measures which Bickel and Lehmann (1979) formulated on an intuitive basis but which are best studied theoretically through reformulation as GL-statistics.

Example 1. Use (12.2.2) with $q(x) = x^2$, $T_t(G) = G^{-1}(t) - G^{-1}(1-t)$, $J^*(t) = (1-2\beta)^{-1}$ for $\beta \le t \le 1 - \beta$ and $= 0$ elsewhere, where β is chosen in $(0, 1/2)$, $D = 0$, and take $h(x) = x$ in defining $\widehat{H}_n$ (i.e., take $\widehat{F}_n$). Then the relevant GL-statistic is essentially

$$(n - 2[n\beta])^{-1} \sum_{k=[n\beta]+1}^{n-[n\beta]} (X_{nk} - X_{n,n-k+1})^2,$$

a *nonparametric* measure of *spread*. Note that in this example the more general functional $T(\cdot)$ is applied to the *classical* empirical df.   □

Example 2. Use (12.2.2) with $q(x) = x^2$, $T_t(G) = G^{-1}(\frac{t+1}{2})$, $J^*(t) = (1 - \alpha - \beta)^{-1}$ for $\alpha \le t \le 1 - \beta$ and $= 0$ elsewhere, where $0 < \alpha < 1/2 < 1 - \beta < 1$, $D = 0$, and take $h(x_1, x_2) = x_1 - x_2$ in defining H_n. Then $T(H_n)$ yields still another nonparametric measure of spread, one which involves *both* the more general functional $T(\cdot)$ and the more general empirical df $\widehat{H}_n$.   □

12.2.3 The Estimation Error

Our general goal is to study the *estimation error*,

$$T(\widehat{H}_n) - T(H_F),$$

where $T(\widehat{H}_n)$ is given by (12.2.1) using a simple L-functional, or, more generally, with $T(\cdot)$ given by a functional of form (12.2.2).

12.3 SOME FOUNDATIONAL TOOLS

We combine *functional analysis* for the functional $T(\cdot)$ with *probabilistic analysis* for the empirical cdf $\widehat{H}_n$. A convenient representation for the latter is

$$\widehat{H}_n = n_{(m)}^{-1} \sum \delta_{h(X_{i_1},\dots,X_{i_m})},$$

where δ_y denotes the cdf placing mass 1 at the point y.

12.3.1 Differentiation Methodology

For some purposes, we require the functional $T(\cdot)$ to be *differentiable*, for which a quite basic form of differential serves very well. For an arbitrary functional $T(\cdot)$ on df's G, the *Gâteaux differential* at G_0 is defined by

$$T'(G_0; G_1 - G_0) = \frac{d}{d\lambda} T(G_0 + \lambda(G_1 - G_0)) \Big|_{0+}.$$

As is well-known [e.g., Serfling (1980)], this yields an *approximation* to $T(G_1) - T(G_0)$, when G_1 is "close" to G_0. To apply this to our object of study, the *estimation error*, we take $G_0 = H_F$ and $G_1 = \widehat{H}_n$, obtaining

$$
\begin{aligned}
T(\widehat{H}_n) - T(H_F) &\doteq T'(H_F; \widehat{H}_n - H_F) \\
&= T'(H_F; n_{(m)}^{-1} \sum \delta_{h(X_{i_1},\dots,X_{i_m})} - H_F) \\
&= n_{(m)}^{-1} \sum T'(H_F; \delta_{h(X_{i_1},\dots,X_{i_m})} - H_F),
\end{aligned}
$$

where in the last step *linearity* of T' in its *second* argument is assumed (to be checked for each specific functional T under consideration). Thus, for *any* functional T whose Gâteaux differential satisfies the above linearity property, the corresponding approximation to the estimation error $T(\widehat{H}_n) - T(H_F)$ has the form of a *U-statistic*, based on the "kernel"

$$T'(H_F; \delta_{h(x_1,\dots,x_m)} - H_F). \tag{12.3.1}$$

That is, under $\boxed{\text{linearity}}$ we have:

$$\boxed{\text{The differential approximation to the estimation error is a U-statistic.}}$$

In particular, for T an *L-functional*, and for the case that the df G_0 has density g_0, we obtain after some manipulations [Serfling (1980)]

$$
\begin{aligned}
T'(G_0; G_1 - G_0) \;=\; & -\int_{\infty}^{\infty} [G_1(y) - G_0(y)] \, J(G_0(y)) \, dy \\
& + \sum_{j=1}^{d} a_j \, \frac{p_j - G_1(G_0^{-1}(p_j))}{g_0(G_0^{-1}(p_j))}.
\end{aligned}
\tag{12.3.2}
$$

More generally, for T given by (12.2.2) with q *differentiable*, we have

$$
\begin{aligned}
T'(G_0; G_1 - G_0) \;=\; & -\int_{0}^{1} q' \circ T_t(G_0) \, T'_t(G_0; G_1 - G_0) \, J^{\star}(t) \, dt \\
& + \sum_{j=1}^{D} A_j \, q' \circ T_{P_j}(G) \, T'_{P_j}(G_0; G_1 - G_0),
\end{aligned}
\tag{12.3.3}
$$

with the quantities $T'_t(G_0; G_1 - G_0)$ being of form (12.3.2). We see that the desired *linearity* of T' indeed holds, whereby we have: *for GL-statistics, the differential approximation to the estimation error is a U-statistic.* For explicit formulation of the relevant kernel given by (12.3.1), see Serfling (1984) and Janssen, Serfling, and Veraverbeke (1984). Here we simply note that the kernel in (12.3.1) has *mean* 0 and we denote its *variance* by

$$
\sigma^2(T, H_F) = \mathrm{Var}\big(T'(H_F; \delta_{h(X_1,\dots,X_m)} - H_F) \big).
$$

12.3.2   The Estimation Error in the U-Empirical Process

The closeness of $T(\widehat{H}_n)$ to $T(H_F)$ is related, of course, to the closeness of $\widehat{H}_n$ to H_F. This becomes manifest in various ways. For example, to establish *asymptotic normality* of $T(\widehat{H}_n) - T(H_F)$, the relevant consideration is the behavior of the normalized difference

$$
n^{1/2}[T(\widehat{H}_n) - T(H_F) - T'(H_F; \widehat{H}_n - H_F)],
$$

for which a precise treatment entails the use of *rates* for the convergence of $\widehat{H}_n$ to H_F in various norms.

On the other hand, to establish the *SLLN* for $T(\widehat{H}_n)$, the relevant considerations are *continuity* rather than differentiability of $T(\cdot)$, combined with convergence of the *quantile functions* $\widehat{H}_n^{-1}$ to H_F^{-1} in various modes of convergence.

Thus the "U-empirical process" which underlies our investigation of GL-statistics becomes itself a target of investigation. The general goal is to establish for $\widehat{H}_n$ the wide collection of results already available for the classical empirical cdf $\widehat{F}_n$.

The first general result for the empirical process of U-statistic structure appears to have been developed by Silverman (1976), in work preceding the appearance of "GL-statistics" and motivated by other considerations. Indeed, treating a larger class of empirical processes, he established *weak convergence* of $n^{1/2}(\widehat{H}_n(\cdot) - H_F(\cdot))$ to a Gaussian process. In Silverman (1983), specifically for the context of GL-statistics, extension with respect to a stronger topology was obtained. One can also treat the the empirical process of U-statistic structure as a special case of "U-process" as introduced by Nolan and Pollard (1987, 1988), for which a general treatment of weak and strong convergence is provided by Arcones and Giné (1993). For a large deviation result for U-processes, see Serfling and Wang (1998).

12.3.3   Extended Glivenko-Cantelli Theory

One class of results for $\widehat{H}_n$ covers the convergence of $\widehat{H}_n$ to H_F in various modes and norms. We call this "Glivenko-Cantelli theory," in a broad sense of the term.

Results for $\|\widehat{H}_n - H_F\|_\infty$

An exponential probability inequality for $\|\widehat{H}_n - H_F\|_\infty$ was established by Helmers, Janssen, and Serfling (1988):

$$P(\|\widehat{H}_n - H_F\|_\infty > d) \leq (1 + 4C[n/m]^{1/2}d)\, e^{-2[n/m]d^2}, \quad d > 0, \ n \geq m,$$

where C is a universal constant and $[\cdot]$ denotes "integer part." This is an analogue of the Dvoretzky, Kiefer, and Wolfowitz (1956) inequality for $\|\widehat{F}_n - F\|_\infty$. In fact, the latter inequality is used as a lemma in Helmers, Janssen, and Serfling (1988) to obtain an exponential bound on the *moment generating function* of $\|\widehat{H}_n - H_F\|_\infty$, thus providing a new tool even for the case $\widehat{F}_n$. As a corollary of the above probability inequality, we readily obtain

$$\|\widehat{H}_n - H_F\|_\infty = O\left(\frac{\log n}{n}\right)^{1/2} \text{ almost surely, } n \to \infty,$$

which gives the "Glivenko-Cantelli Theorem" for $\widehat{H}_n$ along with a rate of convergence. Compare the "in-probability" version,

$$\|\widehat{H}_n - H_F\|_\infty = O_p(n^{-1/2}), \ n \to \infty,$$

for H_F *continuous*, proved in Serfling (1984).

The above probability inequality for $\|\widehat{H}_n - H_F\|_\infty$ also has a *multi-sample* extension, given in Helmers, Janssen, and Serfling (1988). Another variant concerns *weighted* versions of the above sup-norm, i.e.,

$$\|(\widehat{H}_n - H_F)/(w \circ H_F)\|_\infty,$$

where $w(\cdot)$ is some specified weight function. See Silverman (1983) and Helmers, Janssen, and Serfling (1988) for particular results.

Further results

For treatment of $\|\widehat{H}_n - H_F\|_{L_p}$, see Serfling (1984), Helmers, Janssen, and Serfling (1988), and Arcones and Giné (1993), and for $\widehat{H}_n^{-1}(\cdot) - H_F^{-1}(\cdot)$, see Janssen, Serfling, and Veraverbeke (1984) and Helmers, Janssen, and Serfling (1988). Strong approximation of the U-empirical process is treated by Dehling, Denker, and Philipp (1985).

12.3.4 Oscillation Theory, Generalized Order Statistics, and Bahadur Representations

A classical *nonparametric* approach for obtaining a *confidence interval* for a *quantile parameter* $F^{-1}(p)$ is to take as endpoints of the interval a pair of *order statistics*,

$$\left(X_{n,a(n)} \ , \ X_{n,b(n)} \right),$$

with the ranks $a(n), b(n)$ selected to achieve desired confidence. Extension to *sequential fixed-width* nonparametric C. I.'s is obtained by letting n be defined suitably as a random *stopping time* N.

A much more general and interesting class of parameters is defined by retaining the simplicity of the quantile functional,

$$T(G) = G^{-1}(p),$$

with G given by H_F based on various choices of kernel $h(x_1, \ldots, x_m)$. We have seen several examples above. For such parameters we may form nonparametric C. I.'s by taking as endpoints a suitably chosen pair of *generalized order statistics*,

$$\left(W_{n,a(n)} \ , \ W_{n,b(n)} \right),$$

letting n be given by a stopping time N in the case of a *sequential* procedure.

Such applications are based on theoretical results for the behavior of *sequences* of the generalized order statistics, $W_{n,k(n)}$, for certain choices of rank sequence $k(n)$. A key result is a "*Bahadur-type representation*": for $0 < p < 1$, H_F twice differentiable with $H'_F(H_F^{-1}(p)) > 0$, and $k(n)$ satisfying

$$\frac{k(n)}{n_{(m)}} - p = o\left(\left(\frac{\log n}{n}\right)^{1/2}\right),$$

we have that *almost surely* as $n \to \infty$

$$W_{n,k(n)} = H_F^{-1}(p) + \frac{k(n)/n_{(m)} - \widehat{H}_n(H_F^{-1}(p))}{H'_F(H_F^{-1}(p))} + O(n^{-3/4}(\log n)^{-3/4}).$$

In particular, this yields for the (generalized) pth quantile $\widehat{H}_n^{-1}(p)$ a representation as *approximately a sample mean in form*.

A fundamental result on which the above result is based concerns the *oscillation behavior* of the empirical process based on $\widehat{H}_n$. Denote by

$$\omega(g; \delta) = \sup_{|s-t| \leq \delta} |g(s) - g(t)|,$$

the *modulus of continuity* function for a given function g, and by

$$\alpha_n(\cdot) = n^{1/2}[\widehat{H}_n(\cdot) - H_F(\cdot)]$$

the empirical process based on $\widehat{H}_n$. Results on the rate of convergence to 0 of $\omega(\alpha_n; a_n)$ and related quantities, for sequences a_n tending to 0 at appropriate rates, are given in Silverman (1983), Janssen, Serfling, and Veraverbeke (1984), and Choudhury and Serfling (1988). In particular, the latter paper provides a broad treatment including general application to the context of sequential fixed-width nonparametric C. I.'s. The results sharpen and extend previous work of Bahadur (1966) for the case $h(x) = x$ [see also Serfling (1980)] and of Geertsema (1970) for both the cases $h(x) = x$ and $h(x_1, x_2) = \frac{1}{2}(x_1 + x_2)$. For extension to the *multi-sample* case, see Serfling (1992).

12.3.5 Estimation of the Variance of a U-Statistic

The evaluation of the Gâteaux differential of a GL-functional at $\widehat{H}_n - H_F$ was seen to be a U-statistic in form. The variance $\sigma^2(T, H_F)$ of the corresponding kernel (3) is the relevant variance parameter in the asymptotic normality of $T(\widehat{H}_n)$. Some applications require *estimation* of this variance parameter, e.g., for confidence intervals on $T(H_F)$.

General methodology for estimation of the variance of an ordinary U-statistic is available, for example, in Sen (1981). However, in the present case the kernel of our U-statistic involves *unknown parameters*. For GL-statistics which are *quantiles* of $\widehat{H}_n$, estimation of $\sigma^2(T, H_F)$ is treated in Choudhury and Serfling (1988).

12.4   GENERAL RESULTS FOR GL-STATISTICS

12.4.1   Asymptotic Normality and the LIL

Results on asymptotic normality of GL-statistics $T(\widehat{H}_n)$ are developed in Serfling (1984) and Helmers and Ruymgaart (1988) for for $T(\cdot)$ a classical L-functional with bounded scores and unbounded scores, respectively, and in Janssen, Serfling, and Veraverbeke (1984) for $T(\cdot)$ having the more general form (12.2.2). Under moderate regularity conditions, these statistics satisfy

$$n^{1/2}[T(\widehat{H}_n) - T(H_F)] \longrightarrow_d N(0, \sigma^2(T, H_F), \ n \to \infty.$$

For $T(\cdot)$ a simple L-functional, the development parallels the treatment of $T(\widehat{F}_n)$ (ordinary L-statistics) as in Serfling (1980). Briefly, put

$$\Delta_n = n^{1/2}[T(\widehat{H}_n) - T(H_F) - T'(H_F; \widehat{H}_n - H_F)]$$

and decompose this into $\Delta_n = \Delta_{n1} + \Delta_{n2}$, corresponding to the *continuous* (*J*-function) and *discrete* components of the functional T. Then, for Δ_{n1}, establish inequalities of the form

$$|\Delta_{n1}| \le \|W_{\widehat{H}_n, H_F}\|_A \ \|\widehat{H}_n - H_F\|_B, \qquad (12.4.1)$$

where $A = \infty$ and $B = L_p$, or vice versa, and

$$W_{G,F}(y) = \begin{cases} \dfrac{K(G(y)) - K(F(y))}{G(y) - F(y)} - J(F(y)), & G(y) \ne F(y) \\ 0, & G(y) = F(y), \end{cases}$$

with $K(t) = \int_0^t J(u) du$. This sets the stage for an analysis which motivates and exploits some of the *Glivenko-Cantelli* results for $\widehat{H}_n$ in Section 12.3. For the component Δ_{n2}, it turns out that this quantity is precisely that which is treated in the *Bahadur representation* result for $\widehat{H}_n$ as discussed in Section 12.3.4. For $T(\cdot)$ given by the more general functional (12.2.2), the treatment is somewhat more complicated.

For the LIL, a parallel approach works. For the *Berry-Esséen rate* for the convergence in the AN result and its use as a tool in the *bootstrap analysis* of GL-statistics, see Helmers, Janssen, and Serfling (1990).

12.4.2 The SLLN

The classical SLLN states that the sample mean converges almost surely to its expectation, a result that has fundamental and wide application in probability and statistics. Considering now the "statistical setting", we ask

In what generality does the SLLN hold?

For the generality of the class of *L-statistics*, a sharp SLLN was established by van Zwet (1980). This was extended to *GL-statistics* in Helmers, Janssen, and Serfling (1988): under moderate regularity conditions, we have

$$T(\widehat{H}_n) \longrightarrow_{a.s.} T(H_F), \ n \to \infty.$$

In some sense this is a weaker conclusion than asymptotic normality, but, since we thus need to establish it under weaker conditions, the problem can in principle be a harder one (and in fact *is*).

In the development of Helmers, Janssen, and Serfling (1988), the problem was handled by identifying and formulating the functional-analytic and probabilistic components inherent in the problem and then treating these separately. One first investigates the convergence behavior of the functional $T(\cdot)$ evaluated at a *deterministic* sequence of weakly convergent df's G_n. This leads to conditions on $T(\cdot)$ and on $\{G_n\}$, sufficient for convergence of $T(G_n)$ to a limit. Then one establishes, as an extended *Glivenko-Cantelli* property for $\widehat{H}_n$, that with probability 1 the random sequence of *empirical df's* $\{\widehat{H}_n\}$ indeed satisfies the conditions on $\{G_n\}$.

12.4.3 Large Deviation Theory

The *large deviation* problem, specialized to GL-statistics, is to evaluate the limit

$$\lim_{n\to\infty} \frac{\log P(\,|T(\widehat{H}_n) - T(H_F)| \geq d\,)}{n},$$

under appropriate conditions. For ordinary L-statistics as well as other functionals of $\widehat{F}_n$, this has been solved fairly completely in Groeneboom *et al.* (1979). For extension to GL-statistics and other functionals of $\widehat{H}_n$, see Serfling and Wang (1999).

12.4.4 Further Results

Jackknife results were established for U-statistics, by Arvesen (1969) and for L-statistics by Parr and Schucany (1982). For GL-statistics of the simple form (12.2.1), jackknife results have been developed by Shao (1990). It is of interest to extend to the more general form (12.2.2). For *bootstrap* results for GL-statistics, see Helmers, Janssen, and Serfling (1990). *Multi-sample* GL-statistics are treated by Akritas (1986) and Serfling (1992). Generalizing the study of incomplete U-statistics by Blom (1976), *incomplete* GL-statistics based on the form (12.2.1) are investigated by Hössjer (1996). It is also of interest to extend to (12.2.2).

12.5 SOME APPLICATIONS

12.5.1 One-Sample Quantile Type Parameters

A general treatment of GL-statistics having the form $\widehat{H}_n^{-1}(p)$, for some choice of kernel h and $0 < p < 1$, is given by Choudhury and Serfling (1988). Some examples are as follows.

Location estimation

For estimation of the location parameter θ of a symmetric and continuous cdf F, classical nonparametric estimators are provided by the median and by the median of pairwise averages (the Hodges-Lehmann location estimator). More generally, let us consider – as noted in Section 12.1 and proposed in Serfling (1984) – the median of m-wise averages:

$$\text{HL}_{(m)} = \text{median}\left\{\frac{X_{i_1} + \cdots X_{i_m}}{m}\right\}$$

(which for $m > 2$ gives competitors to the classical estimators). With the kernel $h(x_1, \ldots, x_m) = \frac{x_1 + \ldots + x_m}{m}$, this is a *GL-statistic*: $\widehat{H}_n^{-1}(1/2)$. It estimates the *generalized quantile parameter* $H_F^{-1}(1/2)$. Besides the treatment of Choudhury and Serfling (1988) for this example, see also Choudhury (1989, 1990) and, for extension to *multivariate* X_i's, Chaudhuri (1992). In terms of *asymptotic relative efficiency* (ARE) with respect to $\overline{X}$ at the Normal distribution, and *breakdown point* (BP), the estimator $\text{HL}_{(m)}$ exhibits a very favorable trade-off in comparison with other estimators, as shown in the following table.

Estimator	ARE	BP
median	.637	.500
25%-trim	.833	.250
$HL_{(2)}$	.955	.293
$HL_{(3)}$	.981	.206
$HL_{(4)}$	.989	.160

We interpret this finding in the context of robust parametric estimation and arrive at the following principle:

> The use of the median operation, after "smoothing" the data by taking a function of several observations at a time, over all subsets of the data, leads to a statistic which has a favorable combination of efficiency and robustness. I.e., smoothing followed by medianing yields both efficiency and robustness.

A more general type of location estimator is given by taking a kernel of form $h(x_1, \ldots, x_m) = \sum_{i=1}^{m} \alpha_i x_i$ with $\sum_{i=1}^{m} \alpha_i = 1$. See Choudhury and Serfling (1988) for further discussion.

Spread estimation

Among various measures of spread discussed by Bickel and Lehmann (1979) is the median of the distribution of $|X_1 - X_2|$, where X_1 and X_2 are independent r.v.'s having cdf F. This is a generalized quantile parameter, $H_F^{-1}(1/2)$, based on the kernel $h(x_1, x_2) = |x_1 - x_2|$.

More generally, as discussed in Choudhury and Serfling (1988), we might consider the class of spread measures and estimators corresponding to kernels of the form

$$h(x_1, \ldots, x_m) = \left| \sum_{i=1}^{m} \beta_i x_i \right|,$$

with $\sum_{i=1}^{m} \beta_i = 0$. This generalizes the above m-wise average form of kernel and extends an approach studied by Maritz, Wu and Staudte (1977).

Regression slope estimation

Consider the simple linear regression model $Y_i = \alpha + \beta X_i + \epsilon_i$, with $\{\epsilon_i\}$ i.i.d. r.v.'s independent of X_i, and X_i a sequence of *random* regressors. Let F denote the common cdf of the mutually independent pairs (X_i, Y_i), $1 \leq i \leq n$, and let H_F denote the cdf of $h((X_1, Y_1), (X_2, Y_2))$, where

$$h((x_1, y_1), (x_2, y_2)) = \frac{y_2 - y_1}{x_2 - x_1}.$$

For this choice of kernel, the nonparametric estimator of β given by Theil (1950), i.e., the median of the slopes $(Y_i - Y_j)/(X_i - X_j)$, is the corresponding GL-statistic based on the median functional: $\hat{\beta} = \hat{H}_n^{-1}(1/2)$. The results of Choudhury and Serfling (1988) provide sequential nonparametric fixed-width confidence intervals for this classical estimator.

12.5.2   Two-Sample Location and Scale Problems

Location

Suppose $F^{(2)}(x) = F^{(1)}(x - \theta)$, and let F denote $(F^{(1)}, F^{(2)})$. For integer $m \geq 1$ consider the kernel

$$h(x_1^{(1)}, \ldots, x_m^{(1)} ; x_1^{(2)}, \ldots, x_m^{(2)}) = \frac{(x_1^{(1)} + \ldots + x_m^{(1)}) - (x_1^{(2)} + \ldots + x_m^{(2)})}{m}.$$

Assuming $F^{(1)}$ continuous, we have that $H_F(\theta) = 1/2$, i.e., $\theta = H_F^{-1}(1/2)$, and a corresponding estimator is $\hat{\theta}_n = \hat{H}_n^{-1}(1/2)$, where $n = (n_1, n_2)$, the vector of respective sample sizes. The case $m = 1$ is the shift estimator given by Hodges and Lehmann (1963), while the cases $m \geq 2$ represent new competing estimators. Note that under the null hypothesis $\theta = 0$ we have $H_F(0) = 1/2$, and a corresponding *test statistic* is given by $\hat{H}_n(0)$. For the case $m = 2$, this test was proposed by Hollander (1967) [see also discussion in Randles and Wolfe (1979, pp. 96–97]. See Serfling (1992) for a general development.

Scale

Suppose $F^{(2)}(x) = F^{(1)}((x - \theta)/\eta)$, for θ an unknown nuisance parameter and $\eta > 0$ the parameter of interest. With the kernel

$$h(x_1^{(1)}, x_2^{(1)} ; x_1^{(2)}, x_2^{(2)}) = \frac{|x_2^{(1)} - x_1^{(1)}|}{|x_2^{(2)} - x_1^{(2)}|},$$

we have $\eta = H_F^{-1}(1/2)$, and a corresponding estimator is given by $\hat{\eta}_n = \hat{H}_n^{-1}(1/2)$, Under the null hypothesis $\eta = 1$ we have $H_F(1) = 1/2$, and a corresponding *test statistic* is given by $\hat{H}_n(1)$, as proposed by Lehmann (1951). See Serfling (1992) for a general development.

12.5.3   Robust ANOVA

Here we suppose that $F^{(j)}(x) = F_0(x - \Delta_j)$, $1 \leq j \leq c$, and consider estimation of a parameter of form

$$\theta = \sum_{j=1}^{c} d_j \Delta_j,$$

where $d_1, \ldots, d_c$ are specified constants and the Δ_j's are unknown. The problem of nonparametric estimation of θ in the case of a *contrast* ($\sum_1^c d_j = 0$) was initially studied and solved by Lehmann (1963), whose approach consists of expressing θ in the form of a linear combination of the differences $\Delta_i = \Delta_j$ and using nonparametric estimates of these. A rich literature has developed on this approach and its modifications. Using the framework of GL-statistics, however, a straightforward competing estimator may be formulated, based on the kernel

$$h(x_1^{(1)}; \cdots; x_1^{(c)}) = \sum_{j=1}^{c} d_j x_1^{(j)}.$$

We suppose F_0 to be symmetric about 0, in which case we have $\theta = H_F^{-1}(1/2)$, and a natural estimator of θ is thus given by $\hat{\theta}_n = \hat{H}_n^{-1}(1/2)$, where $n = (n_1, \ldots, n_c)$. Surprisingly, this estimator has not been investigated previously in the literature. This formulation also includes the case that θ is *not* a contrast. For testing the null hypothesis $\theta = \theta_0$, a natural test statistic is given by $\hat{H}_n(\theta_0)$.

12.5.4   Robust Regression

Frees (1991) has introduced and investigated a wide class of estimators of β, in which a typical estimator is given by trimming the collection of ordered slopes $(Y_i - Y_j)/(X_i - X_j)$, and then taking a weighted average of the remaining slopes. Using an extended notion of generalized empirical cdf, he represents these as GL-statistics for appropriate choices of kernel.

12.5.5   Robust Estimation of Exponential Scale Parameter

Consider the problem of robust estimation of θ in the two-parameter *exponential* distribution $E(\mu, \theta)$ having cdf

$$G(x) = 1 - e^{-(x-\mu)/\theta}, \quad x \geq \mu, \tag{12.5.1}$$

for $\theta > 0$ and $-\infty < \mu < \infty$, with μ an unknown "nuisance parameter". The maximum likelihood estimator of θ, $\hat{\theta}_{\text{ML}} = \overline{X}_n - X_{n1}$, is *efficient*, being asymptotically normal with mean θ and variance θ^2/n, but is *nonrobust*, having BP $= 0$. Competing *trimmed mean* type estimators $\hat{\theta}_T$ for various choices of trimming level β have been investigated by Kimber (1983a,b) and established to possess relatively high efficiency coupled with favorable robustness. It has been found, however, that these trimmed type estimators are outperformed by generalized median type estimators $\hat{\theta}_{GM}$ based on suitable kernels. This finding illustrates again the general principle stated in Section 12.5.1. As a typical example, $\hat{\theta}_T$ based on 10% upper and lower trimming has ARE $= .85$ and upper BP $= .10$, whereas $\hat{\theta}_{GM}$ for a suitable kernel has ARE $= .94$ and upper BP $= .13$. For full details, see Brazauskas and Serfling (1999). Note that the above exponential scale estimation problem is equivalent, through exponential transformation of the data, to that of tail index estimation in a two-parameter Pareto model.

Acknowledgment This research was supported by NSF Grant DMS-9705209.

REFERENCES

Akritas, M. G. (1986). Empirical processes associated with V-statistics and a class of estimators under random sampling, *Annals of Statistics*, **14**, 619–637.

Arcones, M. A. and Giné, E. (1993). Limit theorems for U-processes, *Annals of Probability*, **21**, 1494–1542.

Arvesen, J. N. (1969). Jackknifing U-statistics, *Annals of Mathematical Statistics*, **40**, 2076–2100.

Bahadur, R. R. (1966). A note on quantiles in large samples, *Annals of Mathematical Statistics*, **37**, 577–580.

Bickel, P. J. and Lehmann, E. L. (1979). Descriptive statistics for nonparametric models, IV. Spread, In *Contributions to Statistics*, Hájek Memorial Volume (Ed., J. Jurecková), pp. 33–40, Academia, Prague.

Blom, G. (1976). Some properties of incomplete U-statistics, *Biometrika*, **63**, 573–580.

Brazauskas, V. and Serfling, R. (1999). Robust estimation of tail parameters for two-parameter Pareto and exponential models via generalized quantile statistics, *Preprint*.

Chaudhuri, P. (1992). Multivariate location estimation using extension of R-estimates through U-statistics type approach, *Annals of Statistics*, **20**, 897–916.

Choudhury, J. (1989). Selecting the median of a special subset of $X + \cdots + X$, *Sesquicentennial Proceedings of American Statistical Association, Section on Statistical Computing*, 212–216.

Choudhury, J. (1990). Sequential confidence intervals based on generalized Hodges-Lehmann location estimators and related statistics, *Communications in Statistics*, **19**, 287–303.

Choudhury, J. and Serfling, R. J. (1988). Generalized order statistics, Bahadur representations, and sequential nonparametric fixed-width confidence intervals, *Journal of Statistical Planning and Inference*, **19**, 269–282.

Dehling, D., Denker, M. and Philipp, W. (1987). The almost sure invariance principle for the empirical process of U-statistic structure, *Annales de l'Institut Henri Poincare, Section B, Calcul des Probabilites et Statistique*, **23**, 121–134.

Dvoretzky, A., Kiefer, J. and Wolfowitz, J. (1956). Asymptotic minimax character of the sample distributon function and of the classical multinomial estimator, *Annals of Mathematical Statistics*, **27**, 642–669.

Frees, E. W. (1991). Trimmed slope estimates for simple linear regression, *Journal of Statistical Planning and Inference*, **27**, 203–221.

Geertsema, J. C. (1970). Sequential confidence intervals based on rank tests, *Annals of Mathematical Statistics*, **41**, 1016–1026.

Groeneboom, P., Oosterhof, J. and Ruymgaart, F. H. (1979). Large deviation theorems for empirical probability measures, *Annals of Probability*, **7**, 553–586.

Helmers, R., Janssen, P. and Serfling, R. (1988). Glivenko-Cantelli properties of some generalized empirical df's and strong convergence of generalized L-statistics, *Probability Theory and Related Fields*, **79**, 75–93.

Helmers, R., Janssen, P. and Serfling, R. (1990). Berry-Esséen and bootstrap results for generalized L-statistics, *Scandinavian Journal of Statistics*, **17**, 65–77.

Helmers, R. and Ruymgaart, F. H. (1988). Asymptotic normality of generalized L-statistics with unbounded scores, *Journal of Statistical Planning and Inference*, **19**, 43–53.

Hodges, Jr., J. L. and Lehmann, E. L. (1963). Estimates of location based on rank tests, *Annals of Mathematical Statistics*, **34**, 598–611.

Hollander, M. (1967). Asymptotic efficiency of two nonparametric competitors of Wilcoxon's two sample test, *Journal of American Statistical Association*, **62**, 939–949.

Hössjer, O. (1996). Incomplete generalixed L-statistics, it Annals of Statistics, **24**, 2631–2654.

Huber, P. J. (1981). *Robust Statistics*, John Wiley & Sons, New York.

Janssen, P., Serfling, R. and Veraverbeke, N. (1984). Asymptotic normality for a general class of statistical functions and applications to measures of spread, *Annals of Statistics*, **12**, 1369–1379.

Kimber, A. C. (1983a). Trimming in gamma samples, *Applied Statistics*, **32**, 7–14.

Kimber, A. C. (1983b). Comparison of some robust estimators of scale in gamma samples with known shape, *Journal of Statistical Computation and Simulation*, **18**, 273–286.

Lehmann, E. L. (1951). Consistency and unbiasedness of certain nonparametric tests, *Annals of Mathematical Statistics*, **22**, 165–179.

Lehmann, E. L. (1963). Robust estimation in analysis of variance, *Annals of Mathematical Statistics*, **34**, 957–966.

Maritz, J. S., Wu, M. and Staudte, Jr., R. G. (1977). A location estimator based on a U-statistic, *Annals of Statistics*, **5**, 779–786.

Nolan, D. and Pollard, D. (1987). U-processes: rates of convergence, *Annals of Statistics*, **15**, 780–789.

Nolan, D. and Pollard, D. (1988). Functional limit theorems for U-processes, *Annals of Probability*, **16**, 1291–1298.

Parr, W. C. and Schucany, W. R. (1982). Jackknifing L-statistics with smooth weight functions, *Journal of American Statistical Association*, **27**, 639–646.

Randles, R. H. and Wolfe, D. A. (1979). *Introduction to the Theory of Nonparametric Statistics*, John Wiley & Sons, New York.

Sen, P. K. (1981). *Sequential Nonparametrics*, John Wiley & Sons, New York.

Serfling, R. J. (1980). *Approximation Theorems of Mathematical Statistics*, John Wiley & Sons, New York.

Serfling, R. J. (1984). Generalized L-, M- and R-statistics, *Annals of Statistics*, **12**, 76–86.

Serfling, R. J. (1992). Nonparametric confidence intervals for generalized quantile parameters in multi-sample contexts, In *Nonparametric Statistics and Related Topics* (Ed., A. K. Md. E. Saleh), pp. 121–139, Elsevier Science Publishers B.V., Amsterdam.

Serfling, R. and Wang, W. (1999). Large deviation results for U- and V-statistics and U- and V-empiricals, *Preprint.*

Serfling, R. and Wang, W. (1998). A large deviation theorem for U-processes, *Preprint.*

Shao, J. (1990). Jackknife variance estimators for generalized L-statistics, *Statistics & Probability Letters*, **11**, 27–32.

Silverman, B. (1976). Limit theorems for dissociated random variables, *Advances in Applied Probability*, **8**, 806–819.

Silverman, B. (1983). Convergence of a class of empirical distribution functions of dependent random variables, *Annals of Probability*, **11**, 745–751.

Theil, H. (1950). A rank-invariant method of linear and polynomial regression analysis, III, *Proceedings Kon. Ned. Akad. v. Wetensch. A*, **53**, 1397–1412.

van Zwet, W. (1980). A strong law for linear functions of order statistics, *Annals of Probability*, **8**, 986–990.

## CHAPTER 13

A CLASS OF ROBUST STEPWISE TESTS FOR MANOVA

DEO KUMAR SRIVASTAVA

St. Jude Children's Research Hospital, Memphis, TN

GOVIND S. MUDHOLKAR

University of Rochester, Rochester, NY

CAROL E. MARCHETTI

Rochester Institute of Technology, Rochester, NY

Abstract: It is well known that Student's t test as well as the ANOVA F test are reasonably validity-robust with respect to moderate departures from normality; see e.g. Mudholkar, Mudholkar and Srivastava (1991), Marchetti, Mudholkar and Mudholkar (1998). However, in the absence of normality substantial power loss is associated with the above procedures. The same holds for Hotelling's T^2 and various normal theory MANOVA procedures in multivariate analysis; see Seber (1984), Mudholkar and Srivastava (1999a,b) and the references therein. Recently, Mudholkar and Srivastava (1999c) have proposed a class of robust stepwise tests as alternatives to Hotelling's problem by incorporating the modification of J. Roy's (1958) step down argument presented in Mudholkar and Subbaiah (1980). In this paper, we extend their reasoning to construct a class of robust tests for the multivariate analysis of variance for the one way classification and examine their robustness properties. The new procedures use relatively familiar univariate tests and avoid any new distributional problems. The robust stepwise tests have a reasonable type I error control and substantially enhanced power at nonnormal alternatives without significant loss of power in the presence of normality.

Keywords and phrases: Decomposition, robust trimmed ANOVA statistics, combining independent P-values

13.1 INTRODUCTION

Consider the problem of testing the simplest multivariate general linear hypothesis, i.e. testing the homogeneity of the means $\mu_1, \mu_2, ..., \mu_K$ of k multivariate normal populations with the common covariance matrix Σ, i.e. testing $H_0 : \mu_1 = \mu_2 = ... = \mu_k$. The properties of the well known invariant tests such as Wilk's likelihood ratio (Λ), Lawley-Hotelling Trace, Bartlett-Nanda-Pillai Trace, Roy's maximum root, have been extensively examined in the literature, e.g. see Anderson (1984), Mardia, Kent and Bibby (1979), Seber (1984). It is well known that none of the invariant tests uniformly dominates the others in terms of power. However, an asymptotic analysis appearing in Hsieh (1979a,b) shows that the likelihood ratio test, which is maximin [Anderson (1984, p. 332)], is superior to Lawley-Hotelling trace and Roy's largest root tests in terms of Bahadur efficiency. For a decision theoretic analysis of multivariate procedures see Kiefer and Schwartz (1965) and Schwartz (1967).

It is generally well recognized that, as in the univariate case, the commonly used multivariate invariant procedures are validity robust for small departures from normality. However, from several robustness studies it is known that, in the context of testing the significance of a mean vector or testing the equality of k ($k \geq 2$) mean vectors, the normal theory based invariant tests are either invalid or very conservative; see Chase and Bulgren (1971), Everitt (1979), Bauer (1981), Srivastava and Awan (1982), Tiku and Singh (1982), Davis (1980), Olson (1974). More seriously, it is believed that the non robustness of these tests is manifested mainly and substantially in loss of power. In the univariate setting, Geary (1947) somewhat flamboyantly remarked that, *"Normality is a myth; there never was, and never will be, a normal distribution"*. Since the multivariate normality entails marginal as well as joint normality of the components, Mudholkar and Srivastava (1998) observe that Geary's provocative comment is a fortiori true in the multivariate case, and the multivariate normality assumption is at best dubious. Thus, in light of this and the few and sketchy studies of efficiency robustness, a need for robust multivariate procedures is strongly indicated. However, the development of multivariate robust methods is in rudimentary stages.

Even in the univariate case robust methods for estimation are better understood and accessible than those for testing the hypotheses. The robust estimation in multivariate setting, including the extensions of well known univariate L, M and R approaches, have been discussed by many including Mood (1941), Bickel (1965), Gnanadesikan and Kettenring (1972), Bebbington (1978), Titterington (1978). Also, as in the univariate case, the literature on multivariate rank tests, which bypass the nuisance of studentization for unknown scale and are related to the R-estimators, is most extensive; see Puri and Sen (1971). However, the justification of these nonparametric multivariate tests is largely asymptotic and the related distribution theory for their implementation in moderate sized samples is not adequately understood. Furthermore, these tests are often regarded as less efficient. M-methods have also been theoretically discussed in the context of univariate and multivariate testing of hypothesis, see Hampel et al. (1986). However, Draper (1988) in his review paper of the robust methods observes that, "the L-methods have historically been the most awkward of the three in generalizing to linear models".

Yet, the L-estimates such as the median, trimmed means and trimeans are the oldest and the most easily motivated estimators of location. The earliest use of trimmed means is in Tukey and McLaughlin's (1963) studentization of trimmed mean followed by its application by Yuen and Dixon (1973) and Yuen (1974) for testing equality of two means. Mudholkar, Mudholkar and Srivastava (1991) note some limitations of these tests and employ the asymptotic distribution of the trimmed means in Huber (1970), together with empirical methods, to construct robust trimmed-t tests valid for samples of size $n \geq 10$. Similar tests based upon quick estimators of location, e.g. trimean and Gastwirth estimator, are given in Patel et al. (1985), and Srivastava, Mudholkar and Mudholkar (1992). These studies demonstrate a dramatic power advantage of the tests based on trimmed means and quick estimators of location. For example, for some nonnormal populations the power of these robust tests can be as high as 70% as compared 14% for the classical t test. More recently, Marchetti (1997), and Marchetti, Mudholkar and Mudholkar (1998) have developed a trimmed ANOVA test based on trimmed means for the one way classification analysis of variance. The purpose of this paper is to combine their arguments and results with those in Mudholkar and Srivastava (1999c) to develop some robust stepwise tests for one way classification for multivariate analysis of variance.

The robust one way ANOVA and other preliminaries are given in Section 13.2. The development of the modified step down procedure and the construction of a class of robust stepwise tests for testing the homogeneity of means of k multivariate samples appears in Section 13.3. Empirical evaluation of the operating characteristics of the procedures are presented in Section 13.4. The final section, Section 13.5, is given to conclusions.

13.2 PRELIMINARIES

In this section we summarize the basic properties of the trimmed means, describe their use in testing of hypothesis, and outline the logic of modified stepwise tests.

13.2.1 Robust Univariate Tests

One Sample. Let $X_1 < X_2 < ... < X_n$ be the order statistics of a random sample from a continuous, symmetric population with distribution function $(d.f.)$ $F(\{x - \mu\}/\sigma)$. Then the g-trimmed mean or δ-trimmed mean, $\delta = g/n$, of the sample is $\widetilde{X} = (X_{g+1} + ... + X_{n-g}) / (n - 2g)$ .Tukey and McLaughlin (1963) used empirical analysis, as Huber (1970) describes, involving "trial" and "error" to propose the studentization of $\widetilde{X}$ by Winsorized variance $S_W^2 / h(h-1)$, where $S_W^2 = [(g+1)(X_{g+1} - \widetilde{X})^2 + (X_{g+2} - \widetilde{X})^2 + ... + (g+1)(X_{n-g} - \widetilde{X})^2]$, and $h = (n - 2g)$. Huber (1970) confirmed the validity of their studentization by showing that as, $n \to \infty$,

$$\sqrt{n}(\widetilde{X} - \mu) \longrightarrow N(0, b_F^2(\delta)\, \sigma^2) \tag{13.2.1}$$

and

$$\sqrt{n-1}(\, \widetilde{s}^2 - b_F^2(\delta)\, \sigma^2\,) \longrightarrow N(0, R_F^2(\delta)\, \sigma^4\,), \tag{13.2.2}$$

where $\widetilde{s}^2 = S_W^2 / n(1-2\delta)^2$. Mudholkar, Mudholkar and Srivastava (1991), fix the normal family as the target population, i.e. take $F=\Phi$, and obtain polynomial approximations,

$$b_\Phi^2(\delta) \approx 1 + 0.48\delta + 1.21\delta^2, \tag{13.2.3}$$

and

$$w_\Phi^*(\delta) = [b_\Phi^4(\delta) / R_\Phi^2(\delta)] \approx 0.5 - 1.62\delta + 1.91\delta^2 - 1.85\delta^3. \tag{13.2.4}$$

They use empirical methods for approximating the small sample null distribution of the trimmed-t statistic proposed by Tukey and McLaughlin (1963).

Two Samples. Now consider random samples of sizes n_1 and n_2 from two symmetric location-scale populations $F((x-\mu_1)/\sigma)$ and $F((x-\mu_2)/\sigma)$. Let $\widetilde{X}_1$ and $\widetilde{X}_2$ denote the δ_1 and δ_2 trimmed means, respectively, and $\widetilde{s}_1$, and $\widetilde{s}_2$ denote the corresponding Winsorized standard deviation estimators defined above. Then, using the asymptotic distribution, as in (2.2), Mudholkar, Mudholkar and Srivastava (1991) propose pooling the Winsorized sample variances, $\tilde{s}_1$, and $\tilde{s}_2$ by

$$\tilde{s}_{pool}^2 = (w_1 \frac{\tilde{s}_1^2}{b_1^2} + w_2 \frac{\tilde{s}_2^2}{b_2^2}) (w_1 + w_2)^{-1}, \qquad (13.2.5)$$

where $b_i^2 = b_\Phi^2(\delta_i)$, $w_i = w_\Phi(\delta_i) = (n_i - 1)w_\Phi^*(\delta_i)$, to construct the two sample trimmed-t statistic for testing $H_0 : \mu_1 = \mu_2$ as,

$$\tilde{t}_p = \frac{(\tilde{X}_1 - \tilde{X}_2)}{\tilde{s}_{pool} \sqrt{b_1^2/n_1 + b_2^2/n_2}}, \qquad (13.2.6)$$

and obtain a scaled Student's t approximation, $\tilde{t}_p \sim A\, t_\nu$, where $\nu = 2*(w_1 + w_2)$ and A, obtained empirically, is given by:

$$A = 1 - 1.3\frac{\delta}{\nu} + 7.5\frac{\delta^2}{\nu} + 16\frac{\delta}{\nu^2} - 150\frac{\delta^3}{\nu^2},$$

where $\delta = (\delta_1 + \delta_2)/2$.

One Way Classification. Now consider the usual one way ANOVA hypothesis, $H_0 : \mu_1 = \mu_2 = ... = \mu_k$, on the basis of samples X_{ij}, $j = 1, 2, ..., n_i$, from k normal populations $N(\mu_i, \sigma^2)$, $i = 1, 2, ..., k$. The above approach used in construction of the one- and two-sample trimmed-t test does not have a straight forward extension for the k-sample case. Hence, Marchetti (1997), and Marchetti, Mudholkar and Mudholkar (1998) begin anew by empirically refining the asymptotic distribution of the trimmed mean and the Winsorized variance given in (13.2.1) and (13.2.2) to make them reasonably applicable in small samples. Specifically, they use 10000 replication Monte Carlo study involving random samples of size n, for various values of n and several values of trimming proportion δ, in order to estimate $r(n, \delta)$, $s(n, \delta)$ and $t(n, \delta)$ such that approximations

$$\sqrt{n}(\tilde{X} - \theta) \approx N(0, r(n, \delta) b_\Phi^2(\delta) \sigma^2), \qquad (13.2.7)$$

and

$$\sqrt{n-1}(\tilde{s}^2 - t(n, \delta) r(n, \delta) b_\Phi^2(\delta) \sigma^2) \approx N(0, s(n, \delta) R_\Phi^2(\delta) \sigma^4), \qquad (13.2.8)$$

hold for values of $n \geq 5$. specifically, Using regression methods, they obtain and recommend the following:

$$r(n, \delta) = 1 - \frac{7\delta}{4\sqrt{n}} + \frac{21\delta^2}{\sqrt{n}} - \frac{57\delta^3}{\sqrt{n}}$$

$$s(n, \delta) \quad = \quad 1 - \frac{45\delta}{n} + \frac{302\delta^2}{n} - \frac{672\delta^3}{n} \tag{13.2.9}$$

$$t(n, \delta) \quad = \quad 1 - \frac{30\delta}{n} + \frac{235\delta^2}{n} - \frac{580\delta^3}{n} \quad ,$$

as the expressions for $r(n, \delta)$, $s(n, \delta)$ and $t(n, \delta)$. Then, for testing the homogeneity of the k means, they propose the following trimmed analog, $\widetilde{F}$, of the normal theory variance ratio:

$$\widetilde{F} \quad = \quad \Sigma Q_i^* (\widetilde{X}_i - \widetilde{X}_{..})^2 \; / \; [(k-1)\widetilde{s}_{pool}^2] \; , \tag{13.2.10}$$

where

$$\widetilde{s}_{pool}^2 \quad = \quad [\Sigma Q_i \widetilde{s}_i^2 \; / \; (r_i t_i b_i^2)] \; / \; \Sigma Q_i \; . \tag{13.2.11}$$

In (13.2.10) and (13.2.11), for the i-th population, $i = 1, 2, ..., k$, $\widetilde{X}_i$, $\widetilde{s}_i^2$ are the trimmed mean and Winsorized variance estimators, b_i^2 is the analog of (13.2.3), $R_i^2 = b_i^4/w_i^*$ where w_i^* is as given in (13.2.4), $Q_i^* = n_i/(r_i * b_i^2)$, $Q_i = [(r_i t_i b_i^2)^2 (n_i - 1)] \; / \; (s_i R_i^2)$, r_i, s_i, t_i are the analogs of (2.9), and $\widetilde{X}_{..} = (\Sigma Q_i^* \widetilde{X}_i)/(\Sigma Q_i^*)$. They suggest that the null distribution of the statistic $\widetilde{F}$ be approximated by the variance ratio F distribution with $(k-1)$ and $\nu \; (= 2\Sigma \, Q_i)$ d.f. It may be noted that in the above development the proportion of trimmings may vary by samples. We will use these results to develop the robust stepwise test for one way multivariate analysis of variance hypothesis.

13.2.2 Combining Independent P-Values

The robust stepwise tests in Mudholkar and Srivastava (1999, c) use the logic in Mudholkar and Subbaiah's (1980) modification of J. Roy's (1958) step down tests, which, in turn, employ classical methods for combining independent tests. The combination of P-values is a meta-analytic tool for an overall judgement regarding a scientific hypothesis. Its investigation often involves conducting several, m, independent studies differing in design and size in which the original hypothesis takes form of possibly different null hypotheses. It is well known that, when the overall hypothesis is true, the P-values of the tests from these studies are independent and uniformly distributed on (0,1). Some of the best known and widely used combination statistics, for combining m independent P-values $P_1, P_2, ...P_m$, are; (i) $\Psi_T = \min(P_i)$ due to Tippett, (ii) $\Psi_F = -2 \, \Sigma \, log(P_i)$ due to Fisher, (iii) $\Psi_N = \Sigma \Phi^{-1}(1 - P_i)$ due to Liptak, and (iv) $\Psi_L = A^{-1/2}\Sigma \, log\{P_i/(1 - P_i)\}$, where $A = \pi^2 m(5m + 2)(15m + 12)$, due to George and Mudholkar (1979). The overall null hypothesis is rejected for small values of Ψ_T and Ψ_L and large values of Ψ_F and Ψ_N. Under the null hypothesis Ψ_T is distributed as a minimum of m uniform variates, Ψ_F has a χ^2 distribution with $2m$ degrees of freedom, Ψ_N is distributed as a $N(0, m)$ variate, and Ψ_L is approximated by a t-distribution with $(5m + 4)$ degrees of freedom. In the following section, we propose robust tests based on the modified step down logic.

13.2.3   Modified Step Down Procedure

The multivariate general linear model, very similar to its better known univariate version, is given by

$$E(\mathbf{Y}) \quad = \quad \mathbf{A}\,\Theta\,, \qquad (13.2.12)$$

where the n rows of the observation matrix $\mathbf{Y}$ (nxp) are independent normal variates with a p.d. covariance matrix Σ, $\mathbf{A}$ (nxm) is a known design matrix of rank m, and Θ (mxp) is the matrix of unknown parameters. The MANOVA hypothesis about the (mxp) matrix Θ of parameters for some matrix $\mathbf{B}$ (txm) of full rank is,

$$H_0 : \ \Psi = \mathbf{B}\,\Theta = 0\,. \qquad (13.2.13)$$

It is well known that the problem of testing the hypothesis in (13.2.13) has an invariant structure and the invariant tests, such as the likelihood ratio (Λ), Hotelling-Lawley trace and Roy's maximum root, all depend upon the characteristic roots of $\mathbf{HE}^{-1}$, where $\mathbf{H}$ and $\mathbf{E}$ are the matrix analogs of the univariate sum of squares due to hypothesis and error; for distributional and other results see Anderson (1984).

A step down procedure for testing H_0 given by J. Roy (1958) involves a sequence of familiar univariate tests. In the first stage of the procedure consideration is restricted to the first components of the p-variate observations, and consists of testing the univariate linear hypothesis obtained by the corresponding restriction of H_0 using the familiar F test at level α_1. Second stage considers the distribution of the second component of the observation vector conditional upon the first component as the covariate, and consists of using an analysis of covariance F test at level α_2, for testing the univariate linear hypothesis implied by H_0 for the conditional distributions; and so on for the further stages of the procedure. The remarkable fact which makes the stepwise procedures simple and convenient is that under H_0 the F tests at the successive stages of the procedure are independent; for details see J. Roy (1958), Anderson (1984) or Seber (1984). The step down procedure involves familiar univariate tests and raises no new distribution problems; but presents the problem of choosing significance levels α_i's for the successive F tests. The modified stepwise tests by Mudholkar and Subbaiah

(1980, 1988) avoid the problem by combining the independent P-values of the p independent step down F tests using one of the combining procedures given in Section 13.2.2. They have shown that if the P-values are combined using either Fisher's or Logit method then the resulting tests are B-optimal and asymptotically equivalent to the likelihood ratio test in terms of Bahadur efficiency. Furthermore, the empirical evidence shows that in the rank one particular case, i.e. the Hotelling's problem, the power functions of these modified stepwise tests are practically indistinguishable from that of Hotelling's T^2 test; see Mudholkar and Subbaiah (1980).

Multivariate One Way Classification. In this particular case we have p-variate random samples $\mathbf{Y_{ij}}$, $j = 1, 2, ..., n_i$, of size n_i, $\Sigma n_i = n$, from k p-variate normal populations $N_p(\mu_i, \boldsymbol{\Sigma})$, $i = 1, 2, ..., k$ and the MANOVA hypothesis reduces to,

$$H_0 : \mu_1 = \mu_2 = ... = \mu_k . \qquad (13.2.14)$$

For simplicity let $p = 2$, and denote the component of the observation vector by U_{ij} and V_{ij}. At the first stage of the stepwise procedure we have to test the homogeneity of the mean of k univariate normal populations on the basis of random sample U_{ij}, $j = 1, 2, ..., n_i$, from them. Let P_1 denote the P-value of the ANOVA test statistic, $F = (n - k)\Sigma n_i(\overline{U}_{i.} - \overline{U}_{..})^2/(k - 1)\Sigma\Sigma(U_{ij} - \overline{U}_{..})^2$.

At the second stage we have k bivariate normal populations of $(U, V)'$ with possibly different means $(\xi_i, \eta_i)'$ but with the same covariance matrix. Consider the conditional distribution of V given U for which

$$E(V_{ij}|U_{ij}) \quad = \quad \eta_i - \beta(U_{ij} - \xi_i) , \; i = 1, 2, .., k. \qquad (13.2.15)$$

So at the second stage the problem of testing the homogeneity of the means at (13.2.15) can be solved by using the standard linear model theory or using the residual like quantities

$$e_{ij} \quad = \quad V_{ij} - bU_{ij} , \qquad (13.2.16)$$

where the regression coefficient b as well as the $s_{2.1}^2$ the estimator of the conditional variance of V_{ij} given U_{ij} are obtained from the pooled sample covariance matrix. Let P_2 denote the P-value corresponding to the analysis of variance F test at the second stage. A similar reasoning can be extended to p-variate populations, and the modified stepwise procedure for the one way MANOVA is completed by combining the P-values corresponding to all the p stages. It may be noted that process could be considerably simplified, with possibly minor loss of efficiency, by using independent estimates of the common regression coefficient of V on U for each sample and regarding e_{ij}'s as approximately independent.

13.3 ROBUST STEPWISE TESTS

A use of the independence properties of the conditional distributions and normal theory regression residual like quantities appears in Tiku and Singh (1982) and Tiku and Balakrishnan (1988) where they use maximum likelihood estimates with censored normal data to test the significance of the equality of the mean vectors of two nonnormal bivariate populations. Mudholkar and Srivastava (1999c) recast and view their work in the framework of the modified step down tests and construct a class of robust stepwise tests for the equality of two nonnormal multivariate populations. Their work was motivated, on the one hand, by the multivariate analog of the Winsor's observation, noted in Mallows and Tukey (1982), that *"all observed distributions are Gaussian in the middle"*, and on the other hand by the belief that for large samples the independence properties would remain approximately valid if the means are replaced by the trimmed means with the associated appropriate changes, especially those involving small sample adaptations of the asymptotic distributions of the trimmed means and the Winsorized variances outlined in Section (13.2.1). The robust analysis of variance by Marchetti, Mudholkar and Mudholkar (1998) developed in Section (13.2.1) is now used in conjunction with the modified stepwise one way MANOVA procedure in Section (13.2.3) to construct the following robust stepwise tests.

Suppose $\mathbf{Y_{ij}}, i = 1, 2, ..., k$ and $j = 1, 2, ..., n_i$ be random samples from k p-variate populations with distribution functions $F(\mathbf{\Sigma}^{-1/2}(\mathbf{y} - \mu_i))$, $i = 1, 2, ..., k$ and suppose it is of interest to test the null hypothesis $H_0 : \mu_1 = \mu_2 = ... = \mu_k$. Let $Y_{ij}^{(l)}$ denote the l-th component, $l = 1, 2, ..., p$, of the p-dimensional observation $\mathbf{Y_{ij}}$.

Step 1. Consider the k univariate samples $Y_{ij}^{(1)}$, $j = 1, 2, ..., n_i$, $i = 1, 2, ..., k$, of the first component of the multivariate samples. Trim a proportion $\delta_i = g_i/n_i$ from each end of the i-th sample and compute the trimmed mean $\widetilde{Y}_i^{(1)}$ and the Winsorized sample standard deviation $\widetilde{s}_i^{(1)}$, $i = 1, 2, ..., k$, and use the trimmed ANOVA statistic $\widetilde{F}^{(1)}$, using equation (13.2.10), to obtain the robust P-value $\widetilde{P}^{(1)}$.

Step 2. Consider regression of the second components $Y_{ij}^{(2)}$ on $Y_{ij}^{(1)}$. Now, in the interest of simplicity regress $Y_{ij}^{(2)}$ on $Y_{ij}^{(1)}$ sample by sample separately and obtain independent estimates b_i^{21} of the common regression coefficient β^{21} and obtain the residual like quantities,

$$e_{ij}^{(2)} = Y_{ij}^{(2)} - b_i^{21} Y_{ij}^{(1)} \quad j = 1, 2, ..., n_i \text{ and } i = 1, 2, ..., k . \quad (13.3.17)$$

Again apply the trimmed ANOVA procedure as discussed in Section 13.2.1

to the residual like quantities $e_{ij}^{(2)}$ in (3.1), and obtain the statistic $\widetilde{F}^{(2)}$ to test the homogeneity of k parameter values $E(Y_{ij}^{(2)} - \beta^{21} Y_{ij}^{(1)})$. Compute the P-value of $\widetilde{F}^{(2)}$ test by replacing n_i by $(n_i - 1)$ in the null distribution approximation discussed in Section 13.2.1 and denote it by $\widetilde{P}^{(2)}$.

Step 3. Now consider the conditional distribution of $Y_{ij}^{(3)}$ given $Y_{ij}^{(2)}$ and $Y_{ij}^{(1)}$. Apply the logic of Step 2 for robust testing of the homogeneity of expected values of $E(Y_{ij}^{(3)} - \beta^{32} Y_{ij}^{(2)} - \beta^{31} Y_{ij}^{(1)})$ across the k populations. That is obtain $\widetilde{F}^{(3)}$ the robust trimmed statistic and the corresponding P-value $\widetilde{P}^{(3)}$ by applying the procedure to the residual like quantities $e_{ij}^{(3)} = (Y_{ij}^{(3)} - b_i^{32} Y_{ij}^{(2)} - b_i^{31} Y_{ij}^{(1)})$ and using $n_i - 2$ instead of n_i in the null distribution approximation. Proceed in a similar manner and obtain the remaining P-values, $\widetilde{P}^{(4)}, ..., \widetilde{P}^{(p)}$, the p robust approximately independent P-values, corresponding to p stepwise trimmed ANOVA tests.

Step 4. Combine the approximately independent P-values $\widetilde{P}^{(l)}$, $l = 1, 2, ..., p$, obtained in Step 3 by using any of the combination methods, namely, Fisher, Logit, Liptak and Tippett, denoted by $\widetilde{T}_F$, $\widetilde{T}_L$, $\widetilde{T}_N$, and $\widetilde{T}_T$, respectively, to obtain the robust P-value for overall MANOVA hypothesis.

Remark. It may be noted that the n_i residual like quantities obtained, for i-th population at the l-th stage of the decomposition, using equations given in Steps 2–4 are not independently distributed. Indeed, their covariance matrix is of rank $(n_i - l)$. One could potentially improve the performance of the proposed statistics using Helmert's transformation and making the residuals independent. However, in view of the convenience of implementation and the fact that, for n_i large in relation to p, this dependence would be negligible one can treat the residual like quantities obtained above as approximately independent.

13.4   A MONTE CARLO EXPERIMENT

13.4.1   The Study

In this section we describe a Monte Carlo experiment conducted in order to understand the operating characteristics of the robust modified stepwise MANOVA tests based on the four classical combination methods. Although, the procedure allows for varying trimming proportions for different samples and different stages, in this experiment, we kept the trimming proportions equal at all stages. That is, we assumed $\delta_{i1} = \delta_{i2} = ... = \delta_{ip} = \delta_i$, $i = 1, 2, ..., k$, where δ_i is the proportion of observations trimmed from the i-th population.

Null Distribution. This part of the simulation experiment, involving 5000 replications, was devoted to the study of Type I error control of the robust stepwise MANOVA tests. Each replication consisted of samples of size n from three bivariate populations with a common covariance matrix Σ, with 1 as the diagonal elements and .5 as the off diagonal element, and mean vectors $\mu'_1 = \mu'_2 = \mu'_3 = (0, 0)$. The robust stepwise MANOVA tests described in Section 13.4, using all four combination methods of Section 13.2.2, were applied to the set of three samples in each replication to test the homogeneity of the population vectors. The proportion of rejections in the 5000 tests was used as the estimate of the Type I error probabilities. The samples were of size n, $n = 10$ (5) 50, and the populations used were: (I) $N_2(\mu, \Sigma)$, (II) $0.8\ N_2(\mu, \Sigma) + 0.2\ N_2(\mu, 9\Sigma)$, (III) $0.8\ N_2(\mu, \Sigma) + 0.2\ N_2(\mu, 16\Sigma)$, (IV) $EC_2(\mu, \Sigma, |\frac{G}{G}|)$, (V) Bivariate T with 3 degrees of freedom and (VI) Bivariate Cauchy. The details of the methods for generating random samples from the above populations are discussed in Mudholkar and Srivastava (1999a) or Johnson (1987). A selection of the results, corresponding to some populations and Fisher combination method, is presented in Table 13.1.

Power Study. The power function of the robust stepwise tests were estimated using the above process in which each set of three samples came from the bivariate populations with a common covariance matrix Σ, as above, and different location vectors. The location vector for the first population was $\mu'_1 = (0, 0)$ and various alternative location vectors for the second and third populations considered were : (A) $\mu'_2 = (\mu_{21}, \mu_{22}) = (0.2, 0.2)$, and $\mu'_3 = (\mu_{31}, \mu_{32}) = (0.2, 0.2)$ (B) $\mu'_2 = (0.5, 0.5)$, $\mu'_3 = (0.5, 0.5)$, and (C) $\mu'_2 = (1.0, 1.0)$ and $\mu'_3 = (1.0, 1.0)$. A selection of the results of the simulation experiments for the power properties is presented in Table 13.2. For a visual depiction of the improvement observed for nonnormal populations a selection of results, corresponding to level $\alpha = .05$, is presented in Figure 13.1. In Figure 13.1, on the alternative axis, 0.2, 0.5, and 1.0 correspond to the alternatives (A), (B) and (C) discussed above and 0 refers to null situation with all three vectors centered about **0**.

Results. From the results of the Monte Carlo experiment, a selection from which appear in Tables 13.1 and 13.2, and Figure 13.1, it is seen that the robust stepwise MANOVA test procedures offer satisfactory Type I error control. For very heavy tailed populations, such as multivariate Cauchy, the accuracy of the null distribution improves with increasing proportion

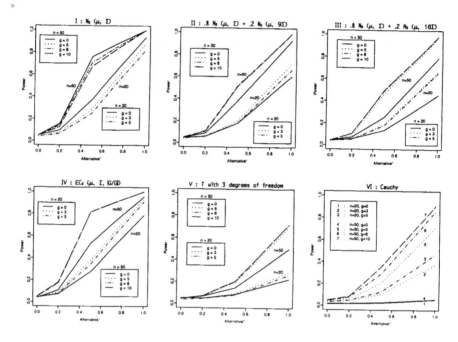

** Three bivariate populations, equal sample size (n), equal trimming
proportion $\delta = (g/n)$
* as discussed in Section 13.4

FIGURE 13.1 Power function of Fisher combination test** of Section 13.3

of trimming. For example, when the underlying distribution is bivariate
Cauchy and the three samples are of equal size 50, then it is seen that the
estimates of 5% nominal probability improve from .0154 to .0386, .0418
and .0476 respectively for 0% , 10%, 16% and 20% trimmings. In terms of
the power functions the robust tests experience minimally lower power at
normal populations. For example, when the three samples of size 50 are
chosen from bivariate normal population, with alternative (B), then the
power declines from .7620 for 0% trimming to .7136, .7094 and .6750 corresponding to 10%, 16% and 20% trimming, respectively. However, their
power advantage is substantial for nonnormal populations, especially when
the populations are very heavy tailed. For example, when three samples
of size 50 are taken from bivariate Cauchy population centered at the alternative (C), discussed above, the power estimates improve from .0558
corresponding to 0% trimming to .7130, .8582, and .9070 corresponding to
10%, 16% and 20% trimming, respectively. In general, we suggest between
15-20% trimming from each end at each stage of the procedure.

13.5 CONCLUSIONS

In general, the multivariate normality assumption, underlying the commonly used normal theory methods, is at best dubious. In this paper, we have presented a class of robust stepwise tests based on the normal theory modified step down tests due to Mudholkar and Subbaiah (1988). They provide satisfactory Type I error control for a broad class of symmetric multivariate populations. However, the real advantage of the robust tests, as in the univariate case [Mudholkar, Mudholkar and Srivastava (1991) and Srivastava, Mudholkar and Mudholkar (1992)], is their remarkably and substantially higher power in case of heavy tailed nonnormal populations without significant loss of power when the normal assumption is satisfied. Hence, they are effective solutions for testing the homogeneity of the means of multivariate samples.

Acknowledgements The research work of Deo Kumar Srivastava was in part supported by the Grant CA 21765 and the American Lebanese Syrian Associated Charities. The work of Carol E. Marchetti was in part supported by a Dean's Summer Followship Grant from the College of Science at RIT.

REFERENCES

Anderson, T. W. (1984). *An Introduction to Multivariate Statistical Analysis*, John Wiley & Sons, New York.

Bauer, P. (1981). On the robustness of Hotelling's T^2, *Biometrical Journal*, **4**, 405–412.

Bebbington, A. C. (1978). A method of bivariate trimming for robust estimation of the correlation coefficient, *Applied Statistics*, **27**, 221–226.

Bickel, P. J. (1965). On some robust estimates of location, *Annals of Mathematical Statistics*, **36**, 847–858.

Chase, G. R., and Bulgren, W. G. (1971). A Monte Carlo investigation of the robustness of T^2, *Journal of the American Statistical Association*, **66**, 499–502.

Davis, A. W. (1980). On the effects of moderate multivariate nonnormality on Wilk's likelihood ratio criterion, *Biometrika*, **67**, 419–427.

Draper, D. (1988). Rank Based Robust Analysis of Linear Models 1. Exposition and Review, *Statistical Science*, **3**, 239–271.

Everitt, B. S. (1979). A Monte Carlo investigation of the robustness of Hotelling's one- and two-sample T^2 tests, *Journal of the American Statistical Association*, **74**, 48–51.

Geary, R. C. (1947). Testing for normality, *Biometrika*, **34**, 209–242.

Gnanadesikan, R., and Kettenring, J. R. (1972). Robust estimates, residuals, and outlier detection with multiresponse data, *Biometrics*, **28**, 81–124.

Hampel, F. R., Ronchetti, E. M., Rousseeuw, R. J., and Stahel, W. A. (1986). *Robust Statistics*, John Wiley & Sons, New York.

Hsieh, H. K. (1979a). On asymptotic optimality of likelihood ratio tests for multivariate normal distribution, *Annals of Statistics*, **7**, 592–598.

Hsieh, H. K. (1979b). Exact Bahadur efficiencies for tests of the multivariate linear hypothesis, *Annals of Statistics*, **7**, 1231–1245.

Huber, P. J. (1970). Studentizing robust estimates, *In Nonparametric Techniques in Statistical Inference* (Ed., K. L. Puri), pp. 453–463, Cambridge University Press, Cambridge, England.

Johnson, M. E. (1987). *Multivariate Statistical Simulation*, John Wiley & Sons, New York.

Kiefer, J. and Schwartz, R. (1965). Admissible Bayes character of T^2-, R^2-, and other fully invariant tests for classical multivariate normal problems, *Annals of Mathematical Statistics*, **36**, 747–770.

Mallows, C. L. and Tukey, J. W. (1982). An overview of techniques of data analysis, emphasizing its exploratory aspects, In *Some Recent Advances in Statistics* (Eds., J. T. De Oliveira and B. Epstein).

Marchetti, C. E. (1997). *Robust Analysis of Variance*, Ph.D. Dissertation, University of Rochester, Rochester, New York.

Marchetti, C. E., Mudholkar, G. S., and Mudholkar, A. (1998). Robust analysis of variance, *Submitted for publication*.

Mardia, K. V., Kent, J. T. and Bibby, J. M. (1979). *Multivariate Analysis*, Academic Press, New York.

Mood, A. M. (1941). On the joint distribution of the median in samples from a multivariate population, *Annals of Mathematical Statistics*, **12**, 268–278.

Mudholkar, A., Mudholkar, G. S. and Srivastava, D. K. (1991). A construction and appraisal of pooled trimmed-t statistics. *Communications in Statistics—Theory and Methods*, **20**, 1345–1359.

Mudholkar, G. S. and George, E. O. (1979). The logit statistic for combining probabilities-an overview, In *Optimizing Methods in Statistics*, (Ed.. J. S. Rustagi), pp. 345–365, Academic Press, New York.

Mudholkar, G. S. and Srivastava, D. K. (1998). The elusive and illusory multivariate normality, *Technical Report*, 98/02, Department of Biostatistics, University of Rochester, Rochester, New York.

Mudholkar, G. S. and Srivastava, D. K. (1999a). Trimmed $\tilde{T}^2$: A robust analog of Hotelling's T^2, *Submitted for publication*.

Mudholkar, G. S. and Srivastava, D. K. (1999b). Robust analogs of Hotelling's two-sample T^2, *Submitted for publication*.

Mudholkar, G. S. and Srivastava, D. K. (1999c). A class of robust stepwise alternatives to Hotelling's T^2, *Submitted for publication*.

Mudholkar, G. S. and Subbaiah, P. (1980). Testing significance of a mean vector - A possible alternative to Hotelling's T^2, *Annals of the Institute of Statistical Mathematics*, **32**, 43–52.

Mudholkar, G. S. and Subbaiah, P. (1988). On a Fisherian detour of the step-down procedure for MANOVA, *Communications in Statistics—Theory and Methods*, **17**, 599–611.

Olson, C. L. (1974). Comparative robustness of six tests in multivariate analysis of variance, *Journal of the American Statistical Association*. **69**, 894–908.

Patel, K. R., Mudholkar, G. S. and Fernando, J. L. I. (1988). Student's t approximations for three simple robust estimators, *Journal of the American Statistical Association*, **83**, 1203–1210.

Puri, M. L. and Sen, P. K. (1971). *Nonparametric Methods in Multivariate Analysis*, John Wiley & Sons, New York.

Roy, J. (1958). Step-down procedure in multivariate analysis, *Annals of Mathematical Statistics*, **29**, 1177–1187.

Schwartz, R. (1967). Admissible tests in multivariate analysis of variance, *Annals of Mathematical Statistics*, **38**, 698–710.

Seber, G. A. F. (1984). *Multivariate Observations*, John Wiley & Sons, New York.

Srivastava, D. K., Mudholkar, G. S. and Mudholkar, A. (1992). Assessing the significance of difference between two quick estimates of location, *Journal of Applied Statistics,* **19**, 405–416.

Srivastava, M. S. and Awan, H. M. (1982). On the robustness of Hotelling's T^2-test and distribution of linear and quadratic forms in sampling from a mixture of two multivariate normal populations, *Communications in Statistics—Theory and Methods*, **11**, 81–107.

Tiku, M. L. and Balakrishnan, N. (1988). Robust Hotelling-type T^2 statistics based on the modified maximum likelihood estimators, *Communication in Statistics - Theory and Methods*, **17**, 1789–1810.

Tiku, M. L. and Singh, M. (1982). Robust statistics for testing mean vectors of multivariate distributions, *Communications in Statistics—Theory and Methods*, **11**, 985–1001.

Titterington, D. M. (1978). Estimation of correlation coefficients by ellipsoidal trimming, *Applied Statistics*, **27**, 227–234.

Tukey, J. W. and McLaughlin, D. H. (1963). Less vulnerable confidence and significance procedures for location based on a single sample: Trimming/Winsorization 1, *Sankhyā, Series A,* **25**, 331–352.

Yuen, K. K. and Dixon, W. J. (1973). The approximate behavior and performance of the two sample trimmed t, *Biometrika,* **60**, 369–374.

Yuen, K. K. (1974). The two sample trimmed t for unequal population variances, *Biometrika*, **61**, 165–170.

TABLE 13.1 Type I error control with Fisher combination statistic of Section 13.3; $k = 3$, $p = 2$, g_i = number and δ_i = % trimmed from the i-th population

						Nominal Probabilites	
				Population I : $N_2(\mathbf{0}, \mathbf{\Sigma})$			
n_1	n_2	n_3	g_1 (δ_1)	g_2 (δ_2)	g_3 (δ_3)	.05	.01
20	20	20	0 (0)	0 (0)	0 (0)	.0510	.0122
20	20	20	3 (15)	3 (15)	3 (15)	.0408	.0076
20	20	20	5 (25)	5 (25)	5 (25)	.0406	.0072
50	50	50	0 (0)	0 (0)	0 (0)	.0492	.0118
50	50	50	5 (10)	5 (10)	5 (10)	.0488	.0094
50	50	50	8 (16)	8 (16)	8 (16)	.0500	.0124
50	50	50	10 (20)	10 (20)	10 (20)	.0514	.0108
20	20	50	0 (0)	0 (0)	0 (0)	.0482	.0126
20	20	50	3 (15)	3 (15)	5 (10)	.0438	.0078
20	50	50	0 (0)	0 (0)	0 (0)	.0542	.0126
20	50	50	4 (20)	6 (12)	10 (20)	.0460	.0082
20	50	50	4 (20)	10 (20)	10 (20)	.0506	.0120

Population II : .8 $N_2(\mathbf{0}, \mathbf{\Sigma}) + .2\ N_2(\mathbf{0}, \mathbf{9\ \Sigma})$

n_1	n_2	n_3	g_1 (δ_1)	g_2 (δ_2)	g_3 (δ_3)	.05	.01
20	20	20	0 (0)	0 (0)	0 (0)	.0464	.0078
20	20	20	3 (15)	3 (15)	3 (15)	.0382	.0072
20	20	20	5 (25)	5 (25)	5 (25)	.0390	.0056
50	50	50	0 (0)	0 (0)	0 (0)	.0470	.0080
50	50	50	5 (10)	5 (10)	5 (10)	.0450	.0096
50	50	50	8 (16)	8 (16)	8 (16)	.0462	.0102
50	50	50	10 (20)	10 (20)	10 (20)	.0510	.0078
20	20	50	0 (0)	0 (0)	0 (0)	.0516	.0088
20	20	50	3 (15)	3 (15)	5 (10)	.0470	.0098
20	50	50	0 (0)	0 (0)	0 (0)	.0488	.0072
20	50	50	4 (20)	6 (12)	10 (20)	.0444	.0094
20	50	50	4 (20)	10 (20)	10 (20)	.0416	.0084

TABLE 13.1 (Cont.)

						Nominal Probabilites	
n_1	n_2	n_3	$g_1\ (\delta_1)$	$g_2\ (\delta_2)$	$g_3\ (\delta_3)$	.05	.01

Population III : $.8\ N_2(\mathbf{0}, \mathbf{\Sigma}) + .2\ N_2(\mathbf{0}, \mathbf{16\ \Sigma})$

n_1	n_2	n_3	$g_1\ (\delta_1)$	$g_2\ (\delta_2)$	$g_3\ (\delta_3)$	.05	.01
20	20	20	0 (0)	0 (0)	0 (0)	.0418	.0060
20	20	20	3 (15)	3 (15)	3 (15)	.0408	.0068
20	20	20	5 (25)	5 (25)	5 (25)	.0364	.0080
50	50	50	0 (0)	0 (0)	0 (0)	.0436	.0084
50	50	50	5 (10)	5 (10)	5 (10)	.0366	.0074
50	50	50	8 (16)	8 (16)	8 (16)	.0432	.0092
50	50	50	10 (20)	10 (20)	10 (20)	.0434	.0078
20	20	50	0 (0)	0 (0)	0 (0)	.0476	.0076
20	20	50	3 (15)	3 (15)	5 (10)	.0422	.0090
20	50	50	0 (0)	0 (0)	0 (0)	.0422	.0078
20	50	50	4 (20)	6 (12)	10 (20)	.0464	.0110
20	50	50	4 (20)	10 (20)	10 (20)	.0460	.0090

Population IV : $EC_2(\mathbf{0}, \mathbf{\Sigma}, |\mathbf{G/G}|\ )$

n_1	n_2	n_3	$g_1\ (\delta_1)$	$g_2\ (\delta_2)$	$g_3\ (\delta_3)$	.05	.01
20	20	20	0 (0)	0 (0)	0 (0)	.0418	.0076
20	20	20	3 (15)	3 (15)	3 (15)	.0398	.0070
20	20	20	5 (25)	5 (25)	5 (25)	.0476	.0102
50	50	50	0 (0)	0 (0)	0 (0)	.0384	.0060
50	50	50	5 (10)	5 (10)	5 (10)	.0446	.0108
50	50	50	8 (16)	8 (16)	8 (16)	.0496	.0104
50	50	50	10 (20)	10 (20)	10 (20)	.0512	.0110
20	20	50	0 (0)	0 (0)	0 (0)	.0464	.0108
20	20	50	3 (15)	3 (15)	5 (10)	.0488	.0116
20	50	50	0 (0)	0 (0)	0 (0)	.0430	.0080
20	50	50	4 (20)	6 (12)	10 (20)	.0454	.0088
20	50	50	4 (20)	10 (20)	10 (20)	.0486	.0096

TABLE 13.1 (Cont.)

Population V : *Bivariate T with 3 Degrees of Freedom*						Nominal Probabilites	
n_1	n_2	n_3	g_1 (δ_1)	g_2 (δ_2)	g_3 (δ_3)	.05	.01
20	20	20	0 (0)	0 (0)	0 (0)	.0506	.0068
20	20	20	3 (15)	3 (15)	3 (15)	.0420	.0074
20	20	20	5 (25)	5 (25)	5 (25)	.0398	.0064
50	50	50	0 (0)	0 (0)	0 (0)	.0458	.0098
50	50	50	5 (10)	5 (10)	5 (10)	.0434	.0066
50	50	50	8 (16)	8 (16)	8 (16)	.0442	.0092
50	50	50	10 (20)	10 (20)	10 (20)	.0432	.0064
20	20	50	0 (0)	0 (0)	0 (0)	.0486	.0088
20	20	50	3 (15)	3 (15)	5 (10)	.0428	.0100
20	50	50	0 (0)	0 (0)	0 (0)	.0440	.0084
20	50	50	4 (20)	6 (12)	10 (20)	.0462	.0102
20	50	50	4 (20)	10 (20)	10 (20)	.0478	.0096

Population VI : *Bivariate Cauchy*

n_1	n_2	n_3	g_1 (δ_1)	g_2 (δ_2)	g_3 (δ_3)	.05	.01
20	20	20	0 (0)	0 (0)	0 (0)	.0182	.0010
20	20	20	3 (15)	3 (15)	3 (15)	.0358	.0064
20	20	20	5 (25)	5 (25)	5 (25)	.0326	.0084
50	50	50	0 (0)	0 (0)	0 (0)	.0154	.0004
50	50	50	5 (10)	5 (10)	5 (10)	.0386	.0070
50	50	50	8 (16)	8 (16)	8 (16)	.0418	.0070
50	50	50	10 (20)	10 (20)	10 (20)	.0476	.0104
20	20	50	0 (0)	0 (0)	0 (0)	.0312	.0026
20	20	50	3 (15)	3 (15)	5 (10)	.0434	.0120
20	50	50	0 (0)	0 (0)	0 (0)	.0396	.0064
20	50	50	4 (20)	6 (12)	10 (20)	.0484	.0148
20	50	50	4 (20)	10 (20)	10 (20)	.0570	.0160

TABLE 13.2 Empirical power functions for Fisher combination statistic of Section 13.3; $k = 3$, $p = 2$, Alternatives (A), (B) and (C) in Section 13.4, g_i = number and δ_i = % trimmed from i-th population

n_1	n_2	n_3	$g_1(\delta_1)$	$g_2(\delta_2)$	$g_3(\delta_3)$	Level $\alpha = .05$ Alternatives (A)	(B)	(C)	Level $\alpha = .01$ Alternatives (A)	(B)	(C)
						Population I : $N_2(-,\Sigma)$					
20	20	20	0 (0)	0 (0)	0 (0)	.0976	.3736	.9342	.0268	.1744	.8154
20	20	20	3 (15)	3 (15)	3 (15)	.0758	.2846	.8812	.0200	.1230	.7122
20	20	20	5 (25)	5 (25)	5 (25)	.0652	.2508	.8226	.0128	.0938	.5944
50	50	50	0 (0)	0 (0)	0 (0)	.1556	.7620	1.000	.0470	.5460	.9988
50	50	50	5 (10)	5 (10)	5 (10)	.1450	.7136	.9998	.0408	.4714	.9984
50	50	50	8 (16)	8 (16)	8 (16)	.1420	.7094	.9998	.0428	.4640	.9978
50	50	50	10 (20)	10 (20)	10 (20)	.1266	.6750	.9994	.0372	.4396	.9960
20	20	50	0 (0)	0 (0)	0 (0)	.1016	.4112	.9684	.0240	.2006	.8954
20	20	50	3 (15)	3 (15)	5 (10)	.0860	.3458	.9444	.0202	.1504	.8298
20	50	50	0 (0)	0 (0)	0 (0)	.0966	.4590	.9808	.0288	.2358	.9204
20	50	50	4 (20)	6 (12)	10 (20)	.0866	.3832	.9530	.0234	.1810	.8660
20	50	50	4 (20)	10 (20)	10 (20)	.0948	.3806	.9516	.0280	.1718	.8496
						Population II : $.8\ N_2(-.\Sigma) + .2\ N_2(-.9\ \Sigma)$					
20	20	20	0 (0)	0 (0)	0 (0)	.0586	.1814	.6046	.0126	.0622	.3792
20	20	20	3 (15)	3 (15)	3 (15)	.0576	.2032	.7190	.0124	.0722	.4842
20	20	20	5 (25)	5(25)	5 (25)	.0610	.1820	.6656	.0154	.0586	.4166
50	50	50	0 (0)	0 (0)	0 (0)	.0880	.3860	.9292	.0268	.1738	.8252
50	50	50	5 (10)	5 (10)	5 (10)	.1094	.5144	.9870	.0290	.2876	.9568
50	50	50	8 (16)	8 (16)	8 (16)	.1092	.5234	.9898	.0314	.2908	.9612
50	50	50	10 (20)	10 (20)	10 (20)	.1028	.5168	.9904	.0284	.2782	.9604
20	20	50	0 (0)	0 (0)	0 (0)	.0666	.1982	.6668	.0162	.0664	.4478
20	20	50	3 (15)	3 (15)	5 (10)	.0618	.2408	.7990	.0128	.0956	.5924
20	50	50	0 (0)	0 (0)	0 (0)	.0676	.2284	.6924	.0200	.0888	.4664
20	50	50	4 (20)	6 (12)	10 (20)	.0724	.2728	.8550	.0158	.1162	.6824
20	50	50	4 (20)	10 (20)	10 (20)	.0774	.2880	.8608	.0218	.1216	.6842

TABLE 13.2 (Cont'd)

n_1	n_2	n_3	$g_1 (\delta_1)$	$g_2 (\delta_2)$	$g_3 (\delta_3)$	Level $\alpha = .05$ Alternatives (A)	(B)	(C)	Level $\alpha = .01$ Alternatives (A)	(B)	(C)
					Population III : .8 $N_2(-\,,\Sigma) + .2\,N_2(-\,,16\,\Sigma)$						
20	20	20	0 (0)	0 (0)	0 (0)	.0560	.1346	.4552	.0098	.0366	.2516
20	20	20	3 (15)	3 (15)	3 (15)	.0520	.1888	.6676	.0114	.0670	.4386
20	20	20	5 (25)	5 (25)	5 (25)	.0510	.1782	.6576	.0112	.0612	.4004
50	50	50	0 (0)	0 (0)	0 (0)	.0714	.2552	.7914	.0152	.1008	.5922
50	50	50	5 (10)	5 (10)	5 (10)	.0870	.4644	.9760	.0210	.2472	.9176
50	50	50	8 (16)	8 (16)	8 (16)	.1054	.4898	.9862	.0276	.2576	.9502
50	50	50	10 (20)	10 (20)	10 (20)	.1040	.4956	.9842	.0284	.2636	.9364
20	20	50	0 (0)	0 (0)	0 (0)	.0536	.1492	.4856	.0096	.0478	.2720
20	20	50	3 (15)	3 (15)	5 (10)	.0606	.2214	.7582	.0158	.0816	.5424
20	50	50	0 (0)	0 (0)	0 (0)	.0600	.1566	.5394	.0124	.0528	.3230
20	50	50	4 (20)	6 (12)	10 (20)	.0646	.2722	.8342	.0148	.1116	.6486
20	50	50	4 (20)	10 (20)	10 (20)	.0762	.2856	.8594	.0224	.1172	.6766
					Population IV : $EC_2(-\,,\Sigma\,,\,|G/G|\,)$						
20	20	20	0 (0)	0 (0)	0 (0)	.0720	.2770	.7862	.0148	.1228	.6142
20	20	20	3 (15)	3 (15)	3 (15)	.0906	.3790	.9360	.0252	.1804	.8246
20	20	20	5 (25)	5(25)	5 (25)	.0760	.3456	.9098	.0206	.1458	.7520
50	50	50	0 (0)	0 (0)	0 (0)	.1080	.5350	.9548	.0302	.3220	.9002
50	50	50	5 (10)	5 (10)	5 (10)	.1724	.8188	.9998	.0600	.6212	.9994
50	50	50	8 (16)	8 (16)	8 (16)	.1726	.8180	1.000	.0618	.6218	.9986
50	50	50	10 (20)	10 (20)	10 (20)	.1794	.8184	1.000	.0594	.6260	.9998
20	20	50	0 (0)	0 (0)	0 (0)	.5362	.6614	.8510	.3038	.4340	.7168
20	20	50	3 (15)	3 (15)	5 (10)	.6974	.8138	.9702	.4002	.5796	.8898
20	50	50	0 (0)	0 (0)	0 (0)	.0818	.3168	.8306	.0258	.1508	.6988
20	50	50	4 (20)	6 (12)	10 (20)	.1036	.5346	.9846	.0322	.3094	.9466
20	50	50	4 (20)	10 (20)	10 (20)	.1162	.5428	.9844	.0336	.3078	.9506

TABLE 13.2 (Cont'd)

n_1	n_2	n_3	$g_1 (\delta_1)$	$g_2 (\delta_2)$	$g_3 (\delta_3)$	Level $\alpha = .05$ Alternatives			Level $\alpha = .01$ Alternatives		
						(A)	(B)	(C)	(A)	(B)	(C)
	Population V : *Bivariate T with 3 Degrees of Freedom*										
20	20	20	0 (0)	0 (0)	0 (0)	.0474	.0836	.2232	.0100	.0188	.0832
20	20	20	3 (15)	3 (15)	3 (15)	.0466	.0906	.2852	.0096	.0210	.1122
20	20	20	5 (25)	5 (25)	5 (25)	.0440	.0786	.2616	.0086	.0194	.0938
50	50	50	0 (0)	0 (0)	0 (0)	.0646	.1436	.5000	.0134	.0408	.2738
50	50	50	5 (10)	5 (10)	5 (10)	.0666	.1938	.6912	.0142	.0752	.4650
50	50	50	8 (16)	8 (16)	8 (16)	.0708	.2006	.7184	.0144	.0696	.4814
50	50	50	10 (20)	10 (20)	10 (20)	.0674	.2044	.7218	.0168	.0726	.4886
20	20	50	0 (0)	0 (0)	0 (0)	.0552	.0982	.2712	.0118	.0262	.1084
20	20	50	3 (15)	3 (15)	5 (10)	.0570	.1102	.3332	.0136	.0316	.1510
20	50	50	0 (0)	0 (0)	0 (0)	.0590	.0938	.2816	.0138	.0270	.1168
20	50	50	4 (20)	6 (12)	10 (20)	.0606	.1082	.3946	.0156	.0306	.1868
20	50	50	4 (20)	10 (20)	10 (20)	.0568	.1214	.4090	.0132	.0374	.1964
	Population VI : *Bivariate Cauchy*										
20	20	20	0 (0)	0 (0)	0 (0)	.0190	.0274	.0564	.0016	.0042	.0130
20	20	20	3 (15)	3 (15)	3 (15)	.0430	.1028	.3766	.0072	.0316	.1886
20	20	20	5 (25)	5(25)	5 (25)	.0454	.1254	.4704	.0098	.0390	.2504
50	50	50	0 (0)	0 (0)	0 (0)	.0166	.0244	.0558	.0012	.0034	.0116
50	50	50	5 (10)	5 (10)	5 (10)	.0580	.2070	.7130	.0176	.0764	.4896
50	50	50	8 (16)	8 (16)	8 (16)	.0746	.2908	.8582	.0184	.1314	.6988
50	50	50	10 (20)	10 (20)	10 (20)	.0816	.3548	.9070	.0244	.1728	.7850
20	20	50	0 (0)	0 (0)	0 (0)	.0314	.0410	.0636	.0034	.0102	.0154
20	20	50	3 (15)	3 (15)	5 (10)	.0558	.1024	.3962	.0152	.0366	.1992
20	50	50	0 (0)	0 (0)	0 (0)	.0360	.0394	.0634	.0044	.0074	.0156
20	50	50	4 (20)	6 (12)	10 (20)	.0588	.1488	.5264	.0190	.0536	.3172
20	50	50	4 (20)	10 (20)	10 (20)	.0712	.2092	.6456	.0202	.0834	.4264

CHAPTER 14

ROBUST ESTIMATORS FOR THE ONE-WAY VARIANCE COMPONENTS MODEL

YOGENDRA P. CHAUBEY   K. VENKATESWARLU

Concordia University, Montreal, Quebec, Canada

Abstract: Since the adaptation of using pseudo–observations generated by M-estimation technique by Rocke (1983, *Biometrika*) for proposing robust estimators of variance components, there have been added quite a few methods in this direction [see the review article by Welsh and Richardson (1997) in *Handbook of Statistics*, Vol. XV]. However, not much extensive comparative studies of these methods are available [see, *e.g.* Richardson and Welsh (1995), *Biometrics*]. In this article, we present our numerical study of the ML and REML procedures and their robust versions for the one-way random effects model. Biases, MSE's and convergence properties of the different procedures are compared.

Keywords and phrases: Variance components, robust methods, unbalanced ANOVA

14.1   INTRODUCTION

Variance component models are used in various fields including sample surveys, animal breeding experiments and quality control procedures. The recent decade has seen several developments towards inference procedures for these models [see Rao (1997), Rao and Kleffé (1988), Searle *et al.* (1992)]. There are various procedures for estimating the variance components. The ANOVA method is still popular in the case of a balanced design. However, for the unbalanced case, the choice is not absolutely clear. With the development of efficient numerical algorithms, there has been a tendency to use the maximum likelihood approach. The maximum likelihood approach, necessarily demands normality assumptions and therefore may not be appropriate when such assumptions are not justified. Even when the data can be considered normal, in general, contamination of the observations and/or gross errors make the maximum likelihood approach [Hartley and Rao (1967)] (which is totally dependent on the model assumptions) inefficient as was demonstrated by Huber (1981) and Hampel et al. (1986). Owing to these practical considerations, there have been attempts to develop robust estimators for variance components.

One possible alternative to the normality assumption is to replace the distributions of error components by longer tailed distributions as investigated in Lange et al. (1989) and Stahel and Welsh (1992). Since the validity of a particular long tailed distribution may not be justified in practice, Rocke (1983) presented robustified version of the ANOVA estimates for the balanced two-component mixed model, which was subsequently extended to the unbalanced data by considering Henderson's Method 3. However, one drawback of this method is its inability to prevent negative estimates. A more general approach was introduced by Fellner (1986) robustifying the Henderson-Harville REML algorithm [see Harville (1977)]. These methods were motivated from the algorithmic nature of Huber's approach to robust estimation in linear models, and the asymptotic properties of the resulting estimators are yet to be established.

A different direction was taken in Huggins (1993), which replaced the quadratic function in the likelihood by a slowly varying function motivated by Huber's original suggestion for the robust estimation of a location parameter. Asymptotic properties of the resulting estimators were established by Richardson (1995). Similar to Huber's proposal I and II, Richardson and Welsh (1995) studied the corresponding ψ functions for the robust estimators for variance component models. These were named Robust ML1 and Robust ML2.

The maximum likelihood estimator is known to be badly biased in small samples; see Swallow and Monahan (1984). This problem was overcome by Patterson and Thompson (1971, 1974) in the classical case by introducing restricted likelihood (likelihood of independent contrasts of the data rather than of the data itself). The resulting likelihood does not involve the fixed parameters and the corresponding estimators of variance components are known as the REML estimators. Richardson and Welsh (1995) also robustified REML algorithm similar to Huber's approach as mentioned above. These are called Robust REML1 and Robust REML2. For a detailed account of various robust methods for mixed linear models the reader may refer to Welsh and Richardson (1997).

There have been numerical studies comparing various estimators in the classical case [such as Hocking and Kutner (1975), Corbeil and Searle (1976), Chaubey (1984) and Swallow, Monahan (1984) and Westfall (1987) and others] for finite samples. For comparing robust estimators there are only a few studies in the literature. Stahel and Welsh (1992, 1996) compared the REML with robust estimators of Fellner (1986) and Rocke (1991) for the balanced one way model. Richardson and Welsh (1995) conducted a small simulation study to compare the ML, REML and their robust versions for a two component mixed model. Gervini and Yohai (1998) compare another robust version of the ML with the robust procedures of Rock (1991) and Fellner (1986).

The purpose of this paper is to provide a detailed comparison of some robust estimators for the unbalanced one way model. Consideration of other procedures is in progress. Section 14.2 outlines the ML and REML methods for estimating the variance components along with their robust versions as proposed in Richardson and Welsh (1995). Section 14.3 gives details of the sampling experiment together with the numerical results and Section 14.4 presents a discussion of these results. The final section presents summary and conclusions.

14.2 MIXED LINEAR MODELS AND ESTIMATION OF PARAMETERS

14.2.1 General Mixed Linear Model

The general mixed linear model is given by

$$y = X\alpha + \sum_{i=1}^{c-1} Z_i \beta_i + \epsilon \qquad (14.2.1)$$

where, y is an n-vector of responses, X and Z_i are known $n \times p$ and $n \times q_i$ design matrices, respectively, α is a p-vector of unknown fixed effects, β_i represent q_i vectors of unobserved random effects, $1 \leq i \leq c-1$; and ϵ is an n-vector of unobserved errors. The random effects β_i are assumed to be independent with mean zero and variance σ_i^2; each component of the error ϵ is assumed to be independent with mean zero and variance σ_c^2; and $\beta_1, \ldots, \beta_{c-1}$ and ϵ are assumed to be independent. Thus, we have

$$E(y) = X\alpha$$

$$var(\beta_i) \;\; = \;\; \sigma_i^2 I_{q_i}; \quad var(\epsilon) = \sigma_c^2 I_n \quad \text{and}$$

$$var(y) \;\; = \;\; V = \sigma_c^2 I_n + \sum_{i=1}^{c-1} \sigma_i^2 Z_i Z_i'. \tag{14.2.2}$$

14.2.2   Maximum Likelihood and Restricted Maximum Likelihood Estimators

Assuming the normality of the random effects and the error term in the model (14.2.1), the log-likelihood function is given by

$$l(\theta) = (-1/2) \log[\|V\|] - (1/2)(y - X\alpha)'V^{-1}(y - X\alpha) \tag{14.2.3}$$

where $\theta = (\alpha, \sigma_1^2, \sigma_2^2, ..., \sigma_c^2)'$. We may write (14.2.3) as

$$l(\theta) = (-1/2) \sum_{j=1}^{g} \log[\|V_j\|] - (1/2)(y_j - X_j\alpha)'V_j^{-1}(y_j - X_j\alpha) \tag{14.2.4}$$

where $y = (y_1', ..., y_g')'$. This partitioning is such that y_j and $y_{j'}$ $(j \neq j')$ are uncorrelated and X and V are partitioned conformably. Now, the maximum likelihood estimator of the variance components $\theta_0 = (\sigma_1^2, \sigma_2^2, ..., \sigma_c^2)'$ can be obtained by solving the following estimating equations

$$\sum_{j=1}^{g} -\text{tr}[V_j^{-1}Z_iZ_i']_j + (y_j - X_j\alpha)'V_j^{-1}[Z_iZ_i']_jV_j^{-1}(y_j - X_j\alpha) = 0,$$

$$i = 1, ..., c, \tag{14.2.5}$$

$$\alpha = (X'V^{-1}X)^{-1}X'V^{-1}y. \tag{14.2.6}$$

Each Z_iZ_i' is block-diagonal with jth block being denoted by $[Z_iZ_i']_j$. These equations can be solved iteratively.

The REML estimator of θ_0 is obtained by maximizing

$$l_R(\theta_0) = (-1/2) \log[\|V\|] - (1/2) \log[\|X'V^{-1}X\|] - (1/2)y'Py. \tag{14.2.7}$$

As shown by Harville (1977), this estimator is obtained from

$$-\text{tr}[PZ_iZ_i'] + (y - X\alpha)'V^1[Z_iZ_i']V^{-1}(y - X\alpha) = 0,$$

$$i = 1, 2, ..., c \tag{14.2.8}$$

where

$$P = V^{-1}[I - X(X'V^{-1}X)^{-1}X'V^{-1}]$$

$$\text{and} \quad \alpha = (X'V^{-1}X)^{-1}X'V^{-1}y. \tag{14.2.9}$$

In the following section we outline the robust versions of (14.2.5) and (14.2.8) as presented by Richardson and Welsh (1995).

14.2.3 Robust Versions of ML and REML Estimators

Huggins (1993) method maximizes the following modified likelihood function with respect to θ_0,

$$l_{rob} = \sum_{j=1}^{g} (-1/2) \log[||V_j||] - \rho_j[V_j^{-1/2}(y_j - X_j\alpha)], \qquad (14.2.10)$$

where the functions rho_j is supposed to diminish the effect of large residuals, providing robust estimates. The resulting estimate are called Robust ML1 and can be obtained by solving the following estimating equations;

$$(1/2) \sum_{j=1}^{g} z_j' V_j^{-1/2}[Z_i Z_i']_j V_j^{-1/2} \psi_j(z_j) - \text{tr}[K_{1j} V_j^{-1}[Z_i Z_i']_j] = 0,$$
$$i = 1, ..., c \qquad (14.2.11)$$

where $K_1 = E[e\psi(e)']$ with $e \sim N(0, I_n)$, $z_j = V_j^{-1/2}(y_j - X_j\alpha)$, $\{\rho_j\}$ are suitable non-negative vector functions and $\psi_j(z) = \partial\rho_j(z)/\partial z$. The choice for ρ being considered for Robust ML1 is $\rho_j(z_j) = \sum_{j=1}^{g} \rho(z_{ji})$, where $\rho(x) = (1/2)x^2$ for $|x| \leq c$ and $\rho(x) = c|x| - (1/2)c^2$ is the usual Huber's ρ function. The second proposal comes from the requirement to bound the values of $z_j'\psi_j(z_j)$ in the above equation rather than bounding only z_j. This results in the following estimating equations:

$$(1/2) \sum_{j=1}^{g} \psi_j(z_j)' V_j^{-1/2}[Z_i Z_i']_j V_j^{-1/2} \psi_j(z_j) - \text{tr}[K_{2j} V_j^{-1}[Z_i Z_i']_j] = 0,$$
$$i = 1, ..., c \qquad (14.2.12)$$

where $K_2 = E[\psi(e)\psi(e)']$. This procedure is called Robust ML2. The equations, (14.2.11) and (14.2.12) also require the value of α, which is obtained by solving

$$\sum_{j=1}^{g} X_j' V_j^{-1} \psi_j[V_j^{-1/2}(y_j - X_j\alpha)] = 0. \qquad (14.2.13)$$

The robust version of REML is obtained by minimizing a robust version of (14.2.7) as in the case of the robust ML method which results in the following proposals:

Robust REML1: Solve

$$\{(y - X\alpha')V^{-1}[Z_i Z_i']V^{-1/2}\psi[V^{-1/2}(y - X\alpha)] - \text{tr}[K_1 P Z_i Z_i']\} = 0,$$
$$i = 1, ..., c. \qquad (14.2.14)$$

Robust REML2: Solve

$$\{\psi[V^{-1/2}(y - X\alpha)]'V^{-1/2}[Z_i Z_i']V^{-1/2}\psi[V^{-1/2}(y - X\alpha)]$$
$$-\text{tr}[K_2 P Z_i Z_i']\} = 0, \quad i = (1,4.2.15)$$

14.2.4 Computation of Estimators for the One Way Model

The one way model is given by

$$y_{ij} = \mu + a_i + \epsilon_{ij} \qquad (14.2.16)$$

where μ is an unknown parameter, a_i denote the random effects and e_{ij} are unobservable errors; $a_i \sim N(0, \sigma_a^2)$, $e_{ij} \sim N(0, \sigma_e^2)$, $j = 1, ..., n_i, i = 1, ..., k$. The constant k is the number of groups and $(n_1, n_2, ..., n_k)$ is called the n-pattern. This model in the matrix form can be written as

$$y = 1_N \mu + Z_1 \alpha_1 + \epsilon \qquad (14.2.17)$$

where 1_N denotes a $N-$vector of one's, $N = \sum n_i$,

$$Z_1 = \begin{bmatrix} 1_{n_1} & 0 & & & \\ & 1_{n_2} & & & \\ & & . & & \\ & & & . & \\ 0 & & & & 1_{n_k} \end{bmatrix}$$

$\alpha_1' = (a_1, a_2, ..., a_k)$, $\epsilon' = (\epsilon_1', \epsilon_2', ..., \epsilon_k')$, ϵ_i denoting the column vector $(\epsilon_{ij}), j = 1, ..., n_i$. Thus y is a vector of observations with mean vector $\mu 1_N$ and variance covariance matrix

$$V = \sigma_a^2 V_1 + \sigma_e^2 V_2. \qquad (14.2.18)$$

With these values the general iterating schemes given in Sections 14.2.2 and 14.2.3 are used in the numerical experiment described in the next section.

14.3 DESCRIPTION OF THE SIMULATION EXPERIMENT

We chose the 10 n-patterns studied in Swallow and Monahan as given below;
$P1 = (3, 5, 7)$
$P2 = (1, 5, 9)$
$P3 = (3, 3, 5, 5, 7, 7)$
$P4 = (1, 1, 5, 5, 9, 9)$
$P5 = (1, 1, 1, 1, 13, 13)$

$P6 = (3, 3, 3, 5, 5, 5, 7, 7, 7)$

$P7 = (1, 1, 1, 5, 5, 5, 9, 9, 9)$

$P8 = (1, 1, 1, 1, 1, 1, 1, 19, 19)$

$P9 = (2, 10, 18)$

$P10 = (3, 15, 27)$

These represent various degrees of unbalancedness as discussed in Swallow and Searle (1978). We generated 1000 trials for each pattern such that the convergence was achieved in a maximum of 200 iterations for different degrees of contamination (CTYPE) in a_i and ϵ_{ij} as described below (see also Rocke (1991)) where N stands for standard normal distribution, LT stands for a mixture distribution with 90% N(0,1) and 10% N(0,9)and VLT stands for a 95% N(0,1) and 5% N(0,100) mixture. Thus, CTYPE (NN) refers to $a_i \sim N$, $\epsilon_{ij} \sim N$ and CTYPE (NLT) *etc.* are defined in a conformal fashion.

The convergence was considered to have been achieved when the estimates at the kth and $(k+1)$th iterations satisfied

$$\frac{|\hat{\sigma}^2_{a,k+1} - \hat{\sigma}^2_{a,k}|}{1 + \hat{\sigma}^2_{a,k}} + \frac{|\hat{\sigma}^2_{e,k+1} - \hat{\sigma}^2_{e,k}|}{1 + \hat{\sigma}^2_{e,k}} < .00001.$$

This criterion is the same as the one used by Swallow and Monahan (1984) except that we have a tighter error bound (.00001 instead of .0001) and the maximum number of iterations was 200 [instead of 20 used by Swallow and Monahan (1984)]. This criterion is based on the premise that relative discrepancy is more meaningful than the absolute discrepancy. Unity is added to each denominator to prevent division by zero. In the case of CTYPE (NN), convergence was reached in less than 200 iterations for all n-patterns. However as contamination increased the number of trials which did not converge in less than 200 iterations also increased (see Table 14.3 for a comparison).

For each combination of CTYPE and n-pattern we computed the estimates of bias and MSE by averaging over all the replications. It should be also pointed out here for clarity that the parameter of interest in this study is the variance component of non-contaminated part. The biases and MSE's for only n-patterns P1, P2 and P3 are displayed in Tables 14.1 and 14.2. A detailed picture is given through graphs for all the patterns. The graphs for absolute biases computed from these simulations for σ^2_a and σ^2_e are presented in Figures 14.1 and 14.2 respectively, where as those for the MSE are presented in Figures 14.3 and 14.4.

14.4   DISCUSSION OF THE RESULTS

14.4.1   Biases of the Estimators of $\hat{\sigma}_a^2$

- For NN case, absolute biases are smaller for REML and its robust versions than for ML and its robust versions. This feature is common with the balanced case [see Swallow and Searle (1978)]. The biases for robust procedures are slightly larger than those for its nonrobust counterpart in the case of REML. However, this trend is reversed in the case of ML.

- If the effects are contaminated, then there is a tendency for increased bias in REML and its robust versions as compared with that of ML and its robust versions. If the effects are uncontaminated, a large contamination in the error can also produce a large absolute bias in REML than that in ML.

- When the random effects are not contaminated but the errors are contaminated, the biases of robust REML are smaller than those of REML.

- When both the components are contaminated, the robust procedures do not seem to have much added (if any) advantage for estimating σ_a^2.

- When the degree of unbalancedness is small (*e.g.*, see P1, P3 and P6), the robust REML procedures have slightly smaller bias than REML, even when both components are contaminated.

14.4.2   Biases of the Estimators of $\hat{\sigma}_e^2$

- The behaviour of the estimators in this case is clearer. For the NN case, biases of all the estimators are small. When there is a contamination in error terms, robust REML2 seems to drastically reduce the bias in REML. Such performance is also visible with respect to ML and its robust versions, especially, for very large contamination.

- The biases in robust REML procedures are generally higher than those for robust ML, especially, when there is a contamination in the error component.

14.4.3   MSE's of Estimators of $\hat{\sigma}_a^2$

- For NN case REML and its robust versions RREML1 and RREML2 have larger MSE's than ML and their robust counterparts.

- When the random effects are not contaminated, but the errors are, the MSE's of robust procedures are reduced when the degree of unbalancedness is small. However, this tendency is reversed as the unbalancedness increases.

- When the random effects are contaminated, robust procedures tend to have larger MSE's than their non robust versions. However, it may depend on the nature of unbalancedness in the data and the degree of contamination.

- Robust ML procedures, generally, seem to be better than the robust REML procedures for estimating σ_a^2.

14.4.4 MSE's of Estimators of $\hat{\sigma}_e^2$

- In the case of no contamination at all, there is no serious loss in efficiency by using any of the robust procedures.

- Significant gains can be achieved by the robust procedures, especially when the degree of contamination is large. In particular, Robust ML2 seems to have smaller MSE's than other procedures.

14.5 SUMMARY AND CONCLUSIONS

It is clear from this numerical study that robust versions of ML and REML methods in case of one way ANOVA model, generally reduce the MSE's for the estimator of the variance component due to the error term, especially, for a large contamination in the error term. However, no such general pattern is evident for the estimators of σ_a^2. Thus, we are of the opinion that better robust versions for estimating σ_a^2 are desired.

REFERENCES

Chaubey, Y. P. (1984). On the comparison of some non-negative estimators for two models, *Communications in Statistics—Simulation and Computation*, **13(5)**, 619–633.

Corbeil, B. R. and Searle, S. R. (1976). A Comparison of variance component estimators, *Biometrics*, **32**, 779–791.

Fellner, W. H. (1986). Robust estimation of variance components, *Technometrics*, **28**, 51–60.

Gervini, D. and Yohai, V. J. (1998). Robust estimation of variance components, *Canadian Journal of Statistics*, **26**, 419–430.

Hampel, F. R., Ronchetti, E. M., Rousseeuw, P. J. and Stahel, W. A. (1986). *Robust Statistics*, John Wiley & Sons, New York.

Harville, D. A. (1977). Maximum likelihood approaches to variance components estimation and related problems, *Journal of the American Statistical Association*, **72**, 320–340.

Hocking, R. R. and Kutner, M. H. (1975). Some analytical and numerical comparisons of estimators for the mixed A.O.V. model, *Biometrics*, **31**, 19–28.

Huber, P. J. (1981). *Robust Statistics*, John Wiley & Sons, New York.

Huggins, R. M. (1993). A robust approach to the analysis of repeated measures, *Biometrics*, **49**, 715–720.

Lange, K. L., Little, R. J. A. and Taylor, J. M. G. (1989). Robust statistical model using t distribution, *Journal of the American Statistical Association*, **84**, 881–896.

Patterson, H. D. and Thompson, R. (1971). Recovery of inter-block information when block sizes are unequal, *Biometrika*, **58**, 545–554.

Rao, C. R. and Kleffe, J. (1988). *Estimation of Variance Components and Applications*, North-Holland, Amsterdam.

Rao, P. S. R. S. (1997) *Variance Components Estimation, Mixed Models, Methodologies and Applications*, Chapman and Hall, London.

Richardson, A. M. and Welsh, A. H. (1995). Robust restricted maximum likelihood in mixed linear models, *Biometrics*, **51**, 1429–1439.

Rocke, D. M. (1983). Robust statistical analysis of inter laboratory studies, *Biometrika*, **70**, 421–431.

Rocke, D. M. (1991). Robustness and balance in the mixed model, *Biometrics*, **47**, 303–309.

Searle, S. R., Casella, G. and McCulloch, C. E. (1992). *Variance Components*, John Wiley & Sons, New York.

Stahel, W. A. and Welsh, A. H. (1992). Robust estimation of variance component, *Research Report*, **69**, ETH Zurich.

Stahel, W.A. and Welsh, A. H. (1997). Approaches to robust estimation in the simple variance component model. *Journal of Statistical Planning and Inference*, **57**, 295–319.

Swallow, W. H. and Monahan, J. F. (1984). Monte Carlo comparison of ANOVA, MIVQUE, REML, and ML estimators of variance components, *Technometrics*, **26**, 47–57.

Swallow, W. H. and Searle, S.R. (1978). Minimum variance quadratic unbiased estimators (MIVQUE) of variance components, *Technometrics*, **20**, 265–272.

Welsh, A. H. and Richardson, A. M. (1997). Approaches to the robust estimation of mixed models, In *Handbook of Statistics* (Eds., G. S. Maddala and C. R. Rao), **15**, pp. 343–384, North-Holland, Amsterdam.

Westfall, P. H. (1987). A comparison of variance component estimates for arbitrary underlying distributions, *Journal of Statistical Planning and Inference*, **82**, 866–874.

TABLE 14.1 Bias of different estimators for σ_a^2 and σ_e^2

Variance Component	Estimation Method	CTYPE						
		NN	NLT	NVLT	LTN	LTLT	LTVLT	VLTVLT
					n-pattern:(3,5,7)			
σ_a^2	ML	-0.3643	-0.3998	-0.3610	0.2238	0.0805	0.1167	1.3721
	RML1	-0.3441	-0.4058	-0.4985	0.2442	0.0782	-0.0309	1.1562
	RML2	-0.3454	-0.4310	-0.4518	0.2286	0.0550	0.0409	1.1237
	REML	0.0451	0.0469	0.2687	0.9338	0.7687	0.9836	2.8706
	RREML1	0.0863	0.0327	-0.1372	0.9852	0.7661	0.5890	2.3624
	RREML2	0.0941	0.0028	-0.0838	0.9790	0.9838	0.6212	2.4935
σ_e^2	ML	-0.0273	0.7637	4.5439	-0.0225	0.7358	4.5046	3.2010
	RML1	-0.0157	0.6060	2.3990	-0.0157	0.5696	2.5001	1.9319
	RML2	-0.0144	0.4248	0.4416	-0.0193	0.3930	0.5182	0.5194
	REML	-0.0151	0.7970	4.7175	-0.0139	0.7667	4.7223	3.3149
	RREML1	0.0046	0.6631	2.6943	0.0020	0.6228	2.7756	2.1686
	RREML2	0.0136	0.4938	0.4909	0.0068	0.4534	0.6123	0.7060
					n-pattern:(1,5,9)			
σ_a^2	ML	-0.4040	-0.3587	0.2329	0.1453	0.1297	0.2273	1.5416
	RML1	-0.3761	-0.3362	0.2145	0.1840	0.1642	0.2284	1.5905
	RML2	-0.3762	-0.3339	0.2415	0.1898	0.1883	0.2326	1.6567
	REML	0.0541	0.2023	1.1355	0.9232	0.9684	1.1252	3.0850
	RREML1	0.1135	0.2517	1.0284	1.0164	1.0450	1.0614	3.1012
	RREML2	0.1213	0.2738	1.3420	1.0358	1.2819	1.2511	3.5126
σ_e^2	ML	-0.0192	0.7382	2.5932	-0.0046	0.7189	2.2748	2.2830
	RML1	-0.0070	0.5870	1.5351	0.0103	0.5580	1.4348	1.3982
	RML2	-0.0050	0.4077	0.4401	0.0128	0.3833	0.4810	0.3919
	REML	-0.0077	0.7723	2.7269	-0.0015	0.7449	2.3867	2.4069
	RREML1	0.0106	0.6334	1.7262	0.0173	0.5951	1.6075	1.5711
	RREML2	0.0197	0.4620	0.5335	0.0273	0.4244	0.6143	0.5130
					n-pattern:(3,3,5,5,7,7)			
σ_a^2	ML	-0.2433	-0.1826	0.0540	0.5216	0.4578	0.5385	3.3712
	RML1	-0.2332	-0.2153	-0.3080	0.5262	0.4280	0.2559	2.9807
	RML2	-0.2339	-0.2298	-0.2272	0.5089	0.4240	0.4613	3.3088
	REML	-0.0478	0.0539	0.4481	0.8702	0.8223	1.0145	4.4079
	RREML1	-0.0313	0.0064	-0.1160	0.8857	0.7822	0.5701	3.8541
	RREML2	-0.0276	-0.0027	0.0681	0.8775	0.7979	1.0208	4.5913
σ_e^2	ML	0.0216	0.7784	3.8570	-0.0047	0.7654	3.6352	3.4797
	RML1	0.0282	0.5703	1.8069	-0.0023	0.5642	1.7844	1.6077
	RML2	0.0294	0.3946	0.4685	-0.0050	0.3940	0.4822	0.4337
	REML	0.0228	0.7830	3.8874	-0.0043	0.7694	3.6651	3.5099
	RREML1	0.0336	0.5849	1.8985	0.0023	0.5775	1.8661	1.6827
	RREML2	0.0387	0.4142	0.5132	0.0034	0.4119	0.5231	0.4743

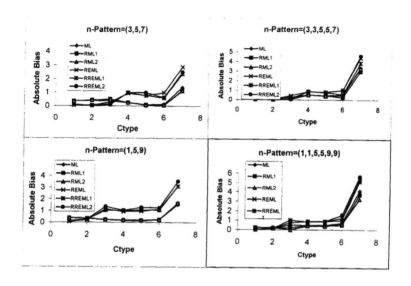

FIGURE 14.1a Absolute bias vs. contamination types for random effect
variance component

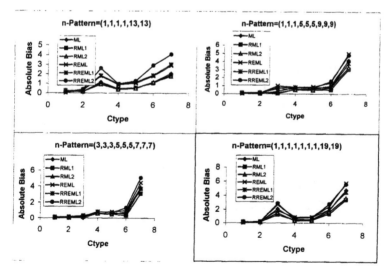

FIGURE 14.1b Absolute bias vs. contamination types for random effect
variance component

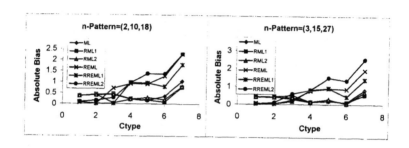

FIGURE 14.1c Absolute bias vs. contamination types for random effect variance component

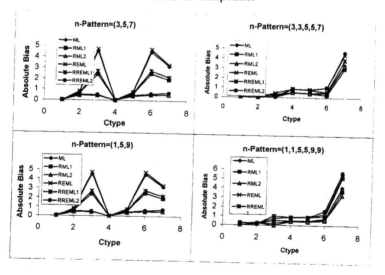

FIGURE 14.2a Absolute bias vs. contamination types for error variance component

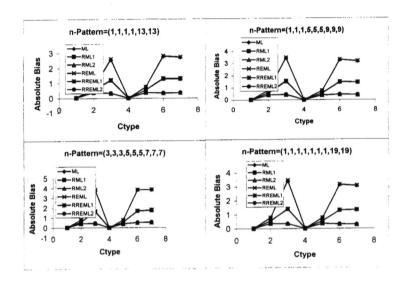

FIGURE 14.2b Absolute bias vs. contamination types for error variance
component

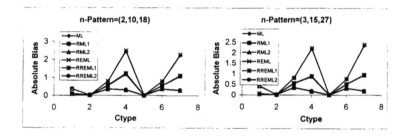

FIGURE 14.2c Absolute bias vs. contamination types for error variance
component

TABLE 14.2 MSE's of different estimators for σ_a^2 and σ_e^2

Variance Component	Estimation Method	CTYPE						
		NN	NLT	NVLT	LTN	LTLT	LTVLT	VLTVLT
		n-pattern:(3,5,7)						
σ_a^2	ML	0.7823	0.8706	1.4176	4.5278	4.0292	4.6480	43.3552
	RML1	0.8146	0.8222	0.9037	4.7067	3.9851	3.6507	42.0383
	RML2	0.8319	0.8100	2.0722	4.7126	4.0790	8.0904	41.5787
	REML	1.5007	1.7339	4.8036	11.0292	9.8250	13.2336	103.9541
	RREML1	1.6385	1.6109	1.7077	11.6758	9.8577	8.9991	98.4896
	RREML2	1.7071	1.6002	2.7501	11.8535	74.4219	11.0710	103.3757
σ_e^2	ML	0.1524	2.0643	103.3943	0.1667	2.1969	107.9137	75.9659
	RML1	0.1602	1.2913	29.9037	0.1670	1.3222	35.8396	37.4485
	RML2	0.1730	0.8072	2.5724	0.1745	0.8540	4.1133	3.0606
	REML	0.1562	2.2007	112.5792	0.1707	2.3180	119.7921	82.1319
	RREML1	0.1657	1.4762	37.9285	0.1739	1.4788	44.6710	46.7762
	RREML2	0.1786	0.9661	3.3544	0.1818	0.9656	7.3851	7.1338
		n-pattern:(1,5,9)						
σ_a^2	ML	0.9704	1.9539	27.8693	4.7500	5.0714	8.5414	51.4415
	RML1	1.0254	2.1142	29.6552	5.1397	5.3396	9.2087	55.1236
	RML2	1.0325	2.2584	28.9563	5.2942	5.9530	9.5551	57.2195
	REML	1.9496	4.6069	65.7224	11.7468	13.0268	21.6921	121.5250
	RREML1	2.1498	5.0550	68.8978	12.8104	14.0980	22.7529	130.1634
	RREML2	2.1979	5.8225	82.1688	13.2992	35.8792	27.5250	145.6212
σ_e^2	ML	0.1518	2.0021	49.6996	0.1732	2.1618	52.0543	52.0543
	RML1	0.1595	1.2685	18.7252	0.1810	1.2956	19.7899	27.2794
	RML2	0.1723	0.7791	2.9569	0.1909	0.9778	3.4047	1.9174
	REML	0.1573	2.1703	54.4292	0.1713	2.2798	43.7717	58.4719
	RREML1	0.1666	1.4263	23.6981	0.1789	1.3987	24.9129	34.1214
	RREML2	0.1802	0.8907	3.8541	0.1912	0.8864	6.3028	4.2974
		n-pattern:(3,3,5,5,7,7)						
σ_a^2	ML	0.4648	0.6055	1.7266	3.5026	3.5523	5.2617	105.4454
	RML1	0.4734	0.5279	0.6582	3.5413	3.4843	3.7267	98.6461
	RML2	0.4827	0.5122	0.6060	3.4890	3.6761	5.2697	117.0506
	REML	0.5886	0.8447	3.3043	5.4102	5.5012	8.6247	155.0287
	RREML1	0.6115	0.7041	0.8882	5.5470	5.4014	5.7320	145.5551
	RREML2	0.6294	0.7121	1.4411	5.5601	5.9078	13.2309	197.8034
σ_e^2	ML	0.0859	1.3380	46.3292	0.0831	1.4843	48.2451	42.3332
	RML1	0.0888	0.7324	11.0704	0.0855	0.8217	14.0499	9.8386
	RML2	0.0946	0.4340	0.8445	0.0917	0.4871	8.8191	0.8707
	REML	0.0863	1.3551	46.9358	0.0834	1.5035	49.3238	43.1850
	RREML1	0.0902	0.7622	12.3410	0.0862	0.8532	15.6365	10.8664
	RREML2	0.0965	0.4598	1.0049	0.0925	0.5121	0.9505	0.9925

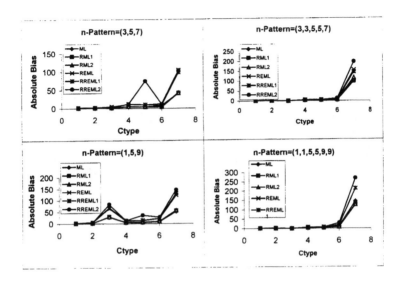

FIGURE 14.3a Mean square error vs. contamination types for random effect variance component

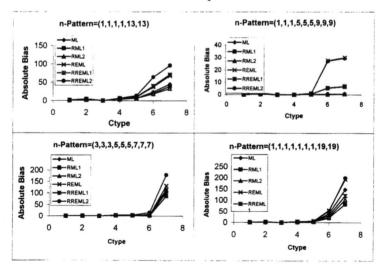

FIGURE 14.3b Mean square error vs. contamination types for random effect variance component

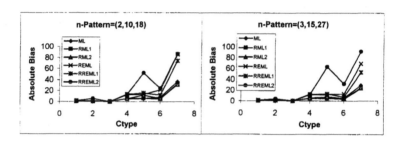

FIGURE 14.3c Mean square error vs. contamination types for random effect
variance component

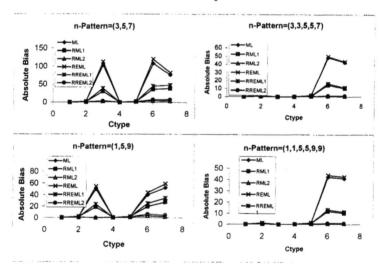

FIGURE 14.4a Mean square error vs. contamination types for error variance
component

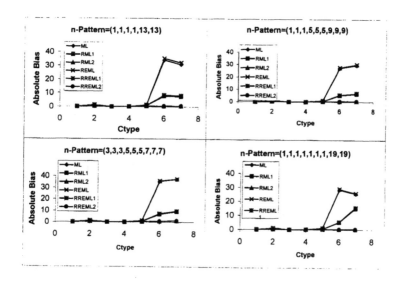

FIGURE 14.4b Mean square error vs. contamination types for error variance component

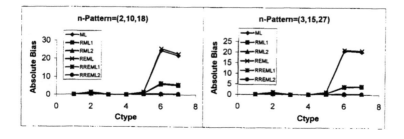

FIGURE 14.4c Mean square error vs. contamination types for error variance component

TABLE 14.3 Number of trials not converged in 200 iterations
(in 1000 trials)

n-Pattern	Estimation Method					
	ML	RML1	RML2	REML	RREML1	RREML2
CTYPE: NVLT						
P1	6	6	52	7	6	86
P2	10	12	67	5	6	118
P3	20	8	65	14	7	83
P4	15	4	98	14	3	139
p5	6	5	97	7	7	142
P6	21	5	64	17	6	96
P7	13	10	99	12	4	139
P8	8	6	93	4	12	124
P9	10	11	155	6	24	243
P10	6	15	258	6	17	365
CTYPE: LTVLT						
P1	10	6	51	9	4	87
P2	8	4	63	4	10	125
P3	20	10	74	11	4	96
P4	15	4	101	11	6	133
P5	7	9	106	6	15	136
P6	14	2	49	14	5	77
P7	6	6	94	12	5	133
P8	3	5	105	4	10	124
P9	4	8	171	3	20	258
P10	6	18	224	4	36	334
CTYPE: VLTVLT						
P1	10	5	44	6	6	90
P2	9	7	70	4	11	119
P3	15	6	77	14	3	100
P4	13	5	97	10	4	135
P5	6	7	105	7	7	156
P6	17	1	58	19	4	94
P7	18	3	104	4	5	143
P8	5	10	92	4	11	110
P9	4	13	183	4	18	265
P10	4	14	223	3	34	331

Part V
Regression and Design

CHAPTER 15

PERFORMANCE OF THE *PTE* BASED ON THE CONFLICTING *W*, *LR* AND *LM* TESTS IN REGRESSION MODEL

Md. BAKI BILLAH

University of Dhaka, Dhaka, Bangladesh

A. K. Md. E. SALEH

Carleton University, Ottawa, Ontario, Canada

Abstract: The problem of estimating the regression coefficients in the usual multiple regression model is considered when it is apriori suspected that the coefficients may be restricted to a subspace. The preliminary test estimator (PTE) based on the Wald (W), Likelihood Ratio (LR), and Lagrangian Multiplier (LM) tests are given. Their bias, mean square error matrix (M), and risk function are derived and compared. In the neighbourhood of the null hypothesis the PTE based on the LM test has the smallest risk followed by the LR based estimator and the estimator based on the W test is the worst. However, the PTE based on the W test performs the best followed by the LR based estimator when the parameter moves away from the subspace of the restriction and the LM based estimator is the worst. A table has been prepared for maximum and minimum guaranteed relative efficiency of the estimators corresponding to the three tests. This table allows one to determine optimum level of significance corresponding to the optimum estimator among the three. It has been shown that the optimum choice of the level of significance becomes the traditional choice by using the W test.

Keywords and phrases: Preliminary test estimators; relative efficiency; level of significance; risk analysis; conflict in tests

15.1   INTRODUCTION

Consider the regression model

$$y = X\beta + e, \tag{15.1.1}$$

where y is an $n \times 1$ vector of the response variable, X is an $n \times p$ matrix of non-stochastic independent variables, β is a p-dimensional column vector of regression parameters and e is the vector of errors associated with y having the same dimension. It is assumed that X is of full rank and $n \geq p$. Also assume that the errors follow normal distribution with mean vector 0 and covariance matrix $\sigma^2 I$.

The null hypothesis to be tested is

$$H_0 : H\beta = h, \tag{15.1.2}$$

where H is a known $q \times p$ matrix of full row rank and h is a known $q \times 1$ vector. The maximum likelihood (ML) estimator for β is the ordinary least squares (OLS) estimator given by

$$\tilde{\beta} = C^{-1}X'y,$$

where $C = X'X$.

Further, the ML estimator of σ^2 is given by

$$\tilde{\sigma}^2 = \frac{1}{n}(y - X\tilde{\beta})'(y - X\tilde{\beta}).$$

The bias and M of the unrestricted estimator of β are 0 and $\sigma^2 C^{-1}$ respectively.

The restricted OLS estimators of β and σ^2 are

$$\hat{\beta} = \tilde{\beta} - C^{-1}H'(HC^{-1}H')^{-1}(H\tilde{\beta} - h) \quad \text{and}$$
$$\hat{\sigma}^2 = \frac{1}{n}(y - X\hat{\beta})'(y - X\hat{\beta}),$$

respectively. The restricted estimator of β has bias,

$$B(\hat{\beta}) = \eta = -C^{-1}H'(HC^{-1}H')^{-1}(H\beta - h)$$

and M, $M(\hat{\beta}) = \sigma^2 C^{-1} - \sigma^2 \Lambda + \eta\eta'$, where $\Lambda = C^{-1}H'(HC^{-1}H')^{-1}HC^{-1}$.

The restricted estimator performs better than the unrestricted estimator when the null hypothesis $H\beta = h$ holds. However, as $H\beta$ differs from h, the restricted estimator may be considerably biased, inefficient and inconsistent while the performance of the unrestricted estimator remains steady over such departure. For this reason, it is desirable to develop an estimator which is a compromise between the unrestricted and the restricted estimators under uncertain prior information $H\beta = h$. This can be done by using the preliminary test approach. The PTE of β is defined by

$$\hat{\beta}^* = \hat{\beta} + [1 - I(\xi^* \le \xi^*_{n,\alpha})](\tilde{\beta} - \hat{\beta}), \qquad (15.1.3)$$

where ξ^* is the general test statistics for testing the hypothesis (15.1.2), $\xi^*_{n,\alpha}$ is the α-level of the critical value of ξ^* and $I(A)$ is the indicator function of the set A. The idea of preliminary test estimator was proposed by Bancroft (1944). The performance of the PTE depends on the size of the test α. The PTE falls in the area of inference with uncertain prior information and have been studied by Bancroft (1944, 1964), Mosteller (1948), Kitagawa (1963), Han and Bancroft (1968) among others. Asymptotic theory together with robustness considerations have been extensively studied by Saleh and Sen (1978, 1984a,b, 1986) among others. Two bibliographies in this area of study are given by Bancroft and Han (1977), and Han, Rao and Ravichandran (1988).

Our main objective of this study is to provide a finite sample theory of the preliminary test estimators (PTE) based on W, LR, and LM tests with normality assumption for the estimation of regression coefficients under general uncertain sub-hypothesis situation stated earlier and to compare the performance of the three estimators. In Section 15.2, we proposed the estimators and test statistics for the null hypothesis $H_0 : H\beta = h$. Section 15.3 contains the bias, M, and risk of the estimators. In Section 15.4, we discuss the relative performance of the estimators. The generalized efficiency is discussed in Section 15.5. Finally, Section 15.6 summarizes the findings.

15.2 THE TESTS AND PROPOSED ESTIMATORS

To test the null hypothesis (15.1.2) the usual F statistic is

$$F = \frac{RRSS - URSS}{URSS} \frac{m}{q}, \qquad (15.2.4)$$

where $m = n - p$, $URSS = (y - X\tilde{\beta})'(y - X\tilde{\beta})$ is the unrestricted residual sum of squares and $RRSS = (y - X\hat{\beta})'(y - X\hat{\beta})$ is the restricted residual sum of squares. Under the alternative hypothesis the distribution of (15.1.2) is a non-central F with $(q, n-p)$ degrees of freedom and with the non-centrality parameter given by

$$\Delta = \frac{1}{2\sigma^2}(H\beta - h)'(HC^{-1}H')^{-1}(H\beta - h). \tag{15.2.5}$$

Three general principles employed for hypothesis testing in econometrics are the W, LR, and LM criteria. The W test was introduced by Wald (1943) and the LM test by Aitchison and Silvey (1958) and Silvey (1959). The LM test is the same as the score test of Rao (1947). Savin (1976) shows that a systematic numerical inequality exists between the test statistics for testing linear restrictions on the coefficients of certain linear models. The inequality relation between the values of the test statistics is $W \geq LR \geq LM$. The three test statistics for testing the hypothesis (15.1.2) are

$$\begin{align}
\xi^W &= (H\tilde{\beta} - h)'(\tilde{\sigma}^2 HC^{-1}H')^{-1}(H\tilde{\beta} - h), \tag{15.2.6}\\
\xi^{LR} &= n[\ln\hat{\sigma}^2 - \ln\tilde{\sigma}^2], \tag{15.2.7}\\
\xi^{LM} &= (H\tilde{\beta} - h)'(\hat{\sigma}^2 HC^{-1}H')^{-1}(H\tilde{\beta} - h). \tag{15.2.8}
\end{align}$$

These test statistics can also be written as follows

$$\xi^W = \frac{nq}{m}F, \tag{15.2.9}$$

$$\xi^{LR} = n\ln\left(1 + \frac{\xi^W}{n}\right), \tag{15.2.10}$$

$$\xi^{LM} = \frac{\xi^W}{1 + \xi^W/n}. \tag{15.2.11}$$

The three test statistics are a function of F statistic since $\xi^W = \frac{nq}{m}F$. Each test statistic has a different exact sampling distribution and hence the critical value for each test statistic is different. When these tests employ exact critical values they are referred to as exacts tests. The PTE defined in terms of exact tests at a given significance level have the same bias, M and risk. However, due to the inequality relation between the value of the test statistics the estimators based on a fixed critical value may have different bias, M and risk.

The exact sampling distributions of the three test statistics is complicated, so that in practice the critical regions of the tests are commonly based on asymptotic approximations. Under the null hypothesis the three test statistics have the same asymptotic chi-square distribution. In most of the econometric applications this asymptotic chi-square distribution is used for the testing purposes. When the exact distribution is approximated by the asymptotic chi-square distribution the critical value for an α level test of H_0 is approximated by the chi-square critical value $\chi_\alpha^2(q)$. The tests based on this approximate critical value are known as large sample tests. The PTE based the large sample tests are defined as follows

$$\hat{\beta}^W = \hat{\beta}I(\xi^W \le \chi_\alpha^2(q)) + \tilde{\beta}I(\xi^W > \chi_\alpha^2(q)), \qquad (15.2.12)$$

$$\hat{\beta}^{LR} = \hat{\beta}I(\xi^{LR} \le \chi_\alpha^2(q)) + \tilde{\beta}I(\xi^{LR} > \chi_\alpha^2(q)), \qquad (15.2.13)$$

$$\hat{\beta}^{LM} = \hat{\beta}I(\xi^{LM} \le \chi_\alpha^2(q)) + \tilde{\beta}I(\xi^{LM} > \chi_\alpha^2(q)). \qquad (15.2.14)$$

15.3 BIAS, M AND RISK OF THE ESTIMATORS

In this section, we give the expressions for the bias, quadratic risk and M matrix of the preliminary test estimators $\hat{\beta}^W$, $\hat{\beta}^{LR}$ and $\hat{\beta}^{LM}$. Direct computation following Judge and Bock (1978) lead to the following results.

The bias of the PTE based on the W, LR, and LM tests are given by (15.3.15)–(15.3.17) respectively:

$$(i) \quad B(\hat{\beta}^W) = -\eta G_{q+2,m}(\ell_1^W; \Delta), \qquad (15.3.15)$$

$$(ii) \quad B(\hat{\beta}^{LR}) = -\eta G_{q+2,m}(\ell_1^{LR}; \Delta), \qquad (15.3.16)$$

$$(iii) \quad B(\hat{\beta}^{LM}) = -\eta G_{q+2,m}(\ell_1^{LM}; \Delta), \qquad (15.3.17)$$

where $\ell_1^W = \frac{m}{n(q+2)}\chi_\alpha^2(q)$, $\ell_1^{LR} = \frac{m}{(q+2)}(e^{\frac{\chi_\alpha^2(q)}{n}} - 1)$, $\ell_1^{LM} = \frac{m\chi_\alpha^2(q)}{(q+2)(n-\chi_\alpha^2(q))}$, $\chi_\alpha^2(q)$ is the critical value of the central χ^2 distribution of q degrees of freedom at α significance level and $G_{q+2,m}(.; \Delta)$ is the cumulative distribution function of a non-central F-distribution with $(q+2, m)$ degrees of freedom and non-centrality parameter Δ.

Note that for $\alpha = 0$, the bias of the three estimators coincide with the bias of the restricted estimator, $\hat{\beta}$, while for $\alpha = 1$, it coincides with that of $\tilde{\beta}$, the unrestricted estimator. Also, as the non-centrality parameter $\Delta \to \infty$, $B(\hat{\beta}^W) = B(\hat{\beta}^{LR}) = B(\hat{\beta}^{LM}) = B(\tilde{\beta}) = 0$ while $B(\hat{\beta})$ becomes unbounded. However, under $H_0 : H\beta = h$, $\Delta = 0$, hence all the estimators are unbiased:

$$B(\hat{\beta}^W) = B(\hat{\beta}^{LR}) = B(\hat{\beta}^{LM}) = B(\tilde{\beta}) = B(\hat{\beta}) = 0. \qquad (15.3.18)$$

The M of the PTE based on the W, LR, and LM tests are given by (15.3.19)–(15.3.21) respectively:

$$(i) \quad M(\hat{\beta}^W) \;=\; \sigma^2 C^{-1} - \sigma^2 \Lambda G_{q+2,m}(\ell_1^W; \Delta)$$
$$+\eta\eta' \left[2G_{q+2,m}(\ell_1^W; \Delta) - G_{q+4,m}(\ell_2^W; \Delta)\right],$$

$$(15.3.19)$$

$$(ii) \quad M(\hat{\beta}^{LR}) \;=\; \sigma^2 C^{-1} - \sigma^2 \Lambda G_{q+2,m}(\ell_1^{LR}; \Delta)$$
$$+\eta\eta' \left[2G_{q+2,m}(\ell_1^{LR}; \Delta) - G_{q+4,m}(\ell_2^{LR}; \Delta)\right],$$

$$(15.3.20)$$

$$(iii) \quad M(\hat{\beta}^{LM}) \;=\; \sigma^2 C^{-1} - \sigma^2 \Lambda G_{q+2,m}(\ell_1^{LM}; \Delta)$$
$$+\eta\eta' \left[2G_{q+2,m}(\ell_1^{LM}; \Delta) - G_{q+4,m}(\ell_2^{LM}; \Delta)\right],$$

$$(15.3.21)$$

where

$$\ell_2^W \;=\; \frac{m}{n(q+4)}\chi_\alpha^2(q),$$

$$\ell_2^{LR} \;=\; \frac{m}{(q+4)}(e^{\frac{\chi_\alpha^2(q)}{n}} - 1),$$

and

$$\ell_2^{LM} \;=\; \frac{m\chi_\alpha^2(q)}{(q+4)(n - \chi_\alpha^2(q))},$$

and $G_{q+4,m}(.;\Delta)$ is the cumulative distribution function of a non-central F-distribution with $(q+4,m)$ degrees of freedom and non-centrality parameter Δ.

If the loss function is $(\beta^* - \beta)'W(\beta^* - \beta)$ for a given non-singular matrix W using the estimator β^*, then the risk is defined by

$$R = E[(\beta^* - \beta)'W(\beta^* - \beta)] = tr(WM),$$

where M is the mean-squared error matrix of β^*. Now direct computation following Judge and Bock (1978) lead to the following theorem.

The risk of the PTE based on the W, LR, and LM tests are given by (15.3.22)–(15.3.24) respectively:

$$(i) \quad R(\hat{\beta}^W) \;=\; \sigma^2 tr(WC^{-1}) - \sigma^2 tr(W\Lambda)G_{q+2,m}(\ell_1^W; \Delta)$$
$$+tr(W\eta\eta') \left[2G_{q+2,m}(\ell_1^W; \Delta) - G_{q+4,m}(\ell_2^W; \Delta)\right],$$

$$(15.3.22)$$

$$(ii) \quad R(\hat{\beta}^{LR}) \;=\; \sigma^2 tr(WC^{-1}) - \sigma^2 tr(W\Lambda)G_{q+2,m}(\ell_1^{LR}; \Delta)$$

$$+tr(W\eta\eta')\left[2G_{q+2,m}(\ell_1^{LR};\Delta)-G_{q+4,m}(\ell_2^{LR};\Delta)\right],$$
$$(15.3.23)$$

$$(iii)\quad R(\hat\beta^{LM}) = \sigma^2 tr(WC^{-1})-\sigma^2 tr(W\Lambda)G_{q+2,m}(\ell_1^{LM};\Delta)$$
$$+tr(W\eta\eta')\left[2G_{q+2,m}(\ell_1^{LM};\Delta)-G_{q+4,m}(\ell_2^{LM};\Delta)\right].$$
$$(15.3.24)$$

The risk of the PTE depend on the matrices C and W. To ease the computations, we let $W=\sigma^{-2}C$. With these substitution in equations (15.3.22)–(15.3.24), the three risks, for simplicity, are

$$(i)\quad R(\hat\beta^W) = p-qG_{q+2,m}(\ell_1^W;\Delta)$$
$$+\Delta\left[2G_{q+2,m}(\ell_1^W;\Delta)-G_{q+4,m}(\ell_2^W;\Delta)\right]$$
$$(15.3.25)$$

$$(ii)\quad R(\hat\beta^{LR}) = p-qG_{q+2,m}(\ell_1^{LR};\Delta)$$
$$+\Delta\left[2G_{q+2,m}(\ell_1^{LR};\Delta)-G_{q+4,m}(\ell_2^{LR};\Delta)\right]$$
$$(15.3.26)$$

$$(iii)\quad R(\hat\beta^{LM}) = p-qG_{q+2,m}(\ell_1^{LM};\Delta)$$
$$+\Delta\left[2G_{q+2,m}(\ell_1^{LM};\Delta)-G_{q+4,m}(\ell_2^{LM};\Delta)\right]$$
$$(15.3.27)$$

15.4 RELATIVE PERFORMANCE OF THE ESTIMATORS

15.4.1 Bias Analysis of the Estimators

The quadratic bias of the PTE based on the W, LR, and LM tests are given by (15.4.28)–(15.4.30) respectively: Here, the quadratic bias is a quadratic form in the bias vectors in equations (15.3.15)–(15.3.17) where the weight matrix is $(1/\sigma^2)C$.

$$(i)\quad QB(\hat\beta^W) = \Delta\left[G_{q+2,m}(\ell_1^W\ \Delta)\right]^2,\qquad (15.4.28)$$

$$(ii)\quad QB(\hat\beta^{LR}) = \Delta\left[G_{q+2,m}(\ell_1^{LR};\Delta)\right]^2,\qquad (15.4.29)$$

$$(iii)\quad QB(\hat\beta^{LM}) = \Delta\left[G_{q+2,m}(\ell_1^{LM};\Delta)\right]^2.\qquad (15.4.30)$$

Since, $\ell_1^{LM}\geq\ell_1^{LR}\geq\ell_1^W$, we have

$$G_{q+2,m}(\ell_1^{LM};\Delta)\geq G_{q+2,m}(\ell_1^{LR};\Delta)\geq G_{q+2,m}(\ell_1^W;\Delta).$$

Therefore, an inequality relation between the quadratic biases of the three estimators is given by

$$QB(\hat\beta^W)\leq QB(\hat\beta^{LR})\leq QB(\hat\beta^{LM}).$$

15.4.2   M Analysis of the Estimators

In this subsection we analyse the M of the PTE based on the three tests and determine their dominance properties.

For the M comparison of $\hat{\beta}^W$ and $\hat{\beta}^{LR}$, we consider the M-difference

$$M(\hat{\beta}^W) - M(\hat{\beta}^{LR}) = \sigma^2 \Lambda \psi - \eta \eta' [2\psi - \psi^*], \qquad (15.4.31)$$

where $\psi = G_{q+2,m}(\ell_1^{LR}; \Delta) - G_{q+2,m}(\ell_1^W; \Delta)$ and $\psi^* = G_{q+4,m}(\ell_2^{LR}; \Delta) - G_{q+2,m}(\ell_2^W; \Delta)$. This M-difference is positive semi-definite whenever for a given non-zero vector $t = (t_1, \cdots, t_p)'$, we have $t'[\sigma^2 \Lambda \psi - \eta \eta' \{2\psi - \psi^*\}]t \geq 0$. That is to say,

$$\psi \sigma^2 t' \Lambda t \geq (2\psi - \psi^*) t' \eta \eta' t.$$

Hence for all $(p \times 1)$ non-zero vectors t, we have

$$\frac{(2\psi - \psi^*)}{\psi} \frac{t' \eta \eta' t}{\sigma^2 t' C^{-1} t} \leq \frac{t' \Lambda t}{t' C^{-1} t},$$

since $t' C^{-1} t > 0$. Now

$$\frac{(2\psi - \psi^*)}{\psi} \max_t \frac{t' \eta \eta' t}{\sigma^2 t' C^{-1} t} \leq \max_t \frac{t' \Lambda t}{t' C^{-1} t},$$

We know that

$$\max_t \frac{t' \eta \eta' t}{\sigma^2 t' C^{-1} t} = \frac{\eta' C \eta}{\sigma^2} = \Delta,$$

and

$$\max_t \frac{t' \Lambda t}{t' C^{-1} t} = ch_{max}[H'(HC^{-1}H')^{-1}HC^{-1}] = 1,$$

since $H'(HC^{-1}H')^{-1}HC^{-1}$ is an idempotent matrix [see Searle (1971)].

Thus (15.4.31) is positive definite if and only if

$$0 \leq \Delta \leq (2 - \psi^*/\psi)^{-1}. \qquad (15.4.32)$$

Within this interval the estimator $\hat{\beta}^{LR}$ performs better than the estimator $\hat{\beta}^W$, and $\hat{\beta}^W$ performs better than $\hat{\beta}^{LR}$ outside the interval i.e. for

$$\Delta \geq (2 - \psi^*/\psi)^{-1}. \qquad (15.4.33)$$

Now, for the M comparison of $\hat{\beta}^{LR}$ and $\hat{\beta}^{LM}$, we consider the M-difference

$$M(\hat{\beta}^{LR}) - M(\hat{\beta}^{LM}) = \sigma^2 \Lambda \psi_1 - \eta \eta' [2\psi_1 - \psi_1^*], \qquad (15.4.34)$$

where

$$\psi_1 = G_{q+2,m}(\ell_1^{LM}; \Delta) - G_{q+2,m}(\ell_1^{LR}; \Delta)$$

and

$$\psi_1^* = G_{q+4,m}(\ell_2^{LM}; \Delta) - G_{q+2,m}(\ell_2^{LR}; \Delta).$$

Proceeding in the similar way as above the M-difference is positive semi-definite whenever

$$\Delta \le (2 - \psi_1^*/\psi_1)^{-1}. \tag{15.4.35}$$

Thus $\hat{\beta}^{LM}$ is superior to $\hat{\beta}^{LR}$ in this range of Δ, otherwise $\hat{\beta}^{LR}$ is superior. Hence, the M of the estimators has the dominance picture

$$M(\hat{\beta}^W) > M(\hat{\beta}^{LR}) > M(\hat{\beta}^{LM}), \tag{15.4.36}$$

for all Δ in the interval

$$\left[0, min\left\{ \left(2 - \frac{\psi^*}{\psi}\right)^{-1}, \left(2 - \frac{\psi_1^*}{\psi_1}\right)^{-1} \right\} \right],$$

while

$$M(\hat{\beta}^W) \le M(\hat{\beta}^{LR}) \le M(\hat{\beta}^{LM}), \tag{15.4.37}$$

within the interval defined by

$$\Delta \in \left(max\left\{ \left(2 - \frac{\psi^*}{\psi}\right)^{-1}, \left(2 - \frac{\psi_1^*}{\psi_1}\right)^{-1} \right\}, \infty \right).$$

15.4.3 Risk Analysis of the Estimators

In this subsection, we provide the risk analysis of the estimators with the general loss function. We study the relative performance of the estimators under the null hypothesis as well as under the alternative hypothesis.

We get from the Courant theorem [see, e.g., Mardia (1979)] that

$$ch_{min}(WC^{-1}) \le \frac{\eta'W\eta}{\eta'C\eta} \le ch_{max}(WC^{-1}),$$

or,

$$\sigma^2 \Delta ch_{min}(WC^{-1}) \le \eta'W\eta \le \sigma^2 \Delta ch_{max}(WC^{-1}),$$

where $ch_{min}(WC^{-1})$ and $ch_{max}(WC^{-1})$ are the minimum and maximum eigenvalues of (WC^{-1}) and $\Delta = \frac{\eta'C\eta}{\sigma^2}$. Now we compare $\hat{\beta}^W$ versus $\hat{\beta}^{LR}$. The risk difference in this case is as follows:

$$R(\hat{\beta}^W) - R(\hat{\beta}^{LR}) = \sigma^2 tr(W\Lambda)\psi - \eta'W\eta(2\psi - \psi^*). \qquad (15.4.38)$$

Thus, we see that the right hand side of (15.4.38) is nonnegative (≥ 0), whenever

$$\Delta \leq \frac{tr(W\Lambda)}{ch_{max}(WC^{-1})(2 - \frac{\psi^*}{\psi})}. \qquad (15.4.39)$$

The length of this interval is bigger than the interval (15.4.32), provided by the M analysis. In this case $\hat{\beta}^{LR}$ performs better than $\hat{\beta}^W$ while $\hat{\beta}^W$ performs better whenever

$$\Delta \geq \frac{tr(W\Lambda)}{ch_{min}(WC^{-1})(2 - \frac{\psi^*}{\psi})}. \qquad (15.4.40)$$

For $W = C$ we note that $tr(W\Lambda) = q$ and the required intervals follows (15.4.39) and (15.4.40). Under H_0, we see that $\hat{\beta}^{LR}$ is superior to $\hat{\beta}^W$ since the difference is positive for all α. We can describe the graph of $\hat{\beta}^{LR}$ as follows: It begins with a value

$$\sigma^2 tr(WC^{-1}) - \sigma^2 tr(W\Lambda)G_{q+2,m}(\ell_1^{LR}; 0)$$

at $\Delta = 0$, then increases crossing the risk of $\hat{\beta}^W$ to a maximum then drops gradually towards $\sigma^2 tr(WC^{-1})$ as $\Delta \to \infty$.

Now we compare $\hat{\beta}^{LR}$ and $\hat{\beta}^{LM}$. Both $\hat{\beta}^{LR}$ and $\hat{\beta}^{LM}$ are superior to $\hat{\beta}^W$ under the null hypothesis $H_0 : H\beta = h$. In general the risk difference is given by

$$R(\hat{\beta}^{LR}) - R(\hat{\beta}^{LM}) = \sigma^2 tr(W\Lambda)\psi_1 - \eta'W\eta(2\psi_1 - \psi_1^*). \qquad (15.4.41)$$

Thus, we obtain that the risk difference is nonnegative (≥ 0) whenever

$$0 \leq \Delta \leq \frac{tr(W\Lambda)}{ch_{max}(WC^{-1})(2 - \frac{\psi_1^*}{\psi_1})}, \qquad (15.4.42)$$

and $\hat{\beta}^{LM}$ performs better than $\hat{\beta}^{LR}$ in this interval. Also, the length of this interval is bigger than the interval (15.4.35), provided by the M analysis.

The risk difference is negative if

$$\Delta \geq \frac{tr(W\Lambda)}{ch_{min}(WC^{-1})(2 - \frac{\psi_1^*}{\psi_1})}.$$ (15.4.43)

Thus, $\hat{\beta}^{LR}$ performs better than $\hat{\beta}^{LM}$ in this interval. Also, the length of this interval (15.4.42) is bigger than the interval (15.4.35), provided by the M analysis.

We can describe the graph of $\hat{\beta}^{LM}$ as follows: It begins with a value

$$\sigma^2 tr(WC^{-1}) - \sigma^2 tr(W\Lambda)G_{q+2,m}(\ell_1^{LM}; \Delta)$$

at $\Delta = 0$, then increases crossing the risk of $\hat{\beta}^{LR}$ and $\hat{\beta}^{W}$ to a maximum then drops gradually towards $\sigma^2 tr(WC^{-1})$ as $\Delta \to \infty$.

Clearly, the risk of the three estimators may be ordered as

$$R(\hat{\beta}^{LM}) < R(\hat{\beta}^{LR}) < R(\hat{\beta}^{W}),$$ (15.4.44)

in the interval

$$0 \leq \Delta \leq min \left[\frac{tr(W\Lambda)}{ch_{max}(WC^{-1})(2 - \frac{\psi^*}{\psi})}, \frac{tr(W\Lambda)}{ch_{max}(WC^{-1})(2 - \frac{\psi_1^*}{\psi_1})} \right],$$

while

$$R(\hat{\beta}^{LM}) \geq R(\hat{\beta}^{LR}) \geq R(\hat{\beta}^{W}),$$ (15.4.45)

in the interval

$$\Delta \in \left(max \left\{ \frac{tr(W\Lambda)}{ch_{max}(WC^{-1})(2 - \frac{\psi^*}{\psi})}, \frac{tr(W\Lambda)}{ch_{max}(WC^{-1})(2 - \frac{\psi_1^*}{\psi_1})} \right\}, \infty \right).$$

Graphical display of relative risks of the three estimators are given in Figures 15.1–15.4.

15.5 EFFICIENCY ANALYSIS AND RECOMMENDATIONS

In this section, we consider the risk efficiency of the three estimators for β and provide a max-min rule for the optimum choice of the level of significance for the preliminary test of the null hypothesis $H_0 : H\beta = h$. Table for relative efficiency (maximum and minimum) and Δ value at which minimum efficiency of a given estimator, $\hat{\beta}^*$ occurs relative to the unrestricted estimator $\tilde{\beta}$ is provided. For discussion we take $W = C$.

The relative efficiency of $\hat{\beta}^*$ compared to $\tilde{\beta}$, is given by

$$RE(\hat{\beta}^* : \tilde{\beta}) = \frac{1}{1 - g(\Delta)} = E(\alpha; \Delta) \qquad (15.5.46)$$

where

$$g(\Delta) = (q/p)G_{q+2,m}(\ell_1^*; \Delta) - (\Delta/q)\left[2G_{q+2,m}(\ell_1^*; \Delta) - G_{q+4,m}(\ell_2^*; \Delta)\right].$$

For a given n, E is a function of α and Δ. The function $E(\alpha, \Delta)$ for $\alpha \neq 0$, has its maximum at $\Delta = 0$ with value $(1 - q/pG_{q+2,m}(\ell_1^*; 0))^{-1}$ (> 1) and decreases crossing the line $E(\alpha, \Delta) = 1$ to a minimum $E(\alpha, \Delta) = E_0$ at $\Delta = \Delta_0$. As Δ increases beyond Δ_0, the relative efficiency increases and approaches 1 as $\Delta \to \infty$. In the case when $\Delta = 0$ and α varies we have $\max_\alpha E(\alpha; 0) = E(0; 0) = (1 - q/p)^{-1}$. The value $E(\alpha; 0)$ decreases as α increases. On the other hand if $\alpha = 0$ and Δ vary, then the curve $E(0; \Delta)$ and $E(1; \Delta) = 1$ intersect at $\Delta = q$. For a general α, $E(0; \Delta)$ and $E(1; \Delta)$ will intersect in the interval $0 \leq \Delta \leq q$; the value of Δ at the intersection decreases as α increases.

In order to choose an estimator with maximum relative efficiency we adopt the following procedure: If it is known that $0 \leq \Delta \leq q$, the estimator $\hat{\beta}$ is chosen since $E(0; \Delta)$ is the largest for all Δ in this range. However, if Δ is unknown, there is no way of choosing a uniformly best estimator. In such case, we pre-assign a value of the efficiency, say E_0, that we are willing to accept then consider the set $B = [\alpha | E(\alpha; \Delta) \geq E_0, \forall \Delta]$ and the estimator is chosen which maximizes $E(\alpha, \Delta)$ over all $\alpha \in B$ and Δ. Thus, we solve for α such that $\max_{\alpha \in B} \min_\Delta E(\alpha, \Delta) = E_0$. Hence, for each estimator we can find optimum level of significance say α_W^*, α_{LR}^* and α_{LM}^* respectively with minimum guaranteed efficiency E_0. Then, we choose $\alpha_W^* = min(\alpha_W^*, \alpha_{LR}^*, \alpha_{LM}^*)$ as the optimum level of significance since $\alpha_W^* < \alpha_{LR}^* < \alpha_{LM}^*$.

Our main objective is to choose the smallest level of significance (α) to yield the best estimator in the sense of highest efficiency. Table 15.1 provides the values of maximum and guaranteed minimum relative efficiency and recommended corresponding size α of the three PTE's for $p = 4$ and $n = 10, 15, 20$. For example, if $n = 10$ and $p = 4$, and the experimenter wishes to have an estimator with a minimum guaranteed efficiency of 0.80, then using Table 15.1, we recommend him to select $\alpha = 0.05$, corresponding to W based PTE, because such a choice of α would yield an estimator with a maximum efficiency of 1.9065 at $\Delta = 0$ and a minimum guaranteed efficiency of 0.8331. Notice that $\alpha = 0.05$ is the traditional level of significance used by experimenters.

15.6 CONCLUSION

In this paper, we studied the effect of the tests on the PTE for the regression parameter when there exists uncertain prior information (UPI) that $H_0 : H\beta = h$ may hold. It is well known that the test statistics satisfies the inequality $W \geq LR \geq LM$. Thus there may exists conflict in the resulting test conclusions when certain fixed critical value is chosen to test the hypothesis.

In this paper, we find that the resulting PTE's of the regression parameter with UPI satisfy the MSE ordering

$$\hat{\beta}^W > \hat{\beta}^{LR} > \hat{\beta}^{LM},$$

for

$$0 \leq \Delta \leq min\left[\left(2 - \frac{\psi^*}{\psi}\right)^{-1}, \left(2 - \frac{\psi_1^*}{\psi_1}\right)^{-1}\right],$$

and the MSE ordering

$$\hat{\beta}^W < \hat{\beta}^{LR} < \hat{\beta}^{LM},$$

for

$$\Delta \geq max\left[\left(2 - \frac{\psi^*}{\psi}\right)^{-1}, \left(2 - \frac{\psi_1^*}{\psi_1}\right)^{-1}\right].$$

Similarly, the risk orderings are

$$\hat{\beta}^{LM} < \hat{\beta}^{LR} < \hat{\beta}^W,$$

in the interval

$$0 \leq \Delta \leq min\left[\frac{tr(W\Lambda)}{ch_{max}(WC^{-1})(2 - \frac{\psi^*}{\psi})}, \frac{tr(W\Lambda)}{ch_{max}(WC^{-1})(2 - \frac{\psi_1^*}{\psi_1})}\right],$$

while the ordering is

$$\hat{\beta}^{LM} > \hat{\beta}^{LR} > \hat{\beta}^W,$$

in the interval

$$\Delta \geq max\left[\frac{tr(W\Lambda)}{ch_{max}(WC^{-1})(2 - \frac{\psi^*}{\psi})}, \frac{tr(W\Lambda)}{ch_{max}(WC^{-1})(2 - \frac{\psi_1^*}{\psi_1})}\right].$$

We have also discussed the method of choosing an optimum level of significance to obtain maxi-mini guaranteed efficient estimators. The W based PTE is found to be performs best in the choice of the smallest level of significance to yield the best estimator in the sense of highest minimum guaranteed efficiency. The most striking feature of the results is the optimum choice of the level of significance becomes the traditional choice by using the W test.

Acknowledgements The research was supported by Natural Science and Engineering Research Council grant A3088 when the first author visited Carleton University, Ottawa, Canada, during the Summer of 1996.

REFERENCES

Aitchison, J. and Silvey, D. (1958). Maximum likelihood estimation of parameters subject to restraints, *Annals of Mathematical Statistics*, **29**, 813–828.

Bancroft, T. A. (1944). On biases in estimation due to the use of preliminary tests of significance, *Annals of Mathematical Statistics*, **15**, 190–204.

Bancroft, T. A. (1964). Analysis and inference for incompletely specified models involving the use of preliminary test(s) of significance, *Biometrics*, **20**, 427–442.

Bancroft, T. A. and Han, C.-P. (1977). Inference based on conditional specification: A note and a bibliography, *International Statistical Review*, **45**, 117–127.

Han, C.-P. and Bancroft, T. A. (1968). On pooling means when variance is unknown, *Journal of the American Statistical Association*, **63**, 1333–1342.

Han, C.-P., Rao, C. V. and Ravichandran, J. (1988). Inference based on conditional specification: A second bibliography. *Communications in Statistics—Simulation and Computation*, **12**, 1–9.

Judge, G. G. and Bock, M. E. (1978). *The Statistical Implication of Pretest and Stein-rule Estimators in Econometrics*, North-Holland, New York.

Kitagawa, T. (1963). Estimation after preliminary tests of significance, *University of California Publications in Statistics*, **3**, 147–186.

Mardia, K. V., Kent, J. T. and Bibby, J. M. (1979). *Multivariate Analysis*, Academic Press, New York.

Mosteller, F. (1948). On pooling data, *Journal of the American Statistical Association*, **43**, 231–242.

Rao, C. R. (1947). Large sample tests of statistical hypotheses concerning several parameters with applications to problems of estimation, *Proceedings of the Cambridge Philosophical Society.* **44**, 50–57.

Saleh, A. K. Md. E. and Sen, P. K. (1984a). Nonparametric preliminary test inference, In *Handbook of Statistics*, Vol. 4, North-Holland, Amsterdam.

Saleh, A. K. Md. E. and Sen, P. K. (1984b). Least-squares and rank-order preliminary test estimation in general multivariate linear models, *Proceedings of the ISI Golden Jubilee International Conference in Statistics: Application and New Directions*, Calcutta, India, 237–253.

Saleh, A. K. Md. E. and Sen, P. K. (1986). On shrinkage least-squares estimation in a parallelism problem, *Communications in Statistics— Theory and Methods*, **15**, 1451–1466.

Savin, N. E. (1976). Conflict among testing procedures in a linear regression model with autoregressive disturbances, *Econometrica*, **44**, 1303–1315.

Silvey, S. D. (1959). The Lagrange Multiplier test, *Annals of Mathematical Statistics*, **30**, 389–407.

Wald, A. (1943). Tests of hypotheses concerning several parameters when the number of observations is large, *Transactions of the American Mathematical Society*, **54**, 426–482.

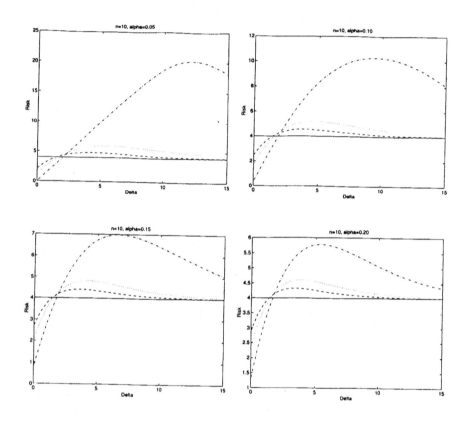

FIGURE 15.1 Risk function of the PTE based on the $W(--)$, $LR(\cdots)$ and $LM(-\cdot-)$ tests for various significance levels

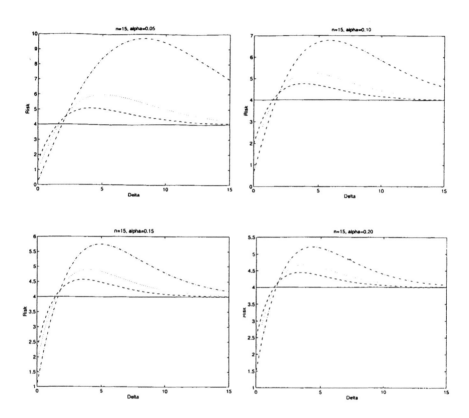

FIGURE 15.2 Risk function of the PTE based on the $W(--)$, $LR(\cdots)$ and $LM(-\cdot-)$ tests for various significance levels

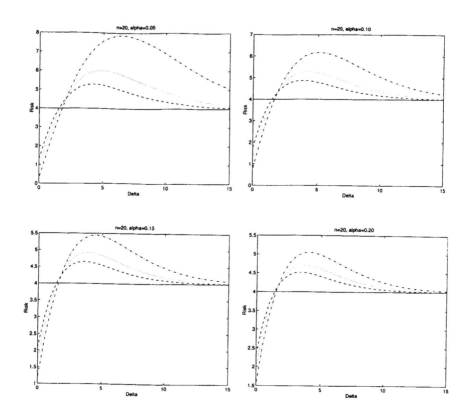

FIGURE 15.3 Risk function of the PTE based on the $W(--)$, $LR(\cdots)$ and $LM(-\cdot-)$ tests for various significance levels

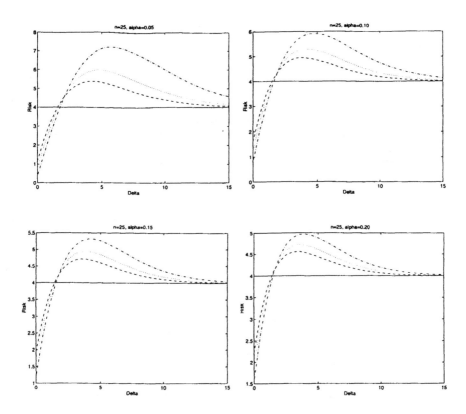

FIGURE 15.4 Risk function of the PTE based on the $W(--)$, $LR(\cdots)$ and $LM(-\cdot-)$ tests for various significance levels

TABLE 15.1 Maximum and minimum guaranteed efficiency of PTE's $(p = 4)$

Test	α :	5%	10%	15%	20%	25%	30%
			$n = 10$				
W	Max	1.90650	1.62379	1.47491	1.37847	1.25309	1.10537
	Min	0.83308	0.87967	0.90612	0.92494	0.93917	0.94999
	Δ	4.00000	4.00000	3.50000	3.50000	3.00000	3.00000
LR	Max	3.38472	2.35811	1.93005	1.68932	1.103767	1.41601
	Min	0.67164	0.77281	0.82963	0.86800	0.89594	0.91749
	Δ	5.50000	5.00000	4.00000	4.00000	3.50000	3.50000
LM	Max	814.66395	13.12121	5.02178	3.13313	2.34099	1.91390
	Min	0.19748	0.38799	0.57301	0.69251	0.77387	0.83199
	Δ	12.50000	9.50000	6.50000	5.50000	4.50000	4.00000
			$n = 15$				
W	Max	2.54892	2.00571	1.74055	1.57678	1.46131	1.37356
	Min	0.78409	0.84151	0.87551	0.89978	0.91849	0.93278
	Δ	4.00000	4.00000	3.50000	3.50000	3.00000	3.0000
LR	Max	4.32222	2.81708	2.22004	1.89372	1.68251	1.53237
	Min	0.66671	0.76046	0.81642	0.85539	0.88377	0.90689
	Δ	5.00000	4.50000	4.00000	3.50000	3.50000	3.50000
LM	Max	21.20441	6.44112	3.75890	2.72890	2.19044	1.86058
	Min	0.41128	0.58923	0.69617	0.76804	0.81968	0.85954
	Δ	8.00000	6.00000	5.00000	4.50000	4.00000	3.50000
			$n = 20$				
W	Max	3.05974	2.28485	1.92521	1.70988	1.56138	1.45049
	Min	0.75864	0.82112	0.85895	0.88608	0.90716	0.92329
	Δ	4.00000	4.00000	3.50000	3.50000	3.00000	3.00000
LR	Max	4.86009	3.30704	2.37686	2.00280	1.76291	1.59349
	Min	0.66721	0.75795	0.81216	0.85018	0.87899	0.90260
	Δ	5.00000	4.00000	4.00000	3.50000	3.50000	3.50000
LM	Max	13.32845	5.38561	3.44083	2.60337	2.13703	1.83932
	Min	0.50846	0.64861	0.73385	0.80375	0.83610	0.86926
	Δ	6.50000	5.00000	4.50000	4.00000	3.50000	3.50000

CHAPTER 16

ESTIMATION OF REGRESSION AND DISPERSION PARAMETERS IN THE ANALYSIS OF PROPORTIONS

SUDHIR R. PAUL

University of Windsor, Windsor, Ontario, Canada

Abstract: Data in the form of proportions arise in Toxicology and other similar fields. These proportions often exibit extra variation than can be explained by a simple binomial distribution. In the analysis of these proportions interest is, generally, in the estimation of the mean or the regression parameters. The dispersion parameter then plays the role of a nuisance parameter. However , in some instances the dispersion parameter or the intraclass correlation parameter is of primary interest. For example, in some binary-data situations in Toxicology the intraclass correlation is interpreted as 'heritability of a dichotomous trait'. In this paper we consider estimation of the dispersion parameter along with the regression parameters by using quadratic estimating functions (QEEs) of Crowder (1987). By varying the coefficients of the QEEs we obtain five sets of estimating equations. We compare large sample relative efficiency of these estimates with the maximum likelihood estimates. Estimated large sample relative efficiencies are also compared for three real life data sets arising from biostatistical practices.

Keywords and phrases: Dispersion parameter, efficiency, Gaussian likelihood, intraclass correlation, joint estimation, optimal estimating equations

283

16.1 INTRODUCTION

Data in the form of proportions arise in toxicology [Weil (1970) and Williams (1975)] and other similar fields [Crowder (1978) and Otake and Prentice (1948)]. These proportions often exibit extra dispersion (over-dispersion or under-dispersion) compared to that predicted by a simple binomial model. In situation like this, interest is often in the estimation of the mean or the regression parameters. The dispersion parameter can then be treated as nuisance parameter. However, in many situations the dispersion parameter or the intraclass correlation parameter may be of interest in its own right, as in some binary-data situations where it is interpreted as 'heritability of a dichotomous trait' [see Elston (1977) and Crowder (1982)]. Estimation of the dispersion parameter is also important for making inference regarding the regression parameters. Marginal or conditional estimation of the dispersion parameter is difficult. So we consider joint estimation of the mean (regression) parameters and the dispersion parameter. The usual procedure is to take a parametric model, such as, the beta-binomial or the extended beta-binomial model to allow over as well as under dispersion and obtain maximum likelihood estimates of the parameters. This procedure may produce inefficient or biased estimates when the parametric model does not fit the data well. Alternatively , more robust estimates, such as moment estimates, quasi-likelihood estimates [Breslow (1990) and Moore and Tsiatis (1991)], extended quasi-likelihood estimates [Nelder and Pregibon (1987)], the Gaussian likelihood estimates [Whittle (1961) and Crowder (1985)], estimates based on the pseudo-likelihood estimating equations of Davidian and Carrol (1987) and estimates based on quadratic estimating functions of Crowder (1987) and Godambe and Thompson (1989) can be considered. Paul and Islam (1998) study six such estimates and compare the small and large sample efficiency and bias properties of these estimates with the maximum likelihood estimates [see Paul and Islam (1998) for details]. Their study show that if interest is only in the mean or the regression parameters then the quasi-likelihood is the method of choice. On the other hand if we need estimation of the dispersion parameter or the joint estimation of the regression and the dispersion paramters then the Guassian likelihood estimates are the estimates of choice. Crowder (1985) finds similar good properties of the Gaussian likelihood estimates. Note that the generalized estimating equations approach of Liang and Ziger (1986) does not apply here as the primary focus of the procedure is to estimate the regression parameters only.

In this paper we consider estimating the regression and the dispersion parameters by the quadratic estimating equations (QEE's) of Crowder

(1987) and Godambe and Thompson (1989). By varying the coefficients of the QEE's we obtain five sets of estimating equations. Some of these five sets of estimating equations coincide with some of those studied by Paul and Islam (1998). We compare the large sample efficiency of the five sets of estimates obtained by the QEE's and the estimates obtained by quasi-likelihood method with the maximum likelihood estimates. We also compare estimated relative efficiencies of the estimates for three sets of real life data arising from biostatistical practices.

Joint estimation of the regression parameters and the dispersion parameter by the seven methods are discussed in Section 16.2. In Section 16.3 we derive and compare large sample relative efficiencies. Estimated relative efficiencies of the estimates for three sets of real life data are compared in Section 16.4. A discussion is given in Section 16.5.

16.2   ESTIMATION

16.2.1   The Extended Beta-Binomial Likelihood

We assume that $Y_i|p \sim$ binomial (n_i, p), for $i = 1, ..., m$. That is,

$$P(Y_i = y_i|p) = \binom{n_i}{y_i} p^{y_i}(1-p)^{n_i-y_i}.$$

Note that the binomial parameter p may not remain constant within a litter. So, we assume that the binomial probability p is a random variable distributed as a beta distribution with parameters α and β having probability density function

$$f(p|\alpha, \beta) = \frac{p^{\alpha-1}(1-p)^{\beta-1}}{B(\alpha, \beta)}, 0 < p < 1, \alpha > 0, \beta > 0,$$

where $B(\alpha, \beta) = \frac{\Gamma(\alpha+\beta)}{\Gamma(\alpha)\Gamma(\beta)}$ is the beta function. Then, the marginal distribution of Y_i is the beta-binomial distribution with probability function

$$P(Y_i = y_i|n_i) = \binom{n_i}{y_i} \frac{B(\alpha + y_i, n_i + \beta - y_i)}{B(\alpha, \beta)}.$$

Reparameterizing α and β as $\pi = \alpha/(\alpha + \beta)$ and $\phi = 1/(1 + \alpha + \beta)$, the unconditional distribution Y_i can be expressed as

$$P(Y_i = y_i|n_i)$$
$$= \binom{n_i}{y_i} \frac{\prod_{r=0}^{y_i-1}\{(1-\phi)\pi + r\phi\} \prod_{r=0}^{n_i-y_i-1}\{(1-\phi)(1-\pi) + r\phi\}}{\prod_{r=0}^{n_i-1}\{1 + \phi(r-1)\}},$$

with $0 \leq \pi_i \leq 1$ and $max(\frac{-1}{n_i-1}) < \phi < 1$ [Prentice (1986)]. The mean structure $\pi = \pi_i$ is given by the logistic model

$$\pi_i(X_i; \beta) = e^{X_i\beta}/(1 + e^{X_i\beta}),$$

where $X_i\beta = X_{i1}\beta_1 + \cdots + X_{ik}\beta_k$ and $X_1, ..., X_k$ are k explanatory variables , $\beta_1, ..., \beta_k$ are the k regression parameters. The mean and variance of the extended beta-binomial variate Y_i are $n_i\pi_i$ and $n_i\pi_i(1 - \pi_i)\{1 + (n_i - 1)\phi\}$. Clearly when $\phi = 0$, the beta- binomial variance coincides with the binomial variance. The parameter ϕ is the dispersion parameter or the intraclass correlation parameter. Maximum likelihood estimate (mle) of $\beta = (\beta_1, ..., \beta_k)$ and ϕ can be obtained by solving the ml estimating equations

$$
\begin{aligned}
\frac{\partial l}{\partial \beta_j} = &\sum_{i=1}^{m}\{\sum_{r=0}^{y_i-1}\frac{1-\phi}{(1-\phi)\pi_i + r\phi} \\
&- \sum_{r=0}^{n_i-y_i-1}\frac{1-\phi}{(1-\phi)(1-\pi_i)+r\phi}\}d_{ij}(\beta) = 0, \quad j = 1, ..., k
\end{aligned}
$$

and

$$
\begin{aligned}
\frac{\partial l}{\partial \phi} = &\sum_{i=1}^{m}\{\sum_{r=1}^{y_i-1}\frac{-\pi_i+r}{(1-\phi)\pi_i+r\phi} + \sum_{r=0}^{n_i-y_i-1}\frac{-(1-\pi_i)+r}{(1-\phi)(1-\pi_i)+r\phi} \\
&- \sum_{r=0}^{n_i-1}\frac{r-1}{(1-\phi)+r\phi}\} = 0,
\end{aligned}
$$

simultaneously, where

$$d_{ij}(\beta) = \frac{\partial \pi_i}{\partial \beta_j} = \pi_i(1-\pi_i)X_{ij}.$$

Now, denote the parameter vector (β, ϕ) by λ. Then the maximum likelihood estimate of λ is denoted by $\hat{\lambda}_{ML}$.

16.2.2 The Quasi-Likelihood Method

The quasi-likelihood [Wedderburn (1974)] is based on the knowledge of the form of first two moments of the random variable $Z_i = Y_i/n_i$, which coincides with those based on the extended beta-binomial model. The quasi-likelihood with the above mean and variance is given by $Q = \sum_{i=1}^{n} Q(z_i, \pi_i, \phi)$,

where

$$Q(z_i, \pi_i, \phi) = \int_{z_i}^{\pi_i} \frac{(z_i - t)n_i}{t(1-t)\{1 + (n_i - 1)\phi\}} dt.$$

Then, given ϕ, the unbiased estimating equation for β_j is

$$U_j(\beta, \phi) = \frac{\partial Q}{\partial \beta_j} = \sum_{i=1}^{m} \frac{(z_i - \pi_i)n_i d_{ij}(\beta)}{\pi_i(1 - \pi_i)\{1 + (n_i - 1)\phi\}} = 0,$$
$$j = 1, ..., k. \qquad (16.2.1)$$

No such estimating equation exists for ϕ. However, an unbiased estimating equation for ϕ can be obtained by using moment method, which, when the k β parameters are estimated is [Breslow (1984, 1990) and Moore and Tsiatis (1991)]

$$U_{k+1}(\beta, \phi) = \sum_{i=1}^{m} \frac{(z_i - \pi_i)^2 n_i}{\pi_i(1 - \pi_i)\{1 + (n_i - 1)\phi\}} - (m - k) = 0. \qquad (16.2.2)$$

Denote the estimates of β and ϕ, obtained by solving equations (16.2.1) and (16.2.2) simultaneously, by $\hat{\lambda}_{QL}$.

16.2.3   Estimation Using Quadratic Estimating Equations

For joint estimation Crowder (1987) proposed a general class of estimating functions called the quadratic estimating functions. Joint estimation of the parameters by the quadratic estimating functions avoid the failure of the maximum quasi-likelihood estimation to give reasonable results. For more details see Crowder (1987). We consider estimating functions quadratic in z_i that has general form (3.1) of Crowder (1987)

$$g_\lambda = \sum_{i=1}^{m} [a_{i\lambda}(z_i - \pi_i) + b_{i\lambda}\{(z_i - \pi_i)^2 - \sigma_{i\lambda}^2\}],$$

where $a_{i\lambda}$ and $b_{i\lambda}$ are specified nonstochastic functions of λ. Thus, the unbiased quadratic estimating equations for $\beta_1, ..., \beta_k$ and ϕ have the form

$$U_j(\beta, \phi) = \sum_{i=1}^{m} [a_{i\beta_j}(z_i - \pi_i) + b_{i\beta_j}\{(z_i - \pi_i)^2 - \sigma_{i\lambda}^2\}] = 0,$$
$$j = 1, ..., k, \qquad (16.2.3)$$

and

$$U_{k+1}(\beta, \phi) = \sum_{i=1}^{m}[a_{i\phi}(z_i - \pi_i) + b_{i\phi}\{(z_i - \pi_i)^2 - \sigma_{i\lambda}^2\}] = 0. \quad (16.2.4)$$

If we take

$$a_{i\beta_j} = [\frac{1}{\sigma_{i\lambda}^2} + \frac{n_i(1 - 2\pi_i)^2}{2(1 - \phi)\pi_i^2(1 - \pi_i)^2}]d_{ij}(\beta),$$

$$b_{i\beta_j} = \frac{-n_i\phi(1 - 2\pi_i)d_{ij}(\beta)}{2(1 - \phi)\pi_i(1 - \pi_i)\sigma_{i\lambda}^2},$$

$$a_{i\phi} = \frac{-(1 - 2\pi_i)n_i}{2(1 - \phi)^2\pi_i(1 - \pi_i)},$$

and

$$b_{i\phi} = \frac{n_i\{1 + (n_i - 1)\phi^2\}}{2(1 - \phi)^2\{1 + (n_i - 1)\phi\}\sigma_{i\lambda}^2}$$

we obtain the Gaussian estimating equations [see Paul and Islam (1998)]. Denote the estimates by $\hat{\lambda}_{GL}$.

If we take $a_{i\beta_j} = d_{ij}(\beta)/\sigma_{i\lambda}^2$, $b_{i\beta_j} = 0$, $a_{i\phi} = 0$ and $b_{i\phi} = \frac{(n_i-1)}{\{1+(n_i-1)\phi\}\sigma_{i\lambda}^2}$, then we obtain the unbiased estimating equations(QEE's) studied by Paul and Islam (1998) which were obtained by combining the quasi-likelihood estimating equations for the regression parameters and the optimal quadratic estimating equations of Crowder (1987) for the dispersion parameter after setting γ_1 and γ_2 to zero. Denote the estimates so obtained by $\hat{\lambda}_{M1}$.

For

$$a_{i\beta_j} = \frac{[-(\gamma_{2i\lambda} + 2 + \gamma_{1i\lambda}(1 - 2\pi_i)\sigma_\lambda/\pi_i(1 - \pi_i)]d_{ij}(\beta)}{\sigma_{i\lambda}^2\gamma_{i\lambda}},$$

$$b_{i\beta_j} = \frac{[\gamma_{1i\lambda} - (1 - 2\pi_i)\sigma_\lambda/\pi_i(1 - \pi_i)]d_{ij}(\beta)}{\sigma_{i\lambda}^3\gamma_{i\lambda}},$$

$$a_{i\phi} = \gamma_{1i\lambda}\pi_i(1 - \pi_i)(n_i - 1)/n_i\sigma_{i\lambda}^3\gamma_{i\lambda}$$

and

$$b_{i\phi} = -\pi_i(1 - \pi_i)(n_i - 1)/n_i\sigma_{i\lambda}^4\gamma_{i\lambda},$$

where, $\gamma_{i\lambda} = \gamma_{2i\lambda} + 2 - \gamma_{1i\lambda}^2$, we obtain the optimal quadratic estimating equations [Crowder (1987)]. Note that the forms of the skewness γ_{1i} and the kurtosis γ_{2i} are not known. We take these based on the second, third and fourth moments of the beta-binomial distribution, which are:

$$\mu_{2i} = \pi_i(1 - \pi_i)\{1 + (n_i - 1)\phi\}/n_i,$$
$$\mu_{3i} = \mu_{2i}(1 - 2\pi_i)\{1 + (2n_i - 1)\phi\}/n_i(1 + \phi)$$

and

$$\mu_{4i} = \mu_{2i}[\frac{\{1 + (2n_i - 1)\phi\}\{1 + (3n_i - 1)\phi\}\{1 - 3\pi_i(1 - \pi_i)\}}{(1 - \phi)}$$
$$+ (n_i - 1)\{\phi + 3n_i\mu_{2i}\}]\frac{1 - \phi}{(1 + \phi)(1 + 2\phi)n_i^2}.$$

Denote the estimates obtained by solving these optimal quadratic estimating equations by $\hat{\lambda}_{M2}$. Further, denote the estimates obtained by solving the optimal quadratic estimating equations with $\gamma_{1i\lambda} = \gamma_{2i\lambda} = 0$ by $\hat{\lambda}_{M3}$. Note, the estimates $\hat{\lambda}_{M3}$ are also obtained by using the pseudo-likelihood estimating equations of Davidian and Carrol (1987).

Finally, if we take $a_{i\beta_j} = d_{ij}(\beta)/\sigma_{i\lambda}^2, b_{i\beta_j} = 0, a_{i\phi} = \frac{n_i\{1 + (n_i - 1)\phi^2\}}{2(1 - \phi)^2\{1 + (n_i - 1)\phi\}\sigma_{i\lambda}^2}$ and $b_{i\phi} = \frac{-(1 - 2\pi_i)n_i}{2(1 - \phi)^2\pi_i(1 - \pi_i)}$, we obtain a set of unbiased estimating equations obtained by combining the quasi-likelihood estimating equations for the regression parameters and the Gaussian likelihood estimating equation for ϕ. Denote the estimates by $\hat{\lambda}_{M4}$.

16.3   ASYMPTOTIC RELATIVE EFFICIENCY

We compare the asymptotic relative efficiency of the estimates $\hat{\lambda}$ obtained by the seven estimation procedures. Asymptotic relative efficiency of $\hat{\lambda}_t$ is $Var(\hat{\lambda}_{ML})/Var(\hat{\lambda}_t)$, where $t = QL, GL, M1, M2, M3, M4$. Expressions for $Var(\hat{\lambda}_{ML})$ and $Var(\hat{\lambda}_{QL})$ have been obtained by Paul and Islam (1998). So, we omit these from presentation in this paper. The estimating equations for all the other five estimates given in this paper have the general form (16.2.1) with specific expressions for $a_{i\beta_j}, b_{i\beta_j}, a_{i\phi}, b_{i\phi}$ for each method given in Section 16.2. So, asymptotic $Var(\hat{\lambda}_t)$ are given in general forms.

From results of Inagaki (1973), the estimators $\hat{\lambda}_t, t = GL, M1, M2, M3, M4$ under conditions similar to those for which standard ML asymptotics hold, is consistent and asymptotically, as $m \to \infty$, normal with covariance matrix

$$Var\hat{\lambda}_t = \{A(\hat{\lambda}_t)\}^{-1}B(\hat{\lambda}_t)[\{A(\hat{\lambda}_t)\}^{-1}]^T,$$

where $A(\lambda), B(\lambda)$ are $(k+1) \times (k+1)$ matrices with

$$
\begin{aligned}
A_{js} &= E\frac{-\partial U_j}{\partial \beta_s} \\
&= \sum_{i=1}^{m}[a_{i\beta_j} + b_{i\beta_j}(1 - 2\pi_i)\{1 + (n_i - 1)\phi\}/n_i]d_{is}(\beta), \\
A_{j,k+1} &= E\frac{-\partial U_j}{\partial \phi} = \sum_{i=1}^{m} b_{i\beta_j}\pi_i(1 - \pi_i)(n_i - 1)/n_i, \\
A_{k+1,j} &= E\frac{-\partial U_{k+1}}{\partial \beta_j} \\
&= \sum_{i=1}^{m}[a_{i\phi} + b_{i\phi}(1 - 2\pi_i)\{1 + (n_i - 1)\phi\}/n_i]d_{is}(\beta), \\
A_{k+1,k+1} &= E\frac{-\partial U_{k+1}}{\partial \phi} = \sum_{i=1}^{m} b_{i\phi}\pi_i(1 - \pi_i)(n_i - 1)/n_i, \\
B_{js} &= E(U_j U_s) = \sum_{i=1}^{m}[a_{i\beta_j} a_{i\beta_s}\mu_{2i} + a_{i\beta_j} b_{i\beta_s}\mu_{3i} \\
&\quad + a_{i\beta_s} b_{i\beta_j}\mu_{3i} + b_{i\beta_j} b_{i\beta_s}(\mu_{4i} - \mu_{2i}^2)], \\
B_{j,k+1} &= E(U_j U_{k+1}) = \sum_{i=1}^{m}[a_{i\beta_j} a_{i\phi}\mu_{2i} + a_{i\beta_j} b_{i\phi}\mu_{3i} \\
&\quad + a_{i\phi} b_{i\beta_j}\mu_{3i} + b_{i\beta_j} b_{i\phi}(\mu_{4i} - \mu_{2i}^2)] = B_{k+1,j}
\end{aligned}
$$

and

$$
B_{k+1,k+1} = E(U_{k+1}^2) = \sum_{i=1}^{m}[a_{i\phi}^2\mu_{2i} + 2a_{i\phi}b_{i\phi}\mu_{3i} + b_{i\phi}^2(\mu_{4i} - \mu_{2i}^2)],
$$

where $\mu_{2i}, \mu_{3i}, \mu_{4i}$ are the second, third and the fourth moments of z_i. The forms of the third and the fourth moments of the z_i are generally unknown. However, for the purpose of comparison, we consider the third and fouth moments of z_i based on the beta-binomial distribution as given in Section 16.2. Note that with no covariate we have only two parameters $\pi = e^{\beta_1}/(1 - e^{\beta_1})$ and ϕ and $d_{ij}(\beta) = 1$.

For numerical relative efficiency comparisons we consider a model with two parameters π and ϕ and a simple logit linear regression model with parameter β_0, β_1 and ϕ. As in Paul and Islam (1998) we consider litter sizes [obtained from a real life experiment by Potthoff and Whittinghill

(1966)] n_i:11, 1, 6, 7, 8, 6, 2, 19, 4, 2, 15, 6, 6, 10, 8, 4, 5, 6, 6, 4, 12, 8, 4, 5, 5, 6, 4, 10, 8, 11, 4, 4, 4, 2, 2, 3.

For the two parameter model we considered four sets of combinations of the parameters π and ϕ : $\phi = .1$, $\pi = .05, .10(.05), ..., .90$; $\phi = .4$, $\pi = .05, .10(.05), ..., .90$; $\pi = .1$, $\phi = .05, .10(.05), ..., .90$; $\pi = .4$, $\phi = .05, .10(.05), ..., .90$. The relative efficiency results for $\hat{\pi}$ are summarized in Table 16.1 and those for ϕ are summarized in Table 16.2.

For the three parameter model we took litter specific covariate values: .005, .01, .015, .02, .025, .03, .005, .01, .015, .02, .025, .03, .005, .01, .015, .02, .025, .03, .005, .01, .015, .02, .025, .03, .005, .01, .015, .02, .025, .03, .005, .01, .015, .02, .025, .03. For the parameter β_0, β_1 and ϕ we considered six sets of values of (β_1, ϕ) =(.1, .1), (1.5, .1), (10, .1), (.1, .4), (1.5, .4), (10, .4) and considered each of these six sets of (β_1, ϕ) with β_0 =.05, .10(.05), ..., .90. The relative efficiency results for $\hat{\beta}_1$ are summarized in Table 16.3 and those for $\hat{\phi}$ are summarized in Table 16.4. The results for $\hat{\beta}_0$ are similar to those for $\hat{\beta}_1$.

From Table 16.1 and Table 16.3 we see that the methods $QL, M1$ and $M4$ provide high efficiency for the mean (regression) parameters. The methods QL and $M1$ provide identical results and for method $M4$ efficiency, sometimes, drops to .92. Note that all these three methods have the quasi-likelihood estimating equations for the mean (regression) parameters. The method $M2$ shows high efficiency (never dropping below .94). The method GL also produces high efficiency, although efficiency drops considerably as all parameter values become large (for example, for $\beta_0 = .8, \beta_1 = 10, \phi = .4$ efficiency drops to .80). The method $M3$ shows inconsistent behavior. Efficiency, some times is very low (for $\pi = .05, \phi = .4$, efficiency is .37) and some times very high (for $\beta_0 = .8, \beta_1 = 10, \phi = .4$ efficeincy is 1.44).

From Table 16.2 and Table 16.4 we see that efficiency of $\hat{\phi}$ by the $QL, M1$ and $M3$ methods are, in general, lower than those given by the methods $GL, M2$ and $M4$. Sometimes the efficiency drops to close to zero. Efficiency of $\hat{\phi}$ by the methods $M2$ is consistently the best. The next best appears to be the GL method. The efficiency of $\hat{\phi}$ by the method $M4$ is close to that by the GL method, slightly lower in some instances. For the joint estimation it appears that the method $M2$ (the optimal quadratic estimating equations) produces the best efficiency results. Both the GL and the $M4$ methods produce good efficiency results, although on the whole, it looks as though the method $M4$ has an edge over GL. Note that the method $M4$ combines the good behaviour of the quasi-likelihood estimating equations for the mean (regression) parameters with the GL estimating equation for the dispersion parameter. Note also that given ϕ, the quasi-likelihood estimating equations for the regression parameters β_j, j=1,...,k are optimal and given β_j, j=1,...,k the GL estimating equation for ϕ, being a likelihood equation, is optimal.

16.4   EXAMPLES

In this section we analyse three sets of real data. The first two sets of data do not involve covariates while the third set involves covariates.

Example 16.4.1 This is an example of a set of biological data. The data given in Table 16.5 from Potthoff and Whittinghill (1966) refer to crossovers in fruit flies. For this data set the estimates of π and ϕ by the different methods and the estimated relative efficiencies are given in Table 16.6. Estimates of π by all methods have high efficiencies. Estimate of ϕ by the QL method has relatively low efficiency. Estimates of ϕ by all other methods have similar and reasonably high efficiency (80%).

Example 16.4.2 This is an example of a set of toxicological data. The data given in Table 16.7 refer to proportion of affected foetuses in litters of mice in the low dose group of Paul (1982). For this data set the estimates of π and ϕ by the different methods and the estimated relative efficiencies are given in Table 16.8. Estimates of π by all methods except method $M3$ have high efficiencies. Estimates of ϕ by the GL, $M2$ and $M4$ methods have very high efficiencies (above 94%); the method $M2$ having the highest efficiency. Estimate of ϕ by the QL method has relatively lower efficiency (86%). Estimates of ϕ by the methods $M1$ and $M3$ have very low efficiencies.

Example 16.4.3 This is an example of an experiment in teratology from Shepard, Mackler and Finch (1980). This was an experiment to study the effects of chemical agents or dietary regimens on foetal development in laboratory rats. Female rats were put on iron-deficient diets and divided into four groups. One group of controls was given weekly injections of iron supplement to bring their iron intake to normal levels, while another group was given only placebo injections. Two other groups were given fewer iron-supplement injections than the controls. The rats were made pregnant, sacrificed 3 weeks later, and the total number of foetuses and the number of dead foetuses in each litter were counted. In addition, the hemoglobin levels of the mothers were measured. The data are presented in Table 16.9. Moore and Tsiatis (1991) analyse these data to select an appropriate model. A complete analysis of the data is not intended here. For illustrative purpose we take the quadratic model relating proportion dead in the logit scale to the hemoglobin level. For this data set the estimates of β_0, β_1 and ϕ by the different methods and the estimated relative efficiencies are given in Table 16.10. Clearly method $M2$ is the best. Next best, in terms of the efficiency of $\hat{\phi}$, is the method GL. In terms of efficiencies of the estimates $\hat{\beta}_0$ and $\hat{\beta}_1$ the performance of the method $M4$ is closer to that of the method $M2$.

16.5 DISCUSSION

In previous studies Crowder (1985) and Paul and Islam (1998) found that the Gaussian likelihood method is best for the estimation of the intraclass correlation ϕ. Those studies did not include estimates based on the optimal quadratic functions. In this paper we study the estimates based on the quadratic estimating functions in an unified manner. We show that the Gaussian likelihood estimates (the GL method), the estimates based on the combination of the quasi-likehood estimates for the regression parameters and the optimal quadratic estimating equation for ϕ after setting the skewness and kurtosis to zero (the $M1$ method) and the estimates based on the pseudo-likelihood estimating equations of Davidian and Carrol (1987) (the $M3$ method) are all special cases of the quadratic estimating equations. The present study of the estimates through large sample efficiency and data analyses show that the estimates based on the optimal quadratic estimating equations with the third and fourth cumulants of the beta-binomial distribution are the best, not only for the dispersion parameter but also for the joint estimation of the regression parameters and the dispersion parameter. The next best, at the cost of some loss of efficiency, are the GL or the $M4$ method. The $M4$ method has simpler estimating equations (the QL) for the regression parameters. Neither of these methods require the knowledge of the third and the fourth cumulants.

Note that the large sample efficiency results are similar for both the simulated data and the real data analysed here. Although, small sample properties of some these estimates have been found by earlier studies [Crowder (1985) and Paul and Islam (1998)] to be similar to the large sample properties it might be worthwhile to conduct a small sample study of the properties of the estimates by the quadratic estimating equations, particularly, because, the optimal quadratic estimating equations involve the third and fourth cumulants of the beta-binomial distribution.

Acknowledgements This research was partially supported by the Natural Sciences and Engineering Research Council of Canada.

REFERENCES

Breslow, N. (1984). Extra poisson variation in log-linear models, *Applied Statistics*, **33**, 38–44.

Breslow, N. (1990). Tests of hypotheses in over-dispersed poisson regression and other quasi-likelihood models, *Journal of the American Statistical Assocation*, **85**, 565–571.

Crowder, M. J. (1982). On weighted least-squares and some variants, *Surrey University Technical Report in Statistics*, No. 13.

Crowder, M. J. (1985). Gaussian estimation for correlated binomial data, *Journal of the Royal Statistical Society, Series B*, **47**, 229–237.

Crowder, M. J. (1987). On linear and quadratic estimating functions, *Biometrika*, **74**, 591–597.

Davidian, M. and and Carroll, R. J. (1987). Variance function estimation, *Journal of the Americal Statistical Association*, **82**, 1079–1091.

Dean, C. and Lawless, J. F. (1989). Discussion on an extension of quasi-likelihood estimation (by V.P. Godambe and M.E. Thompson), *Journal of Statistical Planning and Inference*, **22**, 155–158.

Elston, R. C. (1977). Response to query, Consultants corner, *Biometrics*, **33**, 232–233.

Godambe, V. P. and Thompson, M.E. (1989). An extension of quasi-likelihood estimation, *Journal of Statistical Planning and Inference*, **22**, 137–152.

Inagaki, N. (1973). Asymptotic relations between the likelihood estimating function and the maximum likelihood estimator, *Annals of the Institute of Statistical Mathematics*, **25**, 1–26.

Liang, K. Y. and Zeger, S. L. (1986). Longitudinal data analysis using generalized linear models, *Biometrika*, **73**, 13–22.

Moore, D. F. and Tsiatis, A. (1991). Robust estimation of the standard error in moment methods for extra-binomial and extra-poisson variation, *Biometrika*, **47**, 383–401.

Nelder, J. A. and Pregibon, D. (1987). An extended quasi-likelihood function, *Biometrika*, **74**, 221–232.

Paul, S. R. (1982). Analysis of proportions of affected foetuses in teratological experiments, *Biometrics*, **38**, 361–370.

Paul, S. R. and Islam, A. S. (1998). Joint estimation of the mean and dispersion parameters in the analysis of proportions: a comparison of eficiency and bias, *The Canadian Journal of Statistics*, **26**, 83–94.

Potthoff, R. F. and Whittinghill, M. (1966). Testing for homogeneity. I. The binomial and multinomial distribution, *Biometrika*, **53**, 167–182.

Prentice, R. L. (1986). Binary regression using an extended beta-binomial distribution, with discussion of correlation induced by covariate measurement errors, *Journal of the American Statistical Association*, **81**, 321–327.

Wedderburn, R. W. M. (1974). Quasi-likelihood functions, generalized linear models and the Gauss-Newton method, *Biometrika*, **61**, 439–447.

Weil, C. S. (1970). Selection of the valid number of sampling units and a consideration of their combination in toxicological studies involving reproduction, teratogenesis or carcinogenesis reproduction, teratogenesis, *Food and Cosmetics Toxicology*, **8**, 177–182.

Whittle, P. (1961). Gaussian estimation in stationary time series, *Bulletin of the International Statistical Institute*, **39**, 1–26.

Williams, D. A. (1961). The analysis of binary responses from toxicological experiments involving reproduction and teratogenecity, *Biometrics*, **31**, 949–952.

TABLE 16.1 Asymptotic relative efficiency of $\hat{\pi}$ by the $QL, GL, M1 = (QL$ and QEE combination), $M2 = QEE, M3 = (QEE$ with $\gamma_1 = \gamma_2 = 0)$ and $M4 = (QL$ and GL combination) methods; two parameter model

Parameters		Methods					
π	ϕ	QL	GL	M1	M2	M3	M4
.05	.10	1.02	.98	1.02	1.01	.47	1.02
.05	.40	.97	.93	.97	.96	.37	.97
.10	.10	1.02	.99	1.02	1.01	.77	1.00
.10	.20	1.00	.95	1.00	.98	.72	1.00
.10	.40	.98	.93	.98	.96	.69	.98
.20	.10	1.01	.99	1.01	.98	.96	1.01
.20	.40	.97	.94	.97	.95	.92	.97
.30	.10	1.00	.99	1.00	.99	.99	1.00
.30	.40	.97	.96	.97	.94	.96	.97
.40	.10	1.00	1.00	1.00	.98	.99	1.00
.40	.20	.99	.99	.99	.99	.99	.99
.40	.40	.97	.97	.97	.95	.97	.97
.60	.10	1.00	.99	1.00	.98	.99	1.00
.60	.40	.97	.97	.97	.95	.97	.97
.80	.10	1.00	.98	1.00	.98	.94	1.00
.80	.40	.98	.95	.98	.95	.93	.98

TABLE 16.2 Asymptotic relative efficiency of $\hat{\phi}$ by the $QL, GL, M1 = (QL$ and QEE combination), $M2 = QEE, M3 = (QEE$ with $\gamma_1 = \gamma_2 = 0)$ and $M4 = (QL$ and GL combination) methods; two parameter model

Parameters		Methods					
π	ϕ	QL	GL	M1	M2	M3	M4
.05	.10	.79	1.03	.38	1.04	.37	1.02
.05	.40	.84	.83	.18	.90	.13	.59
.10	.10	.88	1.08	.51	1.09	.73	1.07
.10	.20	.80	.87	.34	.90	.51	.75
.10	.40	.85	.83	.22	.90	.34	.63
.20	.10	.94	1.09	.73	1.09	1.02	1.09
.20	.40	.85	.85	.33	.90	.71	.72
.30	.10	.89	1.02	.85	1.01	1.00	1.02
.30	.40	.86	.86	.51	.91	.86	.79
.40	.10	.80	.90	.86	.90	.89	.90
.40	.20	.83	.89	.82	.90	.89	.88
.40	.40	.86	.86	.77	.93	.91	.84
.60	.10	.61	.69	.65	.68	.68	.69
.60	.40	.87	.87	.77	.92	.92	.85
.80	.10	.53	.61	.42	.62	.52	.62
.80	.40	.87	.86	.33	.92	.72	.73

SUDHIR R. PAUL

TABLE 16.3 Asymptotic relative efficiency of $\hat{\beta}_1$ by the $QL, GL, M1 = (QL$ and QEE combination), $M2 = QEE, M3 = (QEE$ with $\gamma_1 = \gamma_2 = 0$) and $M4 = (QL$ and GL combination) methods; the simple logit linear regression model

Parameters			Methods					
β_0	β_1	ϕ	QL	GL	M1	M2	M3	M4
.10	.10	.10	1.00	1.00	1.00	.99	1.02	1.00
.20			1.00	1.00	1.00	.99	1.03	1.00
.40			1.00	1.00	1.00	.99	1.06	1.00
.60			1.00	.99	1.00	.99	1.11	.99
.80			1.00	.99	1.00	.99	1.18	.99
.10	1.50	.10	1.00	1.00	1.00	.99	1.03	1.00
.20			1.00	1.00	1.00	.99	1.03	1.00
.40			1.00	1.00	1.00	.99	1.06	1.00
.60			1.00	.99	1.00	.99	1.12	.99
.80			1.00	.99	1.00	.99	1.19	.99
.10	10.0	.10	1.00	1.00	1.00	.99	1.04	1.00
.20			1.00	1.00	1.00	.99	1.05	1.00
.40			1.00	.99	1.00	.99	1.10	.99
.60			1.00	.99	1.00	.99	1.17	.99
.80			1.00	.98	1.00	.99	1.27	.98
.10	.10	.40	1.00	.97	1.00	.95	1.03	.97
.20			1.00	.93	1.00	.95	1.04	.97
.40			1.00	.86	1.00	.95	1.09	.95
.60			1.00	.82	1.00	.95	1.17	.94
.80			1.00	.81	1.00	.95	1.30	.92
.10	1.50	.40	1.00	.96	1.00	.95	1.03	.98
.20			1.00	.92	1.00	.95	1.04	.97
.40			1.00	.85	1.00	.95	1.09	.97
.60			1.00	.82	1.00	.95	1.18	.95
.80			1.00	.81	1.00	.95	1.31	.93
.10	10.0	.40	1.00	.90	1.00	.95	1.05	.97
.20			1.00	.87	1.00	.95	1.07	.97
.40			1.00	.82	1.00	.95	1.15	.95
.60			1.00	.80	1.00	.95	1.27	.94
.80			1.00	.80	1.00	.94	1.44	.92

TABLE 16.4 Asymptotic relative efficiency of $\hat{\phi}$ by the $QL, GL, M1 = (QL$ and QEE combination), $M2 = QEE, M3 = (QEE$ with $\gamma_1 = \gamma_2 = 0)$ and $M4 = (QL$ and GL combination) methods; the simple logit linear regression model

Parameters			Methods					
β_0	β_1	ϕ	QL	GL	M1	M2	M3	M4
.10	.10	.10	.83	1.00	.98	.99	.98	1.00
.20			.32	1.00	.85	.99	.96	1.00
.40			.02	.99	.23	.99	.88	.99
.60			.00	.99	.05	.98	.76	.99
.80			.00	.98	.02	.97	.64	.98
.10	1.50	.10	.79	1.00	.98	.99	.98	1.00
.20			.24	1.00	.79	.99	.95	1.00
.40			.02	.99	.20	.99	.86	.99
.60			.00	.99	.05	.98	.75	.99
.80			.00	.98	.01	.97	.62	.98
.10	10.0	.10	.45	.98	.88	.99	.93	.99
.20			.06	.99	.45	.99	.88	.99
.40			.01	.99	.08	.98	.77	.99
.60			.01	.98	.02	.98	.65	.98
.80			.00	.97	.01	.97	.53	.97
.10	.10	.40	.80	.90	.88	.95	.92	.90
.20			.25	.90	.37	.95	.83	.89
.40			.02	.90	.04	.95	.59	.88
.60			.03	.89	.01	.94	.40	.86
.80			.00	.89	.00	.94	.27	.83
.10	1.50	.40	.75	.90	.82	.95	.90	.90
.20			.18	.90	.29	.95	.80	.89
.40			.01	.89	.03	.95	.56	.88
.60			.00	.89	.01	.94	.37	.85
.80			.00	.89	.00	.94	.26	.82
.10	10.0	.40	.37	.90	.44	.95	.72	.89
.20			.04	.89	.09	.95	.61	.88
.40			.00	.89	.01	.94	.41	.86
.60			.00	.89	.00	.94	.28	.83
.80			.00	.88	.00	.94	.20	.80

TABLE 16.5 Number of the cross-over offsprings in $m = 36$ families from Potthoff and Whittinghill (1966). y = number of ++ offsprings, n = total cross-over offsprings

y:	7	1	4	3	5	3	0	11	3	0	10	3	0	4	2	2	3	5
n:	11	1	6	7	8	6	2	19	4	2	15	6	6	10	8	4	5	6
y:	2	1	2	3	1	1	4	5	3	3	5	1	1	3	4	0	1	2
n:	6	4	12	8	4	5	5	6	4	10	8	11	4	4	4	2	2	3

TABLE 16.6 The estimates $\hat{\pi}$ and $\hat{\phi}$ and their estimated relative efficiencies by the $ML, QL, GL, M1 = (QL$ and QEE combination), $M2 = QEE, M3 = (QEE$ with $\gamma_1 = 0, \gamma_2 = 0)$ and $M4 = (QL$ and GL combination) methods for the cross-over data

Methods	Estimates of		Estimated relative	
	π	ϕ	efficiency for	
			$\hat{\pi}$	$\hat{\phi}$
ML	0.4728	0.0950	1.0000	1.0000
QL	0.4742	0.1094	0.9449	0.6655
GL	0.4741	0.0953	0.9940	0.8041
M1	0.4741	0.0944	0.9975	0.8044
M2	0.4741	0.0959	0.9921	0.8022
M3	0.4742	0.0944	0.9973	0.8045
M4	0.4741	0.0959	0.9921	0.8016

TABLE 16.7 The toxicological data of law dose group from Paul (1982). $m = 19$ litters. y = number of live foetuses affected by treatment, n = total of live foetuses

y:	0	1	1	0	2	0	1	0	1	0	0	3	0	0	1	5	0	0	3
n:	5	11	7	9	12	8	6	7	6	4	6	9	6	7	5	9	1	6	9

TABLE 16.8 The estimates $\hat{\pi}$ and $\hat{\phi}$ and their estimated relative efficiencies by the $ML, QL, GL, M1 = (QL$ and QEE combination$)$, $M2 = QEE$, $M3 = (QEE$ with $\gamma_1 = 0, \gamma_2 = 0)$ and $M4 = (QL$ and GL combination$)$ methods for the toxicology data

Methods	Estimates of		Estimated relative efficiency for	
	π	ϕ	$\hat{\pi}$	$\hat{\phi}$
ML	0.1272	0.1054	1.0000	1.0000
QL	0.1275	0.1006	1.0158	0.8556
GL	0.1331	0.1006	0.9709	0.9875
M1	0.1265	0.1225	0.9438	0.4527
M2	0.1273	0.1055	0.9988	0.9282
M3	0.1109	0.1576	0.8034	0.4535
M4	0.1277	0.1055	0.9958	0.9431

TABLE 16.9 Low-iron rat teratology data. N Denotes the litter size, R the number of dead foetuses, HB the hemoglobin level, and GRP the group number. Group 1 is the untreated (low-iron) group, group 2 received injections on day 7 or day 10 only, group 3 received injections on days 0 and 7, and group 4 received injections weekly

N	R	HB	GRP	N	R	HB	GRP
10	1	4.1	1	9	7	3.1	1
11	4	3.2	1	14	14	3.6	1
12	9	4.7	1	12	7	4.1	1
4	4	3.5	1	11	9	4.8	1
10	10	3.2	1	13	8	4.7	1
11	9	5.9	1	14	5	4.8	1
9	9	4.7	1	10	10	6.7	1
11	11	4.7	1	12	10	5.2	1
10	10	3.5	1	13	8	4.3	1
10	7	4.8	1	10	10	3.9	1
12	12	4.3	1	14	3	6.3	1
10	9	4.1	1	13	13	4.4	1
8	8	3.2	1	4	3	5.2	1
11	9	6.3	1	8	8	3.9	1
6	4	4.3	1	13	5	7.7	1

N	R	HB	GRP	N	R	HB	GRP
12	12	5.0	1	14	0	12.6	3
10	1	8.6	2	14	1	9.5	3
3	1	11.1	2	11	0	9.8	3
13	1	7.2	2	3	0	16.6	4
12	0	8.8	3	13	0	14.5	4
14	4	9.3	2	9	2	15.4	4
9	2	9.3	2	17	2	14.5	4
13	2	8.5	2	15	0	14.6	4
16	1	9.4	2	2	0	16.5	4
11	0	6.9	2	14	1	14.8	4
4	0	8.9	2	8	0	13.6	4
1	0	11.1	2	6	0	14.5	4
12	0	9.0	2	17	0	12.4	4
8	0	11.2	3				
11	1	11.5	3				

TABLE 16.10   The estimates $\hat{\beta}_0, \hat{\beta}_1$ and $\hat{\phi}$ and their estimated relative efficiencies by the $ML, QL, GL, M1 = (QL$ and GL combination), $M2 = QEE, M3 = (QEE$ with $\gamma_1 = 0, \gamma_2 = 0)$ and $M4 = (QL$ and GL combination) methods for the low-iron rat teratology data

Methods	Estimates of			
	β_0	β_1	β_2	π
ML	6.0158	-1.2512	0.0457	0.2793
QL	5.5728	-1.1344	0.0363	0.2710
GL	5.8247	-1.3005	0.0491	0.2843
M1	5.5738	-1.1389	0.0363	0.2644
M2	5.8566	-1.2210	0.0433	0.2956
M3	5.4828	-1.1437	0.0339	0.3016
M4	5.5818	-1.1381	0.0365	0.2200

Methods	Estimated relative efficiencies for			
	$\hat{\beta}_0$	$\hat{\beta}_1$	$\hat{\beta}_2$	$\hat{\phi}$
ML	1.0000	1.0000	1.0000	1.0000
QL	0.8043	0.7218	0.6178	0.5588
GL	0.7023	0.6454	0.6273	0.7312
M1	0.2865	0.1610	0.0766	0.0229
M2	0.9012	0.8718	0.8399	0.8618
M3	0.2548	0.1894	0.1440	0.3607
M4	0.9238	0.8280	0.7074	0.5844

CHAPTER 17

SEMIPARAMETRIC LOCATION-SCALE REGRESSION MODELS FOR SURVIVAL DATA

XUEWEN LU

Agriculture and Agri-Food Canada, Guelph, Ontario, Canada

R. S. SINGH

University of Guelph, Guelph, Ontario, Canada

Abstract: In survival analysis, a special class of accelerated failure time models has a log form: $\log T = \psi(Z, X) + \sigma\varepsilon$, where T is a random variable denoting the event time, (Z, X) are covariates, $\psi(\cdot, \cdot)$ is a regression function, ε is a random disturbance term. This type of model is also called location – scale model for log lifetime T, where $\psi(Z, X)$ is the location term and σ is the scale parameter. In this paper, covariates are modeled as $\psi(Z, X) = Z^T\beta + \lambda(X)$, Z may be vector-valued, X is a univariate. β is an unknown parameter vector, λ takes value in a real line and is an unknown smooth function. Hence, the relationship between response and covariates is modeled semiparametrically, the conventional maximum likelihood is not directly applicable to estimate the parametric components. This paper uses the method of Severini and Wong (1992, *Annals of Statistics*, 20, 1768–1802) to construct asymptotically efficient estimators of the parametric component and to specify their asymptotic distributions. An application to the Primary Biliary Cirrhosis Data is provided.

Keywords and phrases: Asymptotic, accelerated failure time model, censored data, generalized profile likelihood, location-scale model, semiparametric regression

305

17.1   INTRODUCTION

The accelerated failure time model is an important class of regression models in survival analysis. A number of these models have a log form:

$$\log T = \psi(X) + \sigma\varepsilon, \tag{17.1.1}$$

where ε is a random disturbance term, T is a random variable denoting the event time. $\psi(X)$ is a regréssion function, σ is an unknown parameter. This model is also called location-scale model for log lifetime T, where $\psi(X)$ is the location term, σ is the scale parameter. When $\psi(X)$ has a parametric from, for example, $\psi(X) = X^T\beta$, it is well known that one can use the maximum likelihood estimation to analyze this model, Kalbfleisch and Prentice (1980) and Lawless (1982) give a detailed introduction on the parametric regression model in the context of survival analysis. Most statistical softwares, for example, SAS and Splus can analyze some standard accelerated failure time models. In SAS, the LIFEREG procedure produces estimates of five types of models, it allows for five distributions for ε: normal, extreme value (2 parameter), extreme value (1 parameter), log-gamma and logistic. the function SURVREG in Splus does the similar things. But these packages lack the ability to analyze the model in which the regression function $\psi(X)$ is not parametrized to a linear form or it contains unknown regression functions.

   In the situation of censored data, when no assumptions are made about the form of the regression relationship and the distribution of ε, in order to estimate the functional form of ψ, Fan and Gijbels (1994) use nonparametric regression techniques to transform the observed data in an appropriate way and then apply a locally weighted least squares regression. In this paper, we introduce the following semiparametric regression model,

$$\psi(Z, X) = Z^T\beta + \lambda(X). \tag{17.1.2}$$

In fact, this is a partly linear model. Using this model, rewrite (17.1.1) as

$$Y = Z^T\beta + \lambda(X) + \sigma\varepsilon, \tag{17.1.3}$$

where $Y = \log T$ is log lifetime. We assume that $\lambda(x)$ is a smooth function of x from $\mathbb{R}$ to $\mathbb{R}$ and σ is an unknown scale parameter. β is an unknown $\rho \times 1$ parameter vector. (Z, X) is a vector of explanatory variables. The variable X is continuous with values in a closed interval $\mathcal{X} \in \mathbb{R}$, and Z is discrete or continuous with values in $\mathbb{R}^\rho$.

   The paper is organized as follows. Section 17.2 of the paper introduces the likelihood for the location-scale model under censoring. Section

17.3 describes the generalized profile likelihood and investigates the finite sample behavior of the maximum generalized profile likelihood estimator. Section 17.4 presents three types of semiparametric location-scale regression models with the computation given in the Appendix. Finally, Section 17.5 illustrates the performance of the proposed procedure via analysis of the Primary Biliary Cirrhosis data.

17.2   LIKELIHOOD FUNCTION FOR THE PARAMETRIC LOCATION-SCALE MODELS

Assume that $S_0(\cdot)$ is the survivor function for ε, $f_0(\cdot)$ is its p.d.f., $\lambda(x)$ is parametrized as $\lambda(x) = \lambda(x; \gamma)$. Then the survivor function for Y, given $(Z, X) = (z, x)$, is of the form $S_0(\omega)$, where $\omega = [y - (z^T\beta + \lambda(x; \gamma))]/\sigma$. The p.d.f. for $Y = \log T$ can be written

$$\sigma^{-1} f_0(\omega).$$

Let $f(t|z, x)$ denote the conditional density function of Y given $(Z, X) = (z, x)$, and let $S(t|z, x) = P\{Y > t|(Z, X) = (z, x)\}$ be its conditional survivor function. The conditional distribution function of censoring random variable C given $(Z, X) = (z, x)$ is denoted by $G(t|z, x)$. Then under independent and *noninformative censoring* (i.e., $G(t|z, x)$ does not involve the parameters σ, β and γ), the conditional likelihood function is given by

$$L = \prod_u f(Y_i|Z_i, X_i) \prod_c S(Y_i|Z_i, X_i), \qquad (17.2.4)$$

where $\prod_u$ and $\prod_c$ denote respectively the products involving the uncensored and the censored observations. Let δ_i represent the censoring indicator, *i.e.* $\delta_i = [T_i \leq C_i]$, and Y_i represent a log lifetime or a log censoring time for i^{th} individual, *i.e.* $Y_i = \min(\log T_i, \log C_i)$, the likelihood function can be written as

$$L(\theta, \gamma) = \prod_1^n [\sigma^{-1} f_0(\omega_i)]^{\delta_i} S_0(\omega_i)^{1-\delta_i}, \qquad (17.2.5)$$

where $\theta = (\beta^T, \sigma)^T$, $\omega_i = [y_i - (z_i^T\beta + \lambda(x_i; \gamma))]/\sigma$. Under the location-scale model (17.1.3) with $\lambda(x) = \lambda(x; \gamma)$, we have the log-likelihood for the sample

$$L_n(\theta, \gamma) = \sum_i [\delta_i \log f(Y_i|Z_i, X_i) + (1 - \delta_i) \log S(Y_i|Z_i, X_i)]$$

$$= \sum_i \{\delta_i[-\log(\sigma) + \log(f_0(\omega_i)] + (1 - \delta_i)S_0(\omega_i)\}.$$

$$(17.2.6)$$

Maximization of (17.2.6) leads to the maximum likelihood estimators of σ, β and γ.

17.3   GENERALIZED PROFILE LIKELIHOOD

If $\lambda(x)$ is not parametrized, to estimate the parameter θ and the nonparametric smooth function $\lambda(x)$, we apply the generalized profile likelihood method of Severini and Wong (1992) and Stein (1956). This method is applied by Severini and Staniswalis (1994) in studying the quasi-likelihood estimation in semiparametric models and by Staniswalis, Thall and Salch (1997) in semiparametric regression analysis for recurrent event interval counts. Generalized profile likelihoods are an extension to semiparametric models of the profile likelihood for parametric models. The resulting estimator of θ converges to the true parameter value at a $\sqrt{n}$ rate and is asymptotically efficient for conditionally parametric models. $\hat{\lambda}_{\hat{\theta}}(x)$ is uniformly consistent for $\lambda(x)$.

17.3.1   Application of Generalized Profile Likelihood to Semiparametric Location-Scale Regression Models

Let Z, X and (Y, δ) be random variables such that the distribution of (Y, δ) conditional on $(Z, X) = (z, x)$ is $f((y, \delta); z, \phi) = [\sigma^{-1}f_0(\omega)]^{\delta}S_0(\omega)^{1-\delta}$, where $\omega = [y - (z^T\beta + \lambda(x))]/\sigma$, $\phi = (\theta, \lambda(x))$, $\theta = (\beta^T, \sigma)^T$ and $\lambda(x)$ is a smooth function of x. It is assumed that the joint distribution of the explanatory variables Z and X does not contain information about θ or λ. Severini and Wong (1992) refer to such a semiparametric model as conditionally parametric because, conditional on $(Z, X) = (z, x)$, the model for (Y, δ) is parametrized by a finite-dimensional parameter ϕ.

Set $l = \log(f)$, Severini and Wong (1992, pp. 1773–1774) show that the least favorable curve λ_{θ} maximizes

$$M(\lambda; \theta, x) = E_0(l((Y, \delta); Z, \theta, \lambda)|X = x)$$

with respect to λ. Here, E_0 denotes expectation using the true parameter value. An estimator of $M(\lambda; \theta, x)$ is constructed using nonparametric smoothing. Severini and Wong (1992) propose the following estimator of

$M(\lambda; \theta, x)$,

$$\hat{M}(\lambda; \theta, x) = [\sum_{i=1}^{n} W_b(X_i - x) l((Y_i, \delta_i); Z_i, \theta, \lambda)] / \sum_{i=1}^{n} W_b(X_i - x), \quad (17.3.7)$$

where $(Z_i, X_i, (Y_i, \delta_i))$; $i = 1, \ldots, n$ are i.i.d. sample from the population $(Z, X, (Y, \delta))$, $W(\cdot)$ is a nonnegative weight function, $W_b(\cdot) = (1/b)W(\cdot/b)$, b is the bandwidth. Then by maximizing $\hat{M}(\lambda; \theta, x)$ with respect to λ for each fixed θ and x, an estimator $\hat{\lambda}_\theta(x)$ of the least favorable curve is obtained. The estimator $\hat{\theta}$ that maximizes the generalized profile likelihood

$$L_n(\theta, \hat{\lambda}_\theta) = \sum_{i=1}^{n} l((Y_i, \delta_i); Z_i, \theta, \hat{\lambda}_\theta(X_i))$$

with respect to θ is obtained as the solution of

$$\begin{aligned}
\frac{d}{d\theta} L_n(\theta, \hat{\lambda}_\theta) &= \sum_{i=1}^{n} \left[\frac{\partial}{\partial \theta} l((Y_i, \delta_i); Z_i, \theta, \lambda) \Big|_{\lambda = \hat{\lambda}_\theta(X_i)} \right. \\
&\quad \left. + \frac{\partial}{\partial \lambda} l((Y_i, \delta_i); Z_i, \theta, \lambda) \Big|_{\lambda = \hat{\lambda}_\theta(X_i)} \frac{d}{d\theta} \hat{\lambda}_\theta(X_i) \right] \\
&= 0. \quad (17.3.8)
\end{aligned}$$

Under certain regularity conditions on the likelihood and the nonparametric smoother, Severini and Wong (1992) establish that this estimator of θ is asymptotically efficient for conditionally parametric models and $\hat{\lambda}_\theta(x)$ is uniformly consistent for $\lambda(x)$. They note that their results can easily be extended to allow for multidimensional X, θ.

17.3.2 Estimation and Large Sample Properties

The empirical version of $M(\lambda; \theta, x)$ given in (17.3.7) is a consistent estimation of $M(\lambda; \theta, x)$, it is proportional to a local likelihood for estimating $\lambda_\theta(x)$. For fixed θ, let $\hat{\lambda}_\theta$ be the maximizer of (17.3.7). Substituting $\hat{\lambda}_\theta$ into the log-likelihood (17.2.6), we obtain the generalized profile likelihood $L_n(\theta, \hat{\lambda}_\theta)$ for θ. Let $\hat{\theta}$ be the maximizer of $L_n(\theta, \hat{\lambda}_\theta)$ with respect to θ. Then $\hat{\theta}$ is an estimator of the true parameter value θ_0.

We now discuss the large sample properties of $\hat{\theta}$. For establishing the consistency and asymptotic normality of $\hat{\theta}$, the estimator $\hat{\lambda}_\theta$ must be a consistent estimator of a least favorable curve and must satisfy the nuisance parameter (NP) conditions of Severini and Wong (1992, p. 1779). These can be verified without difficulties by an application of their Lemma 5.

We now establish the consistency and asymptotic normality of $\hat{\theta}$.

Theorem 17.3.1 *Let $\hat{\theta}$ be any element of Θ satisfying*

$$L_n(\hat{\theta}, \hat{\lambda}_{\hat{\theta}}) = \sup_{\theta \in \Theta} L_n(\theta, \hat{\lambda}_\theta),$$

where Θ is a compact subset of $\mathbb{R}^p \times \mathbb{R}^+$ (since $\sigma > 0$). Then, under the regularity conditions provided by Severini and Wong (1992),

$$\hat{\theta} \xrightarrow{P} \theta_0.$$

Theorem 17.3.2 *Under the regularity conditions given above,*

$$\sqrt{n}(\hat{\theta} - \theta_0) \xrightarrow{D} N(0, i_\theta^{-1}),$$

where i_θ is the marginal Fisher information matrix for θ given by

$$i_\theta = E_0 \left(\frac{\partial l}{\partial \theta}(\theta_0, \lambda) + \frac{\partial l}{\partial \lambda}(\theta_0, \lambda)(v^*) \right)^{\otimes 2},$$

$v^* = \lambda'_{\theta_0}$ is the least favorable direction. i_θ can be consistently estimated by $\hat{i}_\theta = -\frac{1}{n} \frac{\partial^2}{\partial \theta \partial \theta^T} L_n(\theta, \hat{\lambda}_\theta)\big|_{\theta=\hat{\theta}}$.

Theorem 17.3.1 and Theorem 17.3.2 follow from Propositions 1 and 2 of Severini and Wong (1992). From their NP conditions along with the fact that $\hat{\theta} = \theta_0 + O_P(n^{-1/2})$, we obtain following result on the estimation of nonparametric component:

Theorem 17.3.3 $\| \hat{\lambda}_{\hat{\theta}} - \lambda \| = o_P(n^{-1/4})$.

17.4   EXAMPLES OF SEMIPARAMETRIC LOCATION-SCALE REGRESSION MODELS

Example 1: Extreme value and Weibull regression models.

If the *p.d.f* of lifetime T, given $(Z, X) = (z, x)$, is of the form

$$\frac{\kappa}{\alpha(z, x)} \left(\frac{t}{\alpha(z, x)} \right)^{\kappa-1} \exp \left[-\left(\frac{t}{\alpha(z, x)} \right)^\kappa \right] \quad t \geq 0,$$

which is a Weibull model, then the *p.d.f* of $Y = \log T$, given (z, x), is

$$f(y|(z, x)) = \frac{1}{\sigma} \exp \left[\frac{y - \mu(z, x)}{\sigma} - \exp \left(\frac{y - \mu(z, x)}{\sigma} \right) \right] \quad -\infty < y < \infty,$$

where $\mu(z,x) = \log[\alpha(z,x)] = Z^T\beta + \lambda(x)$, $\sigma = 1/\kappa$. This model can be written as

$$Y = Z^T\beta + \lambda(X) + \sigma\varepsilon,$$

where ε has a standard extreme value distribution with $p.d.f$ $f_0(s) = \exp(s - e^s)$, $-\infty < s < \infty$. We have $S_0(s) = \exp(-e^s)$, then

$$a_0(s) = 1 - \exp(s), \quad a_1(s) = a_0(s) - 1, \quad a_2(s) = a_0(s) - 1;$$

and

$$b_0(s) = \exp(s), \quad b_1(s) = b_0(s), \quad b_2(s) = b_0(s),$$

where we define $a_0(s) = \frac{f_0'(s)}{f(s)}$, $b_0(s) = \frac{f_0(s)}{1-S_0(s)}$; $a_1(s) = a_0'(s)$, $b_1(s) = b_0'(s)$; $a_2(s) = a_0''(s)$, $b_2(s) = b_0''(s)$. In this case, equation $\phi_1(\eta;\theta,x) = 0$ in Step a of computations (see Appendix) may be solved explicitly to obtain

$$\hat{\lambda}_\theta(x) = \sigma \log \left\{ \frac{\sum W_b(X_j - x) \exp(\frac{Y_j - Z_j^T\beta}{\sigma})}{\sum W_b(X_j - x)\delta_j} \right\},$$

then an iterative approach of (17.5.9) given in Appendix yields $\hat{\theta}$.

Example 2: Normal and log-normal regression models.

If we consider regression models in which lifetime T is log-normal, then log lifetime $Y = \log T$ is normally distributed, ε has a standard normal distribution, $f_0(s) = \frac{1}{\sqrt{2\pi}} \exp\{-\frac{1}{2}s^2\}$, $S_0(s) = 1 - \Phi(s)$. Hence we have

$$a_0(s) = -s, \quad a_1(s) = -1, \quad a_2(s) = 0;$$

and

$$b_0(s) = \frac{f_0(s)}{1-\Phi(s)}, \quad b_1(s) = -sb_0(s) + b_0^2(s),$$

$$b_2(s) = b_0(s)[(b_0(s) - s)(2b_0(s) - s) - 1].$$

Example 3: Logistic and log-logistic regression models.

If we consider regression models in which the distribution of lifetime T is log-logistic, then distribution of log lifetime $Y = \log T$ is logistic. ε has a logistic distribution with $p.d.f$ and survivor function

$$f_0(s) = \frac{\exp(s)}{(1 + \exp(s))^2}, \qquad S_0(s) = \frac{1}{1 + \exp(s)}.$$

Hence, we have

$$a_0(s) = \frac{1 - \exp(s)}{1 + \exp(s)}, \qquad b_0(s) = \frac{\exp(s)}{1 + \exp(s)};$$

$$a_1(s) = \frac{-2\exp(s)}{(1 + \exp(s))^2}, \qquad b_1(s) = \frac{\exp(s)}{(1 + \exp(s))^2};$$

and

$$a_2(s) = a_0(s)a_1(s), \qquad b_2(s) = a_0(s)b_1(s).$$

17.5   AN EXAMPLE WITH CENSORED SURVIVAL DATA: PRIMARY BILIARY CIRRHOSIS (PBC) DATA

The data set is found in Appendix D of Fleming and Harrington (1991). Between January 1974 and May 1984, the Mayo Clinic collected data on PBC, a rare but fatal Chronic liver disease of unknown cause of the 412 registered patients, the first 312 cases participated in the randomized trial, and contain largely complete data; Our analysis is based on those patients. A more extended discussion can be found in Fleming and Harrington (1991, Chap. 4). In this study, the response T is the time (in days) between registration and death, or liver transplantation or time of the study analysis (July 1986), we study the effects of AGE (in months), Log(ALBUMIN), Log(BILIRUBIN), Log(PROTIME) (PROTIME=Prothrombin time) and EDAME. Fleming and Harrington (1991) find that the model with these five variables is biologically reasonable.

We fit semiparametric Log-normal, Log-logistic and Weibull models respectively. We use the quartic kernel

$$W(v) = (15/16)(1 - v^2)^2 I[-1, 1](v)$$

with the boundary modification of Rice (1984). the bandwidth is chosen by visual inspection. Table 17.1 presents the results for the estimates of parameters. Figures 17.1–17.3 report the functional form for AGE and the linear lines fitted by the fully parametric models. These three models give similar results. Relying on the shapes of the semiparametric model estimates of $\lambda(\cdot)$ in Figure 1-3, we find that a linear function of AGE provides a reasonable fit. Thus, the parametric model $\log T$ is $\psi(Z, X) = \beta_0 + \beta_1 * \text{AGE} + \beta_2 * \log(\text{ALBUMIN}) + \beta_3 * \log(\text{BILIRUBIN}) + \beta_4 * \text{EDEMA} + \beta_5 * \log(\text{PROTIME})$. One can see in the Table 17.1, the numerical values of the estimates of parameters common to both parametric and semiparametric models are very similar. However, the standard errors of the parameter estimates are slightly different, suggesting that we may have not completely succeeded in capturing the true shape of the functional form for AGE. Note the shape of the estimated curve for AGE, we see that there are two peaks and three troughs, which are shifted in time, the overall behave of this curve is that it decreases with aging. This reflects the fact that the log-survival time decreases with aging, but it may increase locally with aging.

REFERENCES

Fan, J. and Gijbels, I. (1994). Censored regression: local linear approximations and their applications, *Journal of the American Statistical Association*, **89**, 560–570.

Fleming, T. R. and Harrington, D. P. (1991). *Counting Processes and Survival Analysis*, John Wiley & Sons, New York.

Kalbfleisch, J. D. and Prentice, R. L. (1980). *The Statistical Analysis of Failure Time Data*, John Wiley & Sons, New York.

Lawless, J. F. (1982). *Statistical Models and Methods for Lifetime Data*, John Wiley & Sons, New York.

Rice, J. (1984). Convergence rates for partially splined models, *Statistics & Probability Letters*, **4**, 203–209.

Severini, T. A. and Staniswalis, J. G. (1994). Quasi–likelihood estimation in semiparametric models, *Journal of the American Statistical Association*, **89**, 501–511.

Severini, T. A. and Wong, W. H. (1992). Generalized profile likelihood and conditionally parametric models, *Annals of Statistics*, **20**, 1768–1802.

Staniswalis, J. G., Thall, P. F. and Salch, J. (1997). Semiparametric regression analysis for recurrent event interval counts, *Biometrics*, **53**, 1334–1353.

Stein, C. (1956). Efficient nonparametric testing and estimation. *Proceedings of the Third Berkeley Symposium on Mathematics, Statistics and Probability*, **1**, pp. 187–196, University California Press, Berkeley.

APPENDIX: COMPUTATION OF THE ESTIMATES

Please note that the method provided in previous sections may be computationally intensive since each evaluation of $L_n(\theta, \hat{\lambda}_\theta)$ requires a separate maximization of $\hat{M}_n(\eta; \theta, x)$ for each $X = x_j$, $j = 1, \ldots, n$. The computation of an estimator $\hat{\theta}$ also involves tedious work. If $\hat{\lambda}_\theta$ has a closed-form, the procedure is very much simplified, this is seen in Section 17.4 from the Extreme Value Regression Model.

We now establish an algorithm for computing the estimates of $\theta = (\beta^T, \sigma)^T$ and λ. Let $l((Y_j, \delta_i); \theta, \eta) = \log f((Y_j, \delta_i); \theta, \eta) = \delta_j[\log f_0(\omega_{\theta j}(\eta)] + (1 - \delta_j) \log[1 - S_0(\omega_{\theta j}(\eta))] - \delta_j \log \sigma$, $\omega_{\theta j}(\eta) = \frac{Y_j - (Z_j^T \beta + \eta)}{\sigma}$. Define

$$\phi_1(\eta; \theta, x) = \sum_j W_b(X_j - x) \frac{\partial}{\partial \eta} l((Y_j, \delta_j); \theta, \eta),$$

and

$$\phi_{2\beta}(\theta) = \sum_j \frac{\partial}{\partial \beta} l((Y_j, \delta_j); \theta, \hat{\lambda}_\theta(X_j)),$$

$$\phi_{2\sigma}(\theta) = \sum_j \frac{\partial}{\partial \sigma} l((Y_j, \delta_j); \theta, \hat{\lambda}_\theta(X_j)),$$

$$\phi_2(\theta, \hat{\lambda}_\theta) = (\phi_{2\beta}^T(\theta), \phi_{2\sigma}(\theta))^T.$$

Then $\theta = (\beta^T, \sigma)^T$ and λ are estimated by the following procedure:

a. For each x, θ, calculate $\hat{\lambda}_\theta(x)$ by solving $\phi_1(\eta; \theta, x) = 0$ for η.

b. Estimate θ by solving $\phi_2(\theta, \hat{\lambda}_\theta) = 0$ for θ.

c. Estimate λ by $\hat{\lambda}_{\hat{\theta}}$.

We consider step b for calculating $\hat{\theta}$. Let $\mathcal{H}$ denote the $(\rho + 1) \times (\rho + 1)$ matrix by

$$\frac{\partial \phi_2(\theta, \hat{\lambda}_\theta)}{\partial \theta^T}.$$

Using Newton-Raphson iterative method, an initial estimate $\hat{\theta}$ can be updated to $\hat{\theta}^*$ using

$$\hat{\theta}^* = \hat{\theta} - \mathcal{H}^{-1}(\hat{\theta}) \phi_2(\hat{\theta}, \hat{\lambda}_{\hat{\theta}}). \tag{17.5.9}$$

This iteration can be continued until convergence. The estimated asymptotic covariance matrix of $\hat{\theta}$ can be determined using $-\mathcal{H}^{-1}(\hat{\theta})$.

Denote by $\hat{\lambda}'_{k\theta} = \partial\hat{\lambda}_\theta(x)/\partial\theta_k$ and $\hat{\lambda}''_{kl\theta}(x) = \partial^2\hat{\lambda}_\theta(x)/\partial\theta_k\partial\theta_l$. Usually, iterative methods are also needed to calculate $\hat{\lambda}_\theta(x)$, $\hat{\lambda}'_{k\theta}(x)$ and $\hat{\lambda}''_{kl\theta}(x)$. For fixed θ and x, consider step a, solving $\phi_1(\eta;\theta,x) = 0$ for $\eta = \hat{\lambda}_\theta(x)$. Because

$$\phi_1(\eta;\theta,x) = \sum_j W_b(X_j - x)\{\delta_j a_0(\omega_{\theta j}(\eta)) - (1-\delta_j)b_0(\omega_{\theta j}(\eta))\}(-\frac{1}{\sigma}),$$

and

$$\frac{\partial\phi_1}{\partial\eta}(\eta;\theta,x) = \sum_j W_b(X_j - x)\{\delta_j a_1(\omega_{\theta j}(\eta)) - (1-\delta_j)b_1(\omega_{\theta j}(\eta))\}(\frac{1}{\sigma})^2.$$

Then an initial estimate η_0 can be updated to η by Newton-Raphson method:

$$\eta = \eta_0 - \phi_1(\eta_0;\theta,x)/\frac{\partial\phi_1}{\partial\eta}(\eta_0;\theta,x). \qquad (17.5.10)$$

As for calculating $\hat{\lambda}'_{k\theta}$ and $\hat{\lambda}''_{kl\theta}(x)$, since $\hat{\lambda}_\theta(x)$ satisfies $\phi_1(\hat{\lambda}_\theta;\theta,x) = 0$ for all θ, x, it follows that

$$\frac{\partial\phi_1(\hat{\lambda}_\theta(x);\theta,x)}{\partial\theta} = 0, \quad \frac{\partial^2\phi_1(\hat{\lambda}_\theta(x);\theta,x)}{\partial\theta\partial\theta^T} = 0, \quad \text{for all } \theta,x. \qquad (17.5.11)$$

Denote $\omega^x_{\theta j} = \omega_{\theta j}(\hat{\lambda}_\theta(x)) = \frac{Y_j - (Z_j^T\beta + \hat{\lambda}_\theta(x))}{\sigma}$, noticing that

$$\frac{\partial\omega^x_{\theta j}}{\partial\beta} = (-\frac{1}{\sigma})(Z_j^T + \frac{\partial\hat{\lambda}_\theta}{\partial\beta}(x))$$

and

$$\frac{\partial\omega^x_{\theta j}}{\partial\sigma} = (-\frac{1}{\sigma})(\omega^x_{\theta j} + \frac{\partial\hat{\lambda}_\theta}{\partial\sigma}(x)),$$

then

$$\phi_1(\hat{\lambda}_\theta;\theta,x) = \sum_j W_b(X_j - x)\{\delta_j a_0(\omega^x_{\theta j}) - (1-\delta_j)b_0(\omega^x_{\theta j})\}(-\frac{1}{\sigma}) = 0.$$

Hence, for $k,l = 1,\ldots,\rho$, we have

$$\frac{\partial\phi_1(\hat{\lambda}_\theta(x);\theta,x)}{\partial\beta_k}$$

$$= \sum_j W_b(X_j - x)\{\delta_j a_1(\omega^x_{\theta j}) - (1-\delta_j)b_1(\omega^x_{\theta j})\}[\frac{Z_{jk} + \hat{\lambda}'_{k\theta}(x)}{\sigma^2}]$$

$$= 0,$$

$$\frac{\partial \phi_1(\hat{\lambda}_\theta(x); \theta, x)}{\partial \sigma}$$

$$= \sum W_b(X_j - x)\{\delta_j a_1(\omega_{\theta j}^x) - (1 - \delta_j)b_1(\omega_{\theta j}^x)\}[\frac{\omega_{\theta j}^x + \hat{\lambda}'_{(r+1)\theta}(x)}{\sigma^2}]$$

$$= 0$$

and

$$\frac{\partial^2 \phi_1(\hat{\lambda}_\theta(x); \theta, x)}{\partial \beta_k \partial \beta_l}$$

$$= \sum W_b(X_j - x)\{\delta_j a_2(\omega_{\theta j}^x) - (1 - \delta_j)b_2(\omega_{\theta j}^x)\}[Z_{jk} + \hat{\lambda}'_{k\theta}(x)]$$

$$\times [Z_{jl} + \hat{\lambda}'_{l\theta}(x)](-\frac{1}{\sigma})^3$$

$$+ \sum W_b(X_j - x)\{\delta_j a_1(\omega_{\theta j}^x) - (1 - \delta_j)b_1(\omega_{\theta j}^x)\}[\hat{\lambda}''_{kl\theta}(x)](\frac{1}{\sigma})^2$$

$$= 0,$$

$$\frac{\partial^2 \phi_1(\hat{\lambda}_\theta(x); \theta, x)}{\partial \beta_k \partial \sigma} = \frac{\partial^2 \phi_1(\hat{\lambda}_\theta(x); \theta, x)}{\partial \sigma \partial \beta_k}$$

$$= \sum W_b(X_j - x)\{\delta_j a_2(\omega_{\theta j}^x) - (1 - \delta_j)b_2(\omega_{\theta j}^x)\}[\omega_{\theta j}^x + \hat{\lambda}'_{(r+1)\theta}(x)]$$

$$\times [Z_{jk} + \hat{\lambda}'_{k\theta}(x)](-\frac{1}{\sigma})^3$$

$$+ \sum W_b(X_j - x)\{\delta_j a_1(\omega_{\theta j}^x) - (1 - \delta_j)b_1(\omega_{\theta j}^x)\}[\hat{\lambda}''_{k(r+1)\theta}(x)](\frac{1}{\sigma})^2$$

$$= 0,$$

$$\frac{\partial^2 \phi_1(\hat{\lambda}_\theta(x); \theta, x)}{\partial \sigma^2}$$

$$= \sum W_b(X_j - x)\{\delta_j a_2(\omega_{\theta j}^x) - (1 - \delta_j)b_2(\omega_{\theta j}^x)\}$$

$$\times [\omega_{\theta j}^x + \hat{\lambda}'_{(r+1)\theta}(x)]^2(-\frac{1}{\sigma})^3$$

$$+ \sum W_b(X_j - x)\{\delta_j a_1(\omega_{\theta j}^x) - (1 - \delta_j)b_1(\omega_{\theta j}^x)\}[\hat{\lambda}''_{(r+1)(r+1)\theta}(x)](\frac{1}{\sigma})^2$$

$$= 0.$$

Solving these equations, we get, for $k, l = 1, \ldots, \rho$,

$$\hat{\lambda}'_{k\theta}(x) = -\frac{\sum W_b(X_j - x)\{\delta_j a_1(\omega_{\theta j}^x) - (1 - \delta_j)b_1(\omega_{\theta j}^x)\}Z_{jk}}{\sum W_b(X_j - x)\{\delta_j a_1(\omega_{\theta j}^x) - (1 - \delta_j)b_1(\omega_{\theta j}^x)\}},$$

$$(17.5.12)$$

$$\hat{\lambda}'_{(r+1)\theta}(x) \;=\; -\frac{\sum W_b(X_j - x)\{\delta_j a_1(\omega^x_{\theta j}) - (1 - \delta_j)b_1(\omega^x_{\theta j})\}\omega_{\theta j}}{\sum W_b(X_j - x)\{\delta_j a_1(\omega^x_{\theta j}) - (1 - \delta_j)b_1(\omega^x_{\theta j})\}},$$

$$(17.5.13)$$

$$\hat{\lambda}''_{kl\theta}(x)$$
$$= \frac{1}{\sigma}\frac{\sum W_b(X_j - x)\{\delta_j a_2(\omega^x_{\theta j}) - (1 - \delta_j)b_2(\omega^x_{\theta j})\}[Z_{jl} + \hat{\lambda}'_{l\theta}(x)][Z_{jk} + \hat{\lambda}'_{k\theta}(x)]}{\sum W_b(X_j - x)\{\delta_j a_1(\omega^x_{\theta j}) - (1 - \delta_j)b_1(\omega^x_{\theta j})\}},$$

$$(17.5.14)$$

$$\hat{\lambda}''_{k(r+1)\theta}(x)$$
$$= \hat{\lambda}''_{(r+1)k\theta}(x)$$
$$= \frac{1}{\sigma}\frac{\sum W_b(X_j - x)\{\delta_j a_2(\omega^x_{\theta j}) - (1 - \delta_j)b_2(\omega^x_{\theta j})\}[\omega^x_{\theta j} + \hat{\lambda}'_{(r+1)\theta}(x)][Z_{jk} + \hat{\lambda}'_{k\theta}(x)]}{\sum W_b(X_j - x)\{\delta_j a_1(\omega^x_{\theta j}) - (1 - \delta_j)b_1(\omega^x_{\theta j})\}},$$

$$(17.5.15)$$

$$\hat{\lambda}''_{(r+1)(r+1)\theta}(x)$$
$$= \frac{1}{\sigma}\frac{\sum W_b(X_j - x)\{\delta_j a_2(\omega^x_{\theta j}) - (1 - \delta_j)b_2(\omega^x_{\theta j})\}[\omega^x_{\theta j} + \hat{\lambda}'_{(r+1)\theta}(x)]^2}{\sum W_b(X_j - x)\{\delta_j a_1(\omega^x_{\theta j}) - (1 - \delta_j)b_1(\omega^x_{\theta j})\}}.$$

$$(17.5.16)$$

We find an interesting phenomenon, if k or l is replaced by $(r + 1)$ in $\hat{\lambda}'_{k\theta}$ and $\hat{\lambda}''_{kl\theta}$, then let the associated term Z_{jk} or Z_{jl} be replaced by $\omega^x_{\theta j}$, we obtain $\hat{\lambda}'_{(r+1)\theta}(x)$, $\hat{\lambda}''_{k(r+1)\theta}(x)$ and $\hat{\lambda}''_{(r+1)(r+1)\theta}(x)$.

After we obtain the estimates of $\lambda_\theta(x)$, $etc.$, we can compute the estimates of θ. Denote by

$$\omega_{\theta j} = \omega_{\theta j}(\hat{\lambda}_\theta(X_j)) = \frac{Y_j - (Z_j^T\beta + \hat{\lambda}_\theta(X_j))}{\sigma}.$$

Recall that

$$L_n(\theta, \hat{\lambda}_\theta) = \sum_j l((Y_j, \delta_j); \theta, \hat{\lambda}_\theta),$$

$$\phi_2(\theta) = \frac{\partial L_n}{\partial\theta}(\theta, \hat{\lambda}_\theta)$$
$$= \left(\begin{array}{c}\frac{\partial L_n}{\partial\beta}(\theta, \hat{\lambda}_\theta)\\ \frac{\partial L_n}{\partial\sigma}(\theta, \hat{\lambda}_\theta)\end{array}\right) = \left(\begin{array}{c}\phi_{2\beta}(\theta)\\ \phi_{2\sigma}(\theta)\end{array}\right),$$

where

$$\phi_{2\beta}(\theta) = (-\frac{1}{\sigma})\sum_j[\delta_j a_0(\omega_{\theta j}) - (1-\delta_j)b_0(\omega_{\theta j})][Z_j + \frac{\partial\hat\lambda_\theta}{\partial\beta}(X_j)],$$

$$\phi_{2\sigma}(\theta) = -\frac{1}{\sigma}\sum_j\{[\delta_j a_0(\omega_{\theta j}) - (1-\delta_j)b_0(\omega_{\theta j})]$$

$$\times[\omega_{\theta j} + \frac{\partial\hat\lambda_\theta}{\partial\sigma}(X_j)] + \delta_j\},$$

and

$$\frac{\partial\phi_{2\beta}(\theta)}{\partial\beta^T} = (-\frac{1}{\sigma})\sum_j[\delta_j a_0(\omega_{\theta j}) - (1-\delta_j)b_0(\omega_{\theta j})][\frac{\partial^2\hat\lambda_\theta(X_j)}{\partial\beta\partial\beta^T}]$$

$$+(\frac{1}{\sigma})^2\sum_j[\delta_j a_1(\omega_{\theta j}) - (1-\delta_j)b_1(\omega_{\theta j})]$$

$$\times[Z_j + \frac{\partial\hat\lambda_\theta}{\partial\beta}(X_j)][Z_j + \frac{\partial\hat\lambda_\theta}{\partial\beta}(X_j)]^T,$$

$$\frac{\partial\phi_{2\beta}(\theta)}{\partial\sigma}$$

$$= (-\frac{1}{\sigma})\phi_{2\beta}(\theta) + (-\frac{1}{\sigma})\sum_j[\delta_j a_0(\omega_{\theta j}) - (1-\delta_j)b_0(\omega_{\theta j})][\frac{\partial\hat\lambda_\theta(X_j)}{\partial\beta\partial\sigma}]$$

$$+(\frac{1}{\sigma})^2\sum_j[\delta_j a_1(\omega_{\theta j}) - (1-\delta_j)b_1(\omega_{\theta j})][\omega_{\theta j} + \frac{\partial\hat\lambda_\theta}{\partial\sigma}(X_j)]$$

$$\times[Z_j + \frac{\partial\hat\lambda_\theta}{\partial\beta}(X_j)],$$

$$\frac{\partial\phi_{2\sigma}(\theta)}{\partial\beta^T} = [\frac{\partial\phi_{2\beta}(\theta)}{\partial\sigma}]^T,$$

$$\frac{\partial \phi_{2\sigma}(\theta)}{\partial \sigma}$$

$$
= \; (-\frac{1}{\sigma})\phi_{2\sigma}(\theta) + (-\frac{1}{\sigma})\sum_{j}[\delta_j a_0(\omega_{\theta j}) - (1 - \delta_j)b_0(\omega_{\theta j})][\frac{\partial^2 \hat{\lambda}_\theta(X_j)}{\partial \sigma^2}
$$

$$
+(-\frac{1}{\sigma})(\omega_{\theta j} + \frac{\partial \hat{\lambda}_\theta}{\partial \sigma}(X_j))] + (\frac{1}{\sigma})^2 \sum_{j}[\delta_j a_1(\omega_{\theta j}) - (1 - \delta_j)b_1(\omega_{\theta j})]
$$

$$
\times [\omega_{\theta j} + \frac{\partial \hat{\lambda}_\theta}{\partial \sigma}(X_j)]^2.
$$

Then

$$
\mathcal{H} \;=\; \frac{\partial^2 L_n(\theta, \hat{\lambda}_\theta)}{\partial\theta\partial\theta^T} = \frac{\phi_2(\theta)}{\partial\theta^T} = \begin{pmatrix} \frac{\phi_{2\beta}(\theta)}{\partial\beta^T} & \frac{\phi_{2\beta}(\theta)}{\partial\sigma} \\ \frac{\phi_{2\sigma}(\theta)}{\partial\beta^T} & \frac{\phi_{2\sigma}(\theta)}{\partial\sigma} \end{pmatrix}.
$$

TABLE 17.1 Estimates of the parameters under the semiparametric and parametric models for PBC data

Parameters	Estimates	SE	χ^2
Model 1: Log-normal model			
Semiparametric model			
σ	0.8534	0.055	240.76
Log (Albumin)	1.6415	0.535	9.41
Log (Bilirubin)	-0.5283	0.069	58.62
Edema	-0.7797	0.233	11.20
Log (Protime)	-3.2561	0.816	15.92
Parametric model			
σ	0.865	0.056	238.59
Log (Albumin)	1.4807	0.528	7.88
Log (Bilirubin)	-0.5371	0.069	59.81
Edema	-0.765	0.230	11.03
Log (Protime)	-3.210	0.821	15.30
Intercept	12.024	2.114	32.35
Age	-0.002	0.0005	16
Model 2: Log-logistic model			
Semiparametric model			
σ	0.4583	0.033	192.87
Log (Albumin)	1.8234	0.514	12.58
Log (Bilirubin)	-0.5195	0.065	63.88
Edema	-0.7226	0.228	10.04
Log (Protime)	-3.0778	0.774	15.81
Parametric model			
σ	0.4654	0.034	187.37
Log (Albumin)	1.679	0.509	10.90
Log (Bilirubin)	-0.5335	0.066	64.71
Edema	-0.6790	0.229	8.80
Log (Protime)	-3.0679	0.780	15.47
Intercept	11.4333	1.964	34.23
Age	-0.0022	0.0005	19.36

TABLE 17.1 (cont'd) Estimates of the parameters under the semiparametric and parametric models for PBC data (cont'd)

Parameters	Estimates	SE	χ^2
Model 3: Extreme value model			
Semiparametric model			
σ	0.6104	0.0427	204.35
Log (Albumin)	2.20	0.443	24.66
Log (Bilirubin)	-0.5147	0.059	76.10
Edema	-0.5042	0.188	7.19
Log (Protime)	-2.4975	0.730	11.70
Parametric model			
σ	0.620	0.0437	201.30
Log (Albumin)	1.8485	0.4240	19.01
Log (Bilirubin)	-0.5259	0.0583	81.37
Edema	-0.4821	0.1808	7.11
Log (Protime)	-2.3774	0.7273	10.68
Intercept	9.865	1.870	27.83
Age	-0.0021	0.0005	17.54

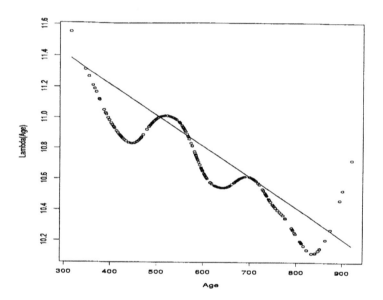

FIGURE 17.1 Fitted function for age using semiparametric log-normal model

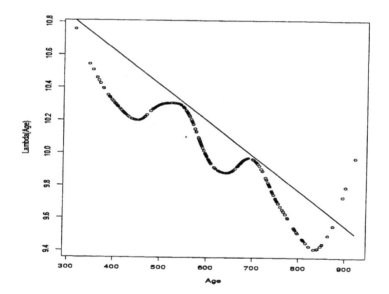

FIGURE 17.2   Fitted function for age using semiparametric log-logistic model

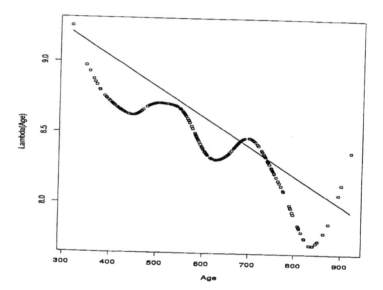

FIGURE 17.3 Fitted function for age using semiparametric extreme value model

CHAPTER 18

ANALYSIS OF SATURATED AND SUPER-SATURATED FACTORIAL DESIGNS: A REVIEW

KIMBERLY K. J. KINATEDER   DANIEL T. VOSS
WEIZHEN WANG

Wright State University, Dayton, OH

Abstract: Various methods have been proposed in the literature for the analysis of saturated and super-saturated factorial designs, but few of these methods are known to provide strong control of error rates. This paper is a review of known results and open problems concerning the strong control of error rates in the analysis of such designs.

Keywords and phrases: Closed test, directional inference, effect sparsity, experiment-wise error rate, orthogonal factorial design, non-orthogonal factorial design, saturated design, step-down test, stepwise test, super-saturated design

18.1 INTRODUCTION

This paper concerns methods of analysis of saturated and super-saturated designs which strongly control error rates for individual or family-wise inference. Methods known to strongly control error rates are reviewed, with discussion of the techniques by which the results have been established. Also, a variety of related open problems are stated.

Substantial progress has been made establishing methods of analysis of orthogonal saturated designs. Methods known to strongly control error rates include: an exact, closed, step-down test for simultaneously testing

325

the hypotheses $H_{0i} : \theta_i = 0$ $(i = 1, 2, \ldots, k)$, applicable for a broad class of statistics [Voss (1988)]; analogous exact individual tests of the each of the hypotheses $H_{0i} : \theta_i = 0$ [Berk and Picard (1991)]; exact individual confidence intervals [Voss (1999)]; and exact simultaneous confidence intervals [Voss and Wang (1999)].

There remain a number of open problems concerning the analysis of orthogonal saturated designs. Strong control of error rates has not been established for any of the following methods: the aforementioned methods of Voss (1988), Berk and Picard (1991), Voss and Wang (1999), and Voss (1999) if the methods are adaptive with respect to the number of terms used to form the denominator of each statistic; adaptive methods along the lines of Lenth (1989); step-down tests using sharper critical values as recommended by Zahn (1975a,b), Venter and Steel (1998), and Langsrud and Naes (1998), and as discussed by Voss (1988); and the step-up tests of Venter and Steel (1998) and Langsrud and Naes (1998). It is also an open problem to show that directional error rate is controlled for the step-down tests of Zahn (1975a,b), Voss (1988), Venter and Steel (1998), and Langsrud and Naes (1998), or for the step-up tests of Venter and Steel (1998) and Langsrud and Naes (1998).

For saturated designs which are non-orthogonal, there has been very little progress developing methods known to strongly control error rates. It is problematic that the estimators are correlated. Kunert (1997) provided a method of transforming the correlated estimators into uncorrelated estimators, with $k!$ possible such transformations to orthogonality. Furthermore, he proposed the use of either a predetermined transformation or the best of the $k!$ transformations to obtain an improved variance estimator—namely, one which is robust to the presence of a few non-negligible effects. Using such variance estimators, he also proposed methods of data analysis but did not establish strong control of error rates. Kinateder, Voss and Wang (1999) obtained exact individual confidence intervals for each of the effects θ_i $(i = 1, 2, \ldots, k)$, using Kunert's (1997) method of transformation to orthogonality. The result depends on making an a priori choice of one of $(k - 1)!$ possible transformations, where the $(k - 1)!$ choices depend on i. However, the method of Kinateder, et.al. has its shortcomings, as will be discussed in Section 18.4.

Concerning the analysis of non-orthogonal saturated designs, there are more open problems than results. It remains an open question whether the confidence intervals of Kinateder, Voss and Wang (1999), with appropriate modification of critical values, would strongly control the error rate if *any* of the $(k - 1)!$ possible transformations is used. Alternatively, tighter confidence intervals might be obtained if a methodology could more directly take into account the correlation structure of the estimators. More generally, there are still no methods of simultaneous inference, either tests or confidence intervals, known to strongly control error rates.

Finally, for the analysis of super-saturated designs, there are still *no* methods which are known to strongly control error rates. It remains an open problem to find individual or simultaneous tests or confidence intervals in this case.

18.2   BACKGROUND

Consider the analysis of data for an unreplicated or fractional factorial design which is saturated or super-saturated. Suppose there are n observations and k parameters of interest. Denote the parameters by θ_i, for $i \in K = \{1, 2, \ldots, k\}$. These are typically treatment contrasts. Throughout this paper we assume *effect sparsity*—namely, that few of the effects are nonzero (or non-negligible).

Also assume throughout the paper that the following linear model is appropriate:

$$\boldsymbol{Y} \sim N_n(\boldsymbol{X\beta}, \boldsymbol{I}\sigma^2),$$

where $\boldsymbol{Y}_{n \times 1}$ is a vector of independent, normally distributed observations with common unknown variance σ^2, $\boldsymbol{X}_{n \times p}$ is the design matrix, $\boldsymbol{\beta}_{p \times 1}$ is the vector of unknown parameters, and $\boldsymbol{I}_{n \times n}$ is the identity matrix. Without loss of generality, the vector $\boldsymbol{\theta}_{k \times 1}$ of effects of interest is of the form $\boldsymbol{\theta} = \boldsymbol{C\beta}$ for some $k \times p$ matrix $\boldsymbol{C}$ of rank k.

18.2.1   Orthogonality and Saturation

For purposes of analysis, we classify the designs of interest into three types: orthogonal saturated designs, non-orthogonal saturated designs, and super-saturated designs. A design is *saturated* or *super-saturated* if the rank of the design matrix $\boldsymbol{X}$ is n. Then all degrees of freedom are consumed by the estimation of model parameters, leaving no degrees of freedom for error. Given a saturated or super-saturated design, the design is saturated if the vector $\boldsymbol{\theta}$ is estimable—otherwise, the design is super-saturated. In other words, a design is *saturated* if the rank of $\boldsymbol{X}_{n \times p}$ is n and the row space of $\boldsymbol{C}$ is in the row space of $\boldsymbol{X}$. A design is *super-saturated* if the rank of $\boldsymbol{X}_{n \times p}$ is n and the row space of $\boldsymbol{C}$ is *not* in the row space of $\boldsymbol{X}$. A design is necessarily super-saturated, for example, if $n < k + 1$, the effects of interest are treatment contrasts, and the model includes an intercept.

For a saturated design, the vector $\hat{\boldsymbol{\theta}}$ of effect (least squares) estimators is of the form

$$\hat{\boldsymbol{\theta}}_{k \times 1} = \boldsymbol{A}'\boldsymbol{Y} \sim N_k(\boldsymbol{\theta}, \boldsymbol{A}'\boldsymbol{A}\sigma^2),$$

where $\boldsymbol{A}_{n \times k}$ is of full rank k. A saturated design is *orthogonal* if the matrix $\boldsymbol{A}'\boldsymbol{A}$ is diagonal, or equivalently, if the columns of $\boldsymbol{A}$ are orthogonal—otherwise, the design is *non-orthogonal*.

For example, consider the regular 2_{III}^{7-4} fractional factorial design shown in Table 18.1, (with defining relation generated by the effects ABD, ACE, BCF and $ABCG$). The $n = 8$ observations allow the independent estimation of the $k = 7$ factorial main effects, the matrix $\boldsymbol{A}$ being $1/4$ times the array in Table 18.1. However, having $n = k + 1$ leaves no error degrees of freedom after adjustment for mean response. The same design would be super-saturated if used to estimate all main effects and two-factor interactions of the seven factors.

TABLE 18.1   A regular 2_{III}^{7-4} fractional factorial design

Obs.	A	B	C	D	E	F	G
1	-1	-1	-1	1	1	1	-1
2	-1	-1	1	1	-1	-1	1
3	-1	1	-1	-1	1	-1	1
4	-1	1	1	-1	-1	1	-1
5	1	-1	-1	-1	-1	1	1
6	1	-1	1	-1	1	-1	-1
7	1	1	-1	1	-1	-1	-1
8	1	1	1	1	1	1	1

Another example is the Plackett-Burman (1946) design in Table 18.2 for $n = 12$ observations. This is an orthogonal saturated design for the estimation of the main effects of $k = 11$ factors each at two levels under a first-order model. The 12 observations allow the independent estimation of the $k = 11$ factorial main effects, the matrix $\boldsymbol{A}$ being $1/6$ times the array in Table 18.2. However, having $n = k + 1$ leaves no error degrees of freedom after adjustment for mean response. The same design would be nearly saturated, but non-orthogonal, if used for any four of the factors to model their four main effects and six two-factor interactions. The same design would be super-saturated if used to estimate the main effects and two-factor interactions of any five of the factors.

TABLE 18.2 The 12-run Plackett-Burman design

Obs.	A	B	C	D	E	F	G	H	I	J	K
1	1	-1	1	-1	-1	-1	1	1	1	-1	1
2	1	1	-1	1	-1	-1	-1	1	1	1	-1
3	-1	1	1	-1	1	-1	-1	-1	1	1	1
4	1	-1	1	1	-1	1	-1	-1	-1	1	1
5	1	1	-1	1	1	-1	1	-1	-1	-1	1
6	1	1	1	-1	1	1	-1	1	-1	-1	-1
7	-1	1	1	1	-1	1	1	-1	1	-1	-1
8	-1	-1	1	1	1	-1	1	1	-1	1	-1
9	-1	-1	-1	1	1	1	-1	1	1	-1	1
10	1	-1	-1	-1	1	1	1	-1	1	1	-1
11	-1	1	-1	-1	-1	1	1	1	-1	1	1
12	-1	-1	-1	-1	-1	-1	-1	-1	-1	-1	-1

The main challenge for a saturated design is that there are no error degrees of freedom for independent estimation of variability. Consequently, the data analysis depends on the relative magnitudes of either the effect estimates or their sums of squares. Such methods of data analysis invariably depend on an assumption of effect sparsity.

18.2.2   Control of Error Rates

This paper concerns the strong control of error rates, individually or familywise, in the analysis of saturated and super-saturated designs.

Individual and familywise control of error rates

Control of individual error rates is control of the Type I error rate for testing each null hypothesis $H_{i0} : \theta_i = 0$ or, equivalently, control of the confidence level associated with each confidence interval (L_i, U_i) for θ_i, $(i \in K)$. *Familywise control of error rates* is control of the chance of making *any* Type I errors while testing the family of k hypotheses $H_{i0} : \theta_i = 0$ $(i \in K)$, or, equivalently, control of the level of confidence that all of the k confidence intervals (L_i, U_i) $(i \in K)$ are simultaneously correct. [See Hochberg and Tamhane (1987, pages 5–12).]

Even in the analysis of a screening experiment, there is value in the use of *both* individual and familywise inference procedures. Then failure to identify a non-negligible effect (a Type II error) may be more detrimental than falsely asserting a nonexistent effect to be statistically significant (a Type I error). Individual inference procedures are generally preferred for the analysis of screening experiments because of their greater power. However, this greater power is at the expense of a higher rate of false positives. In comparison, simultaneous methods of inference are more conservative, but effects found to be significant by such methods are more often real effects. Both individual and simultaneous inference procedures should be used for the analysis of a screening experiment, since they provide different information which is available at no cost!

Strong control of error rates

A procedure provides *strong control of the error rate* if the error rate is controlled over all parameter configurations [Hochberg and Tamhane (1987, p. 3)]. For example, consider an individual confidence interval for θ_1. Then the other effects $\theta_2, \ldots, \theta_k$ are nuisance parameters. It is not enough that the confidence level for capturing θ_1 be as specified only if the other parameters $\theta_2, \ldots, \theta_k$ are all zero, since they almost certainly are not.

For strong control of the individual error rate for a confidence interval for θ_1, the desired probability inequality is of the form

$$P_{\boldsymbol{\theta}}(L_1 < \theta_1 < U_1) \geq 1 - \alpha, \qquad (18.2.1)$$

where the inequality is understood to hold for all $\boldsymbol{\theta}$.

Similarly, for strong control of the familywise error rate for simultaneous confidence intervals for the θ_i $(i \in K)$, the desired probability inequality is of the form

$$P_{\boldsymbol{\theta}}(L_i \leq \theta_i \leq U_i \ : \ \forall \, i \in K) \geq 1 - \alpha. \qquad (18.2.2)$$

If the appropriate probability inequality, (18.2.1) or (18.2.2), holds, and if the infimum over $\boldsymbol{\theta}$ of the corresponding probability is $1 - \alpha$, then the error rate is α, the confidence level is $100(1 - \alpha)\%$, and the confidence intervals are said to be *exact*. If the probability inequality holds but the infimum exceeds $1 - \alpha$, then the confidence intervals are *conservative*.

The notion of an exact confidence interval is equivalent to the notion of a test being of size α, the size of an individual test being

$$\alpha = \sup_{\boldsymbol{\theta} \in H_0} P_{\boldsymbol{\theta}}(\text{reject } H_0).$$

Consider what is meant by the size when simultaneously testing the hypotheses $H_{0i} : \theta_i = 0$ for $i \in K$. Let I be a nonempty subset of $K = \{1, 2, \ldots, k\}$. Then $H_I = \cap_{i \in I} H_{0i}$ denotes the hypothesis that H_{0i} is true for all $i \in I$, viewing each hypothesis as a subset of the parameter space.

For simultaneous testing the k hypotheses H_{0i}, the size of the test is

$$\alpha = \sup_{\boldsymbol{\theta}} P_{\boldsymbol{\theta}}(\exists\, \theta_{i_0} = 0, \text{ but declare } \theta_{i_0} \neq 0)$$

$$= \max_{I \subseteq K} \sup_{\boldsymbol{\theta} \in H_I} P_{\boldsymbol{\theta}}(\text{reject } H_{0i} \text{ for some } i \in I).$$

For strongly controlling error rates in the analysis of saturated and supersaturated designs, the nature of the problem and the results available depend upon the design. The easiest case is when the design is orthogonal and saturated, as discussed in Section 18.3. Non-orthogonal saturated designs may be analyzed by transformation or projection to orthogonality, as will be discussed in Section 18.4. For a super-saturated design, non-estimability of $\hat{\boldsymbol{\theta}}$ poses additional problems for the analysis, as discussed in Section 18.5.

18.3   ORTHOGONAL SATURATED DESIGNS

18.3.1   Background

This section contains a review of known results and open problems in the analysis of orthogonal saturated designs. In this case,

$$\hat{\boldsymbol{\theta}}_{k \times 1} \sim N_k(\boldsymbol{\theta}_k, \boldsymbol{D}s^2),$$

where the covariance matrix $\text{Cov}(\hat{\boldsymbol{\theta}}) = \boldsymbol{D}\sigma^2$ is diagonal and known up to the constant σ^2. Since the estimators can be scaled each to have variance σ^2, assume without loss of generality that $\boldsymbol{D}$ is the identity matrix. Thus, we assume $\hat{\theta}_i \sim N(\theta_i, \sigma^2)$ and independent ($i \in K$). Effect sparsity is also assumed.

There have been many methods proposed for the analysis of orthogonal saturated designs.

Daniel (1959) made a fundamental contribution, using half-normal plots for the subjective analysis of orthogonal saturated designs. Normal probability plots are still widely used for this purpose even today.

At the same time, Daniel (1959) and Birnbaum (1959) considered more formal, objective methods of analysis of such designs. Each considered testing for a nonzero effect, assuming at most one effect is nonzero. Birnbaum provided an optimal decision rule for this case, based on the size of the largest of k independent sum of squares relative to the total sum of squares (or equivalently, to the sum of the rest of the effect sums of squares). An optimal level-α test for the detection of at most one nonzero effect could be iterated to test for multiple nonzero effects, but the iterative procedure would no longer be level-α or optimal. Birnbaum also considered optimal decision rules in the case of at most two nonzero effects, noting that the problem was then already quite complex. Zahn (1975a,b) considered some variations on the iterative methods of Daniel (1959) and Birnbaum (1959), but his results were primarily empirical. The subjective use of normal probability plots remained the standard methodology for the analysis of orthogonal saturated designs until the late 1980's.

Then Box and Meyer (1986, 1993) provided Bayesian methods for obtaining posterior probabilities that effects are active, and there followed a flurry of papers proposing new frequentist methods, making refinements on the methods, and making empirical comparisons of the many variations. See for example papers by Voss (1988), Benski (1989), Lenth (1989), Berk and Picard (1991), Loh (1992), Juan and Peña (1992), Schneider, Kasperski and Weissfeld (1993), Dong (1993), Torres (1993), Haaland and O'Connell (1995), Venter and Steel (1996, 1998), Voss and Wang (1999), and Langsrud and Naes (1998). Hamada and Balakrishnan (1998) provide an extensive review of existing methods, including a Monte Carlo-based comparison of the operating characteristics of the methods.

All of the afore-mentioned methods rely on an assumption of effect sparsity. Most of the methods are heuristically appealing, and in many cases the operating characteristics have been studied empirically or justified in approximation. However, relatively few of the methods are known to provide strong control of error rates, which is the focus of this work.

The objective here is to obtain tests of the hypotheses $H_{i0} : \theta_i = 0$ or confidence intervals for the parameters θ_i ($i \in K$) which strongly control error rates either individually or simultaneously. Strong control of error rates and confidence levels requires establishment of appropriate probability inequalities, which corresponds to a stochastic ordering of distributions. In order to establish this theoretically, the following definition and lemma are useful.

A family of distribution functions $F_\theta(x)$ on the real line, with a real parameter θ, is said to be *stochastically decreasing* if $\theta < \theta'$ implies $F_\theta(x) \leq F_{\theta'}(x)$ for all x [Lehmann (1986, p. 84)].

The following lemma follows from similar results of Alam and Rizvi (1966) and Mahamunulu (1967), used in the ranking and selection literature for identifying least favorable configurations. For related discussion of applications to ranking and selection, see Gupta, Huang and Panchapakesan (1982).

Lemma 18.3.1 *[Stochastic Ordering Lemma, Voss (1999)] Let $F_{i\theta_i}(x)$, with real parameter θ_i, be a stochastically increasing family of distribution* *functions on the real line, for $i = 1, 2, \ldots, k$. Let $X_1, X_2, \ldots, X_k$ be independent random variables, where the distribution function of X_i is $F_{i\theta_i}(x_i)$. For any fixed i, $1 \le i \le k$, if the statistic $t = t(x_1, x_2, \ldots, x_k)$ is a nonincreasing function of x_i when all x_j for $j \ne i$ are held fixed, then the distribution of $T = t(X_1, X_2, \ldots, X_k)$ is stochastically decreasing in θ_i.*

In the rest of this section, some known results and open problems are discussed in some detail.

18.3.2   Simultaneous Stepwise Tests

Closed, step-down tests

The first procedure known to strongly control the error rate in the analysis of an orthogonal saturated design was a closed, step-down testing procedure of Voss (1988) for simultaneously testing the hypotheses $H_{i0} : \theta_i = 0$ ($i \in K$). The following broad class of statistics was considered. Let ϕ be a nonnegative, increasing function on the nonnegative real numbers, and let $X_i = \phi(|\hat{\theta}_i|)$, with corresponding order statistics

$$X_{(1)} < X_{(2)} < \cdots < X_{(k)}.$$

The test statistics are of the form

$$R_{(i)} = X_{(i)}/D, \ i \in K, \tag{18.3.3}$$

where $D = \sum_{i=1}^{k} a_i X_{(i)}$, for nonnegative scalars a_i not all zero.

Included, for example, are the statistics

$$R_{(i)} = SS_{(i)}/QMSE, \ i \in K, \tag{18.3.4}$$

which compare the order statistics of the sums of squares to the *quasi mean squared error QMSE* obtained as the average of the ν smallest sums of squares. These statistics are obtained if $\phi(x) = x^2$ and, for fixed ν ($1 \le \nu \le k$), $a_i = 1/\nu$ for $i \le \nu$ and $a_i = 0$ otherwise.

Also included, for example, are the statistics

$$R_{(i)} = |\hat{\theta}|_{(i)}/D, \ (i \in K), \tag{18.3.5}$$

which compare the order statistics $|\hat{\theta}|_{(i)}$ of the absolute effect estimates $|\hat{\theta}_i|$ to the average of the ν smallest absolute effect estimates, for $\phi(x) = x$ and, for fixed ν ($1 \le \nu \le k$), $a_i = 1/\nu$ for $i \le \nu$ and $a_i = 0$ otherwise.

These test statistics include several statistics proposed in the literature. They include for example the ratio of the largest sum of squares to the total sum of squares considered by Cochran (1941), the modulus-ratio statistics of Daniel (1959), the X and S statistics of Zahn (1975a,b), and (the square of) the ratio statistic of Schneider, Kasperski and Weissfeld (1993).

A variation on the statistics in equation (18.3.3) is for D to be the median or some other quantile of the absolute effect estimates. Then the statistics in (18.3.3) are analogous to adaptive statistics used by Lenth (1989) and others, except these are not adaptive. Adaptive methods are considered in Section 18.3.6.

Voss (1988) obtained the critical values as follows. For fixed function ϕ and scalar vector $\boldsymbol{a} = (a_1, a_2, \ldots, a_k)'$, let $c_\alpha(i, \boldsymbol{a}, k)$ be the upper-α quantile of

$$C_i = \max\{X_1, X_2, \ldots, X_i\}/D, \ (i \in K),$$

under the *null distribution*—namely, when $\theta_i = 0$ for all i. Thus, $\alpha = P_{\boldsymbol{\theta}}(C_i > c_\alpha(i, \boldsymbol{a}, k) \mid \boldsymbol{\theta} = (0, \ldots, 0)')$.

The step-down testing procedure of Voss (1988), illustrated here using the statistics $ss_{(i)}/qmse$ of equation (18.3.4), is as follows. Let $\theta_{(i)}$ denote the parameter corresponding to the ith smallest sum of squares, $ss_{(i)}$. If $ss_{(k)}/qmse > c_\alpha(k, \boldsymbol{a}, k)$, then assert $\theta_{(k)} \neq 0$ and continue; otherwise stop. If $ss_{(k-1)}/qmse > c_\alpha(k - 1, \boldsymbol{a}, k)$, then assert $\theta_{(k-1)} \neq 0$ and continue; otherwise stop. Continue in this fashion, asserting $\theta_{(i)} \neq 0$ for each i such that $ss_{(j)}/qmse > c_\alpha(j, \boldsymbol{a}, k)$ for all $j \geq i$.

Theorem 18.3.1 *[Voss (1988)]   This step-down testing procedure is of familywise size α.*

Proof of the above result was based on the following observation. Denote the (unknown) number of negligible effects by m. Without loss of generality, let the first m effects be negligible. A necessary condition for a false assertion to occur is that

$$\max\{X_1, X_2, \ldots, X_m\}/D > c_\alpha(m, \boldsymbol{a}, k).$$

Under the null distribution, this occurs with probability α. Voss (1988) argued that

$$\max\{X_1, X_2, \ldots, X_m\}/D$$

is stochastically decreasing in $|\hat{\theta}_i|$ for each $i > m$, so

$$P_{\boldsymbol{\theta}}\left(\max\{X_1, X_2, \ldots, X_m\}/D > c_\alpha(m, \boldsymbol{a}, k)\right) \leq \alpha$$

for all $(\theta_{m+1}, \ldots, \theta_k)$ when $\theta_1 = \theta_2 = \cdots = \theta_m = 0$. Application of the Stochastic Ordering Lemma makes the proof rigorous. Because the probability bound is achieved in the null case, the size of the test is α.

Marcus, Peritz, and Gabriel (1976) provided a general method of constructing step-down tests which strongly control the familywise error rate. The method is called the *closure method*. Given the finite family of hypotheses $\{H_{0i} : i \in K\}$, the closure of this family is obtained by taking all non-empty intersections of parameter spaces, $H_I = \cap_{i \in I} H_{0i}$ for $I \subseteq K = \{1, 2, \ldots, k\}$. The method hinges on the existence of a level-α test of each hypothesis H_I. The *closed testing procedure* rejects H_I at level α if and only if H_K is rejected by its associated level-α test for all $K \supseteq I$. [For further details on step-down tests and closed tests, see Hochberg and Tamhane (1987, pp. 53–54).]

The procedure of Voss (1988) is a closed, step-down test. To see this, consider the test of H_I. This hypothesis is rejected if

$$\max_{i \in I}\{X_i\}/D > c_\alpha(|I|, \boldsymbol{a}, k),$$

where $|I|$ is the number of elements in the set I. By definition of the critical value $c_\alpha(|I|, \boldsymbol{a}, k)$, the Type I error rate is exactly α when $\theta_i = 0$ for all $i \in K$. Also, by the Stochastic Ordering Lemma, the distribution of $\max_{i \in I}\{X_i\}/D$ is stochastically decreasing in $|\theta_j|$ for each $j \notin I$. It follows that the test of H_I is of size α.

Iterative methods and sharper critical values: an open problem

In the step-down testing procedure of Voss (1988), the critical value $c_\alpha(i, \boldsymbol{a}, k)$ is determined from the null distribution of the random variable

$$C_i = \max\{X_1, X_2, \ldots, X_i\}/D, \ i \in K.$$

Here C_i is a function of all k random variables X_i, because $D = \sum_{i=1}^{k} a_i X_{(i)}$. Voss (1988) observed that sharper critical values are obtained if C_i is taken to be a function of only i random variables. Specifically, $c_\alpha(i, \boldsymbol{a}, i) < c_\alpha(i, \boldsymbol{a}, k)$ for all i for which the former is well defined, which is the case if $a_j = 0$ for all $j > i$. Use of the sharper critical values corresponds to iteratively testing the effect corresponding to the largest of i estimators, for $i = k, k-1, \ldots$, the ith test statistic being a function of only i effects. Use of these sharper critical values has been advocated for example by Daniel (1959), Zahn (1975a,b), Venter and Steel (1998), and Langsrud and Naes (1998).

In fact, Venter and Steel (1998) considered a more general class of statistics than those in equation (18.3.3). They considered statistics of the form

$$T_{(i)} = X_{(i)}/D_i, \ i \in K, \tag{18.3.6}$$

where $X_{(i)} = |\hat{\theta}|_{(i)}$ and $D_i = \sum_{j<i} a_{ij} X_{(j)}$, for nonnegative scalars a_{ij} not all zero. Thus, the denominators D_i can depend on i. It is reasonable to allow $\phi(x) = x^2$, which would enlarge the class of statistics. Concerning D_i, one could also allow a_{ij} to be nonzero for $j \geq i$. While this in not desirable, the resulting class of statistics (18.3.6) would then generalize those of Voss (1988) in Equation (18.3.3).

For the step-down testing procedure, Venter and Steel (1998) recommend using critical values obtained as the upper-α quantile of

$$C_i = \max\{X_1, X_2, \ldots, X_i\}/D_i$$

under the *null distribution* of $X_1, X_2, \ldots, X_i$, taking the $X_{(j)}$ in D_i to be the order statistics of only $X_1, X_2, \ldots, X_i$, not of $X_1, X_2, \ldots, X_k$. These critical values were chosen so that, for $I \subseteq K$, the test of "$H_I : \theta_i = 0 \ \forall \ i \in I$" has Type I error rate α if θ_i is infinite for each $i \notin I$.

Similarly, Langsrud and Naes (1998) propose forward selection and backward elimination strategies, incorporating, but not requiring, an independent estimator of error variance (σ^2). Their analysis involves the statistics

$$\Psi_j(t) = \frac{SS_{(j)}}{(1/(q_j+t))(ts^2 + \Sigma_{i=1}^{q_j} SS_{(i)})}, \quad j = 2, \ldots, k, \qquad (18.3.7)$$

where s^2 denotes the independent error estimator with t degrees of freedom. They provide a stochastic ordering result to compare different null distributions of interest. Nonetheless, they do not go so far as to rigorously strongly control error rates (termed "protection levels" in their work), but rather establish the protection level in the null case.

Empirical evidence suggests that familywise error rate is strongly controlled if step-down testing is used with the sharper critical values [Zahn (1969, 1975b) and Venter and Steel (1998)], but this result has only been proven for the cases of $k = 2, 3$ [Zahn (1969)]. These tests are *not* closed, step-down tests—the method provides no α-level test of the individual hypotheses $H_{0i} : \theta_i = 0$, for example. It remains an open problem to prove that such methods strongly control the familywise error rate for $k > 3$.

Directional inference: an open problem

Another open problem concerning step-down tests is the following. Suppose a step-down procedure for testing the hypotheses H_{0i} ($i \in K$) controls the familywise error rate to be at most α. Is the error rate still at most α if, for each hypothesis H_{0i} rejected, one infers $\theta_i > 0$ if $\hat{\theta}_i > 0$ and $\theta_i < 0$ if $\hat{\theta}_i < 0$? Shaffer (1980) and Holm (1979) establish such control of *directional error rate* in other scenarios. See Hsu (1996, p. 20) for discussion.

Step-up tests

Holms and Berrettoni (1969) proposed a step-up testing procedure for testing the hypotheses $H_{0i} : \theta_i = 0$ $(i \in K)$, using the test statistics $SS_{(i)}/\sum_{j=1}^{i} SS_{(j)}$ for $i \geq m$, where m is a pre-specified integer, $1 < m \leq k$. More recently, Venter and Steel (1996, 1998) proposed using the same step-up testing procedure but with the more general class of statistics $T_{(i)} = X_{(i)}/D_i$ of equation (18.3.6). Following Venter and Steel (1998), the procedure is as follows. For fixed m $(1 < m \leq k)$, let n be the minimum value in $\{m, m+1, \ldots, k\}$ such that $T_{(i)} > c_\alpha(i, a_i)$, for critical values $c_\alpha(i, a_i)$, $a_i = (a_{i1}, \ldots, a_{i,i-1})$ $(i = m, m+1, \ldots, k)$. The procedure is to infer $\theta_{(i)} \neq 0$ for all $i = n, n+1, \ldots, k$, where $\theta_{(i)}$ is the parameter corresponding to $ss_{(i)}$. If no such n exists, no inferences are made.

Venter and Steel (1998) conjecture that if $c_\alpha(i, a_i)$ is the upper-α critical value of $\max\{X_1, \ldots, X_i\}/D_i$, then their step-up procedure strongly controls the familywise error rate to be at most α. A similar conjecture can be made for the step-up procedure of Langsrug and Naes (1998), corresponding to equation (18.3.7). It remains an open problem to prove these conjectures.

18.3.3   Individual Tests

Individual tests which strongly control error rates follow from the existence of closed, step-down tests, since the closure method requires the existence of an α-level test of each hypothesis $H_{i0} : \theta_i = 0$.

Theorem 18.3.2 *[Berk and Picard (1991)] A size-α test of the hypothesis $H_{i0} : \theta_i = 0$ is to reject H_{i0} if $X_i/D > c_\alpha(1, a, k)$, where X_i, D and $c_\alpha(1, a, k)$ are as defined for Theorem 18.3.1.*

The critical value for testing H_{i0} is obtained as the upper-α quantile of the null distribution of X_1/D, and θ_i is asserted to be nonzero if X_i/D exceeds the critical value.

It is not clear whether directional error rate is controlled if a directional inference is made when the null hypothesis is rejected. If directional inference is desired, it can be obtained with a slight modification of the denominator used to obtain a confidence interval (see Section 18.3.4).

Loughin and Noble (1997) proposed using a permutation test with the test statistic of Birnbaum (1959), comparing the largest sum of squares to the sum of the rest, to test for a single non-negligible effect. They also considered extensions to testing multiple effects, recommending use of a step-up procedure. However, error rates were considered only under the complete null distribution of no active effects.

18.3.4   Individual Confidence Intervals

Exact individual confidence intervals were obtained by Voss (1999). Strong control of the error rate follows from the Stochastic Ordering Lemma, once an appropriate pivotal quantity is identified. To obtain the pivotal quantity, the denominator D used for individual and simultaneous tests is modified to be independent of the estimator of the effect of interest.

To illustrate the method, consider a confidence interval for the first effect, θ_1. Let

$$SS_{(1:1)} < SS_{(2:1)} < \cdots < SS_{(k-1:1)}$$

denote the order statistics of the $k-1$ sums of squares SS_j excluding SS_1. Furthermore, let $QMSE_1$ denote the quasi mean squared error obtained as the average of the ν smallest of these $k-1$ order statistics, for ν a pre-specified integer, $(1 \leq \nu \leq k)$. Then following Voss (1999),

$$(\hat{\theta}_1 - \theta_1)^2 / QMSE_1 \qquad (18.3.8)$$

is a pivotal quantity with respect to θ_1. By the Stochastic Ordering Lemma, the distribution of (18.3.8) is stochastically decreasing in $|\theta_i|$ for all $i \neq 1$. Hence, we have the following result.

Theorem 18.3.3 *An exact confidence interval for θ_1 is*

$$\hat{\theta}_1 \pm \sqrt{q_\alpha(\nu, k) qmse_1}\,,$$

where $q_\alpha(\nu, k)$ is the upper-α quantile of the null distribution of the pivotal quantity (18.3.8).

These confidence intervals are not adaptive. For open problems concerning confidence intervals based on adaptive methods, see Section 18.3.6.

Conservative simultaneous confidence intervals can be obtained by applying the Bonferroni method to these exact individual confidence intervals, but exact simultaneous confidence intervals can also be obtained, as seen next.

18.3.5   Simultaneous Confidence Intervals

Voss and Wang (1999) obtained exact simultaneous confidence intervals for the k parameters θ_i, by consideration of the distribution of the maximum of the pivotal quantities used by Voss (1999) for individual confidence intervals. The method of proof differs, because the Stochastic Ordering Lemma does *not* apply.

Following Voss (1999), let

$$SS_{(1:i)} < SS_{(2:i)} < \cdots < SS_{(k-1:i)}$$

denote the order statistics of the $k - 1$ sums of squares SS_j excluding SS_i, and let $QMSE_i$ denote the quasi mean squared error obtained as the average of the ν smallest of these $k - 1$ order statistics, where ν is a predetermined integer. Consider the distribution of

$$M = \max_{i \in K} \left\{ (\hat{\theta}_i - \theta_i)^2 / QMSE_i \right\} .$$

Voss and Wang (1999) showed that the distribution of M is stochastically decreasing in each of the $|\theta_i|$ if each $\hat{\theta}_i$ has a symmetric unimodal distribution, so that the null distribution can be used to obtain upper-α critical values for exact simultaneous confidence intervals.

Theorem 18.3.4 *[Voss and Wang (1999)]* *If the $\hat{\theta}_i$ are independently distributed, and if $\hat{\theta}_i$ has a symmetric, unimodal distribution with mean θ_i $(i \in K)$, then exact simultaneous confidence intervals are*

$$\theta_i \in \hat{\theta}_1 \pm \sqrt{m_\alpha(\nu, k) \mathrm{qmse}_i} ,$$

where $m_\alpha(\nu, k)$ is the upper-α quantile of the null distribution of M.

The Stochastic Ordering Lemma does not apply to M. Instead, this theorem was established by direct computation of the distribution function of M, with the problem reducing to consideration of the conditional distribution of $\hat{\theta}_1$, for given $\hat{\theta}_2, \ldots, \hat{\theta}_k$.

18.3.6   Adaptive Methods

A challenging open problem is to show strong control of error rates for adaptive methods of inference.

Lenth (1989) proposed use of an estimate of the error standard deviation σ that is adaptive to the number of nonzero effects. Specifically, he computed a *preliminary estimate*

$$\hat{\sigma}_0 = 1.5 \times \mathrm{median} \left\{ |\hat{\theta}_i| : 1 \leq i \leq k \right\}$$

then obtained a second, more robust estimate as 1.5 times the median of those absolute effect estimates not exceeding $2.5\hat{\sigma}_0$—namely,

$$\hat{\sigma} = 1.5 \times \mathrm{median} \left\{ |\hat{\theta}_i| : 1 \leq i \leq k \text{ and } |\hat{\theta}_i| \leq 2.5\hat{\sigma}_0 \right\} .$$

This estimate $\hat{\sigma}$ is *adaptive*, because it is computed from a random number of the smaller effect estimates.

Lenth then based tests on the ratios $\hat{\theta}_i/\hat{\sigma}$ and confidence intervals on the quantities $(\hat{\theta}_i - \theta_i)/\hat{\sigma}$. He recommended use of critical values from the t-distribution with $k/3$ degrees of freedom, based on fitting scaled chi-squared distributions to the empirical distributions of $\hat{\sigma}^2$ for $k = 7$, 15 and 31 by matching the first two moments. For variations on the method of Lenth (1989), see Juan and Peña (1992), Dong (1993), and Haaland and O'Connell (1995).

The estimate $\hat{\sigma}$ and also the quantities $\hat{\theta}_i/\hat{\sigma}$ and $(\hat{\theta}_i - \theta_i)/\hat{\sigma}$ were shown in Voss (1999) to not be monotone in the absolute values of the effect estimates. Hence, the Stochastic Ordering Lemma does not apply, and control of error rates and confidence levels for such adaptive methods remains an important open problem, requiring an alternate method of proof.

Other open problems concern adaptive versions of the step-down tests of Voss (1988), the individual tests of Berk and Picard (1991), the individual confidence intervals of Voss (1999), and the simultaneous confidence intervals of Voss and Wang (1999). Specifically, suppose in each case the denominator is obtained as the average of the ν smallest sums of squares. If the number of sums of squares, ν, used to form the denominator is adaptive—namely, if ν varies from sample to sample—then can the corresponding procedure be shown to strongly control the error rate?

18.4   NON-ORTHOGONAL SATURATED DESIGNS

This section contains a review of known results and open problems in the analysis of non-orthogonal saturated designs. The results are few and the open problems many. The only method known to strongly control error rates is an exact individual confidence interval procedure of Kinateder, Voss and Wang (1999). The confidence intervals could also be used for exact individual tests. The methodology is essentially an extension of the results of Voss (1999) for the case of orthogonal designs, utilizing projections to orthogonality introduced by Kunert (1997).

In the non-orthogonal case, the vector of effect estimators is of the form $\hat{\theta}_{k \times 1} = A'Y \sim N_k(\theta, A'A\sigma^2)$, where $A_{n \times k} = (a_1, \ldots, a_k)$ is of full rank k, and the matrix $A'A$ is non-diagonal. The latter condition corresponds to the columns of A being non-orthogonal.

18.4.1 Individual Confidence Intervals

The method of Kinateder, Voss and Wang (1999) is presented here. Consider obtaining an exact confidence interval for the first effect, θ_1. The fundamental idea is to transform the dependent estimators $\hat{\theta}_i = a_i'Y$ into independent estimators $\hat{\tau}_i = b_i'Y$, to which the Stochastic Ordering Lemma can be applied to construct the desired confidence interval. Following Kunert (1997), the transformation to independence is accomplished by use of projections to orthogonalize the columns a_i of $A_{n \times k} = (a_1, \ldots, a_k)$. Equivalently, one applies the Gram-Schmidt process to the columns of A (without scaling the columns to have norm one) to obtain the matrix $B_{n \times k} = (b_1, \ldots, b_k)$ with orthogonal columns. It follows that $B = AC$, for $C_{k \times k} = (c_{hi})$ an upper-triangular matrix.

Then for $\hat{\tau}_{k \times 1} = (\hat{\tau}_1, \ldots, \hat{\tau}_k)' = B'Y$, $\hat{\tau}_{k \times 1} \sim N(\tau, D\sigma^2)$, where $\tau = C'\theta$ and $D = B'B$. The parameters τ_i are called the *induced effects*. By construction, the columns of B are orthogonal, so $D = (d_{ij})$ is a diagonal matrix. Consequently, the estimators $\hat{\tau}_i$ $(i = 1, \ldots, k)$ are independently distributed, and $\hat{\tau}_1 = \hat{\theta}_1$ is an unbiased estimator of $\theta_1 = \tau_1$.

Because the induced estimators $\hat{\tau}_i$ are independently distributed, the approach of Voss (1999) for an orthogonal design can be applied to the $\hat{\tau}_i$ to obtain an exact confidence interval for θ_1. Specifically, the ν smallest sums of squares of the $k-1$ estimators $\hat{\tau}_i$ for $i \neq 1$ are pooled together into a quasi mean squared error, $QMSE_1$, for pre-specified integer ν, $(1 \leq \nu < k)$. The quantity

$$Q_1^2 = (\hat{\theta}_1 - \theta_1)^2 / (d_{11} \times QMSE_1)$$

is then a pivotal quantity with respect to θ_1, where $d_{11} = \text{Var}(\hat{\theta}_1)/\sigma^2 = \text{Var}(\hat{\tau}_1)/\sigma^2$, and the distribution of Q_1^2 is stochastically decreasing in $|\tau_j|$ for all $j \neq 1$.

Theorem 18.4.1 *[Kinateder, Voss and Wang (1999)] An exact confidence interval for θ_1 is*

$$\hat{\theta}_1 \pm \sqrt{q_\alpha(\nu, k)\, d_{11}\, qmse_1}\,,$$

where $q_\alpha(\nu, k)$ is the upper-α quantile of the null distribution of the pivotal quantity Q_1^2.

Exact confidence intervals for each of the other $k-1$ effects can be obtain analogously. Also, the existence of exact individual confidence intervals implies the existence of exact individual tests.

An apparent shortcoming of the method of Kinateder, Voss and Wang (1999) is that the induced effects $\tau_1, \ldots, \tau_k$ tend to have less effect sparsity than the effects $\theta_1, \ldots, \theta_k$. Specifically, the induced effect τ_i corresponds in a sense to θ_i but is a linear combination of $\theta_1, \ldots, \theta_i$. This *contamination* of τ_i by some θ_j for $j < i$ is a consequence of the nonorthogonality of the design and the projection to orthogonality inherent in the methodology.

18.4.2   Open Problems

As noted previously, there are more open problems than proven methods in the case of non-orthogonal saturated designs. Here are some.

In method of Kinateder, Voss and Wang (1999) just presented, the induced effects depend on the subjective choice of the order of projections, or equivalently, on the order of the columns of the matrix A, the columns being in one-to-one correspondence with the effects θ_i. If the columns of the matrix A were permuted before applying the Gram-Schmidt process, then a different set of induced effects and a different confidence interval width would result. This subjective choice of the order of projections must be made a priori—use of that order which gives the tightest confidence interval would invalidate the procedure, making the procedure liberal. In order to remove this subjectivity, it is of interest to consider the distribution of the pivotal quantity using that permutation of the last $k - 1$ columns of A which minimizes the resulting quasi mean squared error, or equivalently, which minimizes the confidence interval width. It is an open problem to show that this variation on the procedure still provides strong control of the confidence level.

Also concerning the method of Kinateder, Voss and Wang (1999) for constructing individual confidence intervals, a different transformation or projection to orthogonality is needed for each effect. It is desirable to obtain exact methods for which this is not the case. Along these lines, Kunert (1997) used just one such transformation to obtain independent sums of squares with which to construct a single estimate of σ^2, then proposed using this same estimator of σ^2 for the inferences for each of the effects. It is still an open problem to show that his approach strongly controls the error rate.

Finally, there are no known results concerning simultaneous tests or confidence intervals which strongly control error rates.

18.5   SUPER-SATURATED DESIGNS

While there is only one method known to strongly control error rates for the analysis of non-orthogonal saturated designs, there do not exist *any* for the analysis of super-saturated designs. This is an obvious gap in the statistical literature, because design and analysis of experiments are inseparable, and substantial work has been done on the construction of super-saturated designs. Construction of designs has been considered for example by Booth and Cox (1962), Srivastava (1975), Srivastava and Gupta (1979), Anderson and Thomas (1980), Ghosh (1980, 1981), Rosenberger and Smith (1984), Ohnishi and Shirakura (1985), Barnett and Hurwitz (1990), Shirakura (1991), Lin (1993, 1995), Nguyen (1996), Tang and Wu (1997), Yamada and Lin (1997).

Even discussion of the analysis of super-saturated designs in the literature is scarce. This is probably due to the difficulty of the problem. Super-saturated designs not only present the difficult problem of non-orthogonality but also the additional complication of non-estimability. Specifically, if all effects of interest are included in the model, then the model is over-parameterized, so effects are not estimable. Two approaches have been suggested in the literature to circumvent this problem of non-estimability.

The first method of analysis proposed concerns a special class of super-saturated designs called search designs, introduced by Srivastava (1975). Suppose the $k = k_1 + k_2$ effects can be partitioned into two sets of sizes k_1 and k_2, respectively, in such a way that all k_1 effects in the first set may be nonzero so are to be estimated, and at most m of the other k_2 effects are nonzero. A design is a $k_1 + m$ *search design* if it allows the (search and) identification of all nonzero effects under a noiseless model. An equivalent condition is that, for each combination of $2m$ of the k_2 effects in the second set, the submatrix of the design matrix corresponding to the $k_1 + 2m$ effects is of rank $k_1 + 2m$ [Srivastava (1975)]. For the analysis of search design, Srivastava suggested use of the best submodel of given size for the analysis, the best submodel being the one yielding the smallest mean squared error. This appears to be the best starting point for the analysis of super-saturated designs, but no results concerning control of error rates are available.

The second approach suggested for the analysis of super-saturated designs is by Westfall, Young and Lin (1998). They proposed use of forward selection for model building and discussed error control, but they were unable to establish strong control of error rates.

REFERENCES

Alam, K. and Rizvi, M. H. (1966). Selection from multivariate normal populations, *Annals of the Institute of Statistical Mathematics*, **18**, 307–318.

Anderson, D. A. and Thomas, A. M. (1980). Weakly resolvable IV.3 search designs for the p^n factorial experiment, *Journal of Statistical*

Planning and Inference, **4**, 299–312.

Barnett, E. H. and Hurwitz, A. M. (1990). Near orthogonality: systematic supersaturated designs—extension and application, In *Transactions of the 46th Annual Rochester ASQC Conference, University of Rochester*, 27–42, American Society for Quality Control, Milwaukee.

Benski, H. C. (1989). Use of a normality test to identify significant effects in factorial designs, *Journal of Quality Technology*, **21**, 174–178.

Berk, K. N. and Picard, R. R. (1991). Significance tests for saturated orthogonal arrays, *Journal of Quality Technology*, **23**, 79–89.

Birnbaum, A. (1959). On the analysis of factorial experiments without replication, *Technometrics*, **1**, 343–357.

Booth, K. H. V. and Cox, D. R. (1962). Some systematic supersaturated designs, *Technometrics*, **4**, 489–495.

Box, G. E. P. and Meyer, R. D. (1986). An analysis for unreplicated fractional factorials, *Technometrics*, **28**, 11–18.

Box, G. E. P. and R. D. Meyer (1993). Finding the active factors in fractionated screening experiments, *Journal of Quality Technology*, **25**, 94–105.

Cochran, W. G. (1941). The distribution of the largest of a set of estimated variances as a fraction of their total, *Annals of Eugenics*, **11**, 47–52.

Daniel, C. (1959). Use of half-normal plots in interpreting factorial two-level experiments, *Technometrics*, **1**, 311–341.

Dong, F. (1993). On the identification of active contrasts in unreplicated fractional factorials, *Statistica Sinica*, **3**, 209–217.

Ghosh, S. (1980). On main effect plus one plans for 2^m factorials, *Annals of Statistics*, **8**, 922–930.

Gupta, S. S., Huang, D. Y. and Panchapakesan, S. (1982), On some inequalities and monotonicity results in selection and ranking theory, In *Inequalities in Statistics and Probability* (Ed., Y. L. Tong), Vol. 5, pp, 211–227, IMS Lecture Notes—Monograph Series, Hayward, California.

Haaland, P. D. and O'Connell, M. A. (1995). Inference for effect-saturated fractional factorials, *Technometrics*, **37**, 82–93.

Hamada, M. and Balakrishnan, N. (1998). Analyzing unreplicated factorial experiments: A review with some new proposals (with Discussion), *Statistica Sinica*, **8**, 1–41.

Hochberg, Y. and Tamhane, A. C. (1987). *Multiple Comparison Procedures*, John Wiley & Sons, New York.

Holm, S. (1979). A simple sequentially rejective multiple test procedure, *Scandinavian Journal of Statistics*, **6**, 65–70.

Holms, A. G., and Berrettoni, J. N. (1969). Chain-pooling ANOVA for two-level factorial replication-free experiments, *Technometrics*, **11**, 725–746.

Hsu, J. C. (1996). *Multiple Comparisons: Theory and Methods*, Chapman and Hall, New York.

Juan, J. and Peña, D. (1992). A simple method to identify significant effects in unreplicated two-level factorial designs, *Communications in Statistics—Theory and Methods*, **21**, 1383–1403.

Kinateder, K. J., Voss, D. T. and Wang, W. (1999). Exact confidence intervals in the analysis of nonorthogonal saturated designs, *Submitted for publication.*

Kunert, J. (1997). On the use of the factor-sparsity assumption to get an estimate of the variance in saturated designs, *Technometrics*, **39**, 81–90.

Langsrud, O. and Naes, T. (1998). A unified framework for significance testing in fractional factorials, *Computational Statistics & Data Analysis*, **28**, 413–431.

Lehmann, E. L. (1986). *Testing Statistical Hypotheses*, Second edition, John Wiley & Sons, New York.

Lenth, R. V. (1989). Quick and easy analysis of unreplicated factorials, *Technometrics*, **31**, 469–473.

Lin, D. K. J. (1993). A new class of supersaturated designs, *Technometrics*, **35**, 28–31.

Lin, D. K. J. (1995). Generating systematic supersaturated designs, *Technometrics*, **37**, 213–225.

Loh, W. Y. (1992). Identification of active contrasts in unreplicated factorial experiments, *Computational Statistics & Data Analysis*, **14**, 135–148.

Loughin, T. M., and Noble, W. (1997). A permutation test for effects in an unreplicated factorial design, *Technometrics*, **39**, 180–190.

Mahamunulu, D. M. (1967). Some fixed-sample ranking and selection problems, *Annals of Mathematical Statistics*, **38**, 1079–1091.

Marcus, R., Peritz, E. and Gabriel, K. R. (1976). On closed testing procedures with special reference to ordered analysis of variance, *Biometrika*, **63**, 655–660.

Nguyen, N.-K. (1996). An algorithmic approach to constructing supersaturated designs, *Technometrics*, **38**, 69–73.

Ohnishi, T. and Shirakura, T. (1985). Search designs for 2^m factorial experiments, *Journal of Statistical Planning and Inference*, **11**, 241–245.

Plackett, R. L. and Burman, J. P. (1946). The design of optimum multi-factorial experiments, *Biometrika*, **33**, 305–325.

Rosenberger, J. L. and Smith, D. E. (1984). Interruptible supersaturated two-level designs, *Communications in Statistics—Theory and Methods*, **13**, 599–609.

Schneider, H., Kasperski, W. J., and Weissfeld, L. (1993). Finding significant effects for unreplicated fractional factorials using the n smallest contrasts, *Journal of Quality Technology*, **25**, 18–27.

Shaffer, J. P. (1980). Control of directional errors with stagewise multiple test procedures, *Annals of Statistics*, **8**, 1342–1347.

Shirakura, T. (1991). Main effect plus one or two plans for 2^m factorials, *Journal of Statistical Planning and Inference*, **27**, 65–74.

Srivastava, J. N. (1975). Designs for searching non-negligible effects, In *A Survey of Design and Linear Models* (Ed., J. N. Srivastava), pp, 507–519, North-Holland, Amsterdam.

Srivastava, J. N. and Gupta, B. C. (1979). Main effect plan for 2^m factorials which allow search and estimation of one unknown effect, *Journal of Statistical Planning and Inference*, **3**, 259–265.

Tang, B. and Wu, C. F. J. (1997). A method for constructing super-saturated designs and its Es^2 optimality, *The Canadian Journal of Statistics*, **25**, 191–201.

Torres, V. A. (1993). A simple analysis of unreplicated factorials with possible abnormalities, *Journal of Quality Technology*, **25**, 183–187.

Venter, J. H. and Steel, S. J. (1996). A hypothesis-testing approach toward identifying active contrasts, *Technometrics*, **38**, 161–169.

Venter, J. H. and Steel, S. J. (1998). Identifying active contrasts by step-wise testing, *Technometrics*, **40**, 304–313.

Voss, D. T. (1988). Generalized modulus-ratio tests for analysis of factorial designs with zero degrees of freedom for error, *Communications in Statistics—Theory and Methods*, **17**, 3345–3359.

Voss, D. T. (1999). Analysis of orthogonal saturated designs, *Journal of Statistical Planning and Inference*, **78**, 111-130.

Voss, D. T. and Wang, W. (1999). Exact simultaneous confidence intervals in the analysis of orthogonal saturated designs, *Journal of Statistical Planning and Inference* (to appear).

Westfall, P. H., Young, S. S., and Lin, D. K. J. (1998). Forward selection error control in the analysis of supersaturated designs, *Statistica Sinica*, **8**, 101–117.

Yamada, S. and Lin, D. K. J. (1997). Supersaturated design including an orthogonal base, *The Canadian Journal of Statistics*, **25**, 203–213.

Zahn, D. A. (1969). *An empirical study of the half-normal plot*, Ph.D. thesis, Harvard University, Boston.

Zahn, D. A. (1975a). Modifications of and revised critical values for the half-normal plots, *Technometrics*, **17**, 189–200.

Zahn, D. A. (1975b). An empirical study of the half-normal plot, *Technometrics*, **17**, 201–211.

CHAPTER 19

ON ESTIMATING SUBJECT-TREATMENT INTERACTION

GARY GADBURY

University of North Carolina, Greensboro, NC

HARI IYER

Colorado State University, Fort Collins, CO

Abstract: We begin by considering a population of units $\mathbf{U} = (u_1, u_2, \ldots, u_N)$. Our objective is to study the effect of a treatment t on these units with respect to a particular response of interest in the context of a randomized experiment. We make use of a model sometimes referred to as a "Potential response model" or as "Rubin model for causal inference." The model has been used by others to analyze problems associated with estimating a mean treatment effect in both randomized experiments and observational studies.

However, an "average treatment effect" is a meaningful quantity only when it adequately represents the effect of t on each unit. If the effect of t is highly variable from one unit to another, i.e., when the subject-treatment interaction is nonnegligible, then the average treatment effect loses its importance. In fact, a treatment might appear to be a "beneficial" treatment when examining its average effect even though a substantial proportion of the units in the population experience an "unfavorable" effect. This proportion can be calculated or approximated if one knows the variance of the treatment effects along with the mean. Questions concerning the estimation of the variance of treatment effects in a finite population is the subject of this paper.

Keywords and phrases: Rubin model, non-additivity, counterfactual model, potential response model

349

19.1   INTRODUCTION

We begin by considering a population of units $\mathbf{U} = (u_1, u_2, \ldots, u_N)$. Our objective is to study the effect of a treatment t on these units with respect to a particular response of interest in a designed experiment. This effect is to be assessed by comparing the response when the units are subjected to the treatment t with the response when the units are subjected to a control treatment c which is used as a basis of reference. In the ideal situation when the responses to both treatments t and c are known for each unit, the "true" effect of treatment t can be calculated for each unit by taking the difference between the responses of that unit to treatments t and c, respectively. Without loss of generality let us suppose that a positive difference indicates that the treatment effect is "beneficial" and a negative difference indicates an "unfavorable" effect.

In practice it is not possible to obtain the responses of each unit to each of the two treatments at the same time. Instead, what is often done is the following. Suppose that the size of the experimental group is $N = 2n$. Half of these units, chosen randomly, will receive treatment t, with the other half receiving treatment c. The average effect of treatment t is estimated by computing the difference between the average response of the units receiving treatment t and the average response of the units receiving treatment c.

"Average treatment effect" is a meaningful quantity only when it adequately represents the effect of t applied on each unit. If the effect of t is highly variable from one unit to another, then the average treatment effect loses its importance. In fact, a treatment may appear to be "beneficial" when examining its average effect even though a substantial proportion of the units in the population experience an "unfavorable" effect. This proportion can be calculated or approximated if one knows the variance of the treatment effects in addition to its mean. Questions concerning the estimation of the variance of treatment effects in a finite population is the subject of this paper.

Consider the following matrix of responses $X_i, Y_i, i = 1, \ldots, N$ corresponding to the N units in the finite population under consideration.

$$\begin{pmatrix} X_1 & Y_1 \\ X_2 & Y_2 \\ \vdots & \vdots \\ X_N & Y_N \end{pmatrix}. \qquad (19.1.1)$$

Here X_i represents the response of unit u_i if treatment t is applied to it and Y_i denotes its response if treatment c is applied. Even though only one of the two responses can be observed on any given unit, conceptually, the "true" treatment effect for the ith unit is defined as D_i given by

$$D_i = X_i - Y_i. \tag{19.1.2}$$

When the values of D_i are different from one unit to the next, i.e., when a subject-treatment interaction is present, it may be useful to consider the variance S_D^2 of the treatment effects given by

$$
\begin{aligned}
S_D^2 &= \frac{1}{N}\{\sum_i D_i^2 - N\overline{D}^2\} \\
&= S_X^2 + S_Y^2 - 2S_{X,Y}
\end{aligned}
\tag{19.1.3}
$$

where S_X^2 and S_Y^2 are the finite population variances of X and Y respectively, and $S_{X,Y}$ is the finite population covariance. The finite population standard deviation of treatment effects is then S_D and the correlation between X and Y is

$$R_{X,Y} = \frac{S_{X,Y}}{S_X S_Y}.$$

Observe that S_D^2 is zero if and only if there is no subject-treatment interaction.

The key problem of estimating the variance of treatment effects is rooted in the fact that the correlation parameter $R_{X,Y}$ is not estimable from observable data. This nonestimable correlation parameter has a long history going back to Neyman (1935) [the idea of potential responses actually dates to Neyman 1923)]. In the 1935 paper, Neyman demonstrated how in a finite population, estimates of standard errors of the estimated mean treatment effect can be biased due to a subject-treatment interaction, but he did not consider estimation of this interaction.

The assumption throughout this paper is that units only receive one treatment, t or c. We do not consider repeated measures or crossover studies which allow the observation of responses to several treatments on the same subject because these observations are not made under identical conditions and, without several additional unverifiable assumptions, inference about the "true" treatment effect is not possible with these designs. Therefore, it seems reasonable that if we are to proceed with an estimation of S_D^2, the unobserved values (which may be thought of as missing values) may have to be estimated.

Several authors have considered the problem of missing values in various contexts. Estimating missing values in survey data has been referred to as imputation [Rubin (1996), Fay (1996) and Rao (1996)]. Imputing survey data is an important topic of research since missing responses may introduce bias in the estimated mean treatment effect if the responses are not missing at random. In our context, though, the missing values occur as a result of a random treatment assignment. Our motivation for estimating these missing responses is to estimate individual treatment effects thereby facilitating an estimate of S_D^2.

When covariate information is available on all units in the finite population as in Rubin (1978), we may be able to use this information to estimate the unobserved values. Rao (1996), for instance, considers stratification, regression, and ratio imputation in survey data. In each of these cases, the overall objective was the estimation of a mean response and its standard error. Another technique to consider is the matching of subjects by covariate information. The problem of matching units in order to obtain a mean treatment effect and its standard error in observational studies has been considered extensively [Rosenbaum (1989), Rosenbaum (1995) and Rosenbaum and Rubin (1983)].

The present work focuses on estimation of the standard deviation S_D using covariate information to impute unobserved potential responses in a two sample design. In Section 19.2 we consider the use of a single covariate and discuss the bias of the resulting estimator of S_D^2. Section 19.3 includes an illustrative example. We conclude with a summary discussion.

19.2   AN ESTIMATOR OF S_D^2 USING CONCOMITANT INFORMATION

Consider a finite population of size $N = 2n$. In a randomized experiment for estimating the effect of the treatment t, suppose half the units in the finite population, chosen randomly, are subjected to treatment t and the other half are subjected to treatment c. We partition the indices $i = 1, 2, \ldots, N$ of $\mathbf{U}$ into two sets v and w, where $i \in v$ if u_i is assigned to the treatment group, and $i \in w$ if u_i is assigned to the control group. We let $\mathbf{X}_v$ denote the vector of responses to t of the n units in the treatment group, and $\mathbf{Y}_w$ denote the vector of responses to c of the n units in the control group. Once units are randomly assigned to a treatment group and a control group as described above, we observe the responses $\mathbf{X}_v$ and $\mathbf{Y}_w$. The observations may be written as follows:

$$\begin{pmatrix} \mathbf{X}_v & -- \\ -- & \mathbf{Y}_w \end{pmatrix}. \tag{19.2.4}$$

The dashed lines in the above matrix constitute values that were not observed.

Suppose a covariate Z is observed on all the units in the finite population. Further suppose that Z is used to predict the unobserved values for the N units. The collection of all observed values and the combined set of observed as well as predicted values for the N population units may be exhibited as shown below.

$$
\begin{pmatrix}
\mathbf{X} & \mathbf{Y} & \mathbf{Z} \\
\hline
X_1 & -- & Z_1 \\
X_2 & -- & Z_2 \\
\vdots & \vdots & \vdots \\
X_n & -- & Z_n \\
-- & Y_{n+1} & Z_{n+1} \\
\vdots & \vdots & \vdots \\
-- & Y_N & Z_N
\end{pmatrix}
\implies
\begin{pmatrix}
\mathbf{x} & \mathbf{y} & \mathbf{z} \\
\hline
X_1 & \hat{Y}_1 & Z_1 \\
X_2 & \hat{Y}_2 & Z_2 \\
\vdots & \vdots & \vdots \\
X_n & \hat{Y}_n & Z_n \\
\hat{X}_{n+1} & Y_{n+1} & Z_{n+1} \\
\vdots & \vdots & \vdots \\
\hat{X}_N & Y_N & Z_N
\end{pmatrix}.
\qquad (19.2.5)
$$

For convenience, we have labeled the population units such that units that received treatment t are labeled 1 through n. We call the populations in 19.2.5 above the "observed population" (on the left), and the "estimated population" (on the right).

Let the vector of treatment differences in the estimated population be $\mathbf{d} = \mathbf{x} - \mathbf{y}$. An obvious estimator of $S_\mathbf{D}^2$ is the quantity $S_\mathbf{d}^2$ defined by

$$
S_\mathbf{d}^2 = \frac{1}{N}[\sum_i d_i^2 - N\bar{d}^2].
\qquad (19.2.6)
$$

Noting that we can only observe half of the potential responses, X and Y, and, so, must predict the other half, one may be concerned that too many values are being imputed in forming the estimator $S_\mathbf{d}^2$. This is an inherent limitation of the problem and it arises due to the nonestimability of the correlation parameter $R_{\mathbf{X},\mathbf{Y}}$. Certainly we expect $S_\mathbf{d}^2$ to be a biased estimator of $S_\mathbf{D}^2$ in general and the bias could be relatively large in magnitude. The first question we consider is, "how large is the bias when expectations are taken over all possible treatment assignments on the true finite population?" It is reasonable to expect that the bias will depend on how well the prediction function estimates the unobserved values. First, we answer this question without making any assumptions regarding the form of the prediction function or of the true finite population. We only assume a random treatment assignment.

A general prediction approach

We describe the estimated population for a given treatment assignment j (that we refer to as sample j), as follows:

$$\begin{pmatrix} X_1 + e_{1j} & Y_1 + f_{1j} \\ X_2 + e_{2j} & Y_2 + f_{2j} \\ \vdots & \vdots \\ X_{2n} + e_{2n,j} & Y_{2n} + f_{2n,j} \end{pmatrix} \qquad (19.2.7)$$

where e_{ij}, is equal to 0 if X_i is actually observed in sample j. If X_i is not observed, then it is estimated by $\hat{X}_{ij}$ and e_{ij} represents an error term which is defined by

$$e_{ij} = \hat{X}_{ij} - X_i.$$

The term f_{ij} is defined in a similar manner, but with respect to Y. Note that the prediction $\hat{X}_{ij}$ may depend not only on the unit u_i, but also the other units in sample j. Here i belongs to the set $\{1, 2, \ldots, N\}$ and j belongs to the set $\{1, 2, \ldots, k\}$ where $k = \binom{2n}{n}$, the total number of possible treatment assignments. For any unit u_i in sample j, note that at least one of e_{ij}, f_{ij} will be zero.

Recall that $\mathbf{D} = (X_1 - Y_1, X_2 - Y_2, \ldots, X_{2n} - Y_{2n})^T$ and define the $2n \times k$ matrix

$$\mathbf{G} = \{g_{ij}\} \qquad (19.2.8)$$

by defining $g_{ij} = e_{ij} - f_{ij}$. We consider both $\mathbf{D}$ and $\mathbf{G}$ as fixed, but not necessarily known. We can write $\mathbf{G}$ in the form

$$\mathbf{G} = [\mathbf{g_1}, \mathbf{g_2}, \ldots, \mathbf{g_k}] \qquad (19.2.9)$$

where

$$\mathbf{g_j} = (g_{1j}, g_{2j}, \ldots, g_{2n,j})^T.$$

That is, $\mathbf{g_j}$ is the jth column of $\mathbf{G}$. Let

$$S_{\mathbf{g_j}}^2 = \frac{1}{2n}[\sum_i g_{ij}^2 - \frac{(\sum_i g_{ij})^2}{2n}]$$

In other words, $S_{\mathbf{g_j}}^2$ is the variance of the jth column of $\mathbf{G}$. Also let

$$S_{\mathbf{D},\mathbf{g_j}} = \frac{1}{2n}\sum_i (D_i - \overline{D})g_{ij}.$$

That is, $S_{\mathbf{D},\mathbf{g_j}}$ is the covariance between $\mathbf{D}$ and the jth column of $\mathbf{G}$. We then have the following proposition.

Proposition 19.2.1 *The expectation of S_d^2 over all possible random treatment assignments on the true finite population (represented by equation (19.1.1)) is given by*

$$
\begin{aligned}
E(S_d^2) &= S_D^2 + \frac{1}{k}\sum_{j=1}^{k}[S_{g_j}^2 + 2S_{D,g_j}] \\
&= S_D^2 + \overline{S_{g_j}^2} + 2\overline{S_{D,g_j}} \qquad (19.2.10)
\end{aligned}
$$

where the overlined quantities simply represent averages of the quantities over all columns of **G**.

PROOF.   Since the treatment assignment is random we can proceed in a straightforward manner. Note that for any fixed j we have,

$$
d_{ij} = D_i + g_{ij}
$$

It follows that,

$$
\begin{aligned}
\frac{1}{N}\sum_{i=1}^{N}(d_{ij} - \bar{d}_{.j})^2 \\
&= Var\{d_{ij}\}_{i=1,\ldots,N} \\
&= Var(D_i) + Var\{g_{ij}\}_{i=1,\ldots,N} + 2Cov(\{D_i\}, \{g_{ij}\})_{i=1,\ldots,N} \\
&= \frac{1}{N}\sum_{i=1}^{N}(D_i - \bar{D})^2 + S_{g_j}^2 + 2S_{D,g_j}.
\end{aligned}
$$

Therefore, for a fixed j,

$$
S_d^2 = S_D + S_{g_j}^2 + 2S_{D,g_j}.
$$

Taking expectations of the left hand side over all possible random treatment assignments (i.e., over all $j = 1, 2, \ldots, k$) gives the result.   □

We write the bias of S_d^2 as an estimator of S_D^2, over all possible treatment assignments as

$$
bias = \overline{S_{g_j}^2} + 2\overline{S_{D,g_j}}. \qquad (19.2.11)
$$

One can see that, for S_d^2 to be an unbiased estimator of S_D^2, one of the following conditions must hold.

1. the variance of each column of **G** is zero, or

2. $\overline{S^2_{g_J}} = -2\overline{S_{D,g_J}}$.

In practice, neither of the conditions is expected to hold in general. To develop a better understanding of the nature of the bias, we consider the finite population of responses to be a sample of size N from a suitable superpopulation, and investigate the expectation of the bias computed over this superpopulation. We use the symbol $\mathcal{E}$ to denote this expectation operator. We must also choose a prediction function since the distribution of the errors g_{ij} becomes relevant. This is discussed in the next proposition.

Proposition 19.2.2 *Suppose the following assumptions are satisfied:*

1. *$(X_i, Y_i, Z_i), i = 1, \ldots, N = 2n$ is an iid sample from a superpopulation which is normal with mean equal to*

$$\begin{pmatrix} \mu_X \\ \mu_Y \\ \mu_Z \end{pmatrix}$$

   and covariance matrix equal to

$$\begin{pmatrix} \sigma_X^2 & \sigma_{XY} & \sigma_{XZ} \\ \sigma_{XY} & \sigma_Y^2 & \sigma_{YZ} \\ \sigma_{XZ} & \sigma_{YZ} & \sigma_Z^2 \end{pmatrix}.$$

2. *Treatment assignment is random with n units receiving treatment t and the rest receiving treatment c.*

3. *A least squares linear regression function relating X and Z is used to predict the unobserved X values. Also, a least squares linear regression function relating Y and Z is used to predict the unobserved Y values. So, for a given sample s, where indices $i = 1, \ldots 2n$, are partitioned into sets v and w with $i \in v$ if $X_i \in s$, and $i \in w$ if $Y_i \in s$, we have*

$$\hat{\mathbf{X}}_w = b_o(v)\mathbf{1_n} + b_1(v)\mathbf{Z}_w$$
$$\hat{\mathbf{Y}}_v = b'_o(w)\mathbf{1_n} + b'_1(w)\mathbf{Z}_v$$

4. *The population size $N = 2n \geq 8$.*

Then the expectation of this bias over the superpopulation is given by

$$BIAS = \mathcal{E}(bias) = -\frac{n-4}{2(n-3)}(\sigma_{X|Z}^2 + \sigma_{Y|Z}^2) + \frac{2n-1}{n}\sigma_{XY|Z} \quad (19.2.12)$$

where $\sigma_{X|Z}^2 = \sigma_X^2(1 - \rho_{XZ}^2)$ and $\sigma_{Y|Z}^2 = \sigma_Y^2(1 - \rho_{YZ}^2)$ are the conditional variances of X and Y, respectively, and $\sigma_{XY|Z} = \sigma_X\sigma_Y(\rho_{XY} - \rho_{XZ}\rho_{YZ})$ is the conditional covariance of X and Y, given Z. Here ρ_{XZ} is the correlation between X and Z, and ρ_{YZ} is the correlation between Y and Z, in the superpopulation.

The proof of Proposition 19.2.2 follows from standard results concerning expectations of quadratic forms. As an immediate consequence of this proposition, we have the following corollary.

Corollary 19.2.1 *As $\sigma_{X|Z}^2$ and $\sigma_{Y|Z}^2$ approach zero, BIAS in equation (19.2.12) approaches zero.*

This result follows by noting that $\sigma_{XY|Z}$ approaches zero as $\sigma_{X|Z}^2$ and $\sigma_{Y|Z}^2$ approach zero.

Thus, when Z is a good covariate, i.e., it is a good predictor of X and Y, the estimator S_d^2 of S_D^2 may be expected to have a very small bias.

Note that all superpopulation parameters in $BIAS$ in equation (19.2.12) are estimable from the observed data with the exception of $\sigma_{XY|Z}$ or, equivalently, ρ_{XY}. If one were to write the likelihood function based on the observed data, one would note that the parameter ρ_{XY} does not even appear in the likelihood function. In the special case when $\sigma_{X|Z}^2 = \sigma_{Y|Z}^2$ we see that BIAS is zero if and only if

$$\rho_{XY|Z} = \frac{n(n-4)}{(2n-1)(n-3)}. \tag{19.2.13}$$

where $\rho_{XY|Z}$ is the partial correlation coefficient of X and Y conditional on Z. For values of $\rho_{XY|Z}$ larger than $\frac{n(n-4)}{(2n-1)(n-3)}$ $BIAS$ is positive. For large N the condition in (19.2.13) reduces to $\rho_{XY|Z} = 1/2$.

We can get an idea of the sensitivity of $BIAS$ to varying values of ρ_{XY}, by fixing all other parameters and letting ρ_{XY} vary over values such that the trivariate correlation matrix of $(X, Y, Z)^T$ is positive definite. For illustration we fix $\sigma_X = \sigma_Y = \sigma_Z = 1$, and $\rho_{XZ} = \rho_{YZ} = 0.7$. All means are set to zero since the results will not depend on the population means. Figure 19.1 shows the relationship between the true superpopulation variance of treatment effects, $Var(X - Y)$, and the expected value of S_d^2. Under the assumed superpopulation structure $BIAS$ is negative for most values of ρ_{XY}. However $BIAS$ becomes much smaller as ρ_{XY} approaches the values of ρ_{XZ} and ρ_{YZ}. As ρ_{XY} becomes larger than ρ_{XZ} and ρ_{YZ}, $BIAS$ becomes positive indicating that, on average, the estimator will overestimate

$Var(X - Y)$. It is clear that, as the values of ρ_{XZ} and ρ_{YZ} become close to one, $BIAS$ becomes smaller for all allowable values of ρ_{XY}.

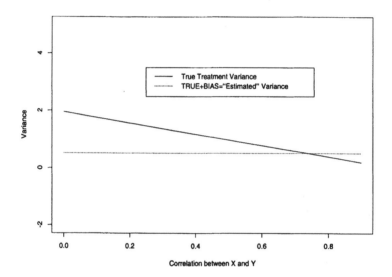

FIGURE 19.1 Illustration of the sensitivity of $BIAS$ to varying values of ρ_{XY}. $\sigma_X = \sigma_Y = \sigma_Z = 1$, and $\rho_{XZ} = \rho_{YZ} = 0.7$. The true treatment variance is $Var(X - Y) = \sigma_X^2 + \sigma_Y^2 - 2\sigma_X\sigma_Y\rho_{XY}$

A related contribution by Rubin and Thayer (1978) and Thayer (1983) shows how, for selected plausible values of the partial correlation $\rho_{X,Y|Z}$, one can average over the covariate Z obtaining limits on the simple correlation ρ_{XY}. Furthermore, under specific conditions, the limits on ρ_{XY} are tighter than the selected plausible limits on $\rho_{X,Y|Z}$. However, this analysis required selecting plausible limits on the partial correlation which may or may not be practical depending on the particular application. Also note that the partial correlation is not estimable from the observed data due to the fact that there is no information on the simple correlation, ρ_{XY}.

Using the fact that the trivariate correlation matrix of the random vector $(X, Y, Z)^t$ must be positive definite one can obtain bounds for S_D^2 and, under the superpopulation model, for σ_D^2. These bounds are estimable based on observable data. Gadbury and Iyer (1999) gave the maximum likelihood estimator for mathematical bounds for σ_D^2 and for the proportion P_- of the population units experiencing an unfavorable treatment effect. They also gave the large sample distributions of these estimators. However, they did not study the characteristics of any estimator for the variance of S_D^2, nor did they consider randomization based inference setup as we do in this article.

Simulations could be used to examine the characteristic of estimated bounds for S_D^2 with respect to a randomization distribution. Results of a simulation study that examined the operating characteristics of estimated bounds may be found in Gadbury and Iyer (1997). The simulation results indicated that, when a suitable covariate Z is available, useful estimates for the bounds on S_D^2 can be calculated from observed data. Furthermore, as the correlations R_{XZ} and R_{YZ} approach one, the estimator S_d^2 approaches the true variance S_D^2.

19.3   AN ILLUSTRATIVE EXAMPLE

We illustrate the ideas reported in this paper using a constructed example so that the true variance of treatment effects, S_D^2, is known. A finite population of potential responses $(X_i, Y_i, Z_i), i = 1, \ldots, N = 40$ was generated as iid observations from a normal superpopulation with mean equal to $(\mu_X, \mu_Y, \mu_Z)^T = (3, 0, 0)^T$, correlation matrix equal to

$$
\begin{pmatrix}
1 & \rho_{XY} & \rho_{XZ} \\
\rho_{XY} & 1 & \rho_{YZ} \\
\rho_{XZ} & \rho_{YZ} & 1
\end{pmatrix}
=
\begin{pmatrix}
1 & 0.5 & 0.75 \\
0.5 & 1 & 0.75 \\
0.75 & 0.75 & 1
\end{pmatrix},
$$

and $\sigma_X = \sigma_Y = 3$ and $\sigma_Z = 1$. The potential responses are shown in Table 19.1. The true treatment effects are shown in the column $D = X - Y$. The finite population mean treatment effect is $\overline{D} = 2.75$, and the variance is $S_D^2 = 7.63$ ($S_D = 2.762$). Note that the standard deviation of treatment effects, S_D, is nearly equal to the average of 2.75. It is the mean treatment effect that is usually estimated from observed data, but knowing something about the magnitude of the variance of treatment effects will help in understanding how the treatment is affecting individuals in the population.

A random treatment assignment produces the observed population shown in Table 19.2. As mentioned earlier, either X or Y can be observed for a unit but not both, so a true treatment effect, D, cannot be observed for any unit. Let $\mathbf{X}_v$ and $\mathbf{Y}_w$ be the observed X and Y, respectively, after

treatment assignment as indicated in Section 19.2. Then it is straightforward to show [Rubin (1974)] that $\overline{X}_v - \overline{Y}_w = 3.425$ (for this example) is
an unbiased estimate for $\overline{D}$.

Estimating S_D^2 requires some knowledge about the individual treatment
effects. So we estimate the D_i $i = 1, \ldots, 40$ as described in Section 19.2 in
Proposition 19.2.2. That is, we use Z to predict missing potential responses
using least squares regression models relating observed X and Z, and relating observed Y and Z. The resulting estimated population is shown in
Table 19.3.

The variance of predicted treatment effects is $S_d^2 = 3.638$ which is a biased estimate of the true variance from results stated in Proposition 19.2.1.
To investigate the nature of this bias we can use Proposition 19.2.2. We
first estimate the parameters that can be estimated from the observable
data. The sample statistics are,

$$\hat{\sigma}_{X|Z} = 1.942 \quad \hat{\sigma}_{Y|Z} = 2.033 \quad \hat{\sigma}_X = 2.74 \quad \hat{\sigma}_Y = 3.31$$

$$\hat{\rho}_{XZ} = 0.724 \quad \hat{\rho}_{YZ} = 0.801.$$

We then let the nonestimable ρ_{XY} range over selected values such that the
estimated 3-dimensional correlation matrix remained positive definite, that
is, $\rho_{XY} = 0.2, 0.3, \ldots, 0.9$. An estimated $BIAS$ using equation (19.2.12)
would depend on these values of ρ_{XY}. The results are,

ρ_{XY}	0.20	0.30	0.40	0.50	0.60	0.70	0.80	0.90
$B\hat{I}AS$	-10.44	-8.67	-6.90	-5.13	-3.36	-1.60	0.17	1.94

These results show that the bias of S_d^2 could be large in magnitude
unless ρ_{XY} is close to 0.80. Specifically, using equation (19.2.13), we see
that the partial correlation $\rho_{XY|Z} = 0.483$ would correspond to $BIAS = 0$.
One could solve for the simple correlation and find that $\rho_{XY|Z} = 0.483$
corresponds to $\rho_{XY} = 0.774$. Since, for our example, $\rho_{XY} = 0.50$, the
estimated $BIAS$ our estimator S_d^2 would be -5.13, but this value could
not be known in practice since it would depend on subjective knowledge
about the value of ρ_{XY}. The $BIAS$ would approach zero as ρ_{XZ} and ρ_{YZ}
approach 1.

19.4   SUMMARY/CONCLUSIONS

In this paper we have used what is often called a Potential Response Model
to define "true" effect of a treatment on any individual in a population.
Using this framework, one can conceive of true treatment effects for individual units and can thereby conceive of a true average treatment effect,

and a true variance of treatment effects over a finite population. We argued that if the variance of treatment effects is large with respect to the magnitude of the average effect, then the average treatment effect alone does not adequately describe how the treatment is affecting the population. Others have considered inference on the true average treatment effect but inference on the variance has not been considered. Various assumptions have been put forward in the literature, for instance, *unit homogeneity* and *constant effect*, under which the variance of treatment effects is zero (or, in practice, very small). Without such strong assumptions, one must use available data themselves to learn about this variance.

Availability of concomitant information was then considered and a biased estimator of S_D^2 was proposed. The finite population bias of this estimator was derived and the sensitivity of the bias to varying values of the nonestimable correlation between X and Y was examined using a superpopulation framework. These results extend in a straightforward manner to the case of multiple covariates. See Gadbury and Iyer (1997). While it is well known that good covariates allow for more efficient estimation of an average treatment effect, the results of this paper demonstrate that they also lead to useful information on the variance of the treatment effects.

Two final comments are noteworthy. The framework used in this paper did not assume any measurement error in the responses. An added degree of complexity arises if measurement errors are suspected. Also, treatment effects were only considered for one point in time. In general, one could conceive of multiple "true potential populations" of responses occurring at different points in time. Even if we can observe all potential responses, it is still possible that the treatment effects differ in time for individual units. One must then consider what is meant by a "true treatment effect" before inference techniques can be employed. Exploring these two issues is a topic of continuing research.

Acknowledgements This research sponsored in part by United States Department of Agriculture, Forest Service contract 28-C5-898, Rocky Mountain Research Station, Fort Collins, CO.

REFERENCES

Fay, R. E. (1996). Alternative paradigms for the analysis of imputed survey data, *Journal of the American Statistical Association*, **91**, 490–498.

Gadbury, G. L. and Iyer, H. K. (1997). Causal inference and the problem of estimating a subject by treatment interaction, *Technical Report*, Department of Statistics, Colorado State University, Ft. Collins.

Gadbury, G. L. and Iyer, H. K. (1999). Unit-treatment interaction and its practical consequences, *Submitted for publication.*

Neyman, J. (1923). On the application of Probability Theory to agricultural experiments. Essay on principles. Section 9, *Annals of Agricultural Sciences*, pp. 1–51. The English translation by D. M. Dabrowska and T. P. Speed appeared in *Statistical Science*, **5**, 1990.

Neyman, J. (1935). Statistical problems in agricultural experimentation (with discussion), *Supplement to the Journal of the Royal Statistical Society, Series B*, **2**, 107–180. With cooperation of K. Iwaskiewics and St. Kolodziejczyk.

Rao, J. N. K. (1996). On variance estimation with imputed survey data, *Journal of the American Statistical Association*, **91**, 499–506.

Rosenbaum, P. R. (1989). The role of known effects in observational studies, *Biometrics*, **45**, 557–569.

Rosenbaum, P. R. (1995). *Observational Studies*, Springer-Verlag, New York.

Rosenbaum, P. R. and Rubin, D. B. (1983). The central role of the propensity score in observational studies for causal effects, *Biometrika*, **70**, 41–55.

Rubin, D. B. (1974). Estimating causal effects of treatments in randomized and nonrandomized studies, *Journal of Educational Psychology*, **66**, 688–701.

Rubin, D. B. (1978). Bayesian inference for causal effects: The role of randomization, *Annals of Statistics* **6**, 34–58.

Rubin, D. B. (1996). Multiple imputation after 18+ years, *Journal of the American Statistical Association*, **91**, 473–489.

Rubin, D. B. and Thayer, D. (1978). Relating tests given to different samples, *Psychometrika*, **43**, 3–10.

Thayer, D. T. (1983). Maximum likelihood estimation of the joint covariance matrix for sections of tests given to distinct samples with application to test equating, *Psychometrika*, **48**, 293–298.

TABLE 19.1 True finite population of potential responses

Unit	Z	X	Y	D	Unit	Z	X	Y	D
1	10.6	16.7	15.0	1.7	21	9.6	14.3	12.8	1.5
2	8.7	10.5	9.7	0.8	22	9.5	12.3	11.5	0.8
3	9.4	10.2	11.7	-1.5	23	10.1	12.0	13.2	-1.2
4	10.3	15.6	11.4	4.2	24	9.8	11.3	10.8	0.5
5	10.5	14.9	10.5	4.4	25	9.7	13.9	8.6	5.3
6	10.8	17.5	9.7	7.8	26	6.9	8.0	1.0	7.0
7	9.9	12.6	9.7	2.9	27	9.8	13.3	9.4	3.9
8	9.3	8.5	9.7	-1.2	28	8.6	10.7	7.6	3.1
9	9.7	15.4	10.7	4.7	29	9.9	12.6	11.6	1.0
10	9.8	14.8	4.8	10.0	30	11.4	16.2	8.7	7.5
11	10.7	12.9	15.0	-2.1	31	10.6	16.9	14.2	2.7
12	10.9	16.1	14.8	1.3	32	10.2	16.1	13.0	3.1
13	9.0	9.9	8.2	1.7	33	9.3	9.5	9.0	0.5
14	8.4	9.8	6.9	2.9	34	8.3	10.6	6.5	4.1
15	9.9	12.5	12.4	0.1	35	9.9	12.0	12.5	-0.5
16	8.6	13.0	8.0	5.0	36	8.7	11.6	8.5	3.1
17	10.7	16.9	12.2	4.7	37	10.5	17.4	11.6	5.8
18	9.5	12.3	9.4	2.9	38	11.1	15.5	13.1	2.4
19	9.6	11.3	13.2	-1.9	39	9.6	15.2	12.6	2.6
20	9.5	12.9	10.5	2.4	40	10.1	17.4	11.4	6.0

TABLE 19.2 Observed responses from the population after treatment assignment

Unit	Z	X	Y	Unit	Z	X	Y
1	10.6	16.7		21	9.6	14.3	
2	8.7		9.7	22	9.5		11.5
3	9.4		11.7	23	10.1		13.2
4	10.3		11.4	24	9.8		10.8
5	10.5	14.9		25	9.7		8.6
6	10.8	17.5		26	6.9		1.0
7	9.9		9.7	27	9.8	13.3	
8	9.3	8.5		28	8.6	10.7	
9	9.7	15.4		29	9.9	12.6	
10	9.8		4.8	30	11.4	16.2	
11	10.7	12.9		31	10.6		14.2
12	10.9		14.8	32	10.2	16.1	
13	9.0	9.9		33	9.3	9.5	
14	8.4		6.9	34	8.3	10.6	
15	9.9		12.4	35	9.9	12.0	
16	8.6		8.0	36	8.7		8.5
17	10.7		12.2	37	10.5		11.6
18	9.5		9.4	38	11.1	15.5	
19	9.6		13.2	39	9.6	15.2	
20	9.5	12.9		40	10.1	17.4	

Table 19.3 Estimated population after treatment assignment and prediction of unobserved responses

Unit	Z	x	y	d	Unit	Z	x	y	d
1	10.6	16.7	13.0	3.7	21	9.6	14.3	10.2	4.1
2	8.7	10.7	9.7	1.0	22	9.5	12.6	11.5	1.1
3	9.4	12.4	11.7	0.7	23	10.1	14.1	13.2	0.9
4	10.3	14.6	11.4	3.2	24	9.8	13.4	10.8	2.6
5	10.5	14.9	12.8	2.1	25	9.7	13.1	8.6	4.5
6	10.8	17.5	13.6	3.9	26	6.9	6.2	1.0	5.2
7	9.9	13.6	9.7	3.9	27	9.8	13.3	10.8	2.5
8	9.3	8.5	9.4	-0.9	28	8.6	10.7	7.5	3.2
9	9.7	15.4	10.5	4.9	29	9.9	12.6	11.1	1.5
10	9.8	13.4	4.8	8.6	30	11.4	16.2	15.3	0.9
11	10.7	12.9	13.3	-0.4	31	10.6	15.3	14.2	1.1
12	10.9	16.1	14.8	1.3	32	10.2	16.1	11.9	4.2
13	9.0	9.9	8.6	1.3	33	9.3	9.5	9.4	0.1
14	8.4	9.9	6.9	3.0	34	8.3	10.6	6.6	4.0
15	9.9	13.6	12.4	1.2	35	9.9	12.0	11.1	0.9
16	8.6	10.4	8.0	2.4	36	8.7	10.7	8.5	2.2
17	10.7	15.6	12.2	3.4	37	10.5	15.1	11.6	3.5
18	9.5	12.6	9.4	3.2	38	11.1	15.5	14.4	1.1
19	9.6	12.9	13.2	-0.3	39	9.6	15.2	10.2	5.0
20	9.5	12.9	10.0	2.9	40	10.1	17.4	11.6	5.8

Part VI
Sample Size and Methodology

CHAPTER 20

ADVANCES IN SAMPLE SIZE METHODOLOGY FOR BINARY DATA STUDIES–A REVIEW

M. M. DESU

State University of New York at Buffalo, NY

Abstract: One of the topics that received lot of attention recently is the sample size requirements for studies undertaken to establish therapeutic equivalence of two treatments. Several papers on this topic appeared in the journal, *Statistics in Medicine*, in the decade of nineties. In this paper we present a review of these advances in the sample size methodology for equivalence studies with binary responses. Both parallel and paired data studies will be discussed. In case of paired data studies the classical problem of testing for equality of treatment effects will also be considered. This review is limited to the work that appeared in Statistics in Medicine, however we also refer to the related work that appeared elsewhere to make the review some what complete.

Keywords and phrases: Binary data, therapeutic equivalence, parallel studies, paired data studies, sample size

20.1 ESTABLISHING THERAPEUTIC EQUIVALENCE IN PARALLEL STUDIES

At times we are interested in comparing the effect of a new treatment with the effect of a standard treatment. In the classical hypothesis testing framework, a null hypothesis of equality of the effects is postulated and we want to reject this null hypothesis in favor of the alternative that the effects are not equal. However in the context of therapeutic equivalence problem it is of interest to establish that the new treatment is as effective as the standard one. It is usually formulated as a hypothesis testing problem. The new treatment is considered as equivalent to the standard one if it is only negligibly inferior. This equivalence is taken as the alternative hypothesis and the nonequivalence case is taken as the null hypothesis. It is a common practice to use a parallel study to gather data for this purpose; however use of paired data studies also have been discussed in literature. So first we will focus our attention on the size of parallel studies. Next we consider a similar problem in relation to paired data studies. The determination of the size of a study, where a confidence bound based test is used, will also be discussed.

Tests for one-sided equivalence or therapeutic equivalence (Tests for non-zero risk difference or non-unity relative risk) in comparative Bernoulli trials

Let $\pi_1(\pi_2)$ be the effect rate of the standard treatment (new treatment). Let Δ be the difference in the effect rates, that is,

$$\Delta = \pi_1 - \pi_2.$$

The effect rates are also known as response probabilities and Δ is also called the risk difference. Usually the two treatments are considered as equivalent if Δ is small. Dunnett and Gent (1977) formulated this (one-sided) equivalence problem as that of testing

$$H_0 : \Delta = \Delta_0 \text{ versus } H_1 : \Delta < \Delta_0, \qquad (20.1.1)$$

where $\Delta_0(> 0)$ is a specified constant. It should be noted that the null hypothesis indicates nonequivalence. This formulation has also been used by Rodary et al. (1989), who derived an expression for the sample size in relation to a parallel study. Blackwelder (1982) considered a slightly different formulation, where one tests

$$H_0 : \Delta \geq \Delta_0 \text{ versus } H_1 : \Delta < \Delta_0. \qquad (20.1.2)$$

This formulation is the one considered by Farrington and Manning (1990). Also, it should be noted that formulation (20.1.1) with $\Delta_0 = 0$ reduces to the usual two sample problem with one-sided alternatives. Such a formulation is used by Rodary et al. (1989) in relation to an efficacy trial. The usual homogeneity testing problem has been discussed extensively in literature. Sahai and Khurshid (1996) gave a comprehensive review of the work on testing for the equality of effect rates.

The formulation (20.1.2) will be referred to as the Δ-formulation. In the following we will summarize the results for this formulation. Then results for the relative risk formulation (ψ-formulation) will follow.

20.1.1 Tests under Δ-Formulation (20.1.2)

Using formulation (20.1.1) and the test statistic of Dunnett and Gent, Rodary et al. (1989) derived an expression for the sample size. These results also hold for the formulation (20.1.2). The various investigations considered tests based on the statistic

$$\hat{\Delta} = p_1 - p_2, \tag{20.1.3}$$

where p's are sample proportions, which are the natural estimates of the effect rates. This statistic is an unbiased estimator of Δ. We assume that a parallel study, with each group of size n, needs to be planned. This n is the sample size we want to determine. This determination is done so as to ensure an α-level test has power $1 - \beta$ at the alternative $\Delta = \Delta_1 < \Delta_0$.

To define the tests, we need an expression for the variance of $\hat{\Delta}$. This variance is given by

$$\text{var}(\hat{\Delta}) = [\pi_1(1 - \pi_1) + \pi_2(1 - \pi_2)]/n \equiv \Sigma^2(\pi_1, \pi_2)/n. \tag{20.1.4}$$

An alternative (equivalent) expression is

$$\begin{aligned} \text{var}(\hat{\Delta}) &= [\pi_1(1 - \pi_1) + (\pi_1 - \Delta)(1 - \pi_1 + \Delta)]/n \\ &\equiv \sigma^2(\pi_1, \Delta)/n. \end{aligned} \tag{20.1.5}$$

The critical region of the test to be used is chosen using the approximate normal distribution of $\hat{\Delta}$. Thus a test with approximate size α

$$\text{rejects } H_0 \text{ in favor } H_1 \text{ if } \hat{\Delta} - \Delta_0 < z_\alpha.v, \tag{20.1.6}$$

where z_p is the pth quantile of the standard normal distribution and v^2 is an estimate of the variance of $\hat{\Delta}$. Three different estimates, proposed earlier, were investigated by Farrington and Manning (1990). The relevant details are given below.

Blackwelder:

The suggested estimate is

$$v_B^2 = \Sigma^2(p_1, p_2).$$

This estimate uses the sample proportions to estimate π's. It is also used at times for testing the homogeneity hypothesis.

Dunnett and Gent:

In this approach π's are estimated under the restriction $\Delta = \Delta_0$, subject to the marginal totals remaining equal to those observed. The resulting estimates are

$$\hat{\pi}_1 = (p_1 + p_2 + \Delta_0)/2, \quad \hat{\pi}_2 = (p_1 + p_2 - \Delta_0)/2.$$

These estimates are used to compute v^2 as

$$v_{DG}^2 = \Sigma^2(\hat{\pi}_1, \hat{\pi}_2)/n.$$

Maximum likelihood estimation:

Restricted maximum likelihood method is used to get the estimates of π's. To obtain these estimates one need to solve a third degree equation and the relevant details are given in Farrington and Manning (1990). Using these estimates the v^2 is computed as

$$v_{ML}^2 = \Sigma^2(\pi_{1ML}, \pi_{2ML})/n.$$

This method has been proposed earlier by Miettinen and Nurminen (1985) and others.

Using the power, which is approximated by a normal probability, the required sample size is obtained. The general formula for the sample size (per group) depends on two quantities A and B. These are

$$\begin{aligned} A &= z_{1-\beta}\sigma(\pi_1, \Delta_1) \\ &\equiv z_{1-\beta}[\pi_1(1 - \pi_1) + (\pi_1 - \Delta_1)(1 - \pi_1 + \Delta_1)]^{1/2}, \end{aligned}$$

where π_1 and Δ_1 are population values under the alternative and

$$B = z_{1-\alpha}\Sigma(\bar{\pi}_1, \bar{\pi}_2),$$

where $\bar{\pi}_1, \bar{\pi}_2$ are the large sample approximations to the estimates of π's used in computing v. The required sample size is

$$n_\Delta = (A + B)^2/(\Delta_1 - \Delta_0)^2. \tag{20.1.7}$$

We now present the B values corresponding to the estimation methods indicated earlier.

Blackwelder:

$$B = z_{1-\alpha}\Sigma(\pi_1, \pi_2); \tag{20.1.8}$$

Dunnett and Gent, Rodary et al.:

$$\begin{aligned} B &= z_{1-\alpha}\sigma(\pi_1, \Delta_0) \\ &\equiv z_{1-\alpha}[\pi_1(1 - \pi_1) + (\pi_1 - \Delta_0)(1 - \pi_1 + \Delta_0)]^{1/2}; \tag{20.1.9} \end{aligned}$$

Restricted ML estimation (Farrington and Manning):

$$\begin{aligned} B &= z_{1-\alpha}\Sigma(\bar{\pi}_{1\Delta}, \bar{\pi}_{2\Delta}) \\ &\equiv z_{1-\alpha}[\bar{\pi}_{1\Delta}(1 - \bar{\pi}_{1\Delta}) + \bar{\pi}_{2\Delta}(1 - \bar{\pi}_{2\Delta})]^{1/2}, \tag{20.1.10} \end{aligned}$$

where $\bar{\pi}'$s are the large sample approximations to the proposed estimates of π's.

Roebruck and Kühn (1995) made a comparative study of the sample size calculations of Blackwelder, Rodary *et al.* and Farrington and Manning. The following are their recommendations.

> "the methods of Dunnett and Gent and Rodary et al. may be ruled out in a first step due to bad performance of sample size formula and to the fact that, in nearly no situations is the test superior to both of the others. The comparison of Blackwelder's formulae with those of Farrington and Manning suggests, that both may be used for $\Delta_0 < \pi_1/2$ with the restriction that for $n_1/n_2 = (3/2)$ and small π_1 only the latter may be used. The Farrington/Manning test (and consequently their sample size formula) should be used for $\pi_1 \leq 0.1$ and $\Delta_0 \geq \pi_1$. For the quite unusual configurations as $(\pi_1, \Delta_0) \in \{(0.2, 0.3), (0.3, 0.3), (0.3, 0.45)\}$, Blackwelder's test performs better and can be used in combination with his sample size formula."

20.1.2 Tests under Relative Risk Formulation (ψ Formulation)

For comparing the two effect rates, another popular measure is the risk ratio or relative risk, which is defined as

$$\psi = (\pi_1/\pi_2). \tag{20.1.11}$$

Using this measure, Farrington and Manning formulated the equivalence problem as that of testing

$$H_0 : \psi \geq \psi_0 \; versus \; H_1 : \psi < \psi_0, \qquad (20.1.12)$$

where ψ_0 is a specified constant. Here the test statistic is

$$T_R = p_1 - \psi_0 p_2. \qquad (20.1.13)$$

It is easy to see that the expectation of this statistic is

$$E[T_R] = \pi_1 - \psi_0 \pi_2 = (\psi - \psi_0)\pi_2,$$

and the variance is

$$w^2(\pi_1, \pi_2) = [\pi_1(1 - \pi_1) + \psi_0^2 \pi_2(1 - \pi_2)]/n.$$

The suggested test of approximate size α

rejects H_0 in favor H_1 if $T_R < z_\alpha.w_0$,

where w_0^2 is an estimate of the null variance of T_R. Farrington and Manning obtained a formula for the sample size. This expression is

$$n_\psi = (A^* + B^*)^2/(\pi_1 - \psi_0\pi_2)^2, \qquad (20.1.14)$$

where

$$A^* = z_{1-\beta}[\psi_0^2 \pi_2(1 - \pi_2) + \pi_1(1 - \pi_1)]^{1/2},$$

and

$$B^* = z_{1-\alpha}[\pi_{1R}(1 - \pi_{1R}) + \psi_0^2 \pi_{2R}(1 - \pi_{2R})]^{1/2}.$$

Here π_{1R} and π_{2R} are the large sample approximations to the corresponding quantities used in the estimate w_0^2.

Blackwelder (1993) considered two other methods and made a comparison. One method is based on a logarithmic transformation and the other one uses a Poisson approximation. These results will be described now.

Logarithmic transformation

Here the test depends on $\log \hat{\psi}$, an estimate of $\log \psi$. This statistic is

$$\log \hat{\psi} = \log(p_1/p_2).$$

The approximate mean of this statistic is $log\psi$ and the approximate variance is

$$v^2(\pi_1, \pi_2) = [(1 - \pi_1)/\pi_1 + (1 - \pi_2)/\pi_2]/n.$$

The distribution of this statistic can be approximated by a normal distribution. So a test of approximate size α

$$\text{rejects } H_0 \text{ if } \log \hat{\psi} - \log \psi_0 < z_\alpha . v_*(p_1, p_2), \qquad (20.1.15)$$

where $v_*^2(p_1, p_2)$ is an estimate of the variance $v^2(\pi_1, \pi_2)$. Now the sample size is seen to be

$$n_L = (z_\alpha + z_\beta)^2 [(1 - \pi_1)/\pi_1 + (1 - \pi_2)/\pi_2]/(\log \psi_0 - \log \psi_1)^2 \, (20.1.16)$$

Poisson approximation

Let x_i be the number of cases which showed effect, so that the sample proportion is $p_i = x_i/n(i = 1, 2)$. For large n and small p_i, x_i is distributed approximately as a Poisson variable with mean np_i. The conditional distribution of x_1, given $X = x_1 + x_2$, is a binomial distribution with parameters X and Π, where

$$\Pi = \psi/(1 + \psi).$$

The hypotheses of (20.1.12) can be stated in terms of Π. Using this equivalent formulation, and approximating the conditional binomial distribution by an appropriate normal distribution, the required number of cases, X, can be seen to be

$$X = [z_\alpha \{\Pi_0 (1 - \Pi_0)\}^{1/2} + z_\beta \{\Pi_1 (1 - \Pi_1)\}^{1/2}]^2/(\Pi_0 - \Pi_1)^2. \ (20.1.17)$$

As X is determined by the data, for planning purposes one usually computes the required sample size as

$$n_P = X/[\pi_1 + \pi_2]. \qquad (20.1.18)$$

Blackwelder's comparative study indicated that the method of Farrington and Manning is generally preferable to the logarithmic transform method. The asymptotic formulation of the score test (method of Farrington and Manning) is more generally applicable, even if the analysis uses the log statistic. The Poisson approximation is appropriate in general for risks up to about 0.05.

20.1.3   Confidence Bound Method for Δ Formulation

Rodary *et al.* (1989) suggested a test based on an upper confidence bound for Δ. One can use the power of this test to find an expression for the sample size. The suggested confidence bound is

$$\begin{aligned} \Delta_U &= \hat{\Delta} + z_{1-\alpha} SE(\hat{\Delta}) \\ &= (p_1 - p_2) + z_{1-\alpha}[p_1(1 - p_1) + p_2(1 - p_2)]^{1/2}/n^{1/2}. \end{aligned}$$

The corresponding test rejects H_0 if this bound is less than Δ_0. Approximating the power function by a normal probability and replacing the p's in SE by population π's the sample size can be seen to be

$$n_C = (z_{1-\alpha} + z_{1-\beta})^2 [\pi_1(1 - \pi_1) + \pi_2(1 - \pi_2)]^2 / (\Delta_1 - \Delta_0)^2, \quad (20.1.19)$$

where $\pi_2 = \pi_1 - \Delta_1$.

Nam (1994) considered the sample size requirements for stratified studies where the null hypothesis of non-unity relative risk is under test. The score method is used to obtain the test and the sample size is derived for this score test. When we consider only one stratum, results of Nam (equation 5) coincides with the results of Farrington and Manning .

Yanagawa et al. (1994) proposed Mantel-Haenszel type tests for testing equivalence in relation to a stratified study. It will be interesting to derive the sample size formula for these tests.

All the above investigations are concerned with comparative binomial trials. In the next section we consider similar problems in relation to paired data studies.

20.2 SAMPLE SIZE FOR PAIRED DATA STUDIES

We consider an experiment where we observe binary responses on N pairs. The data is the set of observations on the N vectors (Y_{1j}, Y_{2j}) for $j = 1, 2, \ldots, N$. The variable $Y_1(Y_2)$ is the response under standard treatment or treatment I(new treatment or treatment II). Since the response variables are discrete variables the data can be summarized as a frequency table. The distinct values are called the outcomes. The frequency distribution and the probability model is displayed in a tabular form as follows.

Table 20.1 Probability model for paired data studies

outcome	(0,0)	(0,1)	(1,0)	(1,1)
frequency	m_{00}	m_{01}	m_{10}	m_{11}
probability	π_{00}	π_{01}	π_{10}	π_{11}

The quantities m_{ij} are the outcome frequencies and π_{ij} are the probabilities associated with the outcomes. Now

$$P(Y_1 = 1) = \pi_{11} + \pi_{10} \equiv \pi_I,$$

and

$$P(Y_2 = 1) = \pi_{11} + \pi_{01} \equiv \pi_{II}.$$

First we consider the problem of testing for the equality of π_I and π_{II}. Then we will discuss the problem of equivalence.

20.2.1 Testing for Equality of Correlated Proportions

The problem of interest is to test the null hypothesis

$$H_0 : \pi_I = \pi_{II} \text{ or } \Delta(= \pi_I - \pi_{II}) = 0. \tag{20.2.20}$$

This problem is usually referred to as testing for *equality of two corre-lated proportions*. It is a common practice to restate the null hypothesis of (20.2.20) as

$$H_0 : \pi_{10} = \pi_{01} \text{ or } \Delta(= \pi_{10} - \pi_{01}) = 0 \text{ or } \psi(= \pi_{10}/\pi_{01}) = 1. \tag{20.2.21}$$

Various procedures for this testing problem have been proposed. A very popular testing procedure is the McNemar's test. May and Johnson (1997a) made a comparative study of the validity and power of various tests proposed for this testing problem. They recommend, "the use of either the modified Wald's test, the mid-P test or McNemar's test if $m_{10} + m_{01} < 40$ and the interest is in holding the nominal 5 percent level." It would be of interest to determine the sample size, N, in relation to the recommended modified Wald's test. Also May and Johnson (1997b) made a study of different methods proposed for constructing a confidence interval for the difference $\Delta = \pi_I - \pi_{II} = \pi_{10} - \pi_{01}$. They recommend the Quesenberry and Hurst confidence intervals.

There are various investigations dealing with the determination of sample size for studies using McNemar's test. Two popular approaches will be described. One approach uses the conditional power function and determines the number of discordant pairs needed. The other approach uses multinomial unconditional power function to determine the sample size.

The McNemar's test uses the standardized statistic

$$Z_M = (m_{10} - m_{01})/[m_{10} + m_{01}]^{1/2} \equiv X/M^{1/2}. \tag{20.2.22}$$

The critical region is determined using the approximate standard normal distribution for Z. For the one-sided alternatives $H_1 : \psi > 1$ or $\Delta > 0$, the test

$$\text{rejects } H_0 \text{ if } Z_M > z_{1-\alpha}. \tag{20.2.23}$$

A method for sample size determination for a case-control study is given by Schlesselman (1982). In this method the conditional distribution of X given M is used to find the power function. Using a normal approximation to the conditional power function Schlesselman obtained an expression for the number of discordant pairs, M, required to provide power $1-\beta$ to detect a specified value $\psi_1(> 1)$ for the ratio $\psi = \pi_{10}/\pi_{01}$. Now the required number of discordant pairs, M, can be seen to be

$$M = [\{z_{1-\alpha}(\psi_1 + 1) + 2z_{1-\beta}\psi_1^{1/2}\}/(\psi_1 - 1)]^2. \qquad (20.2.24)$$

For planning purposes, this value of M is converted to the sample size estimate as

$$N_c = M/\pi_d, \qquad (20.2.25)$$

where $\pi_d(= \pi_{10} + \pi_{01})$ is the probability of a discordant pair. Clearly one needs to input a value for π_d to calculate this sample size. Lachin (1992) refers to this method of calculating the sample size from the M value given by (20.2.25) as *First-order unconditional power function* method.

Using the multinomial model for the frequencies we can get an expression for the variance of X. Approximating the distribution of X by a normal distribution, an approximation to the required sample size is obtained as

$$N_m = [z_{1-\alpha}(\pi_d)^{1/2} + z_{1-\beta}(\pi_d - \Delta_1^2)^{1/2}]^2/\Delta_1^2 \qquad (20.2.26)$$

where Δ_1 is value of the difference, $\Delta = \pi_{10} - \pi_{01}$, under the alternative at which we want to control the power. Here also one needs to use a value for π_d like in the result (20.2.25). This result, when expressed in terms of ψ_1 and π_{01}, reads as

$$N_m = \frac{[z_{1-\alpha}(\psi_1 + 1)^{1/2} + z_{1-\beta}\{(\psi_1 + 1) - (\psi_1 - 1)^2\pi_{01}\}^{1/2}]^2}{[(\psi_1 - 1)^2\pi_{01}]}.$$
$$\qquad (20.2.27)$$

This result appears in Connett *et al.* (1987).

The above discussion is from Lachin (1992), who made a comparison of the various sample size proposals for the McNemar's test, in reference to a matched case-control study. His analysis found that the multinomial based result given by Connor (1987) and Connett *et al.* (1987) is *fairly accurate*. In the context of evaluating M using (20.2.24), Lachin indicated that a referee suggested a sequential approach. Under this approach one recruits matched pairs until M discordant pairs are included. Thus to implement this sampling scheme one need not specify π_d. It may be noted that Lui (1997) did consider such a sampling method and determined M by controlling the power of the associated equivalence test .

Lachenbruch (1992) also investigated the problem of sample size for studies based on McNemar's test. He states that usually investigators can specify the marginal probabilities π_I and π_{II} rather than ψ and π_d. This information on the marginal probabilities leads to some restrictions on the possible values of the cell probabilities , π_{ij}. Thus he proposes a method for computing the sample size needed in these cases and compares them with the values given by the two formulas mentioned earlier. His proposal for the sample size is

$$N_L = 0.25[z_{1-\alpha} + z_{1-\beta}]^2/[(0.5 - s)^2|\pi_I + \pi_{II} - 2\pi_{11}|], \quad (20.2.28)$$

where $s = (\pi_I - \pi_{11})/(\pi_I + \pi_{II} - 2\pi_{11})$. Lachenbruch found that the values given by his method are close to the values found in a Monte Carlo study of Connett *et al.*

20.2.2 Tests for Establishing Equivalence

Nam (1997) considered the problem of testing the clinical equivalence hypothesis. Identifying treatment I as the new treatment and treatment II as the standard treatment, he is interested in testing a hypothesis about

$$\Delta = \pi_I - \pi_{II} = \pi_{10} - \pi_{01}.$$

The problem is posed as testing

$$H_0 : \Delta = \Delta_0 \text{ versus } H_1 : \Delta > \Delta_0, \quad (20.2.29)$$

where $\Delta_0 < 0$. Nam derived a test using the score method . He also considered a test based on the difference between the observed response rates. Lee and Lusher (1991) proposed a test based on an upper confidence bound for Δ. These three methods will be described now.

Score test

The score function is

$$S(\pi_{01}, \Delta) = m_{10}/(\pi_{01} + \Delta) - (m_{00} + m_{11})/(1 - 2\pi_{01} - \Delta). \quad (20.2.30)$$

Let $\nu(\pi_{01}, \Delta)$ be the approximate variance of the score function. The score test is based on the statistic

$$Z_S = S(\bar{\pi}_{01}, \Delta_0)/[\nu(\bar{\pi}_{01}, \Delta_0)]^{1/2}, \quad (20.2.31)$$

where $\bar{\pi}_{01}$ is the MLE of π_{01} under H_0. The test

$$\text{rejects } H_0 \text{ when } Z_S > z_{1-\alpha}. \quad (20.2.32)$$

For $\Delta_0 = 0$, the statistic Z_S reduces to the McNemar's statistic Z_M of (20.2.22). Thus the test (20.2.32) can be seen to be a generalization of the McNemar's test (20.2.23). Approximating the power function by a normal probability and replacing the MLE of π_{01} by its large sample limit π_{01}^*, the sample size is obtained. Let

$$C = [z_{1-\alpha}/(2\pi_{01}^* + \Delta_0 - \Delta_0{}^2)^{1/2} + z_{1-\beta}/(2\pi_{01} + \Delta_1 - \Delta_1{}^2)^{1/2}]^2,$$

and

$$D = [(\pi_{10}(1 - \pi_{01}^*) - \pi_{10}^*(1 - \pi_{01})/\{\pi_{10}^*(1 - \pi_{10}^* - \pi_{01}^*)\}]^2, \quad (20.2.33)$$

where $\pi_{10}^* = \pi_{01}^* + \Delta_0$. Now the required sample size is

$$N_S = C/D. \tag{20.2.34}$$

Test based on sample proportions

Nam also considered a test based on the observed proportions

$$p_{ij} = m_{ij}/N \quad \text{for} \quad i, j = 0, 1,$$

which are the unrestricted ML estimates of the probabilities π_{ij} . In terms of these proportions, estimates of π_I and π_{II} are

$$p_I = (m_{11} + m_{10})/N, p_{II} = (m_{11} + m_{01})/N.$$

Thus an estimate of Δ is

$$\hat{\Delta} = p_I - p_{II} = (m_{10} - m_{01})/N \equiv X/N.$$

A statistic of interest is

$$t = \hat{\Delta} - \Delta_0 = p_I - p_{II} - \Delta_0 = (m_{10} - m_{01} - N\Delta_0)/N \equiv (X - N\Delta_0)/N.$$

It is easy to see that

$$E(t|H_0) = 0,$$

and

$$var(t|H_0) = (\pi_{10} + \pi_{01} - \Delta_0{}^2)/N = (2\pi_{01} + \Delta_0 - \Delta_0{}^2)/N.$$

The estimated null variance v^* is used to construct the test statistic

$$Z_t = t/[v^*]^{1/2} = (X - N\Delta_0)/[N(2\bar{\pi}_{01} + \Delta_0 - \Delta_0{}^2)]^{1/2},$$

where $\bar{\pi}_{01}$ is the MLE of π_{01} under H_0. An approximate α-level test

$$\text{rejects } H_0 \text{ if } Z_t > z_{1-\alpha}. \tag{20.2.35}$$

It is easy to see that, for $\Delta_0 = 0$, the test statistic Z_t reduces to the McNemar's test statistic Z_M of (20.2.22). So the test (20.2.35) can also be viewed as a generalization of the one-sided version of McNemar's test (20.2.23). Nam further showed that the score method test statistic Z_S is equal to the test statistic Z_t . Thus the two tests (20.2.32) and (20.2.35) are equivalent. However the approximate power functions are different and thus the sample size estimates are different. The sample size required to control the power of the test based on Z_t at $\Delta = \Delta_1 (> \Delta_0)$ is

$$N_t = \frac{[z_{1-\alpha}(2\pi_{01}^* + \Delta_0 - \Delta_0{}^2)^{1/2} + z_{1-\beta}(2\pi_{01} + \Delta_1 - \Delta_1{}^2)^{1/2}]^2}{(\Delta_1 - \Delta_0)^2} ,$$

$$\tag{20.2.36}$$

where π_{01}^* is large sample limit of the restricted MLE $\bar{\pi}_{01}$. A continuity correction modification of the test statistic Z_t has also been considered. The sample size for the continuity corrected test is

$$N_{t'} = N_t[1 + \{1 + 4/(N_t|\Delta_1 - \Delta_0|)\}^{1/2}]^2/4. \tag{20.2.37}$$

Nam states, "the sample size calculated by the score method is intermediate between those by corrected and uncorrected ML methods when the required power is greater than 50 percent. These three values are close and comparable." So one can use either of the two methods considered so far.

A Wald type test

Lu and Bean (1995) considered a similar problem and proposed a test based on a Wald-type statistic

$$\begin{aligned} Z_{LB} &= (m_{10} - m_{01} - N\Delta_0)/(m_{10} + m_{01} - N\Delta_0{}^2)^{1/2} \\ &= (X - N\Delta_0)/(M - N\Delta_0{}^2)^{1/2}. \end{aligned}$$

This statistic is approximately a standard normal variable. An approximate α-level test

$$\text{rejects } H_0 \text{ if } Z_{LB} > z_{1-\alpha}. \tag{20.2.38}$$

It is of some interest to note that the statistic Z_{LB} is equal to the statistic Z_M of (20.2.22), when $\Delta_0 = 0$. So the test (20.2.38) of Lu and

Bean can also be viewed as a generalization of McNemar's test. However this test is different from the test derived by Nam (1997). Lu and Bean derived the required sample size as

$$N_{LB} = \frac{[z_{1-\alpha}(2\pi_{01} + \Delta_0 - \Delta_0^2)^{1/2} + z_{1-\beta}(2\pi_{01} + \Delta_1 - \Delta_1^2)^{1/2}]^2}{(\Delta_1 - \Delta_0)^2}.$$

(20.2.39)

This expression is very much like N_t; however they differ in the first terms.

Nam also compared the sample sizes obtained by Lu and Bean with the sample sizes N_S and N_t . He found that Lu and Bean method substantially underestimates sample size because its false-positive error rate is greater than the nominal rate and thereby also inflates power.

Test based on a confidence bound

Lee and Lusher (1991) considered the problem of demonstrating equivalence using a confidence bound for Δ. First a $100(1-\alpha)percent$ upper confidence bound for Δ is computed as

$$
\begin{aligned}
\Delta_U &= \hat{\Delta} + z_{1-\alpha}SE(\hat{\Delta}) \\
&= (m_{10} - m_{01})/N + z_{1-\alpha}[N(m_{10} + m_{01}) - (m_{10} - m_{01})^2]^{1/2}/N^{3/2} \\
&= (X/N) + z_{1-\alpha}[N.M - X^2]^{1/2}/N^{3/2}.
\end{aligned}
$$

(20.2.40)

It may be noted that this bound is similar to the upper end point of the confidence interval that one can derive from Wald's test. The formal test based on this bound

$$\text{rejects } H_0 \text{ if } \Delta_U < \Delta_0. \tag{20.2.41}$$

Approximating the power function by a normal probability and replacing $SE(\hat{\Delta})$ by the population expression, the required sample size is obtained as

$$N_{LL} = [z_{1-\alpha} + z_{1-\beta}]^2 [\pi_d - \Delta_1^2]/[(\Delta_1 - \Delta_0)^2]. \tag{20.2.42}$$

This expression depends on π_d, the probability of a discordant pair. It may be noted that this type of dependence occured in expressions (20.2.25) and (20.2.26).

REFERENCES

Blackwelder, W.C. (1982). Proving the null hypothesis, *Controlled Clinical Trials*, **3**, 345–353.

Blackwelder, W.C. (1993). Sample size and power for prospective analysis of relative risk, *Statistics in Medicine*, **12**, 691–698.

Connett, J. E., Smith, J. A. and McHugh, R. B. (1997). Sample size and power for pair-matched case-control studies, *Statistics in Medicine*, **6**, 53–59.

Connor, R. (1987). Sample size for testing differences in proportions for the paired-sample design, *Biometrics*, **43**, 207–211.

Dunnett, C.W. and Gent, M. (1977). Significance testing to establish equivalence between treatments with special reference to data in the form of 2x2 tables, *Biometrics*, **33**, 483-486.

Farrington, C. and Manning, G. (1990). Test statistics and sample size formulae for comparative binomial trials with null hypothesis of non-zero risk difference or non-unity relative risk, *Statistics in Medicine*, **9**, 1447–1454.

Lachenbruch, P. A. (1992). On the sample size for studies based on McNemar's test, *Statistics in Medicine*, **11**, 1521–1525.

Lachin, J. M. (1992). Power and sample size evaluation for the McNemar test with application matched case-control studies, *Statistics in Medicine*, **11**, 1239–1251.

Lee, M. L. and Lusher, J. M. (1991). The problem of therapeutic equivalence with paired qualitative data: an example from a clinical trial using haemophiliacs with an inhibitor to factor VIII, *Statistics in Medicine*, **10**, 433–441.

Lu, Y. and Bean, J. A. (1995). On the sample size for one-sided equivalence of sensitivities based upon McNemar's test, *Statistics in Medicine*, **14**, 1831–1839.

Lui, K. (1997). Exact equivalence test for risk ratio and its sample size determination under inverse sampling, *Statistics in Medicine*, **16**, 1777–1786.

May, W. L. and Johnson, W. D. (1997a). The validity and power of tests for equality of two correlated proportions, *Statistics in Medicine*, **16**, 1081–1096.

May, W. L. and Johnson, W. D.(1997b). Confidence intervals for the difference in correlated binary proportions, *Statistics in Medicine*, **16**, 2127–2136.

Nam, J. (1994). Sample size requirements for stratified prospective studies with null hypothesis of non-unity relative risk using score test, *Statistics in Medicine*, **13**, 79–86.

Nam, J. (1997). Establishing equivalence of two treatments and sample size requirements in matched-pairs design, *Biometrics*, **53**, 1422–1430.

Rodary, C., Com-Nougue, C. and Tournade, M. (1989). How to establish equivalence between treatments: a one-sided clinical trial in pediatric oncology, *Statistics in Medicine*, **8**, 593–598.

Roebuck, P. and Kühn, A. (1995). Comparison of tests and sample size formulae for proving therapeutic equivalence based on the difference of binomial probabilities, *Statistics in Medicine*, **14**, 1583–1594.

Sahai, H. and Khurshid, A.(1996). Formulae and tables for the determination of sample sizes and power in clinical trials for testing differences in proportions for the two-sample design, *Statistics in Medicine*, **15**, 1–21.

Schlesselman, J. J. (1982). *Case-Control Studies*, Oxford University Press, New York.

Yanagawa, Y. Tango, T. and Hiejima, Y. (1994). Mantel-Haenszel-type tests for testing equivalence or more than equivalence in comparative clinical trials, *Biometrics*, **50**, 859–864.

CHAPTER 21

ROBUSTNESS OF A SAMPLE SIZE RE-ESTIMATION PROCEDURE IN CLINICAL TRIALS

Z. GOVINDARAJULU

University of Kentucky, Lexington, KY

Abstract: One of the central questions that arise in clinical trials is, how many additional observations, if any, are needed beyond those originally planned. Consider a two treatment normal response double-blind clinical experiment. We wish to test the null hypothesis of equality of the means against one-sided alternative when the common variance σ^2 is unknown. We wish to determine the required total sample size when the error probabilities α and β are specified at a predetermined alternative. Shih (1992) provides a two-stage procedure which is an extension of Stein's one-sample procedure. Assuming a preliminary guessed value of σ, he estimates σ^2 by the method of maximum likelihood via the $E - M$ algorithm. Since he introduces indicator variables which are treated as unknown parameters, the $m\ell e$ of σ^2 may not be consistent. Here, we propose an estimator of σ^2 that has a closed-form and derive expressions for the effective level of significance (α^*) and the power of the test at the specified alternative. In particular, it is shown that $\alpha^* - \alpha$ is negligible and that the power exceeds $1 - \beta$ when the initial (total) sample size is large.

Keywords and phrases: Robustness, clinical trials, double-blind experiment, sample size re-estimation

21.1 INTRODUCTION

Estimation of the required sample size is an important issue in most of clinical trials. Fixed-sample size designs use previous data or guess work of the parameters which can be unreliable. The classical sequential designs are limited to situations where outcome assessment can be made only after patients are enrolled in the trial. Group sequential designs have also been used in clinical trials. However, the type I error rate at each analysis stage need to be adjusted so as to control the overall type I error probability at a specified level. In several clinical trials, especially those dealing with nonfatal ailments, investigators would like to come up with a procedure at an interim stage in order to obtain updated information on the adequacy of the planned initial sample size. This often occurs when the natural history of the ailment is not well known or the therapy under study is new. In those cases, investigators are often unsure of the assumed values of the parameters that were initially employed for calculating the sample size at the planning stage. Note that the initial parameters are obtained, invariably, from previous studies conducted on different patient populations, diagnostic criteria etc. Consequently, the initial sample size does not guarantee either the width of the confidence interval in estimation or the desired power in hypothesis-testing setup. Hence it is desirable to monitor the clinical trial so as to assure that the basic assumptions on the design are reasonably satisfied and to construct procedures for estimating the sample size using these observations available at the interim stage. Thus, Shih (1992) makes a compelling case for not unblinding the treatment codes at the interim stage so that the integrity of the trial is maintained and no conscious or unconscious bias enters.

If the goal of the trial is to reestimate the required sample size, the only decision that would be taken is the determination of how many additional observations, if any, are needed beyond those planned earlier. If no further observations are needed the planned sample size is sufficient and the trial will be carried out. Shih (1992) proposes a two-stage procedure which is an extension of Stein's one-sample (two-stage) procedure to the two-sample situation. Suppose the two treatment responses are normally distributed with unknown means μ_1 and μ_2 and unknown common variance, σ^2. We wish to test $H_0 : \mu_1 = \mu_2$ against the alternative $H_1 : \mu_1 < \mu_2$ with specified error probabilities α and β at $\mu_2 = \mu_1 + \delta^*$, where δ^* is specified. The clinical trial is double-blind so that we do not know to which treatment the response belongs. Shih (1992) assigns $n_1 = n_2 = n/2$ patients at random to each of the two treatments where n is the preassigned initial total sample size. Assuming a preliminary guessed value σ^* of σ, he estimates σ^2 by the method of maximum likelihood via the $E - M$ algorithm. He introduces n indicator variables which take the value one if the patients was assigned to treatment with mean μ_1 and zero, otherwise. The indicator functions are treated at unknown parameters which need to be estimated. However, there is no closed-form expression for the $m\ell e$ of σ^2 and it may not be consistent since the number of nuisance parameters is getting large with n. Hence, it's worth while to come up with alternative estimates for σ^2. Cohen (1967) provides a method of moments estimate of σ^2 from a distribution that is a mixture of two normal distributions. Blumenthal and Govindarajulu (1977) studied the robustness of Stein's two-stage procedure for mixtures of two normal populations differing in means. Here, we use a simpler but unbiased and consistent estimate of σ^2 under H_0 and investigate the robustness of the level of significance and power at the specified alternative.

21.2   FORMULATION OF THE PROBLEM

Let U denote the response. Then

$$U = \begin{cases} X & \text{if the observation is on a patient having treatment I} \\ Y & \text{if the observation is on a patient having treatment II} \end{cases}$$

$$(21.2.1)$$

Since we are allocating equal number of patients to each treatment,

$$P\left(U = X\right) = P\left(U = Y\right) = \frac{1}{2}. \qquad (21.2.2)$$

Then it is easy to observe that

$$EU = \left(\mu_1 + \mu_2\right)/2 \quad \text{and} \quad \text{var} U = \sigma^2 + \frac{1}{4}\left(\mu_2 - \mu_1\right)^2, \qquad (21.2.3)$$

since we assume that $X\left\{Y\right\}$ has mean $\mu_1\left\{\mu_2\right\}$ and variance σ^2. If n_1 out n are allocated to treatment I, then n_1 is distributed binomially with parameters n and $\frac{1}{2}$. If $(U_1, \ldots, U_n)$ denotes the response vector, and $\bar{U} = \sum U_i/n$, we can write

$$\sum_1^n \left(U_i - \bar{U}\right)^2 = \sum_1^{n_1} \left(X_i - \bar{X}\right)^2 + \sum_1^{n_2} \left(Y_j - \bar{Y}\right)^2$$
$$+ \frac{n_1 n_2}{n_1 + n_2}\left(\bar{Y} - \bar{X}\right)^2 \qquad (21.2.4)$$

where $n_2 = n - n_1$ and $\bar{X}$ and $\bar{Y}$ denote the sample mean of X's and Y's respectively. Note that in the double blind case, X's and Y's are not observable. Also

$$E\left\{(n-1)^{-1}\sum_1^n \left(U_i - \bar{U}\right)^2\right\} = \text{var} U = \sigma^2 + \left(\mu_2 - \mu_1\right)^2/4. \quad (21.2.5)$$

Shih (1992) takes $n_1 = n_2 = n/2$ and estimates σ^2 by the method of maximum likelihood via the $E - M$ algorithm. Let $\hat{\sigma}_n$ denote an estimate of σ. If α and β are the specified error probabilities at $\mu_2 = \mu_1 + \delta^*$, then, Shih (1992) obtains

$$n_1 = 2\left(z_\alpha + z_\beta\right)^2 \left(\sigma^*/\delta^*\right)^2 \qquad (21.2.6)$$

where σ^* is a preliminary guessed value of σ and $z_\alpha = \Phi^{-1}(1-\alpha)$, Φ denoting the standard normal distribution. Let

$$M = n_1\left(\hat{\sigma}_n/\sigma^*\right)^2, \qquad (21.2.7)$$

and $N = \max\left(n_1, M\right) =$ total # of observations on each treatment. That is

$$N = \begin{cases} n_1 & \text{if } \hat{\sigma}_n < \sigma^* \\ M & \text{if } \hat{\sigma}_n > \sigma^*. \end{cases} \qquad (21.2.8)$$

Draw $N - n_1$ additional observations from each treatment. Then the decision rule is:

$$\begin{array}{ll} \text{Reject } H_0 \text{ when} & N^{\frac{1}{2}}\left(\bar{Y}_N - \bar{X}_N\right)/2\hat{\sigma}_n > t_{n-1,\alpha} \\ \text{and accept } H_0 & \text{otherwise.} \end{array} \qquad (21.2.9)$$

Then one can ask, (i) What is the effective level of significance of this procedure? (ii) What is the effective power at the specified alternative? We shall provide answers to these questions in the next sections.

21.3   THE MAIN RESULTS

Let $\hat{\sigma}_n^2 = (2n_1 - 1)^{-1} \sum_{i=1}^{n} \left(U_i - \bar{U}\right)^2$. Then we have the following lemma.

Lemma 21.3.1 *We have* $E\left(\hat{\sigma}_n^2|H_0\right) = \sigma^2$, *and* $\hat{\sigma}_n^2$ *tends to* σ^2 *with probability one when* H_0 *holds as* n *gets large.*

PROOF. One can write

$$\sum_{i=1}^{n}\left(U_i - \bar{U}\right)^2 = \sum_{1}^{n_1}\left(X_i - \bar{X}\right)^2 + \sum_{1}^{n_1}\left(Y_j - \bar{Y}\right)^2$$
$$+ (n_1/2)\left(\bar{Y} - \bar{X}\right)^2. \qquad (21.3.10)$$

Hence,

$$\begin{aligned} E\left(LHS|H_0\right) &= \{(n_1 - 1) + (n_1 - 1) + 1\}\sigma^2 \\ &= (2n_1 - 1)\sigma^2 = (n - 1)\sigma^2. \end{aligned}$$

The second assertion follows from the strong law of large numbers. □

We can reasonably assume that $n > 30$. Hence we can replace $t_{n-1,\alpha}$ in (2.9) by z_α.

Lemma 21.3.2 *Let* $Z = \left(\frac{n_1}{2}\right)^{\frac{1}{2}} \left(\bar{Y} - \bar{X}\right)/\sigma$. *Then as* n *becomes large*

(i)
$$\frac{Z^2}{n-1} \rightarrow \begin{cases} 0 & \text{in probability when } H_0 \text{ is true} \\ \delta^{*2}/4\sigma^2 & \text{in probability when } \mu_2 - \mu_1 = \delta^*. \end{cases}$$

Hence

(ii)
$$\hat{\sigma}_n^2 \approx \begin{cases} V & \text{when } H_0 \text{ is true} \\ V + \delta^{*2}/4\sigma^2 & \text{when } \mu_2 - \mu_1 = \delta^*, \end{cases}$$

where $V = W/(n-1)$ *and* $W \underline{d} \chi_{n-2}^2$.

PROOF. When H_0 is true Z has a standard normal distribution and hence Z^2 is distributed as chi-square with one degree of freedom. When $\mu_2 - \mu_1 = \delta^*$, one can write

$$Z = \tilde{Z} + \left(\frac{n_1}{2}\right)^{\frac{1}{2}} \delta^*/\sigma$$

where $\tilde{Z}$ is a standard normal variable. Hence

$$\frac{Z^2}{n-1} = \frac{\tilde{Z}^2}{n-1} + \frac{2}{n-1}\left(\sqrt{\frac{n_1}{2}}\frac{\delta^*}{\sigma}\right)\tilde{Z} + \frac{n_1}{2(n-1)}\frac{\delta^{*2}}{\sigma^2}$$

from which (i) follows.
Since

$$\frac{\hat{\sigma}_n^2}{\sigma^2} = \frac{1}{(n-1)\sigma^2}\left[\sum_1^{n_1}(X_i - \bar{X})^2 + \sum_{j=1}^{n_1}(Y_j - \bar{Y})^2 + \frac{n_1}{2}(\bar{Y} - \bar{X})^2\right],$$

(ii) readily follows from (i). □

Lemma 21.3.3 *Let* $\theta = (\sigma^*/\sigma)^2$. *Then as* n *gets large*

$$\frac{M\theta}{n_1} \rightarrow \begin{cases} 1 & \text{in probability when } H_0 \text{ is true} \\ 1 + \delta^{*2}/4\sigma^2 & \text{in probability when } \mu_2 - \mu_1 = \delta^*. \end{cases}$$

PROOF. From (21.2.7) we have

$$\frac{M\theta}{n_1} = \left(\frac{\hat{\sigma}_n}{\sigma}\right)^2 .$$

Now use Lemma 21.3.2(ii) and note that V tends to one in probability.

Let α^* and β^* denote the effective error probabilities and for the sake of simplicity we write t in the place of z_α. $\qquad\square$

Result 21.3.1 For sufficiently large n (say $n > 30$), we have

$$\alpha^* = \alpha + \int_{v=0}^{\theta} \left\{ \Phi(t) - \Phi\left(tv^{\frac{1}{2}}\right) \right\} dF_V(v) \qquad (21.3.11)$$

where $V = W/(n-1)$ and $W \underline{d} \chi^2_{n-2}$.

PROOF. Consider

$$\begin{aligned}
\alpha^* &= P_0\left((\bar{Y}_N - \bar{X}_N)/\hat{\sigma}_n (2/N)^{\frac{1}{2}} > t\right) \\
&= P_0\left(\bar{Y}_N - \bar{X}_N > (2/N)^{\frac{1}{2}} t\hat{\sigma}_n, \quad \hat{\sigma}_n \le \sigma^*\right) \\
&\quad + P_0\left(\bar{Y}_N - \bar{X}_N > (2/N)^{\frac{1}{2}} t\hat{\sigma}_n, \quad \hat{\sigma}_n > \sigma^*\right) \\
&= T_1 + T_2 \text{ (respectively)} .
\end{aligned}$$

Next recall that

$$(n-1)\hat{\sigma}_n^2/\sigma^2 = W + Z^2$$

where $W \underline{d} \chi^2_{n-2}$, $Z \underline{d}$ normal $(0,1)$ when H_0 is true, and W and Z are independent. Hence

$$\begin{aligned}
T_1 &= P_0\left(Z > t(\hat{\sigma}_n/\sigma), (\hat{\sigma}_n/\sigma)^2 \le \theta\right) \\
&= P_0\left(Z > t\left(\frac{W+Z^2}{n-1}\right)^{\frac{1}{2}}, (W+Z^2)(n-1) \le \theta\right) \\
&= \frac{1}{2}P_0\left(Z^2 > t^2(V + Z^2/(n-1)), Z^2 < (n-1)(\theta - V)\right) \\
&= \frac{1}{2}\int_{v=0}^{\theta} P_0\left(t^2v/\{1 - t^2/(n-1)\} < Z^2 < (n-1)(\theta - v)\right) dF_V(v) \\
&= \int_{v=0}^{\theta} P_0\left(t^2v/\{1 - t^2/(n-1)\} < Z < \sqrt{n-1}(\theta - v)^{1/2}\right) dF_V(v)
\end{aligned}$$

$$= \int_{v=0}^{\theta} \left\{ \Phi \left(\{(n-1)(\theta-v)\}^{1/2} \right) \right.$$
$$\left. - \Phi \left(t\sqrt{v} / \left(1 - t^2/(n-1)\right)^{1/2} \right) dF_V(v) \right\} \qquad (21.3.12)$$

As n gets large,

$$T_1 = \int_{v=0}^{\theta} \left\{ 1 - \Phi\left(t\sqrt{v}\right) \right\} dF_V(v) = \int_{v=0}^{\theta} \Phi\left(-t\sqrt{v}\right) dF_V(v), \qquad (21.3.13)$$

where $F_V(v)$ denotes the distribution of V.

Note that in one or more steps above, we have exploited the symmetry of the normal distribution.

Consider T_2

$$T_2 = P_0 \left(\bar{Y}_M - \bar{X}_M > \left(\frac{2}{M}\right)^{1/2} \hat{\sigma}_n t, \quad \hat{\sigma}_n^2 > \sigma^{*2} \right).$$

Recall that

$$\frac{M}{n_1} = \frac{\hat{\sigma}_n^2}{\sigma^{*2}} = \frac{\sigma^2}{\sigma^{*2}} \cdot \frac{\hat{\sigma}_n^2}{\sigma^2}.$$

Hence

$$T_2 = P_0 \left(\bar{Y}_M - \bar{X}_M > \sqrt{\frac{2}{n_1}} \cdot \frac{\sigma^*}{\hat{\sigma}_n} \cdot \hat{\sigma}_n \cdot t, \quad \hat{\sigma}_n^2 > \sigma^{*2} \right)$$
$$= P_0 \left(\sqrt{\frac{n_1}{2}} \left(\frac{\bar{Y}_M - \bar{X}_M}{\sigma} \right) > \sqrt{v}\theta t, \quad \hat{\sigma}_n^2 > \sigma^{*2} \right)$$
$$\doteq P_0 \left(\sqrt{\frac{n_1}{2}} \cdot \left(\frac{\bar{Y}_M - \bar{X}_M}{\sigma} \right) > \sqrt{v}\theta t, \quad V > \theta \right).$$

From Lemma 21.3.3, we have

$$M\theta/n_1 \to 1 \text{ in probability under } H_0.$$

Hence by Anscombe's theorem (1952)

$$T_2 = P_0 \left(\sqrt{\frac{n_1}{2\theta}} \left(\frac{\bar{Y}_M - \bar{X}_M}{\sigma} \right) > t, \quad V > \theta \right)$$
$$= P_0 \left(\sqrt{\frac{n_1}{2\theta}} (\bar{Y}_M - \bar{X}_M)/\sigma > t \right) P_0(V > \theta)$$
$$\doteq \Phi(-t) P_0(V > \theta). \qquad (21.3.14)$$

Combining (21.3.13) and (21.3.14) we obtain

$$
\begin{aligned}
\alpha^* &= \int_{v=0}^{\theta} \Phi\left(-tv^{\frac{1}{2}}\right) dF_V(v) + \Phi(-t) P_0(V > \theta) \\
&= \alpha + \int_{v=0}^{\theta} \left\{ \Phi\left(-tv^{\frac{1}{2}}\right) - \Phi(-t) \right\} dF_V(v) \\
&= \alpha + \int_{v=0}^{\theta} \left\{ \Phi(t) - \Phi\left(tv^{\frac{1}{2}}\right) \right\} dF_V(v). \qquad (21.3.15)
\end{aligned}
$$

Since V will be close to unity as n gets large and Φ is monotonic increasing, it is surmised that $\alpha^* - \alpha$ will be positive. Hence

$$
\alpha^* - \alpha \le \int_0^\infty \left\{ \Phi(t) - \Phi\left(tv^{\frac{1}{2}}\right) \right\} dF_V(v). \qquad (21.3.16)
$$

Or

$$
\alpha^* \le E\Phi\left(-tV^{\frac{1}{2}}\right). \qquad (21.3.17)
$$

In Table 21.1 we tabulate the values of $E\Phi\left(-tV^{\frac{1}{2}}\right)$ for selected values of α and n_1.

Let us make an asymptotic assessment of the upper bound for the error in α^* given by (21.3.15). Expanding in Taylor series we obtain

$$
\Phi\left(tv^{\frac{1}{2}}\right) - \Phi(t) = t\left(v^{\frac{1}{2}} - 1\right)\phi(t) - \frac{1}{2}t^3\left(v^{\frac{1}{2}} - 1\right)^2 \phi(t) + \cdots .
$$

Hence

$$
E\left\{ \Phi\left(tV^{\frac{1}{2}}\right) - \Phi(t) \right\} = t\phi(t) E\left(V^{\frac{1}{2}} - 1\right) - \frac{1}{2}t^3\phi(t) E\left(V^{\frac{1}{2}} - 1\right)^2 + \cdots .
$$

Let
$$
S = V^{\frac{1}{2}} \qquad \text{and} \qquad \nu = n - 2.
$$

Then $S = \left(\chi_\nu^2/\nu\right)^{\frac{1}{2}}$. One can easily compute

$$
\begin{aligned}
ES &= (2/\nu)^{\frac{1}{2}} \Gamma\left(\frac{\nu+1}{2}\right) / \Gamma(\nu/2) \\
&\doteq \exp\left\{ -\frac{1}{2} + \frac{\nu}{2}\log\left(1 + \frac{1}{\nu}\right) \right\} \\
&\doteq \exp(-1/4\nu) \qquad (21.3.18)
\end{aligned}
$$

after using Stirling's approximation to the gamma functions. Now using (21.3.18) we have

$$E (S - 1) \doteq -1/4\nu$$
$$E (S - 1)^2 \doteq 2 (1 - ES) \doteq 1/2\nu.$$

Further one can easily show that $E (S - 1)^3$ and $E (S - 1)^4$ are $0 (\nu^{-2})$. Hence

$$E \left\{ \Phi \left(tV^{\frac{1}{2}} \right) - \Phi (t) \right\} = t (1 + t^2) \phi (t) /4\nu + 0 (\nu^{-2}). \qquad (21.3.19)$$

For example, $\alpha = 0.05$, $n_1 = 11$ and $t = 1.645$, give

$$\alpha^* - \alpha \le 0.0078,$$

and

$$\alpha = 0.05, n_1 = 31 \text{ and } t = 1.645 \text{ give } \alpha^* - \alpha \le 0.0026$$

which are very close to the values given in Table 21.1.

In Table 21.1, we give numerical values of $\alpha^* = E\Phi (-tS)$ for selected values of $n_1 = (\nu/2) + 1$ and α. $\qquad\qquad\qquad\qquad\qquad\qquad\qquad$ □

TABLE 21.1 Numerical values of $\alpha^* = E\Phi (-tS)$

α n_1	0.01	0.05	0.1
11	0.0142	0.0576	0.1074
21	0.0123	0.0539	0.1037
31	0.0116	0.0526	0.1025
51	0.0110	0.0516	0.1015
∞	0.01	0.05	0.1

From Table 21.1 we infer that $\alpha^* - \alpha$ is negligible.

Next, we will turn to type II error probability. Let $\mu_2 - \mu_1 = \delta^*$ where δ^* is specified. For the sake of simplicity we write

$$P_{\delta^*} (A) = P^* (A) \qquad \text{for any event A.}$$

Then

$$\beta^* = P^* \left(\bar{Y}_N - \bar{X}_N \le u\hat{\sigma}_n/ (2/N)^{\frac{1}{2}} \right) \qquad (21.3.20)$$

Now

$$E_{\delta^*}\left((n-1)\hat{\sigma}_n^2\right)$$

$$= E_{\delta^*}\left[\sum_1^{n_1}(X_i - \bar{X})^2 + \sum_1^{n_1}(Y_i - \bar{Y})^2 + \frac{n_1}{2}(\bar{Y} - \bar{X})^2\right]$$

$$= (n_1 - 1)\sigma^2 + (n_1 - 1)\sigma^2 + \frac{n_1}{2}\left(\frac{\sigma^2}{n_1} + \frac{\sigma^2}{n_1} + \delta^{*2}\right)$$

$$= (2n_1 - 1)\sigma^2 + \frac{n_1}{2}\delta^{*2}. \tag{21.3.21}$$

Thus, the bias in

$$\hat{\sigma}_n^2 = \frac{n_1}{2(n-1)}\delta^{*2} \approx (\delta^*/2)^2 \text{ for } n \text{ large.} \tag{21.3.22}$$

For β^* we have the following result.

Result 21.3.2 We have

$$\beta^* = \int_0^{\theta - \delta^{*2}/4\sigma^2} \Phi\left\{\left(v + \delta^{*2}/4\sigma^2\right)^{\frac{1}{2}}t - \sqrt{\theta}(t + z_\beta)\right\}dF_V(v)$$

$$+ \Phi\left(-z_\beta\left(1 + \delta^{*2}/4\sigma^2\right)^{\frac{1}{2}}\right)P\left(V > \theta - \delta^{*2}/4\sigma^2\right). \tag{21.3.23}$$

PROOF. One can write

$$\beta^* = P^*\left(\bar{Y}_N - \bar{X}_N \le (2/N)^{\frac{1}{2}}\hat{\sigma}_n t, \ \hat{\sigma}_n \le \sigma^*\right)$$

$$+ P^*\left(\bar{Y}_N - \bar{X}_N \le (2/N)^{\frac{1}{2}}\hat{\sigma}_n t, \ \hat{\sigma}_n > \sigma^*\right)$$

$$= \tilde{T}_1 + \tilde{T}_2 \quad \text{(say)}.$$

Now,

$$\tilde{T}_1 = P^*\left(\bar{Y}_{n_1} - \bar{X}_{n_1} < (2/n_1)^{\frac{1}{2}}\hat{\sigma}_n t, \ \hat{\sigma}_n \le \sigma^*\right)$$

$$= P^*\left((\bar{Y}_{n_1} - \bar{X}_{n_1} - \delta^*)\sqrt{\frac{n_1}{2\sigma^2}} < \frac{\hat{\sigma}_n t}{\sigma} - \delta^*\sqrt{\frac{n_1}{2\sigma^2}}, \ \hat{\sigma}_n \le \sigma^*\right)$$

$$= P^*\left(Z^* \le t\left(V + \delta^{*2}/4\sigma^2\right)^{\frac{1}{2}} - \frac{\delta^*}{\sigma}(z_\alpha + z_\beta)\frac{\sigma^*}{\delta^*}, \ V < \theta - \delta^{*2}/4\sigma^2\right)$$

$$= P^*\left(Z^* \le t\left(V + \delta^{*2}/4\sigma^2\right)^{\frac{1}{2}} - \sqrt{\theta}(z_\alpha + z_\beta), \ V < \theta - \delta^{*2}/4\sigma^2\right)$$

$$
= \int_0^{\theta - \delta^{*^2}/4\sigma^2} P^* \left(Z^* < t \left(v + \delta^{*^2}/4\sigma^2 \right)^{\frac{1}{2}} - \sqrt{\theta} \left(z_\alpha + z_\beta \right) \right) dF_V(v)
$$

$$
= \int_0^{\theta - \delta^{*^2}/4\sigma^2} \Phi \left\{ \left(v + \delta^{*^2}/4\sigma^2 \right)^{\frac{1}{2}} t - \sqrt{\ } \left(z_\alpha + z_\beta \right) \right\} dF_V(v)
$$

$$
\doteq \ \Phi \left\{ \left(1 + \delta^{*^2}/4\sigma^2 \right)^{\frac{1}{2}} t - \sqrt{\theta} \left(z_\alpha + z_\beta \right) \right\} F_V \left(\theta - \delta^{*^2}/4\sigma^2 \right)
$$

$$
(21.3.24)
$$

where Z^* denotes a standard normal variable.

Next consider

$$
\tilde{T}_2 = P^* \left\{ \left(\bar{Y}_M - \bar{X}_M \right) < \left(\frac{2}{M} \right)^{\frac{1}{2}} \hat{\sigma}_n t, \ \hat{\sigma}_n > \sigma^* \right\}
$$

$$
= P^* \left\{ \left(\bar{Y}_M - \bar{X}_M \right) < 2^{\frac{1}{2}} \frac{\sigma^*}{n_1^{\frac{1}{2}}} t, \ \hat{\sigma}_n > \sigma^* \right\}
$$

$$
= P^* \left\{ \bar{Y}_M - \bar{X}_M - \delta^* < \sqrt{\frac{2}{n_1}} \sigma^* t - \delta^*, \ V > \theta - \delta^{*^2}/4\sigma^2 \right\}
$$

$$
= P^* \left\{ \sqrt{\frac{n_1}{2}} \frac{\left(\bar{Y}_M - \bar{X}_M - \hat{\delta} \right)}{\sigma} < \frac{\sigma^*}{\sigma} t - \frac{\delta^*}{\sigma} \sqrt{\frac{n_1}{2}}, \ V > \theta - \delta^{*^2}/4\sigma^2 \right\}
$$

Note that

$$
\frac{M\theta}{n_1 \left(1 + \delta^{*^2}/4\sigma^2 \right)} \to 1 \quad \text{in probability.}
$$

Hence

$$
\sqrt{\frac{M}{2}} \doteq \sqrt{\frac{n_1}{2\theta}} \cdot \left(1 + \delta^{*^2}/4\sigma^2 \right)^{\frac{1}{2}}
$$

By Anscombe's Theorem

$$
\sqrt{\frac{n_1}{2\theta}} \frac{\left(\bar{Y}_M - \bar{X}_M - \delta^* \right)}{\sigma}
$$

is asymptotically normal. Hence

$$
\tilde{T}_2 = P^* \left\{ \sqrt{\frac{n_1}{2\theta}} \frac{\left(\bar{Y}_M - \bar{X}_M - \delta^* \right)}{\sigma} \left(1 + \delta^{*^2}/4\sigma^2 \right)^{\frac{1}{2}} \right.
$$

$$
\left. < \left(t - \frac{\delta^*}{\sigma^*} \sqrt{\frac{n_1}{2}} \right) \left(1 + \delta^{*^2}/4\sigma^2 \right)^{\frac{1}{2}}, V > \theta - \delta^{*^2}/4\sigma^2 \right\}
$$

$$
\begin{aligned}
&= \Phi\left\{\left(t - \frac{\delta^*}{\sigma^*}\sqrt{\frac{n_1}{2}}\right)\left(1 + \delta^{*2}/4\sigma^2\right)^{\frac{1}{2}}\right\} P\left(V > \theta - \delta^{*2}/4\sigma^2\right) \\
&= \Phi\left\{-z_\beta\left(1 + \delta^{*2}/4\sigma^2\right)^{\frac{1}{2}}\right\} P\left(V > \theta - \delta^{*2}/4\sigma^2\right). \qquad (21.3.25)
\end{aligned}
$$

Therefore, combining (21.3.24) and (21.3.25) we have

$$
\beta^* =
\begin{cases}
\Phi\left\{\left(1 + \delta^{*2}/4\sigma^2\right)^{\frac{1}{2}} t - \sqrt{\theta}\,(t + z_\beta)\right\} & \text{if } \theta - \delta^{*2}/4\sigma^2 \geq 1 \\[2mm]
\Phi\left(-z_\beta\left(1 + \delta^{*2}/4\sigma^2\right)^{\frac{1}{2}}\right) & \text{if } \theta - \delta^{*2}/4\sigma^2 < 1.
\end{cases}
$$

$$(21.3.26)$$

$\theta - \delta^{*2}/4\sigma^2 \geq 1$ implies that $\theta \geq 1 + \delta^{*2}/4\sigma^2$ and hence

$$
\beta^* = \Phi\left[\left\{\left(1 + \delta^{*2}/4\sigma^2\right)^{\frac{1}{2}} - \sqrt{\theta}\right\} t - \sqrt{\theta} z_\beta\right] < \Phi\left(z_\beta\right) = \beta.
$$

So in general, $\beta^* < \beta$, for all values of θ.

In Tables 21.2 and 21.3 we tabulate $(1 - \beta^*)/(1 - \beta)$ as a percent for selected values of the parameters.

TABLE 21.2 Values of ratio (as a percent) of the effective power to the nominal power at the specified alternative $\mu_2 - \mu_1 = \delta^*$ when $\alpha = \beta$ and $\theta \geq 1 + \delta^{*2}/4\sigma^2$

			$1 - \beta$	
$\sqrt{\theta}$	δ^*/σ	0.99	0.95	0.90
1.01	0.2	100.1	100.2	100.3
1.02	0.35	100.2	100.4	100.6
1.04	0.50	100.3	100.8	101.1
1.1	0.75	100.6	102.0	102.9
1.2	1.0	100.9	103.4	105.4

Remark 21.3.1 For fixed δ^* and σ^*, both δ^*/σ and $\sqrt{\theta}$ increase as σ decreases. When $\sigma^{*2} - \sigma^2 \geq \delta^{*2}/4$ the gain in power is higher than when $\sigma^{*2} - \sigma^2 < \delta^{*2}/4$.

TABLE 21.3 Values of ratio (as a percent) of the effective power to the nominal power at the specified alternative $\mu_2 - \mu_1 = \delta^*$ when $\theta < 1 + \delta^{*^2}/4\sigma^2$

$1 - \beta$ δ^*/σ	0.99	0.95	0.90
0.2	100.04	100.05	100.17
0.35	100.09	100.26	100.36
0.50	100.18	100.57	100.73
0.75	100.36	101.14	101.63
1.0	100.53	101.80	102.62

Remark 21.3.2 The values $0.2, 0.35$ and 0.50 for δ^*/σ are of much interest in clinical trials. From Table 21.3 we infer that the percentage of gain in power is positive and is less than 3 percent for all practical values of δ^*/σ.

21.4 FIXED-WIDTH CONFIDENCE INTERVAL ESTIMATION

Suppose we wish to estimate $\eta = \mu_2 - \mu_1$ with a confidence interval having width $2d$ and confidence coefficient γ. Let $t = z_{(1+\gamma)/2}$ be such that $2\Phi(t) - 1 = \gamma$. As before, let σ^* be a preliminary estimate of σ. Then n_1 the number of patients to be assigned to each treatment is given by

$$n_1 = \left(t\sigma^* \sqrt{2}/d\right)^2 . \tag{21.4.27}$$

Let $\hat{\sigma}_n^2$ where $n = 2n_1$ denote an estimate of σ^2 based on the blinded responses $U_1, \ldots U_n$. Then, according to Stein's (1945) two-stage procedure we stop at n_1 if

$$t\hat{\sigma}_n (2/n_1)^{\frac{1}{2}} \le d.$$

Otherwise, allocate $M - n_1$, additional patients to each treatment where

$$M = 2t^2 \hat{\sigma}_n^2/d^2 . \tag{21.4.28}$$

Note that

$$M/n_1 \doteq (\hat{\sigma}_n/\sigma^*)^2 . \tag{21.4.29}$$

That is, N the total number of patients on each treatment is given by

$$N = \begin{cases} n_1 & \text{if } \hat{\sigma}_n \le \sigma^* \\ M & \text{if } \hat{\sigma}_n > \sigma^* \end{cases} \tag{21.4.30}$$

We assume that n_1 is sufficiently large (≥ 30). After we stop, the confidence interval for $\eta = \mu_2 - \mu_1$ is $\bar{Y}_N - \bar{X}_N \pm d$ (we also assume that after total experimentation it is unblinded and we know which are X and Y observations. We are interested in evaluating the effective coverage probability γ^* of the resultant confidence interval. Towards this we have the following result.

Result 21.4.1 For sufficiently large n_1, we have

$$\gamma^* = 2\Phi\left(t\sqrt{\theta}\right) F_V\left(\theta - \eta^2/4\sigma^2\right) + 2\Phi\left(t\left(1 + \eta^2/4\sigma^2\right)^{\frac{1}{2}}\right)$$
$$\times \left\{1 - F_V\left(\theta - \eta^2/4\sigma^2\right) - 1\right.$$

where

$$V = W/(n-1), \quad W \underline{d} \chi^2_{n-2}. \qquad (21.4.31)$$

or

$$\gamma^* \doteq \begin{cases} 2\Phi\left(t\left(1 + \eta^2/4\sigma^2\right)^{\frac{1}{2}}\right) - 1 & \text{when } \theta < 1 + \eta^2/4\sigma^2 \\ 2\Phi\left(t\sqrt{\theta}\right) - 1 & \text{when } \theta \geq 1 + \eta^2/4\sigma^2 \end{cases}$$
$$(21.4.32)$$

PROOF.

$$\gamma^* = P_\eta\left(|\bar{Y}_N - \bar{X}_N - \eta| \leq d\right)$$
$$= P_\eta\left(|\bar{Y}_N - \bar{X}_N - \eta| \leq d, \quad \hat{\sigma}_n^2/\sigma^2 \leq \theta\right)$$
$$+ P_\eta\left(|\bar{Y}_N - \bar{X}_N - \eta| \leq d, \quad \hat{\sigma}_n^2/\sigma^2 > \theta\right)$$
$$= T_1^* + T_2^* \quad \text{respectively}$$

where $\theta = (\sigma^*/\sigma)^2$. Consider

$$T_1^* = P_\eta\left(|\bar{Y}_N - \bar{X}_N - \eta| \leq d, \quad \hat{\sigma}_n^2/\sigma^2 \leq \theta\right)$$
$$= P_\eta\left(|\bar{Y}_N - \bar{X}_N - \eta| \leq d, \quad V + Z^2/(n-1) \leq \theta\right)$$
$$\doteq P_\eta\left((n_1/2\sigma^2)^{\frac{1}{2}}|\bar{Y}_N - \bar{X}_N - \eta| \leq t\sqrt{\theta}, V \leq \theta - \eta^2/4\sigma^2\right)$$
$$= \left\{2\Phi\left(t\sqrt{\theta}\right) - 1\right\} F_V\left(\theta - \eta^2/4\sigma^2\right) \qquad (21.4.33)$$

after using (21.4.27). Next consider

$$T_2^* = P_\eta\left(|\bar{Y}_M - \bar{X}_M - \eta| \leq d, \quad \hat{\sigma}_n^2/\sigma^2 > \theta\right)$$

Now, recall that $M\theta/n_1 \left(1 + \eta^2/4\sigma^2\right)$ tends to 1 in probability as n gets large and hence by Anscombe's theorem (1952) that

$$\left\{n_1 \left(1 + \eta^2/4\sigma^2\right)/2\sigma^{*2}\right\} \left(\bar{Y}_M - \bar{X}_M - \eta\right) \overset{d}{\approx} \text{normal } (0, 1). \quad (21.4.34)$$

Hence, we obtain

$$T_2^* \doteq \left\{2\Phi\left(t\left(1 + \eta^2/4\sigma^2\right)^{\frac{1}{2}}\right) - 1\right\}\left\{1 - F_V\left(\theta - \eta^2/4\sigma^2\right)\right\}. \quad (21.4.35)$$

Now combining (21.4.33) and (21.4.35) we obtain the desired result. Further, V will be a degenerate random variable at unity when n is sufficiently large and thus (21.4.32) follows.

In Tables 21.4 and 21.5 we provide some values of γ^* for selected values of γ, η/σ and $\sqrt{\theta}$. We infer that $\gamma^* \geq \gamma$ and, in Table 21.4 that γ^* increases with η/σ and the increase in γ^* decreases with γ. The increase in the coverage probability is large when $\theta \geq 1 + \eta^2/4\sigma^2$ than when $\theta < 1 + \eta^2/4\sigma^2$.

TABLE 21.4 Values of $\gamma^* = 2\Phi\left(t\left(1 + \eta^2/4\sigma^2\right)^{\frac{1}{2}}\right) - 1$ when $\theta < 1 + \eta^2/4\sigma^2$

γ η/σ	0.99	0.95	0.90
0.2	0.990	0.951	0.901
0.35	0.991	0.953	0.905
0.50	0.992	0.957	0.911
0.75	0.994	0.963	0.921
1.0	0.996	0.971	0.934

TABLE 21.5 Values of $\gamma^* = 2\Phi\left(t\sqrt{\theta}\right) - 1$ when $\theta \geq 1 + \eta^2/4\sigma^2$

γ $\sqrt{\theta}$	0.99	0.95	0.90
1.0	0.99	0.95	0.90
1.1	0.995	0.969	0.93
1.3	1.000	0.989	0.968
1.414	1.000	0.994	0.98

Acknowledgements I thank Dr. Weichung J. Shih of Merck. Sharp & Dohme Research Labs for bringing this problem to my attention and Dr. Alexie Dmitrienko for his help in computing Table 21.1.

REFERENCES

Anscombe, F. (1952). Large sample theory of sequential estimation, *Proceedings of the Cambridge Philosophical Society*, **48**, 600–607.

Blumenthal, S. and Govindarajulu, Z. (1977). Robustness of Stein's two-stage procedure for mixtures of normal populations, *Journal of the American Statistical Association*, **72**, 192–196.

Cohen, A. C. (1967). Estimation of mixtures of two normal distributions, *Technometrics*, **9**, 15–28.

Shih, W. J. (1992). Sample size reestimation in clinical trials. *Biopharmaceutical Sequential Statistical Applications* (Ed., K. E. Peace), pp. 285–301, Marcel Dekker, New York.

Stein, C. (1945). A two-sample test for a linear hypothesis whose power is independent of the variance, *Annals of Mathematical Statistics*, **16**, 243–258.

Kiefer, J. and Wolfowitz, J. (1956). Consistency of the maximum likelihood estimator in the presence of infinitely many incidental parameters, *Annals of Mathematical Statistics*, **27**, 887–906.

Part VII
Applications to Industry

CHAPTER 22

IMPLEMENTATION OF STATISTICAL METHODS IN INDUSTRY

BOVAS ABRAHAM

University of Waterloo, Waterloo, Ontario, Canada

Abstract: Statisticians have devised many tools to collect and analyse data from experimental and observational studies. However, attempts to bridge the gap between the available tools and what are practiced in industry have been very limited. It is very important for statisticians to direct serious attention to this issue if Statistics is to be relevant in the society at large. In this paper we propose some ideas for implementation of Statistical Methods based on our interaction with industry.

Keywords and phrases: Experimental studies, observational studies, industrial applications

22.1 INTRODUCTION

What do we mean by implementation? An industrial organization is instituting Statistical Thinking and implementing statistical tools so that it becomes a part of the every day business. We are not thinking about a statistician consulting with a scientist or an engineer for a one time project even though such activities are important in their own right. Implementation, in the sense used here, is much broader and the associated issues are not trivial.

401

22.2   LEVELS OF STATISTICAL NEED IN INDUSTRY

We envision Statistics to play important roles at three levels of an organization: Strategic Level, Managerial Level and Operational Level. This
classification is general and somewhat arbitrary. However, this identifies
and emphasizes different tools to be directed at different levels.

(i) Strategic Level (Top of an Organization)

At this level the most emphasis should be on Statistical Thinking (ST)
which includes the following: Notion of Process, Measurement and Data
Based Decisions, Understanding and Dealing with Variation, and Systematic Approach. Decisions at the strategic level requires an understanding
of variation and these decisions should be based on facts supported by
data [Deming (1986)]. Absence of these are quite prevalent in many organizations. Embracing any program that comes along is an expression of
decisions not supported by data.

(ii) Managerial Level (Middle)

This is the level at which systems are devised for implementation of the
directions taken by upper management. In particular, systems for process
control and improvement, robust product and process design, and training
are the responsibility of middle management. Understanding of Statistical
Thinking and some statistical tools are required.

(iii) Operational Level

This is the stage at which the methods are implemented through the
system built at the managerial level. Understanding of statistical tools
such as Control Charting, Capability, Design of Experiments (DOE), Measurement System Analysis, Regression Analysis, etc. and the actual use of
these tools must be one of the objectives. Here people in different areas
may not need the details of all the tools. For instance, an operator who is
using a control chart for maintaining stability of a process need not know
a lot about Design of Experiments; on the other hand an engineer responsible for process improvement should be knowledgeable in several aspects
of Statistical Process Control (SPC) and DOE.

22.3   IMPLEMENTATION: GENERAL ISSUES

Commitment of management

For the success of any program affecting the whole organization the full
commitment of senior management is essential. They have to assess the

situation early and decide to allocate the resources needed. If it is really important for the organization then senior people need to be involved in the implementation as well. Decide in advance what role they can and will play. For example, the success of the Six Sigma program at GE corporation is due to the commitment of its chief executive officer.

Expected benefits

It is important to recognize the benefits of implementation in the beginning. This helps to focus on what is needed. Of course it can help solve problems, improve processes and increase customer satisfaction. Another benefit is that a good measurement of performance can be done. An overall benefit is that it helps the organization to be a learning organization; a knowledge based company is going to be successful in the long run.

Systems thinking

As in any other implementation, there are several components involved and these need to be considered as part of the system for implementation. Some of these components are: Statisticians, Other People, Technology, Methodology, Organizational Structure and Culture. These components have to work jointly so that the system yields improvements. We need to recognize that there will be 'effects' of each component and 'interaction effects' among the components. We have to build the system such that interaction effects are positive and that the total effect is more than the effects of the components. For example if there are two components A and B then $EF(A + B) \neq EF(A) + EF(B)$, but $EF(A + B) = EF(A) + EF(B) + EF(AB)$ where $EF(A)$ stands for the effect of A. It is important to make the interaction AB [i.e., $EF(AB)$] positive so that the effect of A and B is more than the sum of the individual effects of A and B. Some guiding principles such as Deming's 14 points for management can be extremely beneficial during the implementation. Such principles help to foster positive interaction between components such as people and technology. Often sophisticated software is used to train without considering the background of the trainees. This can lead to negative interaction.

Implementation of Statistical Methods can be part of other system implementations such as those of the quality systems ISO 9000, QS-9000, and Six Sigma.

Implementation plan

We need to answer a number of questions before the activities can start. How does it start? When do the activities take place? Who is responsible for the tasks? What is the scope of the system, calendar of activities? What are the review points, and the associated expected results? Are resources assigned for the planned activities?

Project implementation systems

There are many implementation systems that one can use. Deming's PDSA circle (Shehart's wheel) - Plan, Do, Study, Act [Deming (1986)] is a well known example. A similar system is used in a Statistics Course (Statistics 231) at the University of Waterloo - Problem, Plan, Data, Analysis, Conclusion (PPDAC for short). The Institute for Improvement in Quality and Productivity (IIQP) uses a 7-step system- Problem, Plan, Data, Solution, Confirmation, Standardization, Follow-up. There are many others used by various organizations.

22.4 IMPLEMENTATION VIA TRAINING AND/OR CONSULTING

All organizations do have existing knowledge and acquiring new knowledge may require changes in thinking and culture of the organization. However, there has to be sensitivity about this issue and an understanding of existing knowledge base before anything is implemented. Also any plans for training should reflect this understanding.

Introduction of new knowledge requires training and the training needs for the different levels of the organization can be very diverse. This distinctive needs should be recognized and training programs should be designed in such a way to suit each of the Strategic, Managerial, and Operational levels.

Training considerations

Trainers should have a thorough statistical background and good industrial experience. They should be aware of the culture and structure of the organization. Also they should have an understanding of the context in which they are working (for instance, interfacing with other Quality System trainings) and the background of the trainees.

The quality of the material presented is very important as it should be relevant to the particular needs of the trainees. Schedule and duration of each module is also important. In addition, presentation of the material in an understandable and enjoyable way requires careful planning. Material needs to be presented with implementable and understandable technology. Communication between the trainer and trainee and that between software and participants should be smooth. The interaction between people and technology should be positive.

Training programmes can be interfaced with other programs such as ISO 9000, QS-9000, and Six Sigma. In this case sequencing should be carefully planned and the interaction between the programmes need to be positive.

22.5   IMPLEMENTATION VIA EDUCATION

Today's students are tomorrow's employees. Industrial organizations need graduates with technical and nontechnical skills. These students have to get the education from the universities and it is difficult for the universities to provide all the skills needed to function in the workplace. However, a university statistics curriculum can be improved so that potential employees have enough statistics and communication skills. Many authors have discussed ideas for enhancing statistical education, see for example Garfield (1995), Hoerl *et al.* (1993), Hogg and Hogg (1995), Snee (1993), Vere-Jones (1995) and Wild (1995).

Undergraduate programme

A Statistics undergraduate program should include the following: Scientific method, Problem solving system, Measurement system analysis (MSA), Control charting, Design and analysis of experiments (DOE), Regression analysis, Sampling, Computing and Mathematics [see ASA (1980)]. In addition, the students should get experience in solving industry related problems and communicate the results to people in other areas. There are different ways of achieving this goal. One method adopted at the University of Waterloo is to enroll the students in a co-operative program. In this system the students cycle between university and industry after each term during their undergraduate program. We will discuss this further later. Another approach may be to have joint programs between Engineering and Statistics; one can major in Statistics with a minor in Engineering or vice versa.

Graduate programme

A useful model to consider is to require undergraduate engineering background for a graduate degree in Statistics or Engineering Statistics. Also one should design the graduate program to enable the students to have internships in industry. This can enhance familiarity with working environments, hands-on experience, and communication skills. Joint projects such as seminars between university and industry will also be very helpful. Seminars by people working in industry, not necessarily research seminars, but seminars with issues can open up project and thesis topics for students.

22.6   UNIVERSITY–INDUSTRY COLLABORATION

Universities seek academic excellence. Industries require that their employees work on relevant issues. These two goals need not be on a collision course. With proper insight universities can provide academic excellence with relevance. Basically a university provides education to students. It can also provide faculty for training in the workplace. It is difficult for a university, by itself, to provide the well rounded education required for students to function in the workplace. Industry can help by providing contexts for relevance, and by their input into education. Collaboration between university and industry is essential to produce graduates for the future who can handle the difficult issues of the work place [Brajac and MacKay (1994), Hoadley and Kettenring (1990), Snee (1990)]. Such collaboration requires carefully designed systems for implementation. Since this is not an isolated problem, this also should be thought in a systems framework.

In a university-industry partnership needs of University and Industry must be clearly defined, and roles of the partners clearly understood. The system should be flexible so that students and faculty can spend time in industry to enhance nontechnical skills and to gain some hands-on-experience. University courses can be modified to include project oriented teaching. The industry should provide opportunities to gain experience in problem formulation, planning of approach and data collection and problem solving. Different models can be used for undergraduate and graduate students. The collaboration system must make sure that the transition between university and industry is smooth for students as well as faculty. Also the systems should be flexible to accommodate student and faculty interests. It should also be important to recognize that long term commitments are required by both partners.

22.7   UNIVERSITY OF WATERLOO AND INDUSTRY

University of Waterloo has been involved with industry in several areas at different levels. Here we focus on the involvement related to Statistical Methods.

(i) Co-operative Programmes

The University of Waterloo has a large (probably the largest in the world) co-op programme involving about 10,000 undergraduate students annually. In this programme a student goes to an industry for a four month "work term" after every four month school term. Each student is expected to write a work term report for each work term, which will then be evaluated by the employer (industry) as well as by the University. Students are expected to finish 4-6 successful work terms during their degree programme. The Engineering Undergraduate programme at Waterloo is only available through the co-op option while other undergraduate programmes including Statistics are available by co-op as well as regular routes. Each "faculty" (college) has its own special requirements. However, all co-op placements in industry are administered through the Department of Co-Operative Education, a large administrative group on campus. In general, placement rates in Engineering and Mathematics are well over 90%. During recession periods, some difficulties in placing first year students were experienced, prompting some adjustments to the timing of work terms and academic terms. There is a representative group from industry called the Waterloo Advisory Council which meets with University administration twice a year to exchange ideas on many issues facing the University and industry. Co-op education and University-industry collaboration is often discussed. This is an opportunity for the University to get input from industry regarding curriculum changes, new courses and programs.

(ii) Institute for Improvement in Quality and Productivity (IIQP)

University of Waterloo has many centres and institutes working with industry. Recognizing the prominent role of Statistical Methods in the QI activities the IIQP was established in 1985 as a liaison between the University and industry to implement Statistical Methods in industry. Its mission statement states:

> *"The Institute for Improvement in Quality and Productivity at the University of Waterloo is a group of individuals and corporate members committed to the development, communication and application of methods for quality and productivity improvement. The Institute's goal is to serve its members, the University and Business Communities."*

Goals and objectives of the Institute are:

- To provide a focus for multidisciplinary consulting and research in technical and managerial methods for improving quality

- To develop a centre offering courses and seminars for business and industry

- To aid in developing undergraduate and graduate programs in technical and managerial methods for quality improvement

- To facilitate experience-exchange programs between university faculty and industry personnel

- To stimulate development of innovative training methods in quality for the work place.

University members of the Institute include faculty members from approximately ten disciplines, and this enhances interaction among several disciplines which is helpful in carrying out technology transfer activities.

IIQP activities – training and consulting

The IIQP has an active program of in-company and public courses spanning a range of topics in Industrial Statistics. To encourage rapid implementation of the methods taught in the classroom, work related projects are usually included as part of the courses. Third or fourth year undergraduate, Masters and PhD students have the opportunity of getting involved in these projects with industry, or serving as teaching assistants in the short courses that are offered.

Research

By promoting closer contact between faculty members and industry, the Institute encourages increased applied research on topics of great interest to business and industry. It plays a direct role in stimulating research in the University community through financial awards for graduate work in areas relating to quality improvement. The Institute also provides direct financial support for faculty research. It publishes a research report series containing the results of current research done at the Institute. Graduate students have benefited from the funding provided but, more importantly, from the problems generated for their research from the industrial collaboration.

Co-op students

The IIQP has been employing one or two undergraduate co-op students (Engineering and/or Statistics) each term. They routinely work with faculty members to help with projects. They have the opportunity of visiting companies initially with and later without faculty members.

Campus course curriculum changes

Faculty members involved in the partnership gained valuable experience and ideas in working with industry. These have helped them to implement substantial changes to the content of Statistics courses at Waterloo.

The first course (second year undergraduate) taught to Mechanical and Systems Design Engineering students now centres around Continuous Process Improvement. These students are exposed to Experimental Design, Statistical Process Control, etc. and they have conducted experiments at industrial partner facilities during their work terms which have resulted in substantial annual savings. Significant changes have been made to a second year Statistics course in the Faculty of Mathematics. In this course, students conduct experiments in a laboratory in groups, deal with Measurement System issues and write laboratory reports as a team. Major changes have also been made to an advanced course in Experimental Design to reflect the applications in industrial partner facilities.

A wide variety of examples including casting, injection moulding, undercoating, etc. have been collected from partner facilities and these appear as examples in lectures and assignments. Students are excited by the fact that these examples are real and often involve thousands of dollars in savings [see Brajac and MacKay (1994)].

IIQP partnership with industry continues to provide many tangible and intangible benefits:

- Enhancements in content and delivery of courses

- Graduate and Undergraduate student involvement in real projects

- Enhancement of applied research of Faculty and Graduate students

- Professional development of the faculty members

- Enhancement of Statistical Thinking at some industrial partner facilities and modest cultural changes

- Application of newly developed methods in partner facilities

- Savings in real dollars for industrial partners

22.8 CONCLUDING REMARKS

Statistical Thinking and Methods need to become part of the knowledge base of an organization. We outlined many issues related to the implementation of Statistical Methods in Industrial Organizations. Implementation can be achieved by well planned and systematic training in the organizations and through the enhancements of university education by changes in course contents and delivery. We discussed the need for universities to form partnerships with industry to provide opportunities for students to enhance their skills. Some may argue that universities should be a place for education and should not be in the business of training. It is important to keep the balance, and all such endeavours should be motivated by "academic excellence with relevance."

It is important that the professional Statistician is equipped with good technical and non-technical skills. This is a challenge the universities have to face and one model for success is to form partnerships with industry as suggested. There is no need to compromise on academic excellence, however building in "relevance" to the programme enhances its value.

Acknowledgements The author would like to thank the Natural Sciences and Engineering Research Council of Canada for their research support.

REFERENCES

ASA Committee on Training of Statisticians for Industry (1980). Preparing Statisticians for Careers in Industry: Report of the ASA Section on Statistical Education Committee on Training of Statisticians for Industry, *The American Statistician*, **34**, 65–80.

Box, G. E. P. (1976). Science and statistics, *Journal of American Statistical Association*, **71**, 791–799.

Brajac, M. and MacKay, R. J.(1994). Industry-University Co-Operation: A Case Study, *IIQP Research Report*, RR-94-02, University of Waterloo, Waterloo, Ontario, Canada, N2L 3G1.

Deming, W. E. (1986). *Out of the Crisis*, MIT Press, Cambridge, Massachusetts.

Garfield, J. (1995). How students learn statistics, *International Statistical Review*, **63**, 25–34.

Hoadley, A. B. and Kettenring, J. R. (1990). Communications between statisticians and engineers/physical scientists, *Technometrics*, **32**, 243–274.

Hoerl, R., Hooper, J., Jacobs, P. and Lucas J. (1993). Skills for industrial statisticians to survive and prosper in the emerging quality environment, *The American Statistician*, **47**, 280–292.

Hogg, R. V. and Hogg, M.C. (1995). Continuous quality improvement in higher education, *International Statistical Review*, **63**, 35–48.

Snee, R. D. (1990). A partnership is needed, *Technometrics*, **32**, 267–269.

Snee, R. D. (1993). What is missing in statistical education?, *The American Statistician*, **47**, 194–154.

Vere-Jones, D. (1995). The coming of age of statistical education, *International Statistical Review*, **63**, 3–23.

Wild, C. J. (1995). Continuous improvement of teaching: A case study in a large statistics course, *International Statistical Review*, **63**, 49–68.

CHAPTER 23

SEQUENTIAL DESIGNS BASED ON CREDIBLE REGIONS

ENRIQUE GONZÁLEZ

Universidad de La Laguna, La Laguna, Spain

JOSEP GINEBRA

Universitat Politècnica de Catalunya, Barcelona, Spain

Abstract: Assume we can control an input $x_n \in C \subset R$, and observe one response y_n such that $E[y_n|x_n, \beta] = f(x_n; \beta)$ and that the objective is to keep all the responses close to a target T. We propose sequential designs that always improve on Bayesian certainty equivalence designs by searching for the best design in a family that contains them. To regulate the distance and direction that they move away from the certainty equivalence choice, the new designs experiment on a credible region for the root of $f(x; \beta) = T$. These heuristics perturb certainty equivalence to incentive 'active' learning about β and improve future control. We also describe how to apply this approach to the response surface bandit, where we need to keep all the responses close to the maximum of $f(x; \beta)$.

Keywords and phrases: Adaptive designs, certainty equivalence, highest posterior density interval, sequential optimization, stochastic control

23.1 INTRODUCTION

Assume that we can control an input $x_n \in C \subset R$, and observe one response y_n such that:

$$y_n(x_n) = f(x_n; \beta) + \epsilon_n,$$

413

with $E[y|x, \beta] = f(x; \beta)$ known up to a distribution on the parameters β, $\pi_0(\beta)$. We consider sequential designs, denoted by d, that select successive design points $x_{n+1}^d \in C$, given the history of input levels $X_n^d = (x_1^d, \ldots, x_n^d)$ and observations $Y_n^{'} = (y_1(x_1^d), \ldots, y_n(x_n^d))$.

In the multiperiod control problem the objective is to find designs that minimize $r_i(d) = E[L_i(Y_N, T)]$, where $L_i(Y_N, T)$ is a loss function that punishes distance between $Y_N^{'}$ and a known target T. In particular we look into the minimization of $r_2(d) = E[\sum_{n=1}^{N}(y_n(x_n^d) - T)^2]$. In the response surface bandit problem defined in Ginebra and Clayton (1995) and denoted as the R.S.B., the objective is to find designs that maximize $w_i(d) = E[U_i(Y_N)]$, where $U_i(Y_N)$ is a utility that values closeness between $Y_N^{'}$ and the maximum of $f(x; \beta)$. In particular we look into the maximization of $w_1(d) = E_d[\sum_{n=1}^{N} y_n(x_n^d)]$. Expectations are always w.r.t. the joint distribution for Y_N and β, that depend on d through x_n^d.

The multiperiod control problem and the R.S.B. model many situations in process control and in clinical trials where there are costs and ethical imperatives to provide the best set of input values for the individual to be treated next at the same time that we learn about the surface to enhance future performance. References on these areas are Astrom and Wittenmark (1995) and Rosenberger (1996).

When selecting x_n we face two conflicting goals; On one hand we need x_n to be such that $(y_n(x_n) - T)^2$ is small in expectation (or that $y_n(x_n)$ is large for the R.S.B.), but on the other hand we use $(x_n, y_n(x_n))$ to learn about β and help improve the performance at later stages. Each one of these two goals by itself typically requires x_n to be in different parts of the experimental region, and we have to care about both goals at once. This trade off is labeled by the adaptive control literature as the dual aspect of control, described in Chapter 7 of Astrom and Wittenmark (1995).

In principle backward induction identifies the optimal designs [see De Groot (1970) and Berry and Fristedt (1985)], but it is too complicated to implement them for these problems. Partial results can be found in Berliner (1983), Srinivasan (1984), Berry and Fristedt (1985) and references therein. Instead of aiming at optimality we define a subset of sequential designs D^l, and search through simulation for designs $d_l \in D^l$ either minimizing $r_i(d_l)$ or maximizing $w_i(d_l)$, in a way analogous to Ginebra and Clayton (1995).

If we knew $f(x; \beta)$ we would know the root θ that solves $f(\theta; \beta) = T$ and the κ that maximizes $f(x; \beta)$ and experimenting at either $x_n^d = \theta$ or $x_n^d = \kappa$ would be optimal for the multi-period control problem or the R.S.B. respectively. Sequential certainty equivalence designs are the ones that would be optimal if $f(x; \beta)$ was some current estimate for it, $\hat{f}_n(x; \beta)$;

For the control problem, certainty equivalence designs observe y_{n+1} at the x_{n+1}^{ce} that solves $\hat{f}_n(x;\beta) = T$ and for the R.S.B. they observe y_{n+1} at the x_{n+1}^{ce} that maximizes $\hat{f}_n(x;\beta)$. Note that certainty equivalence designs, (c.e. from now on), depend on the estimator chosen for $f(x;\beta)$ but do not depend on the loss or the utility for the problem.

C.e. is the approach used in almost all the engineering applications; In particular they assume that $\hat{f}_n(x;\beta) = f(x;\hat{\beta}_n)$, and thus c.e. designs for them are the ones that would be optimal if β was known to be $\hat{\beta}_n$. Berliner (1982) shows that some certainty equivalence rules are not even admissible. Lai and Robbins (1982), Ying and Wu (1997), Hu (1998) and references therein look into the asymptotic properties of some of these rules for specific models. Recently Chen and Hu (1998) prove that the Bayes c.e. designs are asymptotically optimal for simple linear models.

Ginebra and Clayton (1995) and González and Ginebra (1999a) propose heuristics for the R.S.B. and the multi-period control problem respectively that improve on c.e. designs; These heuristics are different for both problems and require either the estimation of the standard deviation of $\hat{f}_n(x;\beta)$ or the estimation of $\hat{f}_n'(x;\beta)$. In this paper we propose new heuristics based on credible sets for θ and κ that apply to both problems at once. In Section 23.2 we illustrate them by tackling the multiperiod control problem through designs supported on highest posterior density (H.P.D.) sets for θ. In Section 23.3 we illustrate this approach with an example involving the simple linear model, exploring how do these designs and their performances depend on the prior, the level of noise, the horizon and the size of the prior support for θ. In Section 23.4 we describe how to deal with the R.S.B. through designs supported on intervals based on posterior quantiles for κ.

23.2 DESIGNS FOR CONTROL BASED ON H.P.D. SETS

The certainty equivalence designs for the control problem observe y_{n+1} at some estimate of the root of $f(x;\beta) = T$, $x_{n+1}^{ce} = \hat{\theta}_n$, without taking into account the uncertainty in those estimates. By experimenting away from x_{n+1}^{ce}, we may incur into extra immediate loss but that might help learning about θ and increase the chances of experimenting closer to θ in later runs. To decide how far do we move away from the c.e. choice, we propose the use of posterior regions for θ. We illustrate the idea through designs that experiment on highest posterior density regions, denoted by d_{hpd}, that perturb the c.e. design that observes y_{n+1} at the mode of the marginal posterior for θ at stage n, $x_{n+1}^{ce} = \theta_n^{pmo}$.

Let $g_n(\theta)$ be proportional to the posterior marginal density for θ after stage n, $\pi_n(\theta|X_n, Y_n, \pi_0)$. Let g_n^{max} and g_n^{min} be the maximum and minimum values for $g_n(\theta)$ on the support of θ. Given $\tau \in [0, 1]$, let $hpd_n(\tau)$ be the set:

$$hpd_n(\tau) = \{\theta : g_n(\theta) \geq g_n^{max} - \tau(g_n^{max} - g_n^{min})\}.$$

Observe that $hpd_n(\tau = 0)$ only contains θ_n^{pmo} and that $hpd_n(\tau = 1)$ is the support of the posterior density for θ. The region $hpd_n(\tau)$ is either an interval or a set of disjoint intervals and its width is monotonically increasing with τ. When the marginal posterior for θ is unimodal, $hpd_n(\tau)$ will be an interval for any τ but when it is multimodal, $hpd_n(\tau)$ is a set of disjoint intervals for some τ's.

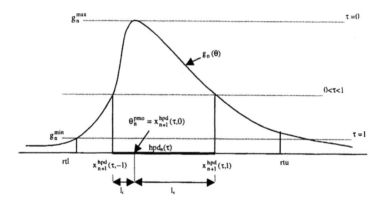

FIGURE 23.1   Example of an $hpd_n(\tau)$ region and of the different ingredients that intervene in the definition of the design $d_{hpd}(\tau, \gamma)$ when the marginal posterior distribution for θ is unimodal. $g_n(\theta)$ is proportional to $\pi_n(\theta|X_n, Y_n, \pi_0)$ and the support for θ is $[rtl, rtu]$

Given that θ_n^{pmo} belongs to $hpd_n(\tau)$ for any τ, it splits that region in two parts; Let l_s and l_i be the lengths of the wider and narrower part respectively, with the convention that when $hpd_n(\tau)$ is a set of disjoint intervals, we evaluate l_i and l_s by only taking into account the points in $hpd_n(\tau)$. Figure 23.1 illustrates all these definitions.

HPD designs are indexed by $(\tau, \gamma) \in [0, 1] \times [-1, 1]$ and are denoted by $d_{hpd}(\tau, \gamma)$. They observe y_{n+1} at a point $x_{n+1}^{hpd}(\tau, \gamma)$ on the set $hpd_n(\tau)$. The index τ regulates the width of the region; The larger τ, the further the design is allowed to stray away from that c.e. choice. The index γ indicates the relative position of $x_{n+1}^{hpd}(\tau, \gamma)$ on the set $hpd_n(\tau)$ and its sign indicates the direction in which we move away from $x_{n+1}^{ce} = \theta_n^{pmo}$. When $\gamma > 0$, $d_{hpd}(\tau, \gamma)$ experiments on the side of θ_n^{pmo} where $hpd_n(\tau)$ is wider, at a distance from x_{n+1}^{ce} equal to γ times the length of that part, l_s. When

$\gamma < 0$ it experiments on the narrower side at a distance from x^{ce}_{n+1} equal to γ times its length, l_i. Thus for fixed τ, the larger $|\gamma|$ the further $x^{hpd}_{n+1}(\tau, \gamma)$ is away from θ^{pmo}_n. When $|\gamma| = 1$, $x^{hpd}_{n+1}(\tau, \gamma)$ is on an end of $hpd_n(\tau)$.

Note that when either $\tau = 0$ or $\gamma = 0$, $d_{hpd}(\tau, \gamma)$ experiments on $x^{hpd}_{n+1} = x^{ce}_{n+1} = \theta^{pmo}_n$. As n increases, $hpd_n(\tau)$ will tend to become narrower and $x^{hpd}_{n+1}(\tau, \gamma)$ will tend to be closer to θ^{pmo}_n, matching the decreasing need to perturb c.e. to learn about θ as we get closer to $n = N$. Observe that analogous designs can be implemented on credible regions other than H.P.D., like with equally tailed intervals based on posterior quantiles for θ.

To find the $d_{hpd}(\tau^*, \gamma^*)$ that minimizes $r_2(\tau, \gamma) = r_2(d_{hpd}(\tau, \gamma)) = E[\sum^N_{n=1}(y_n(x^{hpd}_n) - T)^2]$ we proceed through simulation from the joint parameter and sample space as follows; For each (τ, γ) we repeatedly simulate parameters β^i from $\pi_0(\beta)$ and for each β^i we simulate the use of $d_{hpd}(\tau, \gamma)$ on data from the model for $(y|x, \beta^i)$, estimating $r_2(\tau, \gamma)$ as the average of the observed losses. By searching among various $d_{hpd}(\tau, \gamma)$ we find the $d_{hpd}(\tau^*, \gamma^*)$ that is estimated to be the 'best' for the specific loss, prior and model. We know that $d_{hpd}(\tau^*, \gamma^*)$ always improves on the c.e. design that experiments at $x^{ce}_{n+1} = \theta^{pmo}_n$; Indeed when either τ^* or γ^* are close to 0, that c.e. design is the best HPD design. Müller (1998) and Carlin et al. (1998) present recent uses of simulation on sequential design problems.

23.3   AN EXAMPLE OF THE USE OF HPD DESIGNS

To illustrate the use of HPD designs for control, we explore their performance for the normal linear model, $(y_n|x_n, \beta) \sim N(\beta_0 + \beta_1 x_n, \sigma^2)$, parameterized through $E[y|x, \beta] = T + \beta_1(x - \theta)$. We assume that $\theta = (T - \beta_0)/\beta_1 \sim U(-R, R)$, $\beta_1 \sim N(\mu_{\beta_1} = 1, \sigma^2_{\beta_1})$ and that β_1 is independent of θ, and $T = 0$. To estimate $r_2(\tau, \gamma)$, we simulate 500000 realizations of $(\beta_1, \theta)^i$ from its prior, and for each $(\beta_1, \theta)^i$ we run the $d_{hpd}(\tau, \gamma)$ design on y_n's simulated from $(y_n|x^{hpd}_n, (\beta_1, \theta)^i)$ and average the observed losses; Since the estimated standard deviation for $\hat{r}_2(\tau, \gamma)$ is of the order of .0009, the signal to noise in $\hat{r}_2$ is very large and the (τ^*, γ^*) minimizing $r_2(\tau, \gamma)$ can be easily located through a deterministic optimization process.

To evaluate the relative improvement of $d_{hpd}(\tau^*, \gamma^*)$, with respect to any other HPD design, $d_{hpd}(\tau, \gamma)$, we define $rr(\tau, \gamma)$ to be the relative increase in expected loss when we switch from $d_{hpd}(\tau^*, \gamma^*)$ to $d_{hpd}(\tau, \gamma)$,

$$rr(\tau, \gamma) = \frac{r_2(\tau, \gamma) - r_2(\tau^*, \gamma^*)}{r_2(\tau^*, \gamma^*)}.$$

In particular $rr(0, \gamma)$ is the relative increase in expected loss when we use the c.e. design that experiments on $x_{n+1}^{hpd}(0, \gamma) = x_{n+1}^{ce} = \theta_n^{pmo}$, instead of $d_{hpd}(\tau^*, \gamma^*)$.

Figure 23.2 explores how τ^*, γ^* and $rr(0, \gamma)$ depend on $(\sigma_{\beta_1}, \sigma, N, R)$, where R measures the size of the support for θ. Both τ^* and γ^* increase with increasing σ, while they are rather insensitive to variations in σ_{β_1}. For most of the combinations of $(\sigma_{\beta_1}, \sigma)$ tried, γ^* is positive and thus the best HPD design experiments in the wider part of $hpd_n(\tau)$. When N increases both τ^* and γ^* decrease and thus the larger N, the closer the best HPD design is to the c.e. design. When R increases, τ^* and γ^* first decrease and then they increase.

Figure 23.2 also shows that $rr(0, \gamma)$ increases with increasing σ and with decreasing σ_{β_1}, being more sensitive to changes in σ than in σ_{β_1}. This means that the more we know about β_1 and the smaller the information in $y(x)$ about β, the larger the improvement of $d_{hpd}(\tau^*, \gamma^*)$ over c.e. On the other hand, $rr(0, \gamma)$ decreases with increasing N, consistently with the Bayes c.e. being asymptotically optimal. We have found the behavior for τ^*, γ^* and $rr(0, \gamma)$ with respect to σ_{β_1}, σ, N and R to be similar for all the priors, horizons and sizes of the support for θ that we have tried.

23.4   DESIGNS FOR R.S.B. BASED ON C.P. INTERVALS

The certainty equivalence designs for the R.S.B. observe y_{n+1} at some estimate of κ, the x value maximizing $f(x; \beta)$, $x_{n+1}^{ce} = \hat{\kappa}_n$. Instead, Ginebra and Clayton (1995) perturb c.e. by experimenting on the x_{n+1} maximizing an upper bound for the predicted surface, $\hat{f}_n(x; \beta)$. Here we describe how we could experiment on an equally tailed interval for κ based on its posterior quantiles, perturbing the c.e. design that observes y_{n+1} on the median of the posterior distribution for κ, $x_{n+1}^{ce} = \kappa_n^{pme}$.

Let $q_n^{\alpha/2}$ and $q_n^{1-\alpha/2}$ be the $\alpha/2$ and $(1 - \alpha/2)$ quantiles of the posterior distribution for κ, and let $cp_n(\alpha) = [q_n^{\alpha/2}, q_n^{1-\alpha/2}]$ be a central posterior interval for κ. Note that $cp_n(\alpha = 1)$ only contains κ_n^{pme} and that $cp_n(\alpha = 0)$ is the support of the posterior density for κ. The width of $cp_n(\alpha)$ is monotonically decreasing in α and κ_n^{pme} splits $cp_n(\alpha)$ in two sub-intervals.

The "cp" designs are indexed by $(\alpha, \gamma) \in [0, 1] \times [-1, 1]$ and are denoted by $d_{cp}(\alpha, \gamma)$. They observe y_{n+1} at a point $x_{n+1}^{cp}(\alpha, \gamma)$ on the interval $cp_n(\alpha)$. The smaller α, the further the design is allowed to stray away from that c.e. choice. The index γ indicates the relative position of $x_{n+1}^{cp}(\alpha, \gamma)$ on that interval. When $\gamma > 0$, $d_{cp}(\alpha, \gamma)$ experiments on the side of κ_n^{pme} where $cp_n(\alpha)$ is wider, at a distance from $x_{n+1}^{ce} = \kappa_n^{pme}$ equal to γ times the length of that subinterval. When $\gamma < 0$ $d_{cp}(\alpha, \gamma)$ experiments on

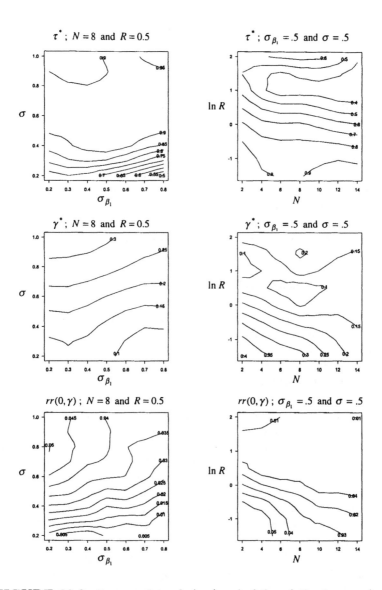

FIGURE 23.2 Contour plots of τ^*, γ^* and of the relative increase in expected loss when c.e. is used instead of $d_{hpd}(\tau^*, \gamma^*)$, $rr_2(\tau = 0, \gamma)$. The model is: $(y_n | x_n, \beta) \sim N(T + \beta_1(x_n - \theta), \sigma^2)$ with $\theta \sim U(-R, R)$, $\beta_1 \sim N(1, \sigma_{\beta_1}^2)$ and β_1 independent of θ. The objective is to minimize $E[\sum_{n=1}^{N} y_n(x_n^d)^2]$

the side of κ_n^{pme} where $cp_n(\alpha)$ is narrower, at a distance from $x_{n+1}^{ce} = \kappa_n^{pme}$ equal to $|\gamma|$ times its length. Thus for fixed α, the larger $|\gamma|$ the further $x_{n+1}^{cp}(\alpha, \gamma)$ is away from κ_n^{pme}. When $|\gamma| = 1$, $x_{n+1}^{cp}(\alpha, \gamma)$ is either $q_n^{\alpha/2}$ or $q_n^{1-\alpha/2}$, depending on which one of these quantiles is closer to κ_n^{pme}.

When either $\alpha = 1$ or $\gamma = 0$, $d_{cp}(\alpha, \gamma)$ experiments on $x_{n+1}^{cp} = x_{n+1}^{ce} = \kappa_n^{pme}$. As n increases, $x_{n+1}^{cp}(\alpha, \gamma)$ tends to be closer to κ_n^{pme}, matching the decreasing need to "actively" learn about κ as we get closer to $n = N$.

To find the design $d_{cp}(\alpha^*, \gamma^*)$ that maximizes $w_i(\alpha, \gamma) = w_1(d_{cp}(\alpha, \gamma)) = E[\sum_{n=1}^{N} y_n(x_n^{cp})]$ we would proceed in a way analogous to the one described in Section 23.2. For each (α, γ) we would repeatedly simulate the use of the corresponding design on data from the assumed model, estimating $w_i(\alpha, \gamma)$. By searching among various $d_{cp}(\alpha, \gamma)$, we find the $d_{cp}(\alpha^*, \gamma^*)$ that is best, improving on the c.e. design that experiments at $x_{n+1}^{ce} = \kappa_n^{pme}$; When α^* is close to 1, or γ^* close to 0, that c.e. design is the best one in this family.

23.5   CONCLUDING REMARKS

The designs here described induce probing actions on certainty equivalence to improve the intermediate estimates for θ, (or κ), and enhance future performance, adding an "active" learning feature on top of the "accidental" learning done by c.e. designs. One way to further improve $d_{hpd}(\tau^*, \gamma^*)$, (or $d_{cp}(\alpha^*, \gamma^*)$), is to re-estimate τ^* (or α^*) and γ^* at some intermediate stages m, based on the posterior distributions of θ (or κ) at those stages.

The main difficulty when implementing the designs proposed is in the repeated computation of the credible regions to estimate (τ^*, γ^*), (or (α^*, γ^*)). Using H.P.D. type designs requires the repeated computation of the profile of the marginal posterior for θ (or κ); When that profile is not easily available we can implement the designs based on central posterior intervals, where the difficulty is the repeated evaluation of the posterior quantiles for θ (or κ). When neither approach is feasible, we can use the designs proposed in González and Ginebra (1999a) and Ginebra and Clayton (1995).

The designs proposed in this paper can presumably be extended to other adaptive design problems with one input variable, but their extensions to problems with more than one input variable is not straightforward; When $x \in R^k$, the solution of $f(x; \beta) = T$ will be an hypersurface and not a real number, and κ will be a point in R^k. González and Ginebra (1999b) presents designs for multiperiod control with multiple control variables that improve both on c.e. as well as on the myopic choice, (that is the one that would be optimal if there was only one observation left to be taken).

APPENDIX: MODEL USED IN SECTION 23.3

In the normal linear model, $(y_n|x_n, \beta) \sim N(\beta_0 + \beta_1 x_n, \sigma^2)$, and $E[y|x, \beta] = \beta_0 + \beta_1 x = T + \beta_1(x - \theta)$, with $\theta = (T - \beta_0)/\beta_1$. To implement the H.P.D. designs in Section 23.3 we use,

Proposition 23.5.1 Let $y_n' = y_n - T$ with $y_n'|x_n, \beta \sim N(\beta_1(x_n - \theta), \sigma^2)$, $\theta \sim U(rtl, rtu)$, $\beta_1 \sim N(\mu_{\beta_1}, \sigma_{\beta_1}^2)$, β_1 independent of θ and σ known. The marginal posterior density for θ is:

$$\pi_n(\theta|Y_n, X_n, \pi_0) \propto$$

$$\frac{I_{[rtl,rtu]}(\theta)}{\sqrt{\sigma_{\beta_1}^2 \sum (x_i - \theta)^2 + \sigma^2}} \exp\left\{ \frac{(\sigma_{\beta_1}^2 \sum y_i(x_i - \theta) + \sigma^2 \mu_{\beta_1})^2}{2\sigma^2 \sigma_{\beta_1}^2 (\sigma_{\beta_1}^2 \sum (x_i - \theta)^2 + \sigma^2)} \right\}.$$

PROOF.   The joint distribution for (Y_n, θ, β_1), (the likelihood times the prior distribution), is:

$$k_n(Y_n, \theta, \beta_1|X_n, \pi_0)$$

$$\propto I_{[rtl,rtu]}(\theta) \exp\left\{ -\frac{1}{2}\left(\frac{(Y_n - \beta_1 X_n \theta_c)'(Y_n - \beta_1 X_n \theta_c)}{\sigma^2} \right. \right.$$

$$\left. \left. + \frac{(\beta_1 - \mu_{\beta_1})^2}{\sigma_{\beta_1}^2} \right) \right\},$$

with $\theta_c' = (-\theta, 1)$. Thus the joint posterior density for (θ, β_1) will be:

$$k_n(\theta, \beta_1|X_n, \pi_0)$$

$$\propto I_{[rtl,rtu]}(\theta) \exp\left\{ \frac{-\beta_1}{2\sigma^2 \sigma_{\beta_1}^2}(\beta_1(\sigma_{\beta_1}^2 \theta_c' X_n' X_n \theta_c) \right.$$

$$\left. -2(\sigma_{\beta_1}^2 Y_n' X_n \theta_c + \sigma^2 \mu_{\beta_1})) \right\}$$

$$= I_{[rtl,rtu]}(\theta) \exp\left\{ \frac{-\beta_1}{2\sigma^2 \sigma_{\beta_1}^2}(\beta_1(\sigma_{\beta_1}^2 \sum_{i=1}^{n}(x_i - \theta)^2 + \sigma^2) \right.$$

$$\left. -2(\sigma_{\beta_1}^2 \sum_{i=1}^{n} y_i(x_i - \theta) + \sigma^2 \mu_{\beta_1})) \right\}.$$

The marginal posterior for θ is obtained integrating out β through using results for normal densities. $\square$

The proof for the next proposition follows by showing that the roots of $d\pi_n(\theta|Y_n, X_n, \pi_0)/d\theta = 0$ are the roots of a third degree polynomial, and thus the marginal posterior for θ has either one or two maxima. We observe instances of bimodality for small n; When that happens, one of the modes tends to be several orders of magnitude larger than the second mode.

Proposition 23.5.2 *Let* $y_n' = y_n - T$ *with* $y_n'|x_n, \beta \sim N(\beta_1(x_n - \theta), \sigma^2)$, *and* $\theta \sim U(rtl, rtu)$, $\beta_1 \sim N(\mu_{\beta_1}, \sigma_{\beta_1}^2)$, β_1 *independent of* θ *and* σ *known. The marginal posterior density for* θ *has at most two modes.*

REFERENCES

Astrom, K. J. and Wittenmark, B. (1995). *Adaptive Control*, Second edition, Addison Wesley, Reading, MA.

Berliner, M. (1982). Uniform improvements on the certainty equivalent rule in a statistical control problem, In *Statistical Decision Theory and Related Topics III, Vol. 1* (Eds., S. S. Gupta and J. Berger), pp. 157–168, Academic Press, New York.

Berliner, M. (1983). Improving on inadmissible estimators in the control problem, *Annals of Statistics*, **11**, 814–826.

Berry, D.A. and Fristedt, B. (1985). *Bandit Problems; Sequential Allocation of Experiments*, Chapman and Hall, London.

Carlin, B. P., Kadane, J. B. and Gelfand, A. E. (1998). Approaches for optimal sequential decision analysis in clinical trials, *Biometrics*, **54**, 964–975.

Chen, K. and Hu, I. (1998). On consistency of Bayes estimates in a certainty equivalence adaptive system, *IEEE Transactions on Automatic Control*, **43**, 943–947.

De Groot, M.H. (1970). *Optimal Statistical Decisions*, McGraw-Hill, New York.

Ginebra, J. and Clayton, M. K. (1995). The response surface bandit, *Journal of the Royal Statistical Society, Series B*, **57**, 771–784.

González, E. and Ginebra, J. (1999a). Bayesian heuristics for the multiperiod control problem; One control variable, *Document de recerca DR 99/03*, Departament d'Estadística i Investigació Operativa, Universitat Politècnica de Catalunya.

González, E. and Ginebra, J. (1999b). Bayesian heuristics for the multi-period control problem; Multiple control variables, *Document de recerca DR 99/05*, Departament d'Estadística i Investigació Operativa, Universitat Politècnica de Catalunya.

Hu, I. (1998). On sequential designs in nonlinear problems, *Biometrika*, **85**, 496–503.

Lai, T. L. and Robbins, H. (1982). Adaptive design and the multiperiod control problem, In *Statistical Decision Theory and Related Topics III, Vol. 2* (Eds. S.S. Gupta and J. Berger), pp. 103–120, Academic Press, New York.

Müller, P. (1998). Simulation based optimal designs. *Bayesian Statistics 6* (Eds., J. M. Bernardo, J. O. Berger, A. P. Dawid and A. F. M. Smith), pp. 459–474, Oxford University Press, Oxford, England.

Rosenberger, W. (1996). New directions in adaptive designs, *Statistical Science*, **11**, 137–149.

Srinivasan, C. (1984). A sharp necessary and sufficient condition for inadmissibility of estimators in a control problem, *Annals of Statistics*, **12**, 927–944.

Ying, Z and Wu, C. F. J. (1997). An asymptotic theory of sequential designs based on maximum likelihood recursions, *Statistica Sinica*, **7**, 75–91.

CHAPTER 24

AGING WITH LAPLACE ORDER CONSERVING SURVIVAL UNDER PERFECT REPAIRS

MANISH C. BHATTACHARJEE

New Jersey Institute of Technology, Newark, NJ

SUJIT K. BASU

National Institute of Management, Calcutta, India

Abstract: Survival distributions that are Laplace order dominated by exponentials is the largest among the standard aging families. In this article, we investigate the subfamily of such distributions for which the asymptotic remaining life under perfect repairs preserves this Laplace ordering.

Keywords and phrases: Survival distributions, perfect repair, Laplace ordering, renewal theory, shock models

24.1 INTRODUCTION

To model the degradation of a repairable equipment over time, it is important to specify not only the 'aging' character of the equipment's initial survival time via a suitable nonparametric specification, but also that of the remaining life of such an equipment in use under a repair discipline. In light of this approach, it is then natural to ask : what are the ramifications of invoking an aging property in conjunction with an appropriate repair scheme? The focus of our present study is on survival distributions which are weakly aging in the sense that they belong to the class $\mathcal{L}$ [Klefsjö (1983)], such that under repeated renewals or replacements, the corresponding asymptotic remaining life of a sufficiently aged unit conserves this property.

Section 24.2 introduces the class $\mathcal{L}_D$ of survival distributions, and its dual, with the above Laplace order conserving property under perfect repairs and briefly reviews some of their properties as a necessary background to investigate their closure under reliability operations in Section 24.3. Negative results on closures are supported by counterexamples. To study their preservation under shock models, Section 24.4 considers the discrete versions of these classes, which are of independent interest as well and clarifies the role of geometric distributions as extreme points therein. In the final section we consider preservation of the class properties under renewal process shock models.

Our motivation for considering the class $\mathcal{L}$ property in this context is that, it is the weakest among all standard aging properties (IFR, DMRL, IFRA, NBUE, [Barlow and Proschan (1975)], HNBUE and $\mathcal{L}$ [Klefsjö (1982, 1983)], and thus constitutes the largest among the survival distribution defined by them. The $\mathcal{L}$ family of distributions is a version of Laplace ordering [see, Stoyan (1983)] relative to the exponential distributions. Note that each of the aging classes above describe some feature of a new equipment's degradation over time until *first* failure, which does not necessarily translate to a corresponding description of its worsening under repair.

24.2   THE CLASS $\mathcal{L}_D$

In what follows, F will generically denote a *distribution function* (d.f.) of the *life* $X \geq 0$ $(F(0-) = 0)$ of a new unit, with *survival function* $\bar{F} := 1 - F$ and Laplace transform

$$L_F(s) := E(e^{-sX}) = \int_0^\infty e^{-st} dF(t), \quad s \geq 0.$$

By $\mu_{r,F}$, we shall denote the r-th moment of F, whenever it is finite, and by η_F its coefficient of variation (c.v.). Klefsjö (1983) defines the class-$\mathcal{L}$ aging (its dual $\bar{\mathcal{L}}$ resp.) property of a survival distribution F, with a finite mean, by

$$\int_0^\infty e^{-st} \bar{F}(t) dt \quad \geq (\leq) \quad \frac{\mu_{1,F}}{1 + \mu_{1,F}}, \quad s \geq 0. \tag{24.2.1}$$

Equivalently, $F \in \mathcal{L}$ if and only if

$$L_F(s) \quad \leq (\geq) \quad (1 + s\mu_{1,F})^{-1}, \quad \forall s \geq 0;$$

i.e., iff the Laplace transform of F is dominated by (dominates, resp.) the Laplace transform of an exponential distribution with the same mean as that of F. For several interesting interpretations of the definition (24.2.1) and its reliability theoretic ramifications, see Klefsjö (*ibid.*).

Returning to the theme of aging relative to renewals, consider an unit with life d.f. F, which is instantaneously replaced (*perfectly repaired*) by a statistically identical and independent copy every time it fails. For the renewal process driven by F, which describes the corresponding point process of failures and repairs, let

$$F_1(x) = \frac{1}{\mu_{1,F}} \int_0^x \bar{F}(t)dt, \quad x \geq 0, \qquad (24.2.2)$$

denote the *first derived distribution* induced by F, i.e.,the d.f. to which the *remaining life* and *age* of an item in use under repeated renewals, converges in distribution. Consider the class of d.f.s such that the asymptotic remaining life distribution conserves the class $\mathcal{L}$ aging (or, its dual) property of a new unit, as defined below.

Definition 24.2.1 A life d.f. F with a finite mean is in the class $\mathcal{L}_D$ ($\bar{\mathcal{L}}_D$, resp.) if F as well as the asymptotic remaining life d.f. F_1 under repeated replacements on failure, are both in $\mathcal{L}$ ($\bar{\mathcal{L}}$, resp.); i.e.,

$$\mathcal{L}_D = \{F \quad : F \in \mathcal{L}, \text{ and } F_1 \in \mathcal{L}\},$$

Replacing $\mathcal{L}$ with $\bar{\mathcal{L}}$ similarly defines the dual class $\bar{\mathcal{L}}_D$.

Bhattacharjee and Sengupta (1996) proved that if $F \in \mathcal{L}$, then F has a finite variance and indeed, like its immediate predecessor class HNBUE among the standard nested aging classes, preserves the property that its coefficient of variation (c.v.) η_F satisfies $\eta_F \leq 1$, the c.v. of exponential distributions. Correspondingly, if $F \in \bar{\mathcal{L}}$, the second moment need not be finite, but if it is, then $\eta_F \geq 1$. It is known [Basu and Bhattacharjee (1984), Bhattacharjee and Sethuraman (1990)] that unit c.v. characterizes the exponentials within all standard nested aging classes up to HNBUE, as well as among survival distributions within all corresponding dual classes up to HNWUE with a finite variance. Bhattacharjee and Sengupta (*ibid*) also provided an example of a d.f. in $\mathcal{L}$, which has a c.v. = 1, but which is not exponential. This counterexample then implies that the preceeding characterization of exponentials does *not* extend to the class $\mathcal{L}$ or its dual, and suggests an immediate question: viz., is there a class smaller than $\mathcal{L}$ ($\bar{\mathcal{L}}$, resp.) not contained within HNBUE (HNWUE, resp.), wherein unit c.v. still characterizes the exponentials? The class $\mathcal{L}_D$ and its dual was originally motivated by Mitra *et al.* (1995) in an effort to answer this question in the affirmative, with the following result.

Theorem 24.2.1 *[Mitra, Basu and Bhattacharjee (1995)] Suppose $F \in \mathcal{L}_D (\bar{\mathcal{L}}_D)$. Then $\eta_F \leq (\geq) 1$. Further, F is exponential if and only if $\eta_F = 1$.*

Mitra *et al.* did not pursue the wider ramifications of the aging property defined by $\mathcal{L}_D$. When we consider the fact that F_1 describes the remaining life in the long run under repeated renewals; the class $\mathcal{L}_D$ acquires a natural setting to explore its properties such as closure under reliability operations, and which describes the motivation for the present work.

The hierarchical relationship of $\mathcal{L}_D$ to the standard aging classes is shown in the following diagram, with a corresponding chain of implication for the respective duals.

Relationship of $\mathcal{L}_D$ to standard aging classes

$$
\begin{array}{c}
\text{IFRA} \longrightarrow \text{NBU} \\
\nearrow \qquad\qquad \searrow \\
\text{IFR} \qquad\qquad\qquad \text{NBUE} \longrightarrow \text{HNBUE} \longrightarrow \mathcal{L} \\
\searrow \qquad\qquad \nearrow \qquad\qquad \nearrow \\
\text{DMRL} \longrightarrow \qquad \longrightarrow \mathcal{L}_D
\end{array}
$$

Except for the nesting relationships of $\mathcal{L}_D$, the other implications are well known. The claim DMRL $\Rightarrow \mathcal{L}_D \Rightarrow \mathcal{L}$ follows, since $\mathcal{L}_D$ is a subset of $\mathcal{L}$ so that, $F \in \text{DMRL} \Rightarrow F \in \mathcal{L}$, and by noting that $F \in \text{DMRL} \Leftrightarrow F_1 \in \text{IFR} \Rightarrow F_1 \in \mathcal{L}$ There are no known relationships between $\mathcal{L}_D$ and NBUE or HNBUE. In fact neither of the last two properties is implied by $\mathcal{L}_D$. This observation follows from our Counterexample 24.3.1 in Section 24.3 which exhibits a d.f. in $\mathcal{L}_D$, but which is not HNBUE and hence is not NBUE.

The moments and Laplace transforms of F and the corresponding *first derived distribution*, which we will find useful for later reference, are related to each other, by

$$\mu_{r,F_1} = \frac{\mu_{r+1,F}}{(r+1)\mu_{1,F}}, \quad r > 0, \tag{24.2.3}$$

$$L_{F_1}(s) = \frac{1 - L_F(s)}{s\mu_{1,F}}, \quad s > 0. \tag{24.2.4}$$

Remark 24.2.1 It follows from the first claim in Theorem 24.2.1 and (24.2.3) that, if $F \in \mathcal{L}_D$, then

$$\mu_{1,F} \geq \frac{\mu_{2,F}}{2\mu_{1,F}} \equiv \mu_{1,F_1} \geq \frac{\mu_{2,F_1}}{2\mu_{1,F_1}} = \frac{\mu_{3,F}}{3\mu_{2,F}}.$$

This chain of inequalities clearly implies,

$$1 \geq \frac{\mu_{2,F}}{2\mu_{1,F}^2} \geq \frac{\mu_{3,F}}{3!\mu_{1,F}^3}, \text{ and } 1 \leq \frac{\mu_{3,F}\mu_{1,F}}{\mu_{2,F}^2} \leq \frac{3}{2}. \qquad (24.2.5)$$

Suppose, the third moment of F agrees with the corresponding exponential moment ($\mu_{3,F} = 3!\mu_{1,F}^3$). Then the first chain in (24.2.5) collapses and immediately implies $\eta_F = 1$, which then guarantees the exponentiality of F by the second claim in Theorem 24.2.1. For higher values of r, an analogous argument can be constructed by repeatedly using (24.2.3). We thus have the following corollary to Theorem 24.2.1, which shows that within $\mathcal{L}_D$, agreement of any moment of order $r = 2, 3, \cdots$ with the corresponding exponential moment also characterizes exponentiality.

Corollary 24.2.1 *Let* $F \in \mathcal{L}_D (\bar{\mathcal{L}}_D)$. *If* $\mu_{r,F} = r!\mu_{1,F}^r$ *for some integer* $r \geq 2$, *then* F *must be exponential.*

24.3   CLOSURE PROPERTIES

This section is devoted to the investigation of preservation properties of $\mathcal{L}_D$ and $\bar{\mathcal{L}}_D$ under reliability operations such as coherent structures and convolutions.

24.3.1   Coherent Structures

We show that neither $\mathcal{L}_D$ nor $\bar{\mathcal{L}}_D$ are closed under coherent structures. The latter is what one would expect (Counterexample 24.3.2), since $\bar{\mathcal{L}}_D$ is *negatively aging*. The first assertion, which requires more effort to demonstrate (Counterexample 24.3.1), is also not unexpected, and is parallel to the corresponding findings that other weakly aging classes such as NBUE and HNBUE are also not closed under formation of coherent structures [Barlow and Proschan (1975) and Klefsjö (1983)].

Counterexample 24.3.1 $\mathcal{L}_D$ is not closed under coherent structures.

Consider the d.f. F with jumps of magnitude 0.3 and 0.7 at $x = 0.3$ and 3 respectively,

$$F(x) = \begin{cases} 0, & \text{for } 0 \leq x < 0.3 \\ 0.3, & \text{for } 0.3 \leq x < 3 \\ 1, & \text{for } x \geq 3 \end{cases}$$

Klefsjö (1983) used this example to conclude that $\mathcal{L}$ is not closed under coherent structures by proving $F \in \mathcal{L}$ and that for the d.f $G := 1 - \bar{F}^2$ of the survival time of a series system of two i.i.d. components with lifetime d.f. F as above, we have $G \notin \mathcal{L}$.

We show $F \in \mathcal{L}_D$. This is enough to demonstrate that $\mathcal{L}_D$ is *not* closed under coherent structures, since the d.f. $G = 1 - \bar{F}^2 \notin \mathcal{L}$ and $\mathcal{L}_D \subset \mathcal{L}$ then implies $G \notin \mathcal{L}_D$.

To check $F \in \mathcal{L}_D$, we only need to show that $F_1 \in \mathcal{L}$, since we already know that $F \in \mathcal{L}$. Now, $F_1 \in \mathcal{L}$ if and only if its Laplace transform satisfies the inequality

$$L_{F_1}(s) \equiv \frac{1 - 0.3e^{-0.3s} - 0.7e^{-3s}}{s\mu_{1,F}} \leq \frac{1}{1 + s\mu_{1,F_1}}, \qquad s > 0,$$

or, equivalently, if and only if

$$\phi(s) := 0.7 + (0.3e^{2.7s} - e^{3s}) + s\{\mu_{1,F_1}(0.7 + 0.3e^{2.7s}) + (\mu_{1,F} - \mu_{1,F_1})e^{3s}\} \geq 0,$$

for all $s \geq 0$. Let $c_r = $ coefficient of s^r in the expansion of $\phi(s)$, $r = 0, 1, 2, \cdots$. Then,

$$c_0 = c_1 = 0,$$

$$c_r = \frac{2.7^{r-1}}{(r-1)!}\left(\frac{0.81}{r} + 0.3\mu_{1,F_1}\right) + \frac{3^{r-1}}{(r-1)!}\left(\mu_{1,F} - \mu_{1,F_1} - \frac{3}{r}\right), \qquad \text{if } r \geq 2.$$

For $r \geq 2$, the first of the two additive terms, defining c_r, is clearly positive. The sign of the second term is that of

$$\mu_{1,F} - \mu_{1,F_1} - \frac{3}{r} > 1.14 - \frac{3}{r} > 0. \qquad \text{for } r \geq 3,$$

since $\mu_{1,F} = 2.19$, $\mu_{2,F} = 6.327$, so that $\mu_{1,F_1} = (6.327/4.38) < 1.05$; and further setting $r = 2$ in the expression of c_r above, we get

$$c_2 > 2.7(0.405 + 0.312) + 3(1.14 - 1.5) = 1.9359 - 1.08 > 0.$$

Thus $c_0 = c_1 = 0$, $c_r > 0$ for all $r \geq 2$, so that $\phi(s) > 0$, for all $s \in (0, \infty)$. Hence $F_1 \in \mathcal{L}$.

Counterexample 24.3.2 The life distribution of a parallel system of two independent exponential components with *different means* is IFRA [Barlow and Proschan (1975, p. 83)] but is *not exponential*, and hence is strictly in $\mathcal{L}$ and thus not in $\bar{\mathcal{L}}$ or, its subset $\bar{\mathcal{L}}_D$. Since, exponentials are trivially in $\bar{\mathcal{L}}_D$, it follows that $\bar{\mathcal{L}}_D$ *is not closed under coherent structures.*

24.3.2 Convolutions

Our main result is that $\mathcal{L}_D$ is closed under formation of cold standby systems of independent components (Theorem 24.3.1), while its dual class is not (Counterexample 24.3.3).

Theorem 24.3.1 *Suppose F and G both $\in \mathcal{L}_D$, and let $H := F * G$ be their convolution. Then $H \in \mathcal{L}_D$.*

PROOF.   Since $F, G \in \mathcal{L}_D$, we have $F, G \in \mathcal{L}$ and the corresponding first derived distributions $F_1, G_1 \in \mathcal{L}$ as well. Hence $H \in \mathcal{L}$, since $\mathcal{L}$ is closed under convolutions [Klefsjö (1983)]. To complete the argument, we further need to show that $H_1 \in \mathcal{L}$.

Since class $\mathcal{L}$ distributions have a finite variance and thus a finite c.v., assume without loss of generality that $\eta_G \leq \eta_F$. Next, note that for any survival distribution S (in $\mathcal{L}$, or not) with a finite mean; using (24.2.4), we see that the corresponding derived distribution $S_1 \in \mathcal{L}$ iff $(s\mu_{1,S})^{-1}\{1 - L_S(s)\} \leq (1 + s\mu_{1,S_1})^{-1}$, i.e.,

$$S_1 \in \mathcal{L} \quad \Leftrightarrow \quad L_S(s) \geq \frac{1 - s(\mu_{1,S} - \mu_{1,S_1})}{1 + s\mu_{1,S_1}}, \quad \text{all } s > 0. \qquad (24.3.6)$$

Since $L_{H_1}(s) = \{1 - L_F(s)L_G(s)\}/\{s(\mu_{1,F} + \mu_{1,G})\}$; by invoking (3.1) for $F, G \in \mathcal{L}_D$, and some routine computations, we see that

$$
\begin{aligned}
L_{H_1}(s) \ &\leq \ \frac{1}{s(\mu_{1,F} + \mu_{1,G})}\left[1 - \left(\frac{1 - s(\mu_{1,F} - \mu_{1,F_1})}{1 + s\mu_{1,F_1}}\right)\right.\\
&\qquad\left.\times \left(\frac{1 - s(\mu_{1,G} - \mu_{1,G_1})}{1 + s\mu_{1,G_1}}\right)\right]\\
&= \ \frac{(1 + sC)}{(1 + s\mu_{1,F_1})(1 + s\mu_{1,G_1})}, \qquad\qquad (24.3.7)
\end{aligned}
$$

where,

$$
\begin{aligned}
C \ &:= \ \frac{\mu_{1,F}\mu_{1,G_1} + \mu_{1,G}\mu_{1,F_1} - \mu_{1,F}\mu_{1,G}}{\mu_{1,F} + \mu_{1,G}}\\
&= \ \alpha\mu_{1,G_1} + \bar{\alpha}(\mu_{1,F_1} - \mu_{1,F}), \qquad\qquad (24.3.8)
\end{aligned}
$$

and

$$\alpha = \mu_{1,F}/(\mu_{1,F} + \mu_{1,G}), \qquad \bar{\alpha} := 1 - \alpha. \qquad (24.3.9)$$

Now, from (24.3.6) and (24.3.7), we see that to conclude $H_1 \in \mathcal{L}$, it is enough to show that

$$(1 + sC)(1 + s\mu_{1,H_1}) \le (1 + s\mu_{1,F_1})(1 + s\mu_{1,G_1}), \text{ all } s > 0. \quad (24.3.10)$$

To verify (24.3.10), we compute the mean of H_1 by using the following easily proved representation [Bhattacharjee et al. (1998)] of H_1 as a mixture,

$$H_1(x) = \alpha F_1(x) + \bar{\alpha}(F * G_1)(x), \qquad (24.3.11)$$

where α is given by (24.3.9). This yields

$$\mu_{1,H_1} = \alpha\mu_{1,F_1} + \bar{\alpha}(\mu_{1,F} + \mu_{1,G_1}). \qquad (24.3.12)$$

Using (24.3.8) and (24.3.12), the coefficient of s on the left hand side of (24.3.10) is,

$$\begin{aligned}
\mu_{1,H_1} + C &= \alpha\mu_{1,F_1} + \bar{\alpha}(\mu_{1,F} + \mu_{1,G_1}) + \alpha\mu_{1,G_1} + \bar{\alpha}(\mu_{1,F_1} - \mu_{1,F}) \\
&= \mu_{1,F_1} + \mu_{1,G_1},
\end{aligned}$$

and thus equals the coefficient of s on the righthand side of (24.3.10). Since $F \in \mathcal{L}$ implies $\eta_F \le 1$ [i.e., $\mu_{1,F_1} \le \mu_{1,F}$, by (24.2.3)]; from (24.3.8), we have $C \le \alpha\mu_{1,G_1}$. Thus, the coefficient of s^2 on the left hand side of (24.3.10) can be bounded above, by

$$C\mu_{1,H_1} \le \alpha\mu_{1,G_1}\mu_{1,H_1}.$$

Hence, we will be done if we can show that $\alpha\mu_{1,H_1} \le \mu_{1,F_1}$, for then the coefficient of s^2 on the left hand side of (24.3.10) will be dominated by its counterpart on right hand side, thereby confirming the inequality. Now, using (24.3.12),

$$\begin{aligned}
\alpha\mu_{1,H_1} \le \mu_{1,F_1} &\Leftrightarrow \alpha\bar{\alpha}(\mu_{1,G_1} + \mu_{1,F}) \le (1 - \alpha^2)\mu_{1,F_1} \\
&\Leftrightarrow \frac{\mu_{1,G_1} + \mu_{1,F}}{\mu_{1,F_1}} \le 1 + \frac{1}{\alpha} = 2 + \frac{\mu_{1,G}}{\mu_{1,F}} \\
&\Leftrightarrow \frac{\mu_{1,G_1}}{\mu_{1,F_1}} \le \frac{2\eta_F^2}{1 + \eta_F^2} + \frac{\mu_{1,G}}{\mu_{1,F}}, \qquad (24.3.13)
\end{aligned}$$

since $\frac{\mu_{1,F_1}}{\mu_{1,F}} = \frac{1}{2}(1 + \eta_F^2)$. However, since $\eta_G \le \eta_F$, we indeed have

$$\frac{\mu_{1,G_1}}{\mu_{1,F_1}} = \frac{\mu_{1,G}}{\mu_{1,F}}\left(\frac{1 + \eta_G^2}{1 + \eta_F^2}\right) \le \frac{\mu_{1,G}}{\mu_{1,F}},$$

which implies that (24.3.13) is satisfied as a strict inequality, and the proof is complete. □

Remark 24.3.1 If $\eta_F < \eta_G$; then use the alternative representation

$$H_1(x) = \bar{\alpha}G_1(x) + \alpha(G * F_1)(x),$$

which follows from (24.3.11) and commutativity of convolutions ($H \equiv F * G = G * F$), and then proceed analogously by reversing the roles of F and G in the subsequent computations following (24.3.11).

Counterexample 24.3.3 Since the convolution of two i.i.d. exponentials (which are trivially in $\mathcal{L}$ as well as in $\bar{\mathcal{L}}_D$) is strictly IFR and thus cannot belong to $\bar{\mathcal{L}}_D$; it follows that $\bar{\mathcal{L}}_D$ *is not closed under convolutions.*

24.3.3 Mixtures

A mixture of exponentials (which are, of course in $\mathcal{L}_D$), being strictly DFR, belongs to $\bar{\mathcal{L}}_D$. Thus $\mathcal{L}_D$ *is not closed under mixtures*, while its dual is, as the next proposition shows.

Theorem 24.3.2 $\bar{\mathcal{L}}_D$ *is closed under arbitrary mixing.*

PROOF. Routine. Let

$$F(t) = \int_\Lambda F_\lambda(t)dP(\lambda), \qquad (24.3.14)$$

be an arbitrary mixture of d.f.s in a family $\{F_\lambda : \lambda \in \Lambda\}$ of $\bar{\mathcal{L}}_D$-distributions with a mixing distribution P on Λ. To show $F \in \bar{\mathcal{L}}_D$, simply note that, if $F_1, F_{\lambda,1}$ denote the first derived distributions corresponding to F, F_1 respectively, then we can write,

$$F_1(t) = \int_\Lambda F_{\lambda,1}(t)dP^*(\lambda), \qquad (24.3.15)$$

where P^* is the probability measure on Λ such that

$$dP^*(\lambda) = \left(\frac{\mu_{1,F_\lambda}}{\mu_{1,F}}\right)dP(\lambda).$$

Note that (24.3.14) implies $\mu_{1,F} = \int_\Lambda \mu_{1,F_\lambda}dP(\lambda)$, which guarantees $P^*(\Lambda) = 1$, so that the measure P^* defined above is indeed a probability on Λ.

Since, $F_\lambda \in \bar{\mathcal{L}}_D$ requires $F_\lambda \in \bar{\mathcal{L}}$ and $F_{\lambda,1} \in \bar{\mathcal{L}}$, all $\lambda \in \Lambda$, and $\bar{\mathcal{L}}$ is closed under mixing [Klefsjö (1983)]; it follows that both the mixtures in (24.3.14) and (24.3.15) must be in $\bar{\mathcal{L}}$, so that $F \in \bar{\mathcal{L}}_D$. □

24.4   THE DISCRETE CLASS $\mathcal{G}_D$ AND ITS DUAL

In order to investigate the preservation of class $\mathcal{L}_D$ property under shock models, in this section we introduce the discrete version $\mathcal{G}_D$ of the class $\mathcal{L}_D$, the corresponding dual class, and note some of their relevant properties. The definition of the class $\mathcal{G}_D$ is prompted by consideration analogous to those which motivated our construction of the class $\mathcal{L}_D$ as a subset of $\mathcal{L}$. To do this, we first need the class $\mathcal{G}$ [Klefsjö (1983)],the discrete counterpart of $\mathcal{L}$.

Definition 24.4.1 [Klefsjö (1983)] The distribution $\boldsymbol{P} = \{p_1, p_2, \cdots\}$ of a positive integer valued random variable N, with mean $EN = m$ and probability generating function (p.g.f.) $\psi_N(z) := Ez^N$, $0 \le z \le 1$ belongs to $\mathcal{G}$ ($\bar{\mathcal{G}}$, resp.), if

$$\psi_N(z) \quad \le (\ge) \quad \frac{z}{z + m(1 - z)} \equiv \psi_{N^*}(z), \quad 0 \le z \le 1, \quad (24.4.16)$$

where, N^* is a geometric r.v. with $EN^* = EN = m$. Assume, $m > 1$, w.l.o.g. Note that (i) N^* has the geometric distribution $\boldsymbol{P}^*$, defined by,

$$P(N^* = k) = \frac{1}{m}\left(1 - \frac{1}{m}\right)^{k-1}, \quad k = 1, 2, \cdots$$

and (ii) the class $\mathcal{G}$ defined above via (24.4.16) is a special case of the partial ordering $<_g$ via generating functions [see, Stoyan (1983)] defined among integer valued r.v.s J, K by $J <_g K$ if $Ez^J \ge Ez^K$, $0 \le z \le 1$; so that,

$$\mathcal{G} = \{N : N^* <_g N\}, \text{ and } \bar{\mathcal{G}} = \{N : N <_g N^*\}.$$

If $\bar{P}_k := P(N > k)$, $k = 0, 1, 2, \cdots$ are the survival probabilities of N, then (24.4.16) can be expressed as,

$$\sum_{k=0}^{\infty} \bar{P}_k z^k = \frac{1 - \psi_N(z)}{1 - z} \quad \ge (\le) \quad \frac{m}{z + m(1 - z)}, \quad 0 \le z < 1.$$

Corresponding to the r.v. N, consider an induced r.v. N_1 with distribution $\boldsymbol{Q} = \{q_1, q_2, \cdots\}$, defined by,

$$q_k \equiv P(N_1 = k) := \frac{P(N \ge k)}{EN} = \frac{\bar{P}_{k-1}}{m}, \quad k = 1, 2, \cdots \quad (24.4.17)$$

with the probability generating function,

$$\psi_{N_1}(z) := Ez^{N_1} = \frac{z}{m}\sum_{k=1}^{\infty} \bar{P}_{k-1} z^{k-1} = \frac{z}{m}\frac{1 - \psi_N(z)}{1 - z}, \quad 0 \le z < 1.$$

$$(24.4.18)$$

Definition 24.4.2 The distribution P of a strictly positive integer valued r.v. N with a finite mean belongs to $\mathcal{G}_D$ $(\bar{\mathcal{G}}_D)$ if the distribution P of N as well as the corresponding induced distribution Q of N_1 belong to $\mathcal{G}$ $(\bar{\mathcal{G}})$, i.e.,

$$\mathcal{G}_D := \{N : N^* <_g N, \text{ and } N_1^* <_g N_1\},$$

where N_1^* is a r.v. with distribution defined by replaying N in (24.4.17) with the geometric r.v. N^* with mean $EN^* = m$. The dual $\bar{\mathcal{G}}_D$ is similarly defined by interchanging the roles of N and N^* in the partial ordering above.

Theorem 24.2.1 has the following analog in the discrete classes $\mathcal{G}$, $\mathcal{G}_D$ (and their duals) which have the family of geometric distributions in their boundary.

Theorem 24.4.1 (i) *If $P \in \mathcal{G}$ $(\bar{\mathcal{G}}$, resp.) with mean m, then*

$$\eta_P \leq (\geq) \ \eta_{P^*} \equiv \left(1 - \frac{1}{m}\right)^{\frac{1}{2}}. \tag{24.4.19}$$

(ii) *Further, if $P \in \mathcal{G}_D$, $(\bar{\mathcal{G}}_D$, resp.), then the extremal value of the c.v. is attained iff $P = P^*$; i.e., if and only if P is geometric.*

PROOF. We indicate the argument when $P \in \mathcal{G}$, $\mathcal{G}_D$ respectively, as the dual case follows similarly.

(i) If the distribution of N is in $\mathcal{G}$, then by using (24.4.18) and (24.4.16), it follows that the distribution of the induced r.v. N_1 must satisfy,

$$E\left(\frac{1 - z^{N_1}}{1 - z}\right) = \frac{1 - \psi_{N_1}(z)}{1 - z} \leq \frac{m}{z + m(1 - z)}.$$

As $z \uparrow 1$, the left hand side converges to EN_1, by the monotone convergence theorem. Hence, as $z \uparrow 1$, the above inequality implies

$$EN_1 \leq m \equiv EN. \tag{24.4.20}$$

From (24.4.18), check that

$$(EN)(EN_1) = m\psi'_{N_1}(1-) = \psi'_N(1-) + \frac{1}{2}\psi''_N(1-)$$
$$= \frac{1}{2}(EN + EN^2). \tag{24.4.21}$$

The inequality (24.4.20) then clearly implies,

$$1 \geq \frac{EN_1}{EN} = \frac{1}{2}\left(\frac{1}{m} + \eta_P^2 + 1\right), \qquad (24.4.22)$$

which leads to the desired conclusion in (24.4.19).

(ii) Suppose $P \in \mathcal{G}_D$. Then its c.v. still satisfies the bound in (24.4.19). Using (24.4.16), (24.4.18) and the defining property of the class $\mathcal{G}_D$, we easily obtain,

$$\frac{z + (1-z)(EN_1 - EN)}{z + (1-z)EN_1} \leq \psi_N(z) \leq \frac{z}{z + (1-z)EN}, \qquad 0 \leq z \leq 1. \qquad (24.4.23)$$

From (24.4.22), we see that η_P attains the extremal value in (24.4.19) if and only if $EN_1 = EN$. When this happens, the bounds on the p.g.f. in (24.4.23) then collapse to show that N must then be a geometric r.v. $\square$

As Theorem 24.4.1 suggests, the class $\mathcal{G}$ ($\bar{\mathcal{G}}$) is strictly larger than $\mathcal{G}_D$ ($\bar{\mathcal{G}}_D$ resp.) since the extreme possible value of the c.v. characterizes the geometric distribution only within the latter classes. The following counterexample shows that attaining the extremal value of the c.v. within the class $\bar{\mathcal{G}}$ does not require a geometric distribution. A similar example can be constructed for the class $\mathcal{G}$.

Counterexample 24.4.1 Consider the distribution P of a r.v. N such that $P(N = 1) = \frac{3}{4}$ and $P(N = 3) = \frac{1}{4}$. Then, $m = \frac{3}{2}$ and $\eta_P = \frac{1}{\sqrt{3}}$, so that η_P attains the bound in (24.4.19). Also, with N^* denoting the geometric r.v. with mean $EN^* = EN$, we have,

$$\psi_N(z) - \psi_{N^*}(z) = \frac{3z + z^3}{4} - \frac{2z}{3-z}, \qquad 0 \leq z \leq 1.$$

Clearly, $\psi_N(z) - \psi_{N^*}(z) \geq 0$, all $z \in [0,1]$ if and only if $z(z-1)^3 \leq 0$, which trivially holds on [0.1]. Thus, $P \in \bar{\mathcal{G}}$, without being geometric, although its c.v. achieves the lower bound in (24.4.19).

24.5 $\mathcal{L}$ AND $\mathcal{L}_D$ AGING WITH SHOCKS

Let S be the standard shock model distribution with survival probability

$$\bar{S}(t) := E\bar{P}_{N(t)} = \sum_{k=0}^{\infty} \bar{P}_k P\{N(t) = k\}, \qquad t \geq 0, \qquad (24.5.24)$$

where $\bar{P}_k$ is the probability of surviving k shocks, and $N(t)$ is the number of shocks occuring up to time t. When shocks arise according to a renewal process, the family of models in (24.5.24) are the so called *renewal process shock models*. We show that when the inter arrival time between shocks is $\mathcal{L}$ ($\mathcal{L}_D$) aging and the number of shocks to failure has the corresponding $\mathcal{G}$ ($\mathcal{G}_D$, resp.) property, then the survival time d.f. S inherits the $\mathcal{L}$ ($\mathcal{L}_D$, resp.) property. A similar finding holds for the dual classes (Theorems 24.5.1 and 24.5.2).

let F be the d.f. driving the renewal process $N(t)$ of shocks, N the number of shocks to failure with $P(N = k) = p_k$ and shock-resistance probabilities $\bar{P}_k = P(N > k), k \geq 0$. Note that when $\{p_k\} \in \mathcal{G}$ or $\mathcal{G}_D$, or their duals; we have $p_0 = 0$; i.e., no failure occurs without shocks.

Theorem 24.5.1 *If $F \in \mathcal{L}(\bar{\mathcal{L}})$ and $\{p_k : k \geq 1\} \in \mathcal{G}(\bar{\mathcal{G}}, resp.)$, then $S \in \mathcal{L}(\bar{\mathcal{L}}, resp.)$.*

PROOF. Whether $p_0 = 0$ or not; when $N(t)$ is a renewal process driven by F, using (24.5.24) we can write

$$S(t) = 1 - \bar{S}(t) = p_0 + \sum_{k=1}^{\infty} p_k F^{*k}(t), \qquad (24.5.25)$$

where F^{*k} is the k-fold convolution of F with itself. Standard computations now yield the Laplace transform of S as, $L_S(s) = \psi_N(L_F(s))$, where ψ_N is the p.g.f. of the number N of shocks to failure. From (24.5.25), we see that,

$$\bar{S}(t) = 1 - p_0 - \sum_{k=1}^{\infty} p_k F^{*k}(t) = \sum_{k=1}^{\infty} p_k \overline{F^{*k}}(t) \qquad (24.5.26)$$

Hence, the mean survival time and the mean time between shocks are related by,

$$\mu_{1,S} = \int_0^{\infty} \bar{S}(t)dt = \sum_{k=1}^{\infty} p_k \int_0^{\infty} \overline{F^{*k}}(t)dt$$

$$= \sum_{k=1}^{\infty} p_k (k\mu_{1,F}) = m\mu_{1,F}, \qquad (24.5.27)$$

where $m = EN$ is the average number of shocks to failure. If $F \in \mathcal{L}(\bar{\mathcal{L}}, resp.)$ and $\{p_k : k \geq 1\} \in \mathcal{G}(\bar{\mathcal{G}}, resp.)$, then using the monotonicity of p.g.f.s and Laplace transforms, the defining properties of the of the classes $\mathcal{L}, \mathcal{G}$ (their duals) and (24.5.25), we get,

$$L_S(s) = \psi_N(L_F(s)) \le (\ge) \psi_N\left(\frac{1}{1 + s\mu_{1,F}}\right) \le (\ge) \frac{1}{1 + s\mu_{1,S}},$$

which proves the desired conclusion. $\square$

Remark 24.5.1 Theorem 24.5.1 supplements the results of Klefsjö (1983), who considered the shock models (24.5.24) where the number of shocks to failure has a distribution in $\mathcal{G}$ (or its dual) but shocks arise as a birth process

Theorem 24.5.2 If $F \in \mathcal{L}_D(\bar{\mathcal{L}}_D)$ and $\{p_k : k \ge 1\} \in \mathcal{G}_D(\bar{\mathcal{G}}_D, resp.)$, then $S \in \mathcal{L}_D(\bar{\mathcal{L}}_D, resp.)$.

PROOF. If S is a renewal process shock model d.f. (24.5.25); the key idea is to derive a suitable representation of the corresponding derived distribution S_1, which is then used to verify its membership in $\mathcal{L}$ or, its dual. From (24.5.25), we have,

$$\int_t^\infty \bar{S}(x)dx = \sum_{k=1}^\infty p_k \int_t^\infty \overline{F^{*k}}(x)dx = \mu_{1,F}\sum_{k=1}^\infty kp_k\bar{G}_k(t),$$

where, for notational convenience, we write G_k for the first derived distribution $(F^{*k})_1$ induced by F^{*k}. It then follows that,

$$\bar{S}_1(t) = \frac{1}{\mu_{1,S}}\int_t^\infty \bar{S}(x)dx = \frac{1}{EN}\sum_{k=1}^\infty kp_k\bar{G}_k(t) = \sum_{k=1}^\infty \pi_k\bar{G}_k(t),$$

where $\{\pi_k : k \ge 1\}$ is the distribution on the positive integers, defined by

$$\pi_k := \frac{kp_k}{EN}, \quad k \ge 1.$$

Consequently,

$$\begin{aligned} \mu_{1,S_1} &= \sum_{k=1}^\infty \frac{\pi_k}{k\mu_{1,F}}\int_0^\infty\int_t^\infty \overline{F^{*k}}(x)dxdt \\ &= \frac{1}{2\mu_{1,F}EN}\sum_{k=1}^\infty p_k\mu_{2,F^{*k}}. \end{aligned} \qquad (24.5.28)$$

Standard computations show that,

$$\mu_{2,F^{*k}} = k\mu_{2,F} + k(k-1)\mu_{1,F}^2 = k\mu_{1,F}^2(1+\eta_F^2) + k(k-1)\mu_{1,F}^2 = k\mu_{1,F}^2(k+\eta_F^2).$$

Plugging this in (24.5.28) and using (24.4.21), we get,

$$\mu_{1,S_1} = \frac{1}{2}\mu_{1,F}\left(\eta_F^2 + \frac{EN^2}{EN}\right) = \frac{1}{2}\mu_{1,F}[2EN_1 + (1 - \eta_F^2)]. \quad (24.5.29)$$

Now, from (24.2.4) and (24.5.25), we get

$$L_{S_1}(s) = \frac{1 - \psi_N(L_F(s))}{s\mu_{1,S}}. \quad (24.5.30)$$

If $F \in \mathcal{L}_D$ and N has a distribution in $\mathcal{G}_D$; in virtue of (24.2.4) and (24.4.23), we note that,

$$L_F(s) = 1 - s\mu_{1,F}L_{F_1}(s) \geq 1 - \frac{s\mu_{1,F}}{1 + s\mu_{1,F_1}}, \quad s > 0, \quad (24.5.31)$$

$$\psi_N(z) \geq 1 - \frac{(1-z)EN}{z + (1-z)EN_1}, \quad 0 \leq z \leq 1. \quad (24.5.32)$$

Observe that the lower bound on the right hand side of (24.5.32) is nondecreasing in z over $[0, 1]$. By monotonicity of L_F and ψ_N, (24.5.30)–(24.5.32) now imply

$$\begin{aligned}
L_{S_1}(s) &\leq \frac{1}{s\mu_{1,S}}\frac{[1 - L_F(s)]EN}{L_F(s) + [1 - L_F(s)]EN_1} \\
&\leq \frac{\mu_{1,F}EN}{\mu_{1,S}}\frac{1}{1 + s[\mu_{1,F_1} + (EN_1 - 1)\mu_{1,F}]} = \frac{1}{1 + s\mu_{1.S_1}},
\end{aligned}$$
$$(24.5.33)$$

where, the last step follows from (24.5.27) and (24.5.29), by noting

$$\begin{aligned}
\mu_{1,F_1} + (EN_1 - 1)\mu_{1,F} &= \frac{1}{2}\mu_{1,F}(1 + \eta_F^2) + \mu_{1,F}(EN_1 - 1) \\
&= \frac{1}{2}\mu_{1,F}[2EN_1 - (1 - \eta_F^2)] = \mu_{1,S_1}.
\end{aligned}$$

This shows $S_1 \in \mathcal{L}$, while Theorem 24.5.1 guarantees $S \in \mathcal{L}$. Hence, $S \in \mathcal{L}_D$. The dual case follows by reversing the inequalities in (24.5.31)–(24.5.33). □

Acknowledgements Research of the second author, as a visitor, was supported by the NJIT Foundation. The hospitality and research support from the Department of Mathematical Sciences, NJIT are gratefully acknowledged.

REFERENCES

Barlow, R. E. and Proschan, F. (1975). *Statistical Theory of Reliability, Probability Models*, Holt, Reinhart and Winston, San Francisco.

Basu, S. K. and Bhattacharjee, M. C. (1984). On Weak convergence within the class of HNBUE life distributions, *Journal of Applied Probability*, **21**, 654–660.

Bhattacharjee, M. C., Abouammoh, A. M., Ahmed, A. N. and Barry, A. M. (1998). Preservation results for survival distributions based on comparisons with asymptotic remaining Life under replacements, *CAMS Technical Report*, Department of Mathematical Sciences, New Jersey Institute of Technology.

Bhattacharjee, A. and Sengupta, D. (1996). On the coefficient of variation for the $\mathcal{L}$ and $\bar{\mathcal{L}}$ classes of life distributions, *Statistics & Probability Letters*, **27**, 177–180.

Bhattacharjee, M. C. and Sethuraman, J. (1990). Families of life distributions characterized by two moments, *Journal of Applied Probability*, **27**, 720–725.

Klefsjö, B. (1982). The HNBUE and HNWUE classes of life distributions, *Naval Research Logistics Quarterly*, **29**, 331–344.

Klefsjö, B. (1983). A useful aging property based on the Laplace transform, *Journal of Applied Probability*, **20**, 615–626.

Mitra, M., Basu, S. K. and Bhattacharjee, M. C. (1995). Characterizing the exponential law under Laplace order domination, *Calcutta Statistical Association Bulletin*, **45**, 171–178.

Stoyan, D. (1983). *Comparison Methods for Queues and Other Stochastic Systems*, John Wiley & Sons, New York.

CHAPTER 25

DEFECT RATE ESTIMATION USING IMPERFECT ZERO-DEFECT SAMPLING WITH RECTIFICATION

NEERJA WADHWA

GE Capital, Stamford, CT

Abstract: An important aspect of any quality control program is estimation of the quality of outgoing products. This paper applies Acceptance Sampling with rectification to the problem of quality assuance when the inspection procedure is imperfect. The objective is to develop effective rectification sampling plans and estimators based on these plans without making the assumption of a perfect inspection procedure. We develop estimators, under two different sampling plans, for the number of undetected defects remaining after a set of lots has been passed. We compare, by extensive simulation, the proposed estimators with existing ones in terms of Root Mean Squares Error (RMSE). One of our estimators, an empirical Bayes estimator, is seen to consistently obtain substantially lower RMSE overall.

Keywords and phrases: Acceptance sampling, rectification, mean squared error, empirical Bayes estimator, imperfection errors

25.1 INTRODUCTION

An increasing number of manufacturers are pursuing high quality standards these days. Manufacturers such as Texas Instruments, Motorola, and General Electric are striving to achieve Six Sigma Quality. This corresponds to

441

a target of no more than 3.4 defects per million products. Given such strin-
gent quality standards, it has become increasingly important to measure
quality reliably, consistently, and accurately. Suppliers, for example, are
frequently required to demonstrate through sampling inspection that their
products meet specified quality standards. Inspection process, therefore,
plays a very crucial role for total quality control in manufacturing. In quality assurance a lot is either accepted outright as satisfactory, or inspection
is done on every item in the lot. Alternately, one may use an acceptance
sampling plan. An acceptance sampling plan is one which indicates conditions for acceptance or rejection of the lot being inspected. In this paper,
we focus on a particular type of sampling inspection, Acceptance Sampling
with Rectification (ASWR). In an acceptance sampling plan, a random
sample is inspected from each lot. The lot is accepted if less than a certain
number, k, of defective units (defectives) are found in the sample; else it is
rejected. Often, to an acceptance sampling plan there is attached a provi-
sion for further inspection of lots rejected by the plan. Rectification calls
for retention of rejected lots and their submission for further inspection.
Commonly, the rejected lot is made to undergo 100% screening operation;
the defectives thus found either discarded or replaced. The process may be
represented as follows:

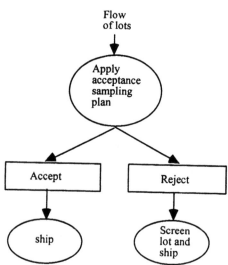

An important aspect of any quality control program is the estimation
of the quality of outgoing products. Extensive research has been conducted
in the area of estimating the proportion of defectives in outgoing lots. For
example, Hahn (1986) investigates two naive estimators and proposes one

empirical Bayes approach to estimating the percentage defectives in accepted lots with zero-defect sampling. Zaslavsky (1988) formally demonstrates Hahn's results and extends them in several ways. Brush, Hoadley, and Saperstein (1990) use a hierarchical Bayes model to estimate the proportion of defectives in accepted lots, based on both accepted and rejected lots. Martz and Zimmer (1990) present an estimator for the percentage of defectives in lots under zero-defect sampling (the sample is deemed acceptable when the number of defectives k, in the initial sample, is zero). Greenberg and Stokes (1992) provide estimators of the number of defectives in a set of T outgoing lots under zero-defect sampling with rectification. To date, almost all research in ASWR has been conducted under the restrictive assumption of the inspection procedure being perfect. This is often an unrealistic assumption. In this paper we consider two rectification sampling plans when the inspection procedure is imperfect. We estimate the defective rates in the lots after zero defect sampling, when the inspection environment is not 100% accurate. We then develop estimators for the number of undetected defectives remaining in a set of accepted lots. Recently, in a working paper, Greenberg and Stokes (1996) have proposed an adjustment to their estimator taking into account imperfections in the inspection procedure. We compare the performance of our estimators with those proposed by Greenberg and Stokes (1992, 1996). We also compare the two sampling plans on the basis of RMSE and cost.

25.2 SAMPLING PLAN A

25.2.1 Model

Consider a set of T lots, each of size n units. A random sample of size m units is selected from each lot and inspected. If no defectives are found in this sample, the lot is accepted. If at least one defective is found, the entire lot undergoes inspection. The defectives detected are discarded and the lot is accepted. The notation in this paper follows Greenberg and Stokes (1992). Let

D_{i1}: The number of defectives in sample i.

D_{i2}: Additional defectives among the un-sampled units in lot i.

D_i: Total number of defectives in lot i. $(D_i = D_{i1} + D_{i2})$.

Y_{i1}: Number of defectives detected in sample i.

Y_{i2}: Additional defectives detected in the remaining $(n - m)$ units in lot i.

Y_i: Number of defectives detected in lot i. $(Y_i = Y_{i1} + Y_{i2})$.

$$Y_i = \begin{cases} D_i & \text{if } D_{i1} > 0 \\ 0 & \text{otherwise} \end{cases}$$

U_i: Number of undetected defectives in lot i. $(U_i = D_i - Y_i)$.

The objective is to estimate the number of defectives, U, in the T outgoing lots; where $U = \sum_{i=1}^{T} U_i$. In the next section we discuss the Greenberg and Stokes (1992, 1996) estimators, and modify the Greenberg and Stokes (1996) estimator.

25.2.2 Modification of Greenberg and Stokes Estimators

In this section we first discuss the Greenberg and Stokes (1992) estimator for the number of undetected defectives remaining in the outgoing lots. We then discuss the Greenberg and Stokes (1996) estimator for the imperfect machine, and then propose a modification to it. The estimator proposed by Greenberg and Stokes (1992) is a non-parametric estimator which allows for general variability of the defective rates across lots. The estimator is defined as:

$$\hat{U}_{GS,1} = \sum_{Y_{i1} > 0} \frac{Y_i}{P_i} - \sum_{i=1}^{T} Y_i, \tag{25.2.1}$$

where

$$P_i = \begin{cases} 1 - \left[\dfrac{\binom{n-Y_i}{m}}{\binom{n}{m}} \right] & \text{if } n - m \leq Y_i, \\ 0 & \text{otherwise,} \end{cases} \tag{25.2.2}$$

P_i is the probability that $Y_{i1} > 0$, and $\sum_{i=1}^{T} Y_i$ is subtracted in (25.2.1) because identified defectives are rectified.

When the inspection process is not error free, two kinds of errors may occur: a defective unit is declared non-defective, or a non-defective unit is declared defective. Let,

$$p = \Pr[\text{unit declared defective} \mid \text{unit is defective}]$$

$$p' = \Pr[\text{unit declared defective} \mid \text{unit is } not \text{ defective}].$$

We will assume that p and p' are known by previous calibration.

An initial adjustment proposed by Greenberg and Stokes (1996) for the estimate of $U = \sum_{i=1}^{T} U_i$, in the case of imperfect inspection procedure is:

$$\hat{U}_{GS,2} \equiv \sum_{Y_{i1}>0} \frac{Y_i - np'}{p - p'}\left(\frac{1}{P_i} - p\right) \tag{25.2.3}$$

where P_i is defined in (25.2.2). Please note that, recently, Greenberg and Stokes (1996) have further updated this estimator to

$$\hat{U}_{GS,2} \equiv \sum_{Y_{i1}>0} \frac{Y_i}{(p - p')P_i} - \sum_{i=1}^{T} \frac{np'}{p - p'}$$
$$- \sum_{i=1}^{T} \frac{(Y_{i1} - mp')p}{(p - p')} - \sum_{Y_{i1}>0} -\frac{(Y_{i2} - (n - m)p')p}{(p - p')}$$

This estimator is an unbiased estimate of U. The original $\hat{U}_{GS,2}$, however, seems to have better RMSE properties for most levels of machine imperfection.

We now propose an estimator for $U = \sum_{i=1}^{T} U_i$. This estimator is a modification of the one proposed by Greenberg and Stokes (1996) in (25.2.3). In our estimator, the probability P_i, i.e., the probability of detecting at least one defective in the sample, has been improved to take into account the imperfections of the inspection procedure. The estimator is,

$$\hat{U}_{new,1} \equiv \sum_{Y_{i1}>0} \frac{Y_i - np'}{p - p'}\left(\frac{1}{P_i} - p\right) \tag{25.2.4}$$

where

$$
\begin{aligned}
P_i' &= P[\text{at least one declared defective in the sample of lot } i] \\
&= 1 - P[\text{no defectives in the sample of lot } i] \\
&= 1 - P[\text{defectives pass in the sample}] \\
&\quad \times P[\text{non-defectives pass in the sample}] \\
&\approx 1 - (1 - p)^{\frac{Y_i - np'}{p - p'}\frac{m}{n}}(1 - p')^{m - \frac{Y_i - np'}{p - p'}\frac{m}{n}} \tag{25.2.5}
\end{aligned}
$$

In the next section, we first discuss the model used, and then propose an Empirical Bayes estimator for the number of undetected defectives remaining in the outgoing lots.

25.2.3   An Empirical Bayes Estimator

The model considered is a modification of the one proposed by Greenberg and Stokes (1992). A set of T lots, each of size n units is considered. A random sample of size m units is selected from each lot and inspected. If no defectives are found in this sample, the lot is accepted. If at least one defective is found, the entire lot undergoes inspection. The defectives detected are discarded and the lot is accepted. We assume that each unit in lot i is independently defective with probability ω_i, and ω_i varies from lot to lot. The model is then specified as:

$$\omega \sim \left(\begin{array}{ll} \text{beta}(a, b) & \text{with probability } \pi \\ 0 & \text{with probability } 1 - \pi \end{array} \right.$$

$$D_{i1} \mid \omega_i \sim \text{binomial}(m, \omega_i)$$

$$D_{i2} \mid \omega_i \sim \text{binomial}(n - m, \omega_i)$$

To account for inspection error, we overlay the above model with the following:

$$Y_{i1} \sim \text{binomial}(D_{i1}, p) + \text{binomial}(m - D_{i1}, p')$$

$$Y_{i2} \mid Y_{i1} > 0 \sim \text{binomial}(D_{i2}, p) + \text{binomial}(n - m - D_{i2}, p')$$

Y_{i1} represents the number of detected defectives in the sample of lot i.

Next, we propose an empirical Bayes estimator for the number of undetected defectives remaining in the outgoing lots. In order to estimate the number of undetected defectives in outgoing lots, researchers have historically based their information only on those lots where at least one defective is found. The estimators discussed above confer to this rationale. It is, however, debatable on how much information about rejected lots one can directly derive from an accepted lot.

Assuming reasonable parameter values, $P(\omega = 0 \mid Y_{i1} = 0)$ is substantially greater than $P(\omega > 0 \mid Y_{i1} = 0)$ (see appendix A1.2). Thus, if no units are declared defective in a lot, the probability that $\omega = 0$ for that lot is very high. In other words, most of the lots where no defectives are found come from the path with probability $1 - \pi$. Conversely, if at least one unit is declared defective, the probability that $\omega > 0$ for that lot is very high. These lots are, thus, more likely to come from the π path. The rejected lots should thus be treated separately from the accepted lots, and one should not directly extrapolate information about one from the other. The empirical Bayes estimator is thus based on the following expression:

$$U_{\text{new},2ab \text{ known}} \equiv \sum_{Y_{i1}>0} \frac{Y_i - np'}{p - p'}(1 - p) + \sum_{Y_{i1}=0} E(D_i \mid Y_{i1} = 0)$$

where $E(D_i \mid Y_{i1} = 0)$ is a constant whose calculation is discussed in appendix A1.1. In $U_{\text{new},2ab \text{ known}}$, we seek to estimate separately the number of undetected defectives in accepted versus rejected lots. For accepted lots, we take the expectation of the number of defectives, given the information that no defectives were observed as an estimate of U.

The reader should note that $U_{\text{new},2ab \text{ known}}$ uses population parameters a, b, and π, and thus cannot be compared fairly with other estimators. Therefore, in the expression for $U_{\text{new},2ab \text{ known}}$ we substitute the parameters a, b, and π with their estimates $\hat{a}, \hat{b}$ and $\hat{\pi}$ respectively. The empirical Bayes estimator thus is,

$$\hat{U}_{\text{new2}} \equiv \sum_{Y_{i1}>0} \frac{Y_i - np'}{p - p'}(1 - p) + \sum_{Y_{i1}=0} \hat{E}(D_i \mid Y_{i1} = 0). \qquad (25.2.6)$$

We estimate a and b using the method of moments approach. As discussed in Appendix A1.2, if zero defectives are observed in the initial sample, the probability of a lot being defective is very small. We thus consider lots where at least one defective is observed in the sample. Using the rationale of hypotheses testing, we consider the null hypothesis to be that of no defectives present in a lot, versus the alternative that there exists at least one defective in the lot. Under the null hypothesis, Y_i is distributed as Binomial(n, p'). Thus for a level of significance of 2.5%, the rejection criterion is to reject a lot if $(Y_{i1} > 0$ and $Y_i > (np' + 2\sqrt{np'(1 - p')})$). Note that, clearly, higher the value of $\mu = a/(a + b)$, the greater the power of the test. Consider a simple example: let $p = 0.99$, $p' = 0.01$, $n = 5000$, and $m = 125$. Even for the smallest value of μ considered, that is $\mu = 0.01$, Pr[Type I Error] = 2.5%, and Pr[Type II Error] ≈ 0.

The mean may be equated as follows:

$$\frac{a}{a + b} = \frac{\sum \# \text{ lots } \hat{\omega}_i}{\# \text{ lots}}, \qquad (25.2.7)$$

where $\hat{\omega}_i = \dfrac{\left(\frac{Y_i - np'}{p - p'}\right)}{n}$, and $\#$ lots = Number of lots satisfying the condition

$$Y_{i1} > 0 \text{ and } Y_i > \left(np' + 2\sqrt{np'(1 - p')}\right).$$

Similarly, we equate the variance term using the method of moments as follows:

$$\frac{ab}{(a + b)^2(a + b + 1)} = \frac{\text{variance}(\hat{\omega}_i)}{\# \text{ lots}} \qquad (25.2.8)$$

We obtain the estimates of a and b by solving the Equations (25.2.7) and (25.2.8).

We estimate π by solving the following equation for $\hat{\pi}$.

$$\hat{\pi} = \frac{\# \text{ lots} + (T - \# \text{ lots}P(\omega > 0 \mid Y_{i1} = 0)}{T} \qquad (25.2.9)$$

The intuition for the above estimate of π, the proportion of defective lots, is as follows: Using the above mentioned rationale of hypotheses testing, the first term in the numerator of Equation (25.2.9) indicates the number of rejected lots, or in other words, the number of defective lots given that at least one defective is observed in the sample. The second term in the numerator accounts for the expected number of defective lots given that no defectives are observed in the sample. This term is included because even when zero defectives are observed in a lot, there exists a possibility of the lot being defective.

25.2.4   Comparison of Estimators

In this section we present simulation evidence that when the inspection procedure is imperfect, the proposed estimators have a lower MSE than $\hat{U}_{GS,1}$ and $\hat{U}_{GS,2}$. The difference in the MSE of the various estimators is usually very large, which makes it difficult to plot the MSE of the estimators on the same scale. Thus, the measure of comparison used is Root Mean Squared Error (RMSE). Appendix A2.2 derives the bias and MSE for estimators $\hat{U}_{GS,1}$, $\hat{U}_{GS,2}$ and $\hat{U}_{new,1}$. These expressions are very general and can be applied to any estimator of the form $\sum_{y_{i1}>0}(\cdot)$. However, a limitation of these expressions is their computational difficulty due to the large number of calculations involved. We, therefore, resort to simulation to get the RMSE of these estimators for large lot/sample sizes (Large lot/sample sizes refer to $n = 5000$ and $m = 1254$. Small lot/sample sizes refer to $n = 15$ and $m = 3$). The expressions for bias and MSE are useful when the lot/sample sizes are small.

The performance of the four estimators is compared for the same values of n, m and T as used in Greenberg and Stokes (1992). Comparisons are done by way of several RMSE figures. The figures are based on 200 simulations, each consisting of 300 lots; one parameter is made to vary in each figure. In Figures 25.1A and 25.1B the RMSE of the estimators is plotted against μ. We take $\pi = 0.1$, and $r = 0.34$ for Figures 25.1A and 25.1B. ρ is the within-lot correlation of defectives, defined in Greenberg and Stokes (1992). They define ρ as follows: Let X_i (or X_j) $= 1$ if the ith (or jth) unit in a lot is defective and 0 otherwise. Then $\rho = \text{corr}(X_i, X_j)$ ($\rho =$

$\frac{(1-\pi)a(a+b+1)+b}{[a(1-\pi)+b](a+b+1)}$ [i]). The RMSE of the estimators is plotted against the average proportion of defectives, $\mu = a/(a + b)$ [ii]. The values considered for population parameters a and b in the simulations are calculated using simultaneous equations [i] and [ii]. The maximum standard error of the RMSE across different levels of μ is stated for these figures. $\hat{U}_{GS,1}$ and $\hat{U}_{GS,2}$ are referred to as GS1 and GS2 in the figures. $\hat{U}_{new,1}$ and $\hat{U}_{new,2}$ are new1 and new2 respectively. $\hat{U}_{GS,1}$ and $\hat{U}_{GS,2}$ and are identical when the inspection procedure is perfect, and their performance lies between $\hat{U}_{new,1}$ and $\hat{U}_{new,2}$. However, when there is a slight imperfection, $\hat{U}_{GS,1}$ rises very fast. Note that this estimator seems to be changing shape for different levels of machine imperfection. Thus, the performance of $\hat{U}_{GS,1}$ seems to be rather poor when the inspection procedure is imperfect.

The RMSE performance of the second estimator, $\hat{U}_{GS,2}$ is substantially better than $\hat{U}_{GS,1}$. It almost always performs better than that of $\hat{U}_{GS,1}$ and its shape is consistent for different levels of p and p'. The standard error of this estimator is almost two orders of magnitude smaller than that of $\hat{U}_{GS,1}$.

$\hat{U}_{new,1}$ performs worse than $\hat{U}_{GS,1}$ and $\hat{U}_{GS,2}$ when the machine is perfect. The reason is that for the perfect machine, P'_i reduces to 1, and thus $\hat{U}_{new,1}$ is always zero. However, with even slight imperfection, the performance of $\hat{U}_{new,1}$ becomes much better. Since we are concerned with the cases when the inspection procedure is imperfect, the overall performance of $\hat{U}_{new,1}$ seems to dominate that of $\hat{U}_{GS,1}$ and $\hat{U}_{GS,2}$.

The fourth estimator i.e., $\hat{U}_{new,2}$, performs well for all levels of p and p'. It has a lower RMSE than the other three estimators at all level of machine imperfection, including the case when the inspection procedure is perfect. In fact, the maximum RMSE of $\hat{U}_{new,2}$ is less than the minimum RMSE of all the other estimators.

In Figure 25.2 we plot the RMSE of the estimators against ρ. We take $\pi = 0.1$ and $\mu = 0.1$. ρ is made to vary from 0.1 to 0.9 in increments of 0.2. In Figure 25.3 we take $\mu = 0.1$, and $\rho = 0.3$. The RMSE of the estimators is plotted against π. Note that π can vary widely depending on whether the process is in control or not. The behavior of the estimators is similar to that described for the case when μ varies.

As is clear from the preceding figures, the performance of $\hat{U}_{new,2}$ is better than the other estimators under all circumstances. We shall, therefore, restrict attention to this estimator.

The reader should note that there is minimal difference in the RMSE performance of $\hat{U}_{new,2ab\ known}$ and $\hat{U}_{new,2}$ for various levels of p and p'. Therefore, for simplicity, we base Figure 25.4 on analytical values of

$\hat{U}_{\text{new},2ab}$ known instead of simulated values of $\hat{U}_{\text{new},2}$. In Figure 25.4, the performance of $\hat{U}_{\text{new},2ab}$ known is studied for different levels of machine imperfection as μ varies. The other parameters are taken to be constant at $n = 5000$, $m = 125$, $\rho = 0.3$ and $\pi = 0.1$. The values of μ range from 0.01 to 0.3.

When the machine is near perfect, the RMSE is low for small and very large values of μ, but is relatively high for intermediate values. To see this, consider a simple example where the machine is perfect and $\mu = 0$ (i.e., there are no defectives in the lots). The RMSE for the undetected defectives is zero. Similarly, when $\mu = 1$, (i.e., either a lot has all defectives, or all non-defectives), the RMSE is again zero. For all other values of μ, the RMSE would be non-zero because the number of undetected defectives would not always be zero. As the machine becomes more imperfect, the RMSE increases monotonically with μ. Looking at the cross section for a particular μ, for small and medium values of μ, the RMSE decreases and later increases as the level of machine imperfection increases. Note that the level of machine imperfection where RMSE starts increasing (approximately $p = 0.95$, $p' = 0.05$) is not shown on the figure. Cases such as this where m and p' are such that all units are almost always sampled are uninteresting and hence were not considered.

The intuitive explanation for this is as follows: As the machine goes from perfect to highly imperfect, there are two counteracting effects. First, the number of lots sampled increases since false errors are detected in the sample. This tends to decrease the RMSE, since more sampling is being done. Second, due to machine imperfection the RMSE increases. When the machine goes from perfect to slightly imperfect, the first effect dominates, thereby reducing the RMSE. However, as the machine approaches severe imperfection, the probability of sampling a lot approaches one. Therefore, the decrease in the RMSE due to the first effect approaches zero. The second effect dominates, thereby increasing the RMSE. Hence, the RMSE first decreases and later increases.

When the value of μ is high, we see that the RMSE increases monotonically as the machine goes from perfect to highly imperfect. The probability of a defective showing up in a sample is high, and therefore, the probability of lot being sampled is high. Hence the first effect is minimal.

25.2.5 Example

To illustrate the computation of the estimators, we simulate a data set based on the model discussed in Section 25.2.3. The data, shown in Tables 25.1 through 25.4 below, is generated by taking 125 samples from each of 300 lots of size 5000. The values for μ, ρ, and π considered in the simulations are 0.1, 0.3, and 0.1 respectively. p and p' are taken to be 0.999 and 0.001. Tables 25.1 and 25.2 present the actual number of defectives and the observed defectives in the 300 lots respectively.

For generation of Table 25.1, the actual number of defectives, D_i, follows Binomial(n, ω). Specifically, $D_{i1} \sim$ Binomial(m, ω), and $D_{i2} \sim$ Binomial$(n - m, \omega)$. Table 25.2 is generated from Table 25.1. The declared number of defectives in the sample, Y_{i1}, is generated in two steps. Defectives that are declared defective are simulated as Binomial(D_{i1}, p). Non-defectives declared defective are simulated as Binomial(D_{i2}, p'). Thus, $Y_{i1} =$ Binomial$(D_{i1}, p) +$ Binomial(D_{i2}, p'). The declared number of defectives in the remaining lot, Y_{i2}, is simulated similarly. Y_i is obtained by adding Y_{i1} and Y_{i2}. The computation of the estimators and their comparison to the actual number of undetected defectives is shown below.

TABLE 25.1 $(p = 0.999, \ p' = 0.001)$

Actual # of defectives (D)	0	1	4	5	13	14	16	25
# of lots	275	1	2	1	1	1	1	1
Actual # of defectives (D)	43	45	60	65	91	120	122	182
# of lots	1	1	1	1	1	1	1	1
Actual # of defectives (D)	249	371	840	841	853	1166	1253	
# of lots	1	1	1	1	1	1	1	
Actual # of defectives (D)	2061	2640						
# of lots	1	1						

TABLE 25.2 $(p = 0.999, \ p' = 0.001)$

# of defectives (Y)	0	3	4	5	6	7	8	9	10
# of lots	248	3	1	14	4	4	3	1	2
# of defectives (Y)	11	12	17	30	47	67	70	101	126
# of lots	1	1	1	1	1	1	1	1	2
# of defectives (Y)	129	188	258	375	841	842	855	1169	1256
# of lots	1	1	1	1	1	1	1	1	1
# of defectives (Y)	2061	2641							
# of lots	1	1							

Actual number of undetected defectives = 100.

$$\hat{U}_{GS,1} = \frac{3}{0.073} + \frac{4}{0.096} + \cdots + \frac{2641}{1} - 11148 \approx 1367$$

$$\hat{U}_{GS,2} = \frac{3 - 5000 \star 0.001}{0.999 - 0.001} \left(\frac{1}{0.073} - 0.999 \right)$$

$$+ \frac{4 - 5000 \star 0.001}{0.999 - 0.001}\left(\frac{1}{0.096} - 0.999\right)$$

$$+ \cdots + \frac{2641 - 5000 \star 0.001}{0.999 - 0.001}\left(\frac{1}{1} - 0.999\right) \approx 1380$$

$$\hat{U}_{new,1} = \frac{3 - 5000 \star 0.001}{0.999 - 0.001}\left(\frac{1}{0.1175} - 0.999\right)$$

$$+ \frac{4 - 5000 \star 0.001}{0.999 - 0.001}\left(\frac{1}{0.117} - 0.999\right)$$

$$+ \cdots + \frac{2641 - 5000 \star 0.001}{0.999 - 0.001}\left(\frac{1}{1} - 0.999\right) \approx 62$$

$$\hat{U}_{new,2} = \frac{11148 - 52 \star 5000 \star 0.001}{0.999 - 0.001}(1 - 0.999) + 248 \star 0.21977 \approx 66$$

As can be seen, the estimators are highly disparate. $\hat{U}_{new,2}$ seems to perform the best amongst all estimators considered; its value of 66 is closest to the actual value (100) of the number of undetected defectives in the accepted lots.

25.3   SAMPLING PLAN B

In this section we present a plan which is referred to as Sampling Plan B. First, we present estimators for the number of undetected defectives left in outgoing lots for this sampling plan in Section 25.3.1. We do so by modifying the proposed estimators for sampling plan A. Next, we compare this sampling plan with plan A on the basis of RMSE in Section ??.

25.3.1   Estimators

We consider a sampling plan where the initial sample is inspected three times instead of once in each lot. A unit is declared defective if it fails at least twice. If no defectives are found, the lot is accepted. If at least one defective is found, the entire lot is screened, and all the defectives are removed. Note that this sampling plan is especially important when p and p' are not known. This is because in order to estimate p and p', it is necessary to inspect some items at least three times [see Johnson, Kotz and Wu (1991)].

   The probability of correctly classifying a defective as well as incorrectly classifying a non-defective for this sampling plan may easily be stated in terms of p and p' respectively (It is easy to show that $q* < (>)q$ if $q <$

($>$)0.5. Since p is usually greater than 0.5, and p' less than 0.5, screening the initial sample three times is equivalent to screening it once but with a inspection procedure that is more perfect.)

The probability of correctly classifying a defective may now be stated in terms of p and p' as follows:

$$
\begin{aligned}
p* &= P(\text{Unit is declared defective} \mid \text{Actually defective}) \\
&= P(\text{Unit declared defective twice} \mid \text{Actually defective}) \\
&\quad + P(\text{Unit declared defective thrice} \mid \text{Actually defective}) \\
&= 3p^2 \left(1 - \frac{2}{3}p\right).
\end{aligned}
$$

Similarly, the probability of incorrectly classifying a non-defective as defective is defined as

$$
p'^* = 3p'^2 \left(1 - \frac{2}{3}p'\right).
$$

Note that the error probabilities for the initial sample will be different from the ones associated with the remaining lot. This is so because the sample is screened three times, whereas the remaining units are screened at most once. Therefore,

$$
\begin{aligned}
E[Y_{i1}] &= D_{i1}p^* + (m - D_{i1})p'^*, \\
E[Y_{i2}] &= D_{i2}p + (n - mD_{i2})p'.
\end{aligned}
$$

Thus, D_i can be estimated by

$$
\frac{Y_{i1} - mp'^*}{p^* - p'^*} + \frac{Y_{i2} - (n - m)p'}{p - p'}.
$$

The estimator, $\hat{U}_{\text{new},1}$, for the number of undetected defectives is modified as,

$$
\begin{aligned}
\hat{U}^*_{\text{new},1} &= \sum_{Y_{i1}>0} \left(\frac{Y_{i1} - mp'^*}{p^* - p'^*} \left(\frac{1}{P_i^*} - p^* \right) \right. \\
&\quad \left. + \frac{Y_{i2} - (n - m)p'}{p - p'} \left(\frac{1}{P_i^*} - p^* \right) \right)
\end{aligned}
\qquad (25.3.10)
$$

where

$$
P_i^* = 1 - (1 - p^*)^{z\frac{m}{n}} (1 - p')^{m - z\frac{m}{n}}
\qquad (25.3.11)
$$

such that

$$z = \frac{Y_{i1} - mp'}{p^* - p'^*} + \frac{Y_{i2} - (n-m)p'}{p - p'} .$$

The second estimator, $\hat{U}_{\text{new},2}$, can be modified for this sampling plan as follows:

$$\begin{aligned}
\hat{U}^*_{\text{new},2} &= \sum_{Y_{i1}>0} \left(\frac{Y_{i1} - mp'^*}{p^* - p'^*} (1 - p^*) \right. \\
&\quad + \left. \frac{Y_{i2} - (n-m)p'}{p - p'} (1 - p) \right) \\
&\quad + \sum_{Y_{i1}=0} \hat{E}^*(D_i \mid Y_{i1} = 0)
\end{aligned}$$

where $\hat{E}^*(D_i \mid Y_{i1} = 0)$ is a constant calculated in the same way as $E^*(D_i \mid Y_{i1} = 0)$ for sampling plan A.

25.4   SUGGESTIONS FOR FURTHER RESEARCH

As part of future research, the proposed estimators can be extended to the case of c-defect acceptance sampling. The estimators can be further modified for situations where lot/sample sizes and/or acceptance numbers vary from lot to lot. Also, the empirical Bayes estimator is based on the beta-binomial model. It can easily be computed for other models, such as gamma-poisson et al.

In this paper, misclassification error probabilities, that is p and p', are assumed to be known. In many practical situations, p and p' will be unknown, so that one would not be able to compute the proposed estimators. However, estimates of p and p' can be substituted in the estimators. Blischke (1964) proposed several estimators for the misclassification error probabilities. Estimation of the probabilities p and p' has also been considered by Johnson et al. (1991). Since computational burden is insignificant these days, maximum likelihood estimators can also be computed [see Greenberg and Stokes (1996)].

APPENDIX A1: CALCULATION OF THE SECOND TERM IN $\hat{U}_{\text{new},2}$

Appendix A1.1: Calculation of $E(D_i \mid Y_{i1} = 0)$

$$E(D_{i1} \mid Y_i = 0)$$

$$= \int_0^1 E(D_{i1} \mid Y_{i1} = 0, \omega) f(\omega \mid Y_{i1} = 0) \partial \omega$$

$$= \frac{m(1-p) \frac{\pi}{Be(a,b)} \int_0^1 \frac{(1-\delta)^m \omega^a (1-\omega)^{b-1}}{\omega(1-p)+(1-\omega)(1-p')} \partial \omega}{(1-p')^m (1-\pi) + \frac{\pi}{Be(a,b)} \int_0^1 (1-\delta)^m \omega^{a-1} (1-\omega)^{b-1} \partial \omega}$$

$$(25.4.12)$$

$$E(D_{i2} \mid Y_i = 0)$$

$$= \int_0^1 E(D_{i2} \mid Y_{i1} = 0, \omega) f(\omega \mid Y_{i1} = 0) \partial \omega$$

$$= \frac{(n-m) \frac{\pi}{Be(a,b)} \int_0^1 (1-\delta)^m \omega^a (1-\omega)^{b-1} \partial \omega}{(1-p')^m (1-\pi) + \frac{\pi}{Be(a,b)} \int_0^1 (1-\delta)^m \omega^{a-1} (1-\omega)^{b-1} \partial \omega}$$

$$(25.4.13)$$

where $\delta = \omega p + (1-\omega) p'$.

The values of integrals in expressions (25.4.12) and (25.4.13) are found by numerical integration. Adding equations (25.4.12) and (25.4.13), we obtain $E(D_i \mid Y_{i1} = 0)$.

Appendix A1.2: Calculations of Probabilities

$$f(\omega > 0 \mid Y_{i1} = 0)$$

$$= \frac{f(Y_{i1} = 0 \mid \omega > 0) f(\omega > 0)}{f(Y_{i1} = 0)}$$

$$= \frac{\frac{\pi}{Be(a,b)} \int_0^1 (1-\delta)^m \omega^a (1-\omega)^{b-1} \partial \omega}{(1-p')^m (1-\pi) + \frac{\pi}{Be(a,b)} \int_0^1 (1-\delta)^m \omega^{a-1} (1-\omega)^{b-1} \partial \omega}$$

$$f(\omega = 0 \mid Y_{i1} > 0)$$

$$= \frac{f(Y_{i1} > 0 \mid \omega = 0) f(\omega = 0)}{f(Y_{i1} > 0)}$$

$$= \frac{(1 - (1 - p')^m)(1 - \pi)}{1 - \left((1 - p')^m(1 - \pi) + \frac{\pi}{\beta e(a,b)} \int_0^1 (1 - \delta)^m \omega^{a-1}(1 - \omega)^{b-1} \partial \omega \right)}$$

APPENDIX A2: ANALYTICAL EXPRESSIONS FOR THE BIAS AND MSE

Appendix A2.1: Bias and the MSE Derivation for $\hat{U}_{\text{new},2ab\text{known}}$

Bias calculations for $\hat{U}_{\text{new},2ab\text{known}}$

$$\text{Bias} = E(U_i - \hat{U}_i) = E(E(U_i - \hat{U}_i \mid \omega))$$

where

$$E(U_i - \hat{U}_i \mid \omega)$$
$$= E\left(\left(\frac{Y_i - np'}{p - p'}\right)(1 - p)I(Y_{i1} > 0) + KI(Y_{i1} = 0) \mid \omega\right)$$
$$\quad - E(D_i - Y_{id}I(Y_{i1} > 0) \mid \omega)$$
$$= \frac{n(1 - p)(\delta - p')}{p - p'} - n\omega(1 - p)$$
$$\quad + (1 - \delta)^m \left(\frac{(1 - p)}{p - p'}((n - m)\delta - np') - K + (n - m)\omega p\right).$$

Note that K represents $E(D_i \mid Y_{i1} = 0)$.

MSE for $\hat{U}_{\text{new},2ab\text{known}}$

$$\text{MSE} = E(\hat{U}_i - U_i)^2$$
$$= E(\hat{U}_i)^2 + E(U_i)^2 - 2E(\hat{U}_i U_i)$$
$$= E\left(E(\hat{U}_i^2) \mid \omega\right) + E\left(E(U_i^2) \mid \omega\right) - 2E\left(E(\hat{U}_i U_i) \mid \omega\right)$$

where

$$E(U_i^2 \mid \omega)$$
$$= E(D_i - I(Y_{i1} > 0)Y_{id} \mid \omega)^2$$
$$= E(D_i^2 \mid \omega) + E(I(Y_{i1} > 0)Y_{id}^2 \mid \omega) - 2E(D_i I(Y_{i1} > 0)Y_{id} \mid \omega)$$

where

$$E(D_i^2 \mid \omega) = n\omega(1 - \omega) + n^2\omega^2$$

$$E(Y_{id}^2 I(Y_{i1} > 0) \mid \omega)$$
$$= n\omega p(1 - \omega p) + (n - m)^2 \omega^2 p^2 (1 - \delta)^m (1 - (1 - \delta)^m)$$
$$+ (n - m)\omega p(1 - \delta)^m (\omega p(2m + 1) - 1)$$
$$+ (n\omega p - (n - m)\omega p(1 - \delta)^m)^2$$

$$E(Y_{id} D_i I(Y_{i1} > 0) \mid \omega)$$
$$= 2p \Big\{ (m\omega(1 - \omega) + m^2 \omega^2)$$
$$+ ((n - m)\omega(1 - \omega) + (n - m)^2 \omega^2)(1 - (1 - \delta)^m)$$
$$+ 2m(n - m)\omega^2 \Big\}$$

$$E(\hat{U}_i^2 \mid \omega) = E \left(\left(\frac{Y_i - np'}{p - p'} \right) (1 - p) I(Y_{i1} > 0) + K I(Y_{i1} = 0) \mid \omega \right)^2$$

$$E(\hat{U}_i^2 \mid \omega)$$
$$= \left(\frac{1 - p}{p - p'} \right)^2 \{ n\delta(1 - \delta) - (n - m)\delta(1 - \delta)^m (\delta(n - m - 1) + 1) \}$$
$$+ (1 - (1 - \delta)^m) \left(\frac{np'(1 - p)}{p - p'} \right)^2 + K^2(1 - \delta)^m$$
$$- 2 \left(\frac{1 - p}{p - p'} \right) \left(\frac{np'(1 - p)}{p - p'} \right) (n\delta - (n - m)\delta(1 - \delta)^m)$$

$$E(U_i \hat{U}_i \mid \omega)$$
$$= E \left(\left[\left(\frac{Y_i - np'}{p - p'} \right) (1 - p) I(Y_{i1} > 0) + K I(Y_{i1} = 0) \right] \right.$$
$$\left. \times [D_i - Y_{id} I(Y_{i1} > 0) \mid \omega) \right)$$

$$E(U_i \hat{U}_i \mid \omega)$$
$$= \left(\frac{1 - p}{p - p'} \right) \left(\frac{\omega(1 - p)}{\omega(1 - p) + (1 - \omega)(1 - p')} \right)$$
$$\times \{ m(m - 1)\delta(1 - \delta) + (n - m)(n - m - 1)\delta(1 - \delta)(1 - (1 - \delta)^m) \}$$

$$+ 2 \left(\frac{1-p}{p-p'} \right) (n-m)m\delta\omega(1-p) + K(n-m)\omega(1-\delta)^m$$

$$- \left(\frac{np'(1-p)}{p-p'} \right)^2 (n\omega - (n-m)\omega(1-\delta)^m)$$

Appendix A2.2: MSE of estimators, $\hat{U}_{GS,1}$, $\hat{U}_{GS,2}$ or $\hat{U}_{new,1}$

MSE calculations

$$
\begin{aligned}
\text{MSE} &= E(\hat{U}_i - U_i)^2 \\
&= E(\hat{U}_i)^2 + E(U_i)^2 - 2E(\hat{U}_iU_i) \\
&= E(E(\hat{U}_i^2) \mid \omega) + E(E(U_i^2) \mid \omega) - 2E(E(\hat{U}_iU_i) \mid \omega)
\end{aligned}
$$

$$E(\hat{U}_i^2) = \sum_{u=0}^{n} P(\hat{U}_i = \hat{u}) \star \hat{u}^2$$

$$E(U_i^2) = \sum_{u=0}^{n} P(U_i = u) \star u^2$$

$$E(\hat{U}_iU_i) = \sum_{k_1=1}^{m} \sum_{k_2=0}^{n-m} \sum_{u=0}^{n} P(Y_{i1} = k_1 \ Y_{i2} = k_2 \ U_i = u)\hat{u}_iu$$

$$P(Y_{i1} = k_1 \ Y_{i2} = k_2 \ U_i = u)$$

$$= \sum_{D_{i1}=0}^{m} \sum_{D_{i2}=0}^{n-m} P(Y_{i1} = k_1 \ Y_{i2} = k_2 \ U_i = uD_{i1}D_{i2})$$

$$= \int_0^\infty \sum_{D_{i1}=0}^{m} \sum_{D_{i2}=0}^{n-m} \{P(U_i = u \mid Y_{i1} = k_1 \ Y_{i2} = k_2 \ D_{i1}D_{i2}\omega)$$

$$P(Y_{i1} = k_1 \ Y_{i2} = k_2 \ D_{i1}D_{i2}\omega)P(D_{i1}D_{i2} \mid \omega)f(\omega)d\omega\}$$

$$= \binom{n}{k_1+k_2}\binom{m}{k_1}\binom{n-m}{k_2}p'^{2(k_1+k_2)}(1-p')^{2(n-k_1-k_2)}$$

$$\times \left((1-\pi) + \pi\frac{\text{Beta}(a, n+b)}{\text{Beta}(a, b)} \right)$$

$$+ \pi \sum_{D_{i1}=0}^{m} \sum_{\substack{D_{i2}=0 \\ D_{i1} \neq D_{i2}=0}}^{n-m} \sum_{1=0}^{k_1} \sum_{1'=0}^{k_2} \{A_1 A_1' B_1 B_1' C_1' C_1 X_1\}$$

where

$$A_1 = \binom{n - D_{i1} - D_{i2}}{k_1 + k_2 - D_{i1} - D_{i2} + u} p'^{k_1 + k_2 - D_{i1} - D_{i2} + u}(1 - p')^{n - k_1 - k_2 - u}$$

$$A'_1 = \binom{D_{i1} + D_{i2}}{D_{i1} + D_{i2} - u} p^{D_{i1} + D_{i2} - u}(1 - p)^{u}$$

$$B_1 = \binom{m}{D_{i1}}\binom{n - m}{D_{i2}}\binom{m - D_{i1}}{1} p'^{l}(1 - p')^{m - D_{i1} - 1}$$

$$B'_1 = \binom{D_{i1}}{k_1 - 1} p^{k_1 - 1}(1 - p)^{D_{i1} - k_1 - +1}$$

$$C_1 = \binom{n - m - D_{i1}}{1'} p'^{1'}(1 - p')^{n - m - D_{i1} - 1'}$$

$$C'_1 = \binom{D_{i2}}{k_2 - 1'} p^{k_2 - 1'}(1 - p)^{D_{i1} - k_2 + 1'}$$

$$X_1 = \frac{\text{Beta}(D_{i1} + D_{i2} + a, n - D_{i1} - D_{i2} + b)}{\text{Beta}(a, b)}$$

REFERENCES

Blischke, W. R. (1964). Estimating the parameters of mixtures of binomial distributions, *Journal of the American Statistical Association Journal*, **59**, 510–527.

Brush, G. G., Hoadley, B. and Saperstein, B. (1990). Estimating outgoing quality using the quality measurement plan, *Technometrics*, **32**, 31–41.

Greenberg, B. S. and Stokes, S. L. (1992). Estimating non conformance rates after zero defect sampling with rectification, *Technometrics*, **34**, 203–213.

Greenberg, B. S. and Stokes, S. L. (1996). Working paper.

Hahn, G. J. (1986). Estimating the percent non conforming in the accepted product after zero defect sampling, *Journal of Quality Technology*, **18**, 182–188.

Johnson, N. L., Kotz, S. and Wu, X. (1991). *Inspection Errors for Attributes in Quality Control*, Chapman & Hall, London.

Martz, H. F. and Zimmer, W. J. (1990), A non-parametric Bayes empirical Bayes procedure for estimating the percent non conforming in accepted lots, *Journal of Quality Technology*, **22**, 95–104.

Zaslavsky, A. (1988). Estimating defective rates in c-defect sampling,
    Journal of Quality Technology, **20**, 248–259.

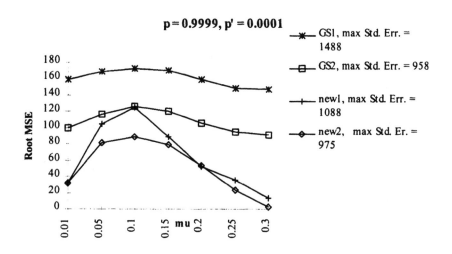

FIGURE 25.1A

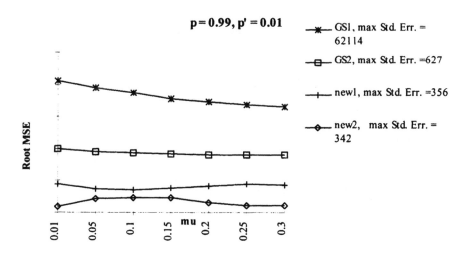

FIGURE 25.1B

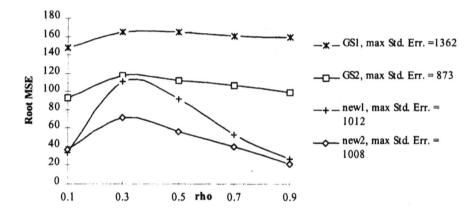

FIGURE 25.2

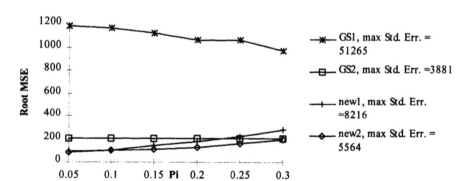

FIGURE 25.3

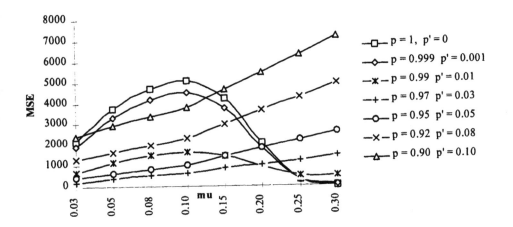

MSE comparisons for different levels of p, p'
n = 5000 m = 125 T = 300 Pi = 0.1 Rho = 0.3

FIGURE 25.4

CHAPTER 26

STATISTICS IN THE REAL WORLD— WHAT I'VE LEARNT IN MY FIRST YEAR (AND A HALF) IN INDUSTRY

REKHA AGRAWAL

GE Corporate Research & Development, Schenectady, NY

It's been a year and a half since I graduated with my Ph.D. in statistics from the University of Waterloo (Canada). Since then, I have been working at GE Corporate Research and Development in the Applied Statistics Program. I thought I would take this opportunity to reflect on my time in industry, and discuss my experiences, the surprises I had coming to industry, and my feelings on being an industrial statistician. The objective is to describe to graduating students the types of experiences they may encounter in their first job and to give them a feel for how they might be spending their time. A secondary objective is to describe those same things to academic departments, so that they can better prepare their students.

26.1   THE GE ENVIRONMENT

The GE Company has 11 businesses: Aircraft Engines, Appliances, Capital Services, Industrial Systems, Information Services, Lighting, Medical Systems, NBC, Plastics, Power Systems and Transportation Systems. This

465

is an incredibly diverse company, unlike any other in the world. This puts us statisticians in the fortunate position of being able to apply statistical methods to such varied applications as evaluating promotional effectiveness on NBC, developing a methodology for evaluating cost on a long term service agreement for locomotives, and helping to isolate the source of defective particles in a thermoplastic resin.

The role of the Corporate Research and Development Center (CRD) is to provide technology and leadership to all of these businesses. In this role, CRD develops "game-changing" technology for the businesses (e.g. Lexan resin, medical cat-scanner), invests in multi-generational product development plans and serves as a key resource in crisis situations, among other things. The businesses must fund activity that is going on at the center on their behalf, although some funding does come to CRD from the corporate office. Therefore, before we can spend almost any time working on an issue with a business, we must have a funding source to charge the effort to. This constant search for charge numbers tends to ensure that we are working on the most vital projects, where someone is willing to pay for our involvement.

CRD is divided into 12 areas of specialization, which we call laboratories. Some examples are the ceramics lab, the polymer and inorganic systems lab and the manufacturing and business process lab. There are also business interface managers for each business, whose role is to co-ordinate all activities between CRD and their business. CRD employs about 1500 people, about half of which have doctoral degrees.

The atmosphere at the center has changed substantially in the past 15 years. Whereas it used to be mostly an academic-like research environment, the emphasis now is on providing immediate value to the businesses. This does not imply that we are always in "fire-fighting" mode we do get involved with longer-term strategic initiatives such as improving the reliability of new product introductions. The emphasis, however, is on providing direct benefit.

It used to be that work time was given for people at CRD to pursue their own research interests. Rumor has it, in fact, that the placemats in the cafeteria had graph paper on them, for people to do quick calculations as they were eating lunch. Now, we are measured on how we can help businesses become more productive and profitable in a competitive global environment the more definable those benefits, the better. If we are interested in pursuing our own academic research interests, we have to do that on our own time. These changes are a significant contributing factor, I feel, to why CRD is one of the few remaining industrial research labs and to our organization becoming a vital part of GE.

The Applied Statistics Program is one of six groups in the Information Technology Laboratory. There are 16 full-time members in our group, a mix of people with master's and Ph.D. degrees. Most of the members of our group have degrees in either economics or engineering, as well as statistics. Our group works with essentially all of the different businesses, and this diversity is one of the things that I find most exciting about my job. This variety of applications necessitates the inclusion of a variety of statistical techniques in our work. We often form part of larger CRD teams, working closely with scientists to solve company problems. As a result of this integration with the businesses, we tend to do quite a bit of traveling, some of which is international.

The people in the Applied Statistics Program are the very best part of my job. I feel very fortunate to have the opportunity to work with and learn from them, because they are some of the best statisticians anywhere.

The mission statement for our group is the following: To provide leadership in helping GE businesses achieve their strategic objectives by developing and implementing tools, methodologies and programs utilizing our unique technical strengths in:

- Applied statistical methods,

- Holistic approaches to problem definition and resolution, and

- Enhancing the businesses' internal capability in the previous two fields.

This mission statement emphasizes a proactive role for our group. Such a role requires close contact and understanding of the businesses, to be able to anticipate their needs.

26.2 SIX SIGMA

A few years ago, GE adopted the Six Sigma quality initiative, originally developed by Motorola and brought to GE with the help of Mikel Harry of the Six Sigma Academy. This program is a disciplined and highly quantitative approach to implementing quality, with a large emphasis on statistics. It defines five major steps to approaching every problem: Define, Measure, Analyse, Improve and Control. At GE, every employee, including company officers, must do Six Sigma training, and demonstrate that they are using the techniques in their jobs. See Hahn and Hoerl (1998) and Hoerl (1998).

At GE, Six Sigma has been led by the very top levels of management. Our chairman, Jack Welch, has made it clear that the key players in driving this initiative, known as Master Black Belts, should be the company's best people and that they will be rewarded with "big jobs" if they are successful. In contrast to the way some other companies have implemented Six Sigma, GE is using it not just as a tool for manufacturing quality, but also in commercial quality (i.e., service, finance, marketing, etc) and design.

In some businesses, Six Sigma training can be up to 13 days, spread over four months. This training is heavily based in statistics (and to some extent statistical thinking), and can involve everything from basic plots, to the normal distribution, to hypothesis testing, to response surface methodology. It provides a step by step approach to problem solving rather than a collection of statistical tools. In almost all cases, the instructors leading the training have not had any formal education in statistics prior to their own Six Sigma training. In some businesses, the trainees must pass a test at the end as well, to demonstrate their ability to use the tools.

Clearly, all of this has some consequences for the statisticians in our company, though these statisticians had nothing to do with the decision of adopting Six Sigma. I find my customers to be well versed in some basic statistical concepts, which makes it easier to communicate with them. They all have access to some standard statistical software, which means they don't ask me to do routine calculations for them. I generally get involved in cases where the statistical issues are non-trivial. I think in general that my customers' familiarity with statistics helps them to understand what a statistician does, although we still need to work hard to help them to understand the added value that we can bring to the table.

Overall, I think Six Sigma has been positive for the statisticians at GE. Since I got to CRD, our group has constantly been in a mode of recruiting and expansion. It does, however, leave some uncertainty about how our roles might change once this initiative is no longer at the forefront. I see the Six Sigma initiative as giving us valuable exposure in which we have an opportunity to show how we add value. It is up to us to make the most of that opportunity so that when Six Sigma does fade, we are still valuable members of teams addressing key company issues.

26.3   THE PROJECTS THAT I'VE WORKED ON

26.3.1   Introduction

In this section, I will describe a sample of projects that I've worked on, to give a flavor for the types of situations I encounter. Most of the projects that I've worked on so far have been related to reliability. I had to learn this subject quickly, because I lacked the foresight to take courses in it while I was in graduate school.

26.3.2 New Product Launch

On a Friday afternoon three weeks after I had arrived at CRD, my boss came into my office and asked "How do you feel about going to Louisville for Monday?" My reply "Sure, where the heck is Louisville?" (My American geography was not yet up to snuf). 48 hours later I was in Louisville, which shows how early I learned one of the crucial lessons in industry the need for being responsive.

The reason for being called to Louisville, KY was a new product introduction. The design team for this new product had been working for months, and they were now close to launch. They had just discovered a potential failure mode, and they wanted some help to project the failure rates associated with that mode. They wanted to avert any potential field failure issues, while a solution to the problem was being sought. The timing was critical, since any delay on a new product launch is extremely expensive.

The biggest issue that we confronted from a statistical point of view on this project was the lack of sufficient data. We developed a comprehensive program to collect data from a variety of sources, including in-house tests, surveys and field tests. In most cases, however, the data available was scanty, with questionable measurement accuracy. Given this, we felt it important to convey the variability associated with any projected estimates of failure. This was difficult to do by the engineers themselves, however, since the techniques we used to assess the variability were somewhat more sophisticated than those they had seen in their Six Sigma training (e.g. Jack-knifing).

26.3.3 Reliability Issue with a Supplied Part

The next project I was involved with was one on a manufactured product, related to the reliability of a supplied part that we used. For convenience, let's say that we are talking about a 'knob" in a "toaster". We manufacture these "toasters" in large volume, and so while the "knob" had a relatively low failure rate, it was still affecting a substantial number of toasters. It was also the single largest failure mode of the toaster. Another thing that compounded the problem was the fact that the failures occurred quickly in the field, which contributed to customer dissatisfaction. Also, failures that occurred in the field were quite expensive to fix, even though the actual cost of the knob was not high: the fix cost 35 times the cost of the knob. All of these factors resulted in high visibility of this project by upper level management.

There were a few things about this project that made it difficult to
handle. The first was the fact that we were unable to correlate any process
variables with the field failure. At the end of the line at the supplier factory,
various characteristics of the knobs were measured, and we would have liked
to show that any one of these characteristics was an indication of a weak
knob, one that would fail later in the field. Unfortunately, this was not the
case. Thus, the only way to determine whether a particular knob was going
to fail was to do some type of cycle testing. Our suppliers were unequipped
to handle this type of extensive testing. We were able to do this testing
at CRD but that was expensive, especially in light of the relatively small
failure rate of the knobs. All of this made it difficult to approach the
problem using a designed experiment.

At the supplier plant, there was a functional test done on 100knobs. We
were able to demonstrate a high correlation between the defect rate that
was found on the knobs at the supplier in that test, and the field failure
rates that we later found in our toasters. In constructing the model for this
relationship we accounted for such factors as our manufacturing period, the
lag between when knobs were manufactured at the supplier and when they
were installed in our toasters, and the amount of time our toasters had
seen in the field. The existence of this relationship meant that we could use
in-house supplier defect levels as a monitoring tool on the quality of knobs
we were installing in our toasters.

Perhaps not surprisingly, the results of this analysis had a negative effect
on the way we interacted with our supplier. They become reluctant to share
any of their in-house data with us, which made it much more difficult to
make progress on this project. Eventually, we were able to reduce the
number of failures due to this failure mode, but progress was slow.

26.3.4   Constructing a Reliability Database

A GE manufacturing business requested our help in constructing a reliability database. They manufactured a series of products, and sold those
products with a specific warranty period. This business had a good understanding of what happened to their products in the warranty period, i.e.
what the failure rates were, and to some extent what the causes of those
failure were. They now wanted to estimate failure rates beyond that warranty period, to extend to the useful life of the product. My assignment
was to find and assess any data that was available to estimate long term
failure rates, and then develop an appropriate methodology. Because of the
nature of the assignment, it was set up as a four month "bridge" program,
which meant that for those four months, I spent 80% of my time at the

client site. The overall goal here was to use these estimates of failure rates to feedback to design and improve long term reliability.

Again, the biggest obstacle in this project was the lack of complete data. In this case, we used service contracts that were sold on the product as the biggest source of information. Clearly, there were some problems with this. For example, were there enough service contracts sold to provide a sufficient sample of the population? Was this particular segment of the population biased in any way relative to the total population? We eventually concluded that while this source of information did not completely meet our needs, it was much better then what we had before. There is now a database and analysis methods set up that estimates failure rates on all of the business products, on all of the known failure modes. These were constructed as a direct result of our recommendations. Any engineer in the business can access this database from his or her desktop computer.

At the beginning of this project, I had a concern about having the value of the statistician confused for the value of the data. In other words, if there had been no data from which long-term failure rates could have been estimated, would I have been considered a failure in this assignment? Fortunately, I didn't have to find out the answer to that question.

Another thing that came across strongly from this project was the importance of having a champion in the business. A large amount of the success that this project had was because my customer in the business was a strong advocate.

26.4 SOME SURPRISES COMING TO INDUSTRY

Maybe the biggest surprise to me in coming to industry is the amount of learning that I've been able to do, and the different opportunities I've been exposed to. That learning has been both technical, and also non-technical things like good and bad ways of interacting with non-statisticians and communicating ideas. It's also been really interesting to me to see the variety of different roles that can be taken in projects everything from statistical consultant to a leadership role, to one of looking for new opportunities.

One thing that I have struggled with at CRD are project funding issues. For example, I was told in October of 1998 to stop working with a particular business, because of a shift in business priorities. This was frustrating, because I had put a lot of time and energy into developing relationships with many people at this business. Although this was clearly a business decision, it is sometimes hard not to take things personally.

The following are some conclusions that I've come to in the time I've spent so far in industry:

Required Soft Skills

1. Flexibility—the ability to adapt to travel, to diverse business environments, to numerous (and sometimes conflicting) customer requirements.

2. Communication—the ability to communicate technical ideas to non-statisticians, from an operator in manufacturing to the CEO of the business.

3. Balance—the ability to balance dual roles of being responsive to the customer and acting as a "change agent" to promote better, newer ways of doing things.

4. Learning—the ability to learn quickly – about application areas and technical areas – from everyone around you.

5. Team skills—the ability to interact positively with people at all levels and from all sorts of different backgrounds.

Required Hard Skills

1. Knowledge of reliability—It was a big mistake for me not to take courses in this at graduate school. While taking such courses would have helped to give me a base in the subject, many of the techniques I have used have been quite specialized (e.g.. Analysis of truncated data or repairable systems) or tailored specifically for the data that was available.

2. Powerpoint engineering—A large part of the communication at GE does not take place in formal, written reports, but rather in presentations. Success in this area, I have found, is dependent on being able to communicate ideas in a concise and creative manner.

Recommendations to Students

1. Take advantage of all opportunities to work with non-statisticians on their problems. One of the hardest things I found to do initially was to talk to a non-statistician about his or her work, and then translate that into a statistical problem. This was especially true when I talked to someone for a long time I would get so engulfed in the details, I found it hard to sort out the information relevant to the statistical problem from the other information. I found that I got better at this with practice, and by watching other people do it.

2. Learn from the experiences of the people around you in handling consulting types of situations. I was fortunate in graduate school to have people around me who frequently worked with industry. This allowed me to get their viewpoint on the "do's" and "don'ts" of working with industry. Of course, this didn't stop me from making some major errors once I got to the job. My favorite story occurred when I told an engineer that the statistical method in question was intuitive, even though he hadn't seen it before. He picked up a contraption that went into a washing machine, and told me that to him, that contraption was intuitive. Hopefully, I made fewer of these mistakes as a result of learning from the people around me in graduate school, and continuing to learn while I'm on the job.

3. Develop a broad base of technical knowledge. I used to get very nervous when I was working at a business and someone would start a conversation with "You are a statistician, let me ask you about..." I was always worried that they would be asking me something that I didn't know about, and that by not being able to answer their question, I would destroy my credibility as an expert. There are many different ways of addressing this problem, but it will always help to have a broad technical base from which you are responding.

4. Be interested in the world outside statistics. I think it helps a lot when working with people if they can sense that you are genuinely interested in what they do, not just your piece of the bigger picture. It is also a great opportunity to learn about all sorts of different areas that you might otherwise not have known about, as exemplified for me when I got to witness the manufacturing process of a street lamp fixture.

Recommendations to Departments

1. Develop contacts with local industry, and allow students direct interactions with those contacts, through term projects, internships or any other available method.

2. Encourage students to take a problem in context and turn it into a statistical one, leaving them to decide what information from the context is relevant to the statistical problem.

3. When presenting standard techniques and methods, encourage "out-of-the- box" thinking. Within two weeks of arriving at this job, someone in the business asked me, "I ran an expensive fractional factorial

experiment, but didn't randomize one of the factors. What can I do now with the results?" This question unnerved me substantially because the area of design of experiments, unlike reliability, was one I felt I knew something about. Unfortunately, I had never considered this engineer's question before.

26.5   GENERAL COMMENTS

I love my job! It's been a truly rich experience, with a plethora of opportunities. The environment I work in is dynamic and exciting, and I've had the opportunity to learn about everything from locomotives to refrigerators; from circuit breakers to supply chain modeling. I can't think of a better thing to be doing.

Acknowledgements   I would like to thank Necip Doganaksoy, Evelyn Eigo, Gerry Hahn, Roger Hoerl, Jock MacKay and Bill Wunderlin for providing excellent feedback on drafts of this paper.

REFERENCES

Hahn, G and Hoerl, R. (1998). Key challenges for statisticians in business and industry (with discussion), *Technometrics*, **40**, 195–213.

Hoerl, R. (1998). Six sigma and the future of the quality profession, *Quality Progress*, **31**, 35–42.

Part VIII

Applications to Ecology, Biology and Health

CHAPTER 27

CONTEMPORARY CHALLENGES AND RECENT ADVANCES IN ECOLOGICAL AND ENVIRONMENTAL SAMPLING

G. P. PATIL   C. TAILLIE

Pennsylvania State University, University Park, PA

Abstract: Surveys for monitoring changes and trends in our environment and its resources involve some unusual conceptual and methodological issues pertaining to the observer, the observed, and the observational process. In this paper, we briefly introduce some of the novel methods of ecological and environmental sampling and present some of the relevant research in progress at the Center for Statistical Ecology and Environmental Statistics in these innovative methods with some emphasis relating to the remote sensing satellite imagery. Keith (1996) may be a good source of additional information.

Keywords and phrases: Adaptive sampling, composite sampling, distance sampling, guided transect sampling, spatial sampling

27.1 CERTAIN CHALLENGES AND ADVANCES IN TRANSECT SAMPLING

As discussed in Patil, Taillie, and Talwalker (1993), the method of line transect sampling has been used to estimate the abundance of plants or animals of a particular species in a given region. The line transect method consists of drawing a baseline across the region to be surveyed and then drawing a line transect through a randomly selected point on the baseline. The surveyor looks around while walking along the line transect and includes the sighted objects of interest in the sample.

It is obvious that the nearer the object or the larger its size, the higher is the probability of sighting the object. Similarly, when the individuals cluster in groups, such as schools or herds, then it is appropriate to regard clusters as the basic sampling units with their encounter probabilities being affected by cluster size. Estimates of cluster abundance can be adjusted to individual abundance using the recorded cluster sizes.

Encounter probabilities in transect sampling can be influenced by numerous other factors such as varying terrain and vegetation cover, weather conditions, time of day, systematic responsive movement toward or away from the transect, etc. Some of these factors are characteristics of the objects themselves and will vary from object to object. "Size" is an example of such a factor. Other factors, notably environmental features, are survey characteristics but can vary from segment to segment in a multi-segmented survey. We refer to members of these two classes as *object-factors* and *survey-factors*, respectively.

It is clearly desirable to account for as many of these factors as possible, and much of the recent transect sampling literature has been concerned with this issue. Section 27.1.1 summarizes some of our own work involving a seabed transect survey for red crabs in which monotonicity of the sighting function was flagrantly violated. The cause was eventually identified and accounted for by putting an additional multiplicative factor in the sighting function. Section 27.1.2 surveys the work of Ramsey, Wildman, and Engbring (1987) and Drummer and McDonald (1987). Each of these papers incorporates the extra-distance factors into the scale parameter of the sighting function. Ramsey *et al.* are concerned with the effects of survey-factors, in which case a purely conditional, regression-like, analysis is adequate. Drummer and McDonald examine an object-factor, specifically size, and need to consider the probability distribution of that factor as well as the visibility bias that occurs in the recorded sizes. Finally, Section 27.1.3 introduces guided transect sampling due to Stahl, Ringvall, and Lamas(1997) in progress in collaboration with the Penn State Center.

27.1.1   Deep-Sea Red Crab

Patil, Taillie, and Wigley (1979, 1980) describe a photographic survey in which features of the optical geometry partially masked the sighting-distance bias by inflating the recorded counts at larger distances from the transect. The survey's purpose was to determine the abundance of the deep-sea red crab, *Geryon quinquedens* Smith, in continental slope waters off the northeastern United States. Water depth at 33 sampling stations ranged from about 200 to 1500 meters. The sampling device was an underwater camera system mounted on a 1200 kg steel sled that was lowered to the ocean floor at each station and towed for 30 to 75 minutes depending upon local conditions. Several hundred non-overlapping photographs were obtained for each station; about half, representing the best quality, were selected for quantitative analysis.

Roughly one crab was sighted for every four frames analyzed. Even so, determination of the perpendicular sighting distance to each crab proved excessively labor-intensive, and an alternative was devised in which every photograph was divided into five zones or strips running parallel to the transect and representing widths of 1.22 meters on the ocean floor. [For technical reasons, the zone closest to the camera was only half as wide, requiring appropriate adjustments; see Patil, Taillie, and Wigley (1979).] The zones were delineated on an overlay and the number of crabs in each zone was counted.

The sighting frequency had been expected to fall off rapidly with distance because of factors such as turbidity, increasingly diffused lighting, and roughness of the seabed. (The camera was angled in a way that exposed the bottoms of seabed depressions provided they were close to the sled.) But, in fact, the histograms of sighting-frequency versus distance showed only a very gradual decline and, at small distances, suggested that frequency might even increase with distance. The explanation was eventually found when it was noted that what appeared in the photograph as a rectangular zone actually represented a trapezoid on the ocean floor with the shorter side closest to the camera. Thus, at sighting-distance x, the photograph exposes a length $a + bx$ where the constants a and b could be determined from the geometry. The data were analyzed using a composite weight function of form

$$w(x) = (a + bx) \cdot v(x; \theta),$$

where the sighting function $v(x; \theta)$ represents the pure sighting-distance bias and would typically involve unknown parameters θ. For the red crab study, the exponential-power sighting function gave a reasonable fit to the data and yielded abundance estimates that were consistent with the results of other survey methods [Wigley, Theroux, and Murray (1975)].

27.1.2   Bivariate Sighting Functions

We let L be the length of the transect and Ω the width of the sighting strip on either side of the transect. Sometimes Ω is established by the sampling design. More often, Ω is taken to be the visible (or audible) horizon, in which case it is convenient to formally let Ω go to infinity and to estimate abundance as a density instead of a count (see below).

With $w(x)$ as the sighting function, the mean detection probability is

$$P = \frac{1}{\Omega} \int_0^\Omega w(x)dx,$$

or

$$\omega = P\Omega, \tag{27.1.1}$$

where

$$\omega = \int_0^\Omega w(x)dx \tag{27.1.2}$$

is known as the *effective half-width*. Writing N and n for the population and sample counts, one has $E[n] = PN$ so that abundance can be estimated as

$$\hat{N} = n/\hat{P} = n\Omega/\hat{\omega}, \tag{27.1.3}$$

provided an estimator $\hat{\omega}$ is available for ω. The awkward quantity Ω can be eliminated if (27.1.3) is divided by the survey area $2L\Omega$ to obtain the density estimator

$$\hat{N}/2L\Omega = n/2L\hat{\omega}. \tag{27.1.4}$$

The recorded right angle distances x have their probability density function given by $w(x)/\omega$, $0 < x < \Omega$, so that ω can, in principle, be estimated from these distances. See Burnham, Anderson and Laake (1980) or Seber (1993) for further details on line transect sampling in general.

When Ω can be taken to be infinite, Ramsey (1979) has suggested a general technique for constructing parametric visibility functions in which the effective half width ω appears explicitly as a scale parameter. Starting with a *kernel* $h(t; \theta)$, $0 < t < \infty$, which is monotone decreasing in t and satisfies $h(0; \theta) = 0$ and $\int_0^\infty h(t; \theta)dt = 1$, Ramsey's family of sighting functions is defined by

$$w(x; \omega, \theta) = h(x/\omega; \theta). \tag{27.1.5}$$

Here, θ is a vector of nuisance parameters that regulate the shape of the sighting function. A common choice of kernel is the exponential-power form,

$$h(t; \gamma) = \exp(-\Gamma(1 + \gamma^{-1})t^\gamma), \tag{27.1.6}$$

which includes the negative exponential and the half-normal as special cases.

Ramsey *et al.* (1987) propose that covariate information $\mathbf{y} = (\mathbf{y_1}, \ldots, \mathbf{y_p})$ be incorporated into the sighting function by letting ω in (27.1.5) be a parametric function of $\mathbf{y}$. Notice that this effectively yields a multivariate sighting function $w(x, \mathbf{y})$. Ramsey et al. suggest the specific form

$$\ln(\omega(\mathbf{y})) = \beta_0 + \Sigma_{j=1}^{P} \beta_j \mathbf{y_j}, \qquad (27.1.7)$$

where the covariates may need to be transformed before inclusion in (27.1.7). Ramsey et al. develop the maximum likelihood estimators conditional upon the recorded covariate values. This conditional approach is appropriate for what we have referred to as survey factors in a multi-segmented survey. Here one wants to pool data from the various segments in order to obtain precise estimates of the visibility functions. However, density is estimated separately for each segment using the segment-specific estimate of ω.

Ramsey *et al.* give an example involving tropical birds with time of day as the covariate. The raw counts suggested a declining abundance through the course of the day; this apparent effect could be accounted for as due to declining visibility (audibility) over time.

Drummer and McDonald (1987) have considered the problem of estimating minke whale abundance where the group size, an object-specific factor, becomes important. They take the group as the basic sighting object and include the group size y as a scaling factor in the exponential-power sighting function

$$w(x, y) = w(x, y; \lambda, \alpha, \gamma) = \exp\{\Gamma(1 + \gamma^{-1})(x/\lambda y^\alpha)^\gamma\}. \qquad (27.1.8)$$

Drummer and McDonald employ a slightly different parametrization. Notice that (27.1.8) falls into the framework of (27.1.7) with the log of group size as covariate since

$$\ln(\omega) = \ln(\lambda) + \alpha \ln(y). \qquad (27.1.9)$$

Since y is an object-factor, its random variation from object to object needs to be taken into account, and the conditional approach of Ramsey et al. is no longer sufficient. Let $f(y)$ be the "natural" distribution of group size. The joint distribution of recorded (x, y) is then proportional to

$$w(x, y)f(y), \qquad (27.1.10)$$

with normalizing constant given by

$$\int \int w(x,y)f(y)dxdy = \int \omega(y)f(y)dy = \lambda \int y^\alpha f(y)dy = \lambda \mu_\alpha = \bar{\omega}.$$
$$\qquad (27.1.11)$$

This argument also shows that recorded y follow the size-biased distribution of order α, $f_\alpha^*(y) = y^\alpha f(y)/\mu_\alpha$. It is readily seen that equations 27.1.1–27.1.4 remain valid when ω is replaced by $\bar{\omega}$.

If one were prepared to assume a parametric form for $f(y)$ then estimation could proceed via maximum likelihood in conjunction with (27.1.10) and (27.1.11). Drummer and McDonald prefer a two-stage approach in which λ and α (as well as the nuisance parameter γ) are estimated using the conditional procedure of Ramsey et al. The moment μ_α in (27.1.11) is then estimated nonparametrically using $\mu_\alpha = 1/E_{f_\alpha^*}[y^{-\alpha}]$, a variant of the Cox (1969) identity for size-biased distributions. Similarly, $\mu_1/\mu_\alpha = E_{f_\alpha^*}[y^{1-\alpha}]$, so that the mean group size, μ_1, can be estimated for the purpose of adjusting group abundance to individual whale abundance.

27.1.3   Guided Transect Sampling

Introduction

As discussed in Stahl, Ringvall, and Lamas (1997), guided transect sampling is primarily intended for the sampling of sparse, geographically scattered, populations for which there exist no list of the units. Basically, it consists of a two-stage design, using wide strips in the first stage and a subsampling procedure in each strip in the second stage. The subsampling is guided by prior information, e.g. in the form of remote sensing image data. Different strategies can be used for the guidance, resulting in different probabilities of inclusion of population units, and consequently in slightly different estimators.

The general principle for second stage subsampling guidance can be coupled with a number of methods for how the samples should be selected along the "guided route." Strip sampling, line transect sampling, adaptive cluster sampling, and plot sampling are examples of methods that can be used. However, in the theoretical set-up of the method, it is assumed that all objects in grid-cells passed by the survey transect are found. The grid-cells, covering the entire area under study, contain the covariate data that are used for directing the sampling effort. The method has some similarities with the covariate-directed sampling approach proposed by Patil, Grigoletto, and Johnson (1996).

The method

An overview of the method, in its basic form, is given in Figure 27.1 below. In the forest area delineated, strips too wide to be entirely surveyed are first randomly laid out. Secondly, a route for the subsampling within each strip is guided by prior information. The details of this guidance are described below.

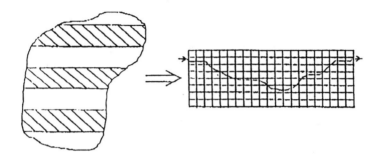

FIGURE 27.1 A general outline of guided transect sampling. A first stage sampling of wide strips (left) is followed by a second stage guided subsampling within each strip (right)

The entire area of interest is partitioned into grid cells of some suitable size, e.g. 20 by 20 meters (or possibly rectangular to simplify the field work). For each such cell, a covariate value is assessed prior to the sampling. For example, the covariate could be the estimated volume of deciduous trees in case the population under study is known to prefer deciduous forest to coniferous forest. Such prior volume estimates can be obtained by, e.g., using satellite data and the kNN-method [e.g. Nilsson (1997)].

In order to facilitate the theoretical description of the method, an assumption is made that all sampling units are detected and counted/measured once the surveyor enters the grid cell they are situated in. Also, the method relies on use of GS, differential in real time, for the guidance of the surveyor through the forest. However, in simple cases it should also be possible to use a compass and a measuring tape.

The first stage strips are laid our randomly, with the restriction that they should match with the system of grid cells. The second stage is a subsampling of grid-cells along a survey line within each first stage strip. Many different strategies can be used for determining where the second-stage transect should be located. One basic idea is, however, that the grid-cells in some manner should be selected with probabilities proportional to their covariate values. Another basic idea is that the field-work should not be too complicated, implying that the survey transect should be some, more or less, connected curve from the beginning to the end of a strip. This is also the reason for introducing the first stage strips. Without them, the survey lines within the forest area would tend to be complicated. Finally, the idea is also to use some line-based inventory rather than plots in order to obtain a more efficient search for individuals of the sparse population in this theoretical description of the method, a strip survey is approximated by a continuous survey of neighboring grid-cells. The surveyor is assumed to perform an entire search for objects in all grid-cells entered.

Conforming to all this, many different strategies for the subsampling within each strip can still be identified. Some straightforward possibilities are:

(i) Random walk (Markovian) with the probability to enter a neighboring cell, in the direction of the survey line, given by the cell's covariate value [Figure 27.2(a)]

(ii) As (i) but allowing the surveyor to step from a particular cell to any of the grid-cells in front. That is, "big steps" are allowed, since the surveyor in this case may go directly from one side of the strip to the other. The strip will no longer be connected [Figure 27.2(b)]

(iii) Random simulation of entire transects through a strip (without considering the covariate data at this stage). Transition is only allowed to neighboring cells. A large number of transects are simulated. For each one, the sum of cell-wise covariate values is calculated. This sum, or some transformation of it, is used for selecting one particular transect by PPS [Figure 27.3(c)].

To make the method useful from a practical point of view, the grid-cells should generally be rectangular (very elongated in the direction of the strips) in order to avoid too much zigzagging for the surveyor. An alternative to this would be to assign higher probabilities for straight continuation than for changing to another row in the grid-cell system. However, in all figures in this theoretical description of the method, square grid-cells are used.

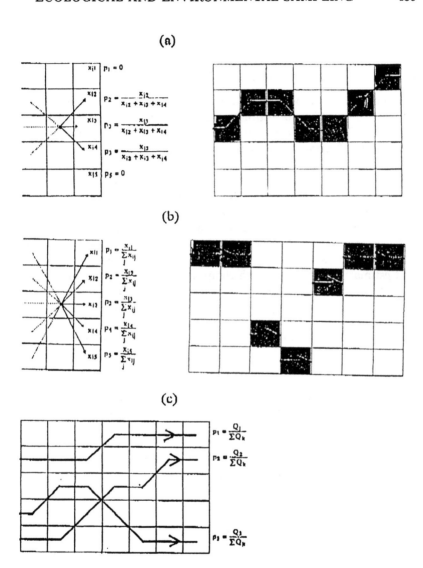

FIGURE 27.2 Different principles for guiding the subsampling. In (a) transition is only allowed to neighboring cells, in (b) transition is allowed to any onward cell, while in (c) entire transects are simulated. In (a) and (b), the probabilities of transition (the p-values) are determined from the covariate values (x-values) in the next stage, denoted i. In (c), entire transects are determined from the sum(Q-values) of covariates in grid-cells visited

27.2 CERTAIN CHALLENGES AND ADVANCES IN COMPOSITE SAMPLING

27.2.1 Estimating Prevalence Using Composites

Consider a trait whose prevalence in a population is denoted by p. We take up the problem of estimating p on the basis of composite samples of size k. Since p is a population mean, this might appear as merely a special case of estimating population means with composite samples. Here, however, a composite sample drawn from the population has the trait in question exactly when one or more of the individual samples making up the composite has the trait. This implies that the measured value on a composite is the indicator function for the composite and this is different from the average of the indicator values for the constituent samples. In fact, the composite indicator is the *maximum* of the individual indicators. The maximum is a nonlinear function, and this nonlinearity is the reason that estimating prevalence with composites requires special attention.

For simplicity, we suppose that the composite sample size k is constant and we let $\pi = \pi_k$ be the prevalence of the trait across all possible composites of size k. We also limit ourselves to (effectively) infinite populations. Now, π and p are related by the formula $1 - \pi = (1 - p)^k$ or

$$p = 1 - (1 - \pi)^{\frac{1}{k}} \equiv H(\pi) . \qquad (27.2.12)$$

We call $H(\cdot)$ the *prevalence transformation*. Its graph is depicted in Figure 27.3 for several values of k. From the Figure, we see that when $k > 1$ the prevalence transformation is monotone increasing, convex, and becomes highly nonlinear when π is large.

The maximum likelihood estimate of π is the sample proportion $\hat{\pi}$ so that the maximum likelihood estimate of p is

$$\hat{p} = H(\hat{\pi}) .$$

Since H is nonlinear, the MLE $\hat{p}$ is biased.

We examine the following issues:

- What is the performance of composite sampling for estimating p as compared with individual sampling?

- What is the optimal value, k_{opt} of the composite sample size k?

- The optimal k will depend on the true, but unknown, value of p. What is the sensitivity and robustness of k_{opt} to misspecification of p?

- Is the bias in $\hat{p}$ important and, if so, how can it be reduced?

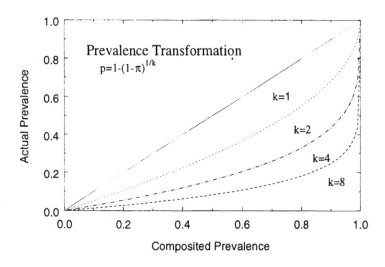

FIGURE 27.3 The prevalence transformation $H(\pi)$ for $k = 1, 2, 4, 8$

Asymptotic performance of compositing

Using statistical differentials, the asymptotic variance of $\hat{p}$ is given by

$$\mathrm{var}_c(\hat{p}) = [H'(\pi)]^2 \, \mathrm{var}(\hat{\pi})$$

$$= \frac{1 - (1-p)^k}{nk^2(1-p)^{k-2}},$$

where $n = n_c$ is the number of composites analyzed. For individual sampling, the variance is

$$\mathrm{var}_i(\hat{p}) = p(1-p)/n,$$

where $n = n_i$ is the number of individuals analyzed. When quantification rather than sample acquisition is the primary cost factor, the relative cost of the two sampling designs can be measured by the ratio of the sample sizes needed to achieve the same variance for the two designs. By the above equations, the *asymptotic relative cost* of compositing compared with individual sampling is

$$\mathrm{RC} = \frac{n_c}{n_i} = \frac{1 - (1-p)^k}{k^2 p(1-p)^{k-1}}.$$

Here, values of RC that are less than unity favor compositing over individual sampling. It is not hard to see that the relative cost satisfies the following:

- RC → $1/k$ as $p → 0$, and

- RC → ∞ as $p → 1$ (unless $k = 1$).

These two properties indicate that neither sampling design is uniformly better than the other and that compositing tends to be better for small p while individual sampling is better for large p. The second property also shows that the use of compositing with an inappropriate choice of k can have disastrous consequences for the relative performance.

Optimal composite sample size and its robustness

The asymptotic relative cost RC is plotted against p in Figure 27.4 for selected values of k. The lower envelope of these curves determines the optimal value of RC as well as the corresponding optimal composite sample size k_{opt}. Notice that the curves in this Figure are steeply rising as p becomes large. This implies that compositing will perform poorly compared to individual sampling if one determines k_{opt} by using a prior value of p that is much smaller than the actual value of p. Thus, it is better to err in the direction of overestimating p and underestimating k_{opt}.

When p is sufficiently large ($p > 2/3$), Figure 27.4 shows that $k_{opt} = 1$ which means that individual sampling is better than compositing for $p > 2/3$. For smaller values of p, k_{opt} increases in discrete jumps as p decreases from 2/3 to 0. We have calculated the values of $p = p_k$ where k_{opt} jumps from the value k to the value $k+1$. These transitional values of p are shown in Table 27.1.

The table shows that k_{opt} gets large very fast as $p → 0$. In fact, $k_{opt} \sim 1.594/p$ when p is small. Often, it may happen that the optimal composite sample size is too large for practical implementation. This may be contrasted with group testing with the Dorfman procedure where $k_{opt} = O(1/\sqrt{p})$ as $p → 0$. However, we have seen above that it is safer to use a smaller than optimal composite sample size.

The table also reveals that the values of π_k are remarkably constant and usually fall in the range $0.7 < \pi_k < 0.8$. From Figure 27.3, we see that this is the range where nonlinearity in the prevalence transformation starts to become pronounced. Consequently, $\hat{p}$ may be significantly biased when optimal composite sample sizes are used.

TABLE 27.1 Values of $p = p_k$ where the optimal composite sample size makes a transition from $k_{opt} = k$ to $k_{opt} = k + 1$. When p is slightly larger than p_k, the optimal composite sample size is k; when p is slightly smaller than p_k the optimal composite sample size is $k + 1$. The composite prevalence $\pi_k = H^{-1}(p_k)$ corresponding to p_k is also tabulated

k	p_k	π_k
1	.667	.667
2	.475	.725
3	.367	.747
4	.299	.759
5	.252	.776
6	.218	.771
7	.192	.774
8	.171	.777
9	.155	.779
50	.031	.794
100	.016	.795
500	.003	.796

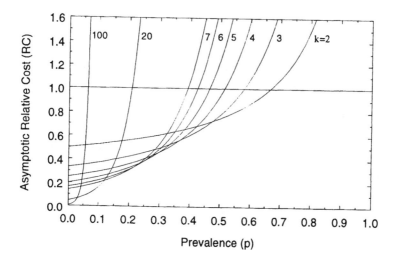

FIGURE 27.4 The asymptotic cost of compositing relative to individual sampling as a function of the true prevalence p for $k = 2(1)7, 20, 100$

Bias reduction

Since the prevalence transformation $p = H(\pi)$ is convex, Jensen's inequality implies that $\hat{p}$ is positively biased,

$$E[\hat{p}] \geq p,$$

where the inequality is strict when $k > 1$. The bias can be quite severe when π is close to unity, e.g., when an optimal composite sample size is used.

Now, H is also monotone increasing, so one way of reducing the bias in $\hat{p}$ is to shrink $\hat{\pi}$ toward zero before applying the prevalence transformation. The *shrinking method* thus uses

$$\hat{p}^* = H(\alpha\hat{\pi}) \tag{27.2.13}$$

to estimate p, where α is a suitably chosen constant satisfying $0 \leq \alpha \leq 1$. Since the bias disappears with large sample sizes, the constant α has to depend upon the sample size n and should go to unity as n becomes large. A natural choice is

$$\begin{aligned}
\alpha &= 1 - \frac{b}{n+c} \\
&= 1 - \frac{b}{n} + \frac{bc}{n^2} + O\left(\frac{1}{n^3}\right),
\end{aligned}$$

where b and c are suitable chosen constants that do not depend upon the sample size.

The choice $b = \frac{k-1}{2k}$ eliminates the first order term from the asymptotic expansion of the bias. The constant c is undetermined by first order considerations. Although one could attempt to eliminate the second order bias term, it is probably better to examine the mean square error. With the above choice of b the shrinking method gives a one parameter family of estimators that are first order unbiased.

A particular member of the shrinking family of estimators has been proposed by Burrows who suggests using

$$\begin{aligned}
\hat{\pi}_{\mathrm{B}} &= \frac{n}{n+b}\hat{\pi} \\
\hat{p} &= H(\hat{\pi}_{\mathrm{B}}),
\end{aligned}$$

where $b = \frac{k-1}{2k}$. Thus, the Burrows estimator is the special case of the shrinking estimator with $c = b$.

Jackknifing is another way of eliminating the first order bias from an estimator. In fact, if we let $\hat{p}_J$ be the jackknifed version of $\hat{p}$, then

$$E\left[\hat{p}_J\right] = p \qquad -\frac{\delta_2}{n^2} + O\left(\frac{1}{n^3}\right)$$

when

$$E\left[\hat{p}\right] = p + \frac{\delta_1}{n} + \frac{\delta_2}{n^2} + O\left(\frac{1}{n^3}\right).$$

Thus, jackknifing eliminates the first order bias without increasing the magnitude of the second order bias.

It is natural to ask how these three methods of bias reduction compare with respect to their effect on the mean square error. For each of the three methods, we find that the mean square error has the form

$$\mathrm{MSE} = \frac{A}{n} + \frac{B}{n^2} + O\left(\frac{1}{n^3}\right),$$

where A is the same for all three methods (A is the asymptotic variance). The shrinking method and jackknifing have different B, but the form of B does not involve the constant c of the shrinking method. Further, B depends upon p and neither the shrinking method nor jackknifing is uniformly better than the other in minimizing the magnitude of B.

Discrimination among the various shrinking estimators would require looking at the small sample properties.

27.2.2   Two-Way Compositing

Compositing of individual samples is a cost-effective method for estimating a population mean, but at the expense of losing information about the individual sample values. The largest of these sample values (*hotspot*) is sometimes of particular interest. Sweep-out methods [Gore and Patil (1994), Gore, Patil, and Taillie (1996)] attempt to identify the hotspot and its value by quantifying a (hopefully, small) subset of individual samples as well as the usual quantification of the composites. Sweep-out design is concerned with the sequential selection of individual samples for quantification on the basis of all earlier quantifications (both composite and individual). The design-goal is for the number of individual quantifications to be small (ideally, minimal).

Sweep-out procedures were originally developed for conventional one-way composites, but Gore, Patil, and Taillie (1995) considered two-way composites using a heuristic method for sequentially selecting samples for quantification. Aragon, Patil, and Taillie (1995) have proposed a more formal and rigorous method for making the selections. In two-way compositing, the individual samples are arranged in a rectangular array and a composite is formed from each row and also from each column. At each step, the procedure employs all available measurements (composite and individual) to form the best linear unbiased predictions for the currently unquantified cells. The cell corresponding to the largest predicted value is chosen next for individual measurement. The procedure continues iteratively and terminates when the largest individual value has been identified with certainty.

The following important and interesting issues arise:

(a) Comparative performance of the two algorithms, perhaps in terms of the mean and variance of the total number of measurements. Least squares prediction appears to encounter the maximum sooner, but takes about the same number of measurements to identify the maximum.

(b) Design questions, such as: Is a square layout better than a rectangular one for a given number of cells? Does a two-way layout provide any benefits over the one-way layout? Is a large layout better than a replication of small layouts?

(c) Distributional questions, such as: How is performance affected by the distribution, especially skewness, of the individual cell values? How is performance affected by correlation among individual cell values as it might arise from sampling in time or space? Is an initial randomization of the cell values advantageous?

27.2.3   Compositing and Stochastic Monotonicity

The basic principle behind the sweep-out methods is that larger individual values are more likely to be found in composites having larger composite values. We have recently attempted to give this intuitively plausible assertion a more rigorous formulation. Initially, we expected that it would be an easy exercise. However, the problem has proven to be surprisingly difficult and the answer—to the extent that we have an answer—has proven to be surprising.

We limit ourselves to composites of size two ($k = 2$). Let X and Y be *iid* random variables that are positive. These are the individual values comprising the composite. Let

$$T = X + Y,$$

so that T is the composite total. The question to be considered is the following: "Is it the case that $X|T = t$ is stochastically increasing in t?"

When the answer to the preceding question is in the affirmative, we say that the distribution of X is *stochastically monotone with respect to (two-fold) convolutions* (SMC). One simple result is the following: A positive random variable is SMC if it has a log-concave density. In fact, this is enough to establish that the family of distributions, $X|T = t$, has monotone likelihood ratio with respect to t.

In the next sections, we give three standard parametric families of distributions which— for varying values of their parameters—are respectively always, never, and sometimes SMC.

Gamma distribution

Let X and Y follow a gamma distribution with index parameter β. Then X is SMC for all values of β. This is easy to see when $\beta \geq 1$ since the gamma density is then log-concave. A direct argument is needed for $\beta < 1$.

Lognormal distribution

Suppose X and Y follow a lognormal distribution with parameters μ and σ^2 (these are the mean and variance on the log-scale). Then, there are no values of μ and σ^2 for which X is SMC. The proof is not easy and will be given elsewhere. However, it turns out that there is a value $t_0 = t_0(\mu, \sigma^2)$ such that X is SMC on the interval $(0, t_0)$, i.e., $X|T = t$ is stochastically increasing in t for $0 < t < t_0$. A natural measure for the size of the interval $(0, t_0)$ is $\Pr(T < t_0)$ and this probability turns out to be at least 0.94 regardless of the values of μ and σ^2 (see Figure 27.5). Thus, the lognormal distribution is SMC with "high probability." Note that the "high probability" refers to the frequency of occurrence of the composites.

Pareto distribution

The Pareto that we have in mind is shifted so that its lower bound is at the origin. The density function is given by

$$\frac{\beta}{(1 + x)^{\beta+1}}, \qquad x > 0,$$

where $\beta > 0$. This distribution is SMC provided $\beta \leq 1/2$. If $\beta > 1/2$ the distribution is not SMC but, as in the lognormal case, there is a $t_0 = t_(\beta)$ such that the Pareto distribution is SMC on the interval $(0, t_0)$. Figure 27.6 plots $\Pr(T < t_0)$ versus β. Here, the probability of stochastic monotonicity is never less than 0.87.

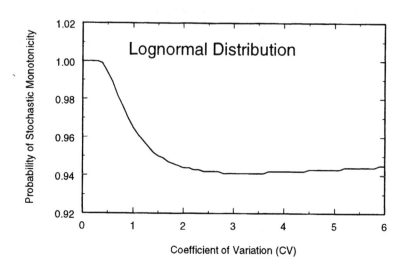

FIGURE 27.5 Probability of stochastic monotonicity for the lognormal distribution. The horizontal axis is the coefficient of variation, given by $\sqrt{\exp(\sigma^2) - 1}$

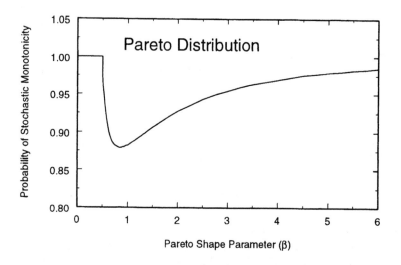

FIGURE 27.6 Probability of stochastic monotonicity for the Pareto distribution

Stochastic monotonicity of the order statistics

Up until now, we have been looking at stochastic monotonicity for an arbitrary one of the individual samples comprising a composite. If the interest is in the largest individual value, then it is more natural to study stochastic monotonicity of the *ordered* individual values. With the same notation as above, let

$$L = \min(X, Y) \qquad \text{and} \qquad U = \max(X, Y).$$

We have established the following result:

Theorem 27.2.1 *The random variable X is SMC if and only if both order statistics are stochastically increasing in t, i.e.,*

- $L|T = t$ *is stochastically increasing in t for all t, and*

- $U|T = t$ *is stochastically increasing in t for all t.*

Our proof applies only to the case of two components. It would be of considerable interest to extend the result to an arbitrary number of components.

The lognormal distribution and the Pareto distribution with $\beta > 1/2$ are not SMC. Therefore, at least one of the order statistics must fail to be stochastically increasing for these distributions. We have been able to show that the failure is for the smaller order statistic and that the larger order statistic is stochastically increasing in t for both distributions (and for many other distributions as well). This is perhaps heartening for the sweep-out paradigm.

27.3   CERTAIN CHALLENGES AND ADVANCES IN ADAPTIVE CLUSTER SAMPLING

27.3.1   Adaptive Sampling and GIS

Several ecological and environmental populations are spatially distributed in a clumped manner. They are not very efficiently sampled by conventional probability based sampling designs. Adaptive sampling is therefore introduced [Thompson (1990)] as a multistage design in which only the initial sample is obtained using a conventional probability based procedure. When the variable of interest for a sampling unit satisfies a given criterion, however, additional units in the neighborhood are selected in the next sampling stage. This procedure is repeated until no new units satisfy the criterion, or the conditions of a stopping rule are satisfied. For methods of unbiased estimation and related statistical inference, see Thompson and Seber (1995).

Consider the point process in Figure 27.7. Here, a sampling frame is delineated by square sampling units laid over the area which contains the population of interest. One may draw a simple random sample of n units using a random number generator for choosing the coordinates for each unit to be included in the initial sample. The variable of interest, Y, in this case may be the number of points per unit, such as the population density for a given species. After obtaining the set of measurements, $\{y_i : i = 1, 2, \cdots, n\}$, each measurement is compared to the given criterion to decide if neighboring units should be sampled. A criterion is typically to include neighboring units in the next sampling stage if $y_i > c$ for some constant c, and otherwise do not include neighboring units in the next stage. If monitoring for an animal or plant species, the criterion may simply be to sample the neighbors of any unit from the initial sample which contains at least one individual. If monitoring contaminant concentrations, the criterion may be to sample neighboring units if a measured concentration in the initial sample exceeds an action level or cleanup standard.

Viewing Figure 27.7, we see that an initial random sample from the grid of sampling units can result in several units that do not satisfy the criterion and would therefore not have any neighboring units sampled in the second stage. For those units in the initial sample which do contain points, each neighboring unit would be sampled in the second stage, and this would be repeated for subsequent stages until clusters are delineated. For our example, the final selection of sample units is shown in Figure 27.8. A cluster is defined as a group of adjacent sampling units that includes all the edge units that do not satisfy the criterion. A network is defined as the group of adjacent units which all satisfy the criterion. Therefore each network set is contained within a cluster set.

While the simple arithmetic mean of the initial random sample of n units is unbiased, the arithmetic mean of the final sample obtained through adaptive cluster sampling is biased upwards. Thompson (1992) presents two "probability proportional to size" estimators that are unbiased for the population mean or total.

With the recent growth of geographic information systems (GIS), spatial data coverages for landscapes are becoming almost universal. Such information, obtained mainly from digitized maps and remotely sensed sources, may provide a powerful aid to adaptive cluster sampling for increasing the efficiency of sampling clustered populations from across a two-dimensional surface.

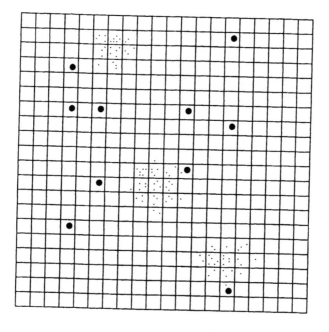

FIGURE 27.7 A grid of sample units, superimposed on a clustered population of point objects, along with a random sample of 10 initial observations

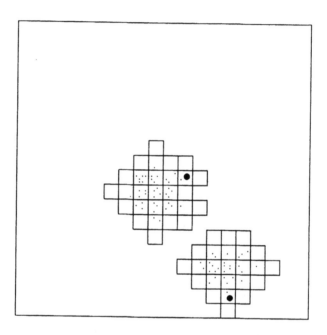

FIGURE 27.8 The final adaptive cluster sample. Two clusters were intercepted by the initial random sample

On one extreme, GIS-based information may dictate where the actual clusters are, thus excluding the need for "adaptive" cluster sampling. However, when clusters must be sought through an initial probability based sample, then GIS based information may be exploited to help decide which neighbors of the initial random sample should in turn be sampled.

Once a measurement is obtained, its corresponding location can be referenced to a GIS database for obtaining auxiliary information about that location and its neighbors. Such information may aid in deciding which neighboring units should be sampled. Analytical results from a GIS may suggest that some neighbors not be sampled, thus saving on sampling and analysis costs. For neighbors that are recommended for sampling, the probability of inclusion may be conditional on the auxiliary information. If inclusion probabilities are affected, then the estimators presented above may require modification, which is an area requiring research.

For one example, consider a wildlife monitoring situation where we detect a colony of ground shrews in an initial random sample. After referring this location to a GIS, we may determine that one neighboring unit is expected to have soils far too wet for suitable shrew habitat. Another location may be marginal, therefore we would only sample it if the budget allows. Meanwhile, GIS may indicate that habitat in the other neighboring units is suitable enough to assign relatively high prior probabilities of shrew presence.

Similar ideas can be applied to pollution monitoring. For example, consider assessing ground water contamination by agricultural pesticides. Any locations revealing a measurement that satisfies the inclusion criterion can be referenced to a GIS for determining things like proximity to farms and geologic formations in order to estimate probabilities of neighboring units being contaminated.

27.3.2 Using Covariate-Species Community Dissimilarity to Guide Sampling

Estimating species richness over a large geographic area based on a subsample of the area presents a special sampling challenge due to the non-additivity of species richness. The basic problem is one of estimating the number of classes in a multinomial population. Various sample-theoretic approaches have been reviewed by Bunge and Fitzpatrick (1993); however, these methods all apply to data that include abundance measurements for each species that is encountered. Data from sampling a large geographic area is most likely to be available as presence/absence recordings for each species encountered within each sample unit.

An approach to this problem has been suggested by Johnson and Patil (1995), which utilizes the species-area relationship. As discussed in Patil, Johnson and Grigoletto (1996), the number of species in an area is expected to grow with increasing area that is sampled, according to a power function, presented as $S = kA^z$, where S is the species richness and A is the area, while z and k are population specific parameters. Since $0 \leq z \leq 1$, this model implies that a "point of diminishing returns" is approached as the area A increases. The objective then becomes to sample enough of the area of interest in order to either encounter all of the species in the area or (more realistically) encounter the plateau region of the true species-area curve. This is desirable for a sample which is intended for estimating the true species richness, whether the estimate is simply the number of species encountered in the sample [Johnson and Patil (1995)], extrapolation of a fitted species-area curve [Patil, Johnson and Grigoletto (1996)] or a bounded monotonically increasing curve [Bunge and Fitzpatrick (1993)], or even when abundance measurements are available so that sample-theoretic methods may apply [Bunge and Fitzpatrick (1993) and Bunge, Fitzpatrick and Handley (1995)].

Fundamentally, we want to encounter as many of the species that occur in the area of interest with an affordably small sampled area. This objective therefore provides the criterion for comparing efficiency of various sampling plans.

Postulating that maximizing the habitat heterogeneity within a selected sample will maximize the number of species encountered, Johnson and Patil (1995) investigated covariate-directed sampling. Using a GIS database that provided synoptic coverage of both breeding bird and tree species in Pennsylvania, they retrospectively sampled the statewide breeding bird community, using incremental tree species richness as a covariate to direct sampling. While this approach performed somewhat better than random sampling, it still missed the northern tier of the state which contained some of the highest bird richness. Indeed, the northern tier of Pennsylvania is amongst the least disturbed regions of the state and therefore has more forest interior, but relatively low tree richness. Johnson, Patil and Rodriguez (1997) attempted to overcome this problem by investigating some other approaches to maximizing forest community dissimilarity as an investigator moves from one sample unit to the next. They further tested these approaches with other covariate-species data that were available.

The dataset was based on a tesselation of Pennsylvania in hexagons, each being 635 km^2. Covariate species and the response variable, breeding bird species, were listed within each hexagon. For more description, see Johnson and Patil (1995). The sampling frame of hexagons, along with a thematic presentation of breeding bird species richness, is also seen in Figure 27.9.

The general protocol for community dissimilarity-directed sampling using known covariate species follow the steps listed below.

1. Select a sample unit with the highest covariate species richness; then enumerate the species of interest that are encountered in this unit.

2. Select the next unit as the one revealing maximum community dissimilarity, compared to either the most recently sampled unit or to all accumulatively sampled units (ties may be broken by choosing at random).

3. Add the number of newly encountered species of interest to the overall species richness.

4. Repeat until an affordable sample size has been reached.

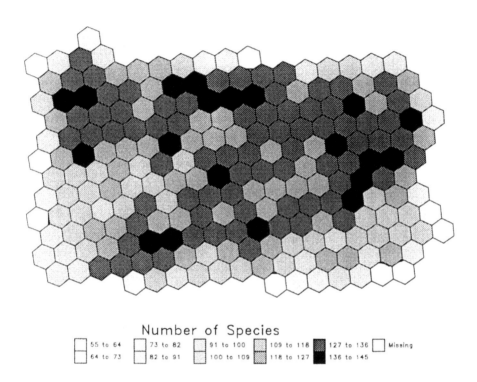

FIGURE 27.9 Bird richness in the hexagons

For example, Johnson, Patil and Rogriguez (1997) used the Jaccard dissimilarity index based on pairwise comparisons between the current and prospective sample units. For a covariate species that occur in both sample units, b that occur exclusively in one unit and c that occur exclusively in the other,

$$D = \frac{b+c}{a+b+c}.$$

They call D the Jaccard dissimilarity index because it is the compliment of the classic Jaccard similarity measure [Ludwig and Reynolds (1988, p. 131)], which is designed for presence/absence data. It is simply the compliment

G. P. PATIL and C. TAILLIE

of the intersection of two finite sets expressed as a proportion of the full union.

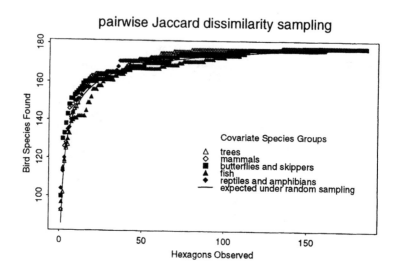

FIGURE 27.10 Species area curves from pairwise Jaccard sampling which selects hexagons in order of maximum covariate-species community dissimilarity based on pairwise comparison of the most recently sampled hexagon with each remaining unsampled one

Using a variety of covariate species that were available, these authors investigated dissimilarity directed sampling by retrospectively sampling the entire state of Pennsylvania for enumerating breeding bird species. The resulting state-wide species area curves are reported in Figure 27.10, along with the expected curve under random sampling, as computed by Equation 27.3.14.

$$E[S_n] = \sum_{i=1}^{s} \left[1 - \frac{\left(\begin{array}{c} N - A_i \\ n \end{array} \right)}{\left(\begin{array}{c} N \\ n \end{array} \right)} \right] \qquad (27.3.14)$$

The "pairwise Jaccard" protocol worked well for trees as well as reptiles and amphibians, but did not consistently outperform what is expected with random sampling for all covariate species that were tested.

This initial exploratory data analysis allows the use of a synoptic database for a whole state to be used to entertain some new ideas about sampling for the enumeration of a non-additive variable such as species richness. For other datasets we may see very different results; therefore, these protocols should continue to be tested with other datasets.

Acknowledgements   Prepared with partial support from the United States Environmental Protection Agency, Environmental Monitoring and Assessment Program, EMAP Design and Statistics Group under a Cooperative Agreement Number CR-821783. The contents have not been subjected to Agency review and therefore do not necessarily reflect the views of the Agency and no official endorsement should be inferred.

REFERENCES

Aragon, M. E. D., Patil, G. P., and Taillie, C. (1995). Use of best linear unbiased prediction for hot spot identification in two-way compositing, *Technical Report 94-0416*, Center for Statistical Ecology and Environmental Statistics, Pennsylvania State University, University Park, PA.

Bunge, J. and Fitzpatrick, M. (1993) Estimating the number of species: a review, *Journal of the American Statistical Association*, **88**, 364–373.

Bunge, J., Fitzpatrick, M. and Handley, J. (1995) Comparison of three estimators of the number of species, *Journal of Applied Statistics*, **22**, 45–59.

Burnham, K. P., Anderson, D. R., and Laake, J. L. (1980). Estimation of density for line transect sampling of biological populations, *Wildlife Monograph*, 72. Supplement to *Journal of Wildlife Management*, **44**.

Burrows, P. M. (1987). Improved estimation of pathogen transmission rates by group testing, *Phytopathology*, **77**, 335–343.

Cox, D. R. (1969). Some sampling problems in technology, In *New Developments in Survey Sampling*, (Eds., N. L. Johnson and J. R. Smith), pp. 506–527, John Wiley & Sons, New York.

Drummer, T. D. and McDonald, L. L. (1987). Size-bias in line transect sampling, *Biometrics*, **13**, 13–21.

Gore, S. D. and Patil, G. P. (1994). Identifying extremely large values using composite sample data, *Environmental and Ecological Statistics*, **1**, 227–245.

Gore, S. D., Patil, G. P., and Taillie (1995). Identifying the largest individual sample value from a two-way composite sample design, *Technical Report 95-0102*, Center for Statistical Ecology and Environmental Statistics, Pennsylvania State University, University Park, PA.

Gore, S. D., Patil, G. P., and Taillie (1996). Identification of the largest individual sample value using composite sample data and certain modifications of the sweep-out method, *Environmental and Ecological Statistics*, **3**, 219-234.

Johnson, G.D. and Patil, G.P. (1995) Estimating statewide species richness of breeding birds in Pennsylvania, *Coenosis*, **10**, 81–87.

Johnson, G. D., Patil, G. P., and Rodriguez, S. (1997). Using covariate-species community dissimilarity to guide sampling for estimating breeding bird species richness, (Invited plenary paper presented at the Conference on Assessment of Biodiversity for Improved Forest Planning, Monte Verita, Switzerland.) *Proceedings of the Conference*, Kluwer Academic Publishers, The Netherlands (To appear).

Keith, L. H. (1996). *Principles of Environmental Sampling*, American Chemical Society, Washington, DC.

Ludwig, J. A. and Reynolds, J. F. (1988). *Statistical Ecology: A Primer on Methods and Computing*, John Wiley & Sons, New York.

Nilsson, M. (1997). Estimation of forest variables using satellite image data and airborne lidar, Swedish University of Agricultural Sciences, Silvestria 17.

Patil, G.P., Grigoletto, M. and Johnson, G.D. (1996). Using covariate-directed sampling of EMAP hexagons to assess the statewide species richness of breeding birds in Pennsylvania, *International Journal of Ecology and Environmental Sciences*, **22**, 177–200.

Patil, G.P., Johnson, G.D. and Grigoletto, M. (1996). Covariate-directed sampling for assessing species richness, In *Spatial Accuracy Assessment in Natural Resources and Environmental Sciences* (Eds., H. T. Mowrer, R. L. Czaplewski, and R. H. Hamre), Second International

Symposium. May 21–23, 1996. General Technical Report RM-GTR-277. U. S. Department of Agriculture, Forest Service, Rocky Mountain Forest and Range Experiment Station, Fort Collins, CO. pp. 569-576.

Patil, G. P., Taillie, C., and Talwalker, S. (1993). Encounter sampling and modeling in ecological and environmental studies using weighted distribution methods, In *Statistics for the Environment* (Eds., V. Barnett and K. F. Turkman), pp. 45-69.

Patil, G. P., Taillie, C., and Wigley, R. L. (1979). Transect sampling methods and their application to the deep-sea red crab, In *Environmental Biomonitoring, Assessment, Prediction, and Management—Certain Case Studies and Related Quantitative Issues* (Eds., J. Cairns, G. P. Patil, and W. E. Waters), pp. 51-75, International Co-operative Publishing House, Fairland, Maryland.

Patil, G. P., Taillie, C., and Wigley, R. L. (1980). Transect sampling methods and their applications to the deep-sea red crab, In *Advanced Concepts in Ocean Measurements for Marine Biology*, (Eds., F. P. Diemer, F. J. Vernberg, and D. Z. Mirkes), pp. 493-505, University of South Carolina Press.

Ramsey, F. L. (1979). Parametric models for line transect surveys, *Biometrika*, **66**, 505-512.

Ramsey, F. L., Wildman, V., and Engbring, J. (1987). Covariate adjustments to effective area in variable area wildlife surveys, *Biometrics*, **43**, 1-11.

Stahl, G., Ringvall, A., and Lamas, T. (1997). Guided transect sampling, Swedish University of Agricultural Sciences, Silvestria 19.

Thompson, S. K. (1990). Adaptive cluster sampling, *Journal of the American Statistical Association*, **85**, 1050-1059.

Thompson, S. K. (1992). *Sampling*, John Wiley & Sons, New York.

Thompson, S. K., and Seber, G. A. F. (1995). *Adaptive Sampling*, John Wiley & Sons, New York.

Wigley, R. L., Theroux, R. B., and Murray, H. E. (1975). Deep-sea red crab, *Marine Fisheries Review*, **37**, 1-21.

CHAPTER 28

THE ANALYSIS OF MULTIPLE NEURAL SPIKE TRAINS

SATISH IYENGAR

University of Pittsburgh, Pittsburgh, PA

Abstract: A common experimental method in neuroscience involves the recording of the activity of a single neuron. However, studies of the functional connectivity of collections of neurons and their behavior require the simultaneous recording of their activity. Current technology permits such recordings of over a hundred neurons. These recordings yield large data sets that present challenging problems in their analysis and their interpretation in biological terms. In this paper, we describe various techniques for detecting functional connections between neurons and describing the nature of those connections.

Keywords and phrases: Coherency, cross-intensity, diffusion process, gravitational clustering, Markovian interval process, multivariate point process, snowflake plot

28.1 INTRODUCTION

The brain consists of many cells called neurons which are the fundamental units that process information. One way that information is transmitted between neurons is through changes in their electrical activity. Most notable is the nerve impulse or action potential, which is a large fluctuation in voltage that is propagated towards other cells. For many purposes, the action potential is usually short enough to be considered a spike and modeled as a point event; a typical recording will yield a sequence of spikes, or a spike train. A common experimental method in neuroscience that dates

507

back to its earliest days is based on the recording of the activity of a single neuron. It has been especially useful for studies of the effects of sensory inputs.

However, a substantial part of recent research in neuroscience involves the study of the cognitive or attentional state of subjects. Such studies require the simultaneous recording of many neurons. In addition, studies of the functional connectivity of collections of neurons require similar data. Current technology permits such recordings of over a hundred neurons. The resulting large data sets present challenging problems in their analysis and their interpretation in biological terms. In this paper, we describe various techniques for detecting functional connections between neurons and describing the nature of those connections.

In Section 28.2 we describe the requisite neurophysiology. In Section 28.3 we describe the various methods that have been proposed and used for the analysis of simultaneously recorded spike trains. In Section 28.4 we conclude with a discussion of avenues for further research.

28.2 PHYSIOLOGICAL BACKGROUND

We begin with a brief account of the physiology of the neuron. This is drawn from more extensive treatments intended for mathematics, signal processing, and statistics audiences are in the books by MacGregor (1987) and Tuckwell (1988, 1989); from neurophysiology texts by Levitan and Kaczmarek (1991) and Shepherd (1994); and the research literature, some of which is cited below.

Neurons differ widely according to their location and function, but they do share certain basic features. In short, a neuron gathers electrochemical signals at its soma or cell body and at its dendrites, which form a treelike projection emanating from the soma. The neuron integrates these signals. When the result of that integration exceeds a certain threshold, the neuron emits an impulse or action potential down another projection called the axon. The axon branches repeatedly, ending at swollen terminals called knobs. These knobs are adjacent to another neuron's cell body or its dendritic tree, to which they send a signal across a small gap known as the synapse. The cell that sends the signal is called presynaptic and the cell that receives it postsynaptic.

The soma contains a nucleus with its genetic materials, and organelles for the synthesis of proteins and their turnover, and to meet its energy needs. The entire cell is encased by a plasma membrane, a lipid double layer which prevents the mixing of the cell contents with that of the extracellular space. The membrane also has important electrical properties.

Typically, its interior is negatively charged and its exterior is positively charged because of different ionic concentrations inside and outside the neuron. The movement of ions across the membrane is largely determined by the voltage and ion concentration gradients, and the permeability of the membrane to the various ion species, such as calcium, chloride, potassium, and sodium. When there are no external signals, the membrane potential is at or near its resting state, which is in the range of -40 to -90 millivolts. Much of the cell's energy is expended in maintaining this potential difference by activating certain proteins at the membrane that transfer ions through channels across the membrane.

Signals received by the postsynaptic neuron move the membrane potential away from rest. The synaptic connections are either excitatory or inhibitory. The response of the postsynaptic cell to an excitatory input is called an excitatory postsynaptic potential or EPSP. This response is due to the release of a chemical neurotransmitter by the presynaptic cell's knob during the spike, which in turn increases the permeability of the postsynaptic membrane to certain ions, including sodium and potassium. The inhibitory postsynaptic potential or IPSP is similar, but with the permeability of potassium and chloride ions increasing while that of sodium ions remaining low. As their names indicate, EPSPs (IPSPs) move the membrane potential toward (away from) the firing threshold.

The dendritic tree and postsynaptic sites on the soma are characterized, at least to a first approximation, by passive electrical properties. That is, the inputs to them are summed linearly with the resting potential, both temporally and spatially. In contrast, a special location known as the trigger zone (usually at the base of the axon, or axon hillock) is characterized by active electrical properties, meaning that there are voltage-dependent ion channels. When the membrane potential at the trigger zone exceeds the firing threshold, a fast depolarization occurs. The axon's membrane is also active; hence the depolarization propagates down the axon and its branches without change. That signal is known as a nerve impulse, action potential, or spike. The spike for a neuron has a characteristic shape; however, the duration of the spike is sufficiently short to be regarded as a point event for many purposes. Finally, a sequence of spikes from a neuron is called a spike train.

Multiple spike train recordings can arise from a single microelectrode that is implanted in the extracellular space near several axons. Current technology also allows for extended recordings from multiple extracellular microelectrodes; hence, simultaneous spike trains from over a hundred units are now possible. Regardless of how the multiple spike trains are obtained, they must first be sorted before any meaningful analysis of the data are done. That is, each spike must be identified with a particular neuron that generated it. This step requires detailed measurements on the shape of the spike itself. The extracellular action potential waveforms differ in detail according to the neuron and the relative position of the measuring micro-electrode. These differences provide a basis for attributing the spikes to the neurons that generate them. There are two general classes of algorithms to do this: feature clustering and template matching. Feature clustering takes several properties of a waveform such as its height and duration and clusters them. Template matching classifies a spike on the basis of the overlap of its waveform with a set of previously determined template waveforms. For a more detailed description of these methods and a recent automatic sorting procedure, see Fee, Mitra, and Kleinfeld (1996).

28.3   METHODS FOR DETECTING FUNCTIONAL CONNECTIONS

In this section, we describe the following approaches that have been used to study multiple spike trains: moment methods, intensity function based methods, frequency domain methods, graphical methods, and parametric methods. The categories are somewhat arbitrary, for there is some overlap among them and most of these methods have their roots in the theory of point processes: see, for example, Daley and Vere-Jones (1988), Karr (1991), and Snyder and Miller (1991). One exception is the gravitational clustering method below which draws upon the cluster analysis and pattern recognition literature.

28.3.1   Moment Methods

We start with the case of two neurons A and B which have firing times $A_1 < A_2 < \cdots < A_{n_A}$ and $B_1 < B_2 < \cdots < B_{n_B}$ over some recording period $[0, T]$. Also, let $N_i(s, t)$ be the number of spikes from neuron i occurring in the interval $(s, t]$ for $i = A, B$. Assuming regularity, the joint spike intensity function is

$$\lambda(t, u) = \lim_{h, k \to 0} \frac{1}{hk} P\{N_A(t, t+h) > 0 \text{ and } N_B(u, u+k) > 0\}.$$

When the two neurons are independent, $\lambda(t, u) = \lambda_A(t)\lambda_B(u)$, where $\lambda_i(t)$ is the marginal intensity function for neuron i; and there is further simplification when the two processes are independent stationary Poisson processes. To estimate $\lambda(t, u)$, we generally need M replicates of the process over a time interval, say $[0, T]$. Subdivide the square $[0, T]^2$ into smaller squares of length h, let S_{ij} be the subsquare with center $(ih/2, jh/2)$, and estimate λ there by

$$\hat{\lambda}\left(\frac{ih}{2}, \frac{ij}{2}\right) = \frac{1}{Mh^2} \sum_{m=1}^{M} I\,(\text{data from trial } m \text{ in } S_{ij}),$$

where I is the indicator function. Hypotheses about the joint intensity can then be assessed using χ^2 goodness-of-fit tests.

Assuming (at least second order) stationarity, one measure of the dependence between A and B is the cross-intensity function

$$\lambda_{AB}(u) = \lim_{h,k\to 0} \frac{1}{hk} P\left\{N_A(u+t, u+t+h) > 0, N_B(t, t+k) > 0\right\}.$$

When the two neurons are independent, $\lambda_{AB}(u)$ is the product of the individual intensities $\lambda_A \lambda_B$, so that large values of $|\,\lambda_{AB}(u) - \lambda_A \lambda_B\,|$ indicates dependence. In the neuroscience literature $\lambda_{AB}(u)$ is also called the cross-correlation function even though it is not bounded. Brillinger (1976) proposed the following estimate of $\lambda_{AB}(u)$:

$$\hat{\lambda}_{AB}(u) = \sum_{i=1}^{n_B} \sum_{j=1}^{n_A} I\left\{A_j - B_i \in (u - h, u + h)\right\},$$

where I is an indicator function. Under suitable regularity, the estimates $\hat{\lambda}_{AB}(u)$ at a finite number of points $\{u_l : 1 \le l \le L\}$ are approximately independent Poisson random variables. Hence, the cross-intensity can be used to estimate $\lambda_{AB}(u)$ at those points. This estimate is also called the cross-correlation histogram or cross-correlogram: for examples of its use and interpretation, see Aersten and Gerstein (1985), Knox (1974), and Moore et al. (1970).

Taking a different viewpoint, Doss (1989) studied the function $K(t_1, t_2)$, which was proposed by Ripley (1976, 1977) in the context of spatial statistics:

$$
\begin{aligned}
K(t_1, t_2) &= \frac{1}{\lambda_A} E\left[N_A(t_1, t_2) \mid B \text{ fires at } t = 0\right] \\
&= \frac{1}{\lambda_B} E\left[N_B(-t_2, -t_1) \mid A \text{ fires at } t = 0\right].
\end{aligned}
$$

When the two neurons are independent, $K(t_1, t_2) = (t_2 - t_1)$, regardless of the marginal intensities λ_A and λ_B. Given data over a time interval $[0, T]$, Ripley proposed the estimate

$$\hat{K}(t_1, t_2) = \frac{T}{n_A n_B} \sum_{i=1}^{n_B} \sum_{j=1}^{n_A} I\left[A_j - B_i \in (t_1, t_2)\right].$$

Assuming certain regularity conditions, Doss showed that as $n_B \to \infty$, $\hat{K}(t_1, t_2)$ is consistent and asymptotically normal with a variance that can also be consistently estimated from the data.

Doss also noted that K and λ_{AB} are related thus:

$$K(t_1, t_2) = \frac{1}{\lambda_A \lambda_B} \int_{t_1}^{t_2} \lambda_{AB}(u)\, du.$$

Even though these two measures of dependence are mathematically equivalent, in practice they are quite different. The regularity conditions imposed by Doss and Brillinger do not imply each other. And while $\hat{\lambda}_{AB}$ is similar to a density estimate with variance on the order of $(hT)^{-1}$, the estimate $\hat{K}(t_1, t_2)$ is similar to that of a distribution function with variance on the order $1/n_B$. Hence, the two methods are complementary.

In practice, the cross-correlogram is the more common of the two. Typically, the analysis for more than two neurons proceeds in a pairwise fashion, although it is widely recognized that such an analysis can miss important three-way or higher order interactions. For one early attempt at dealing with this problem see Gerstein, Perkel, and Subramanian (1978). The next three sections describe several methods that attempt to deal directly with them.

28.3.2 Intensity Function Based Methods

A more elaborate but related approach models the intensity functions of the counting processes of several neurons as functions of each other. The approach is due to Cox and Lewis (1972), who used it in a reliability context. Applications to neuroscience are due to Borisyuk *et al.* (1985), Chornoboy, Schramm, and Karr (1988), and Utikal (1997a,b). In this section, we largely follow Utikal's treatment.

For $(p + 1)$ neurons consider the $(p + 1)$-variate counting process $\mathbf{N} = (\mathbf{N}, \mathbf{N}^{(1)}, \ldots, \mathbf{N}^{(p)})$ that monitors the spikes which occur at the random times $\{T_1, T_2, \ldots\}$, $\{T_1^{(1)}, T_2^{(1)}, \ldots\}$, ..., $\{T_1^{(p)}, T_2^{(p)}, \ldots\}$, respectively. As the notation indicates, this discussion assumes a target neuron with counting process N and trigger neurons with counting processes $N^{(i)}$ for $i = 1, \ldots, p$; however, in principle the neurons could be treated symmetrically. Under certain regularity conditions, the counting processes can be decomposed into the sum of an intensity process $\lambda(t)$ and a martingale $M(t)$:

$$
\begin{aligned}
dN(t) &= \lambda(t)dt + dM(t) \\
dN^{(i)}(t) &= \lambda^{(i)}(t)dt + dM^{(i)}(t), \text{ for } i = 1, \ldots, p.
\end{aligned}
$$

The models of neural interaction say that the intensity functions depend upon the time passed since the most recent firings. Specifically, define the backward recurrence times of $N^{(i)}$ at time t to be $(t - T^{(i)}_{N^{(i)}(t)})$. Then the models for $\mathbf{N}$ have the form above with

$$\lambda(t) = \alpha(t - T_{N(t)}, t - T^{(1)}_{N^{(1)}(t)}, \ldots, t - T^{(p)}_{N^{(p)}(t)})$$
$$\lambda^{(i)}(t) = \alpha^{(i)}(t - T_{N(t)}, t - T^{(1)}_{N^{(1)}(t)}, \ldots, t - T^{(p)}_{N^{(p)}(t)}),$$

where the functions α and α^i are assumed to be unknown; any further information about them about could be incorporated in specific instances.

Cox and Lewis (1972) called such a process a Markov interval process for the case $p = 1$. Chornoboy *et al.* (1988) used the following additive model for $\lambda(t)$:

$$\lambda(t) = \alpha\left(t - T_{N(t)}\right) + \sum_{i=1}^{p} \alpha_i \left(t - T^{(1)}_{N^{(1)}(t)}\right).$$

Utikal (1997) argues that the additive model is not appropriate for the rather sudden jumps in the intensity function that are due to inhibitory or excitatory inputs. He therefore proposes the multiplicative or proportional hazards model

$$\lambda(t) = \lambda_0(t - T_{N(t)})\alpha(t - T^{(1)}_{N^{(1)}(t)}, \ldots, t - T^{(p)}_{N^{(p)}(t)})$$
$$\lambda^{(i)}(t) = \alpha^{(i)}(t - T_{N(t)}, t - T^{(1)}_{N^{(1)}(t)}, \ldots, t - T^{(p)}_{N^{(p)}(t)}),$$

which allows the use of standard statistical packages to compute certain test statistics. At this point, specific parametric models may be imposed to test the different kinds of interactions. Utikal gives an explicit example involving inhibition and excitation, including a delay term for the transmission of a signal from a trigger neuron to the target neuron. Utikal proposes several test statistics to assess the parameters of this semiparametric model.

28.3.3   Frequency Domain Methods

Measures of dependence in the frequency domain are based on the Fourier transform of the processes. As in the spectral theory of stationary Gaussian processes, they provide an alternative to time domain methods, which have certain disadvantages. For instance, the cross-intensity function is analogous to a covariance function; thus, it is unbounded and is a dimensional quantity; Kirkwood (1979) discusses further limitations and other disadvantages of cross-correlation functions. On the other hand, regression techniques involve correlational measures which are bounded; in the Fourier domain, they involve the coherence and phase as measures of association and partial association: see Rosenberg *et al.* (1989). They also provide system identification techniques for studying synaptic interactions: see Brillinger (1975) and Brillinger, Bryant, and Segundo (1976). In this section, we describe some of these techniques.

Assume once again that we have stationary processes with counting measures $N_i(t)$ for neuron $i = A, B$. In that case, the joint spike density $\lambda_{AB}(t, u)$ is a function of the time difference $(u - t)$; we call that $\lambda_{AB}(u - t)$ also. In most cases of interest, the processes become independent as the lag increases:

$$\lim_{|u| \to \infty} \lambda_{AB}(u) = \lambda_A \lambda_B,$$

so that the processes are mixing. Bartlett (1963) defined the cross-spectrum between two point processes at frequency ω as the Fourier transform of the cross-covariance density $q_{AB}(u) = \lambda_{AB}(u) - \lambda_A \lambda_B$,

$$f_{AB}(\omega) = \frac{1}{2\pi} \int_{-\infty}^{\infty} q_{AB}(u) e^{-iu\omega} \, du.$$

The auto-spectrum for the single process N_A is

$$f_{AA}(\omega) = \frac{\lambda_A}{2\pi} + \frac{1}{2\pi} \int_{-\infty}^{\infty} q_{AA}(u) e^{-iu\omega} \, du,$$

where the first term is due to the singularity of the autocovariance at zero. Bartlett's cross-spectrum is related to the empirical Fourier transform of the process itself,

$$\hat{N}_i(\omega; T) = \int_0^T e^{-it\omega} \, dN_i(t)$$

thus:

$$f_{AB}(\omega) = \lim_{T \to \infty} \frac{1}{2\pi T} E\left[\hat{N}_A(\omega; T) \overline{\hat{N}_B(\omega; T)} \right] \text{ for } \omega \neq 0,$$

where the bar indicates the complex conjugate. This expression leads immediately to an estimate of the cross-spectrum.

The cross-spectrum leads to a measure of dependence called the coherence, which comes from the problem of predicting a linear functional of one process from that of the other. Specifically, consider the linear combinations $\int a(t) dN_A(t)$ and $\int b(t) dN_B(t)$, where $a(t)$ and $b(t)$ have Fourier transforms $\hat{a}(\omega)$ and $\hat{b}(\omega)$, respectively. Then the mean-squared error of prediction

$$E \mid \int a(t)dN_A(t) - \mu - \int b(t)dN_B(t) \mid^2$$

is minimized by

$$\hat{b}(\omega) = \hat{a}(\omega)\frac{f_{AB}(\omega)}{f_{AA}(\omega)}.$$

The minimum value achieved is

$$\int \mid \hat{a}(\omega) \mid^2 \left(1- \mid R_{AB}(\omega) \mid^2\right) f_{AA}(\omega)d\omega,$$

where

$$\mid R_{AB}(\omega) \mid^2 = \frac{\mid f_{AB}(\omega) \mid^2}{f_{AA}(\omega)f_{BB}(\omega)}$$

is the coherence of the process at frequency ω. In addition,

$$\mid R_{AB}(\omega) \mid^2 = \lim_{T\to\infty} \mid \text{corr} \left(\hat{N}_A(\omega; T), \hat{N}_B(\omega; T)\right) \mid^2 .$$

Thus, the coherence is between zero and one; when it is zero, one process is of no use in linearly predicting the other, and when it is one, one process gives a perfect linear prediction of the other. Next, the phase spectrum is

$$\theta_{AB}(\omega) = \arg R_{AB}(\omega) = \arg f_{AB}(\omega)$$

is useful in assessing timing relations between the processes. In the simple case that B is a lagged version of A with lag h ($A_j = B_j + h$), then $\theta_{AB}(\omega) = -h\omega$.

For estimating the coherence and phase from observations over a time period $[0, LT]$, Brillinger proposed breaking up the time interval into L disjoint sections of length T, computing the empirical Fourier transform of the counting process, $\hat{N}i(\omega; T, l)$ in each interval $l = 1, \ldots, L$, and estimating the cross-spectrum thus

$$\hat{f}_{AB}(\omega) = \frac{1}{2\pi LT}\sum_{l=1}^{L} \hat{N}_A(\omega; T, l)\overline{\hat{N}_B(\omega; T, l)},$$

to get the following estimate of the coherence:

$$\mid \hat{R}_{AB}(\omega) \mid^2 = \frac{\mid \hat{f}_{AB}(\omega) \mid^2}{\hat{f}_{AA}(\omega)\hat{f}_{BB}(\omega)}$$

The estimate of the phase spectrum follows similarly. Rosenberg *et al.* (1989) contains the details of inference using these measures (for example, using the transform $\tanh^{-1} \mid \hat{R}_{AB}(\omega) \mid^2$ instead of the coherence to improve the normal approximation, and setting critical values). They also show how to use the partial coherence $\mid \hat{R}_{AB|C}(\omega) \mid^2$, which measures the improvement by neuron B when neuron C is already included in the prediction of neuron A. See also Brillinger and Villa (1994) for applications of these methods to the study of neurons from *Aplysia Californica*, a sea hare that has been the subject of many neurophysiological investigations.

28.3.4   Graphical Methods

All of the approaches discussed in this paper use graphical displays. However, we have isolated the two methods in this section because they are graphical procedures that have not been studied systematically from a theoretical standpoint.

The first is called a snowflake plot. Perkel *et al.* (1975) developed it to study networks of three neurons. They postulated that the analysis of relationships between the neurons depends upon the differences between their firing times. If T_{ij} is the jth firing time of neuron i, then the cyclic differences $T_{1j} - T_{2k}$, $T_{2k} - T_{3l}$, and $T_{3l} - T_{1j}$ sum to zero. Hence, they can be displayed in two dimensions. Perkel used a triangular coordinate system which treats the three neuron pairs symmetrically. The axes make a 120° angle with each other. The vertical axis corresponds to an interval between the firings of the first two neurons; the signed distance along that axis is given by $T_{1j} - T_{2k}$ (upwards if the sign of this difference is positive). The other two axes correspond to the other two pairs of neurons, and are treated similarly. Thus, the point on the snowflake plot corresponding to T_{1j}, T_{2k}, T_{3l} is given by drawing the three perpendiculars to the axes at the given distances and plotting the point where they meet. The Cartesian coordinates of that intersection is $([(T_{3l} - T_{1j}) - (T_{2k} - T_{3l})]/\sqrt{3}, T_{2k} - T_{1j})$.

Next, let the span L be the time between the first firing and the last firing among the three neurons. The original proposal for snowflake plot suggested that all such points with span less than L (therefore all points) be plotted. While this is feasible for three neurons, it becomes less so with larger numbers of neurons, so it becomes necessary to define smaller spans and only plot points that fall within them. Of course, the choice of appropriate span depends on an assessment of the length of time during which neural interactions can be sustained: for example, can the firing of the first neuron now affect another neuron ten seconds hence ? Thus, any implementation should experiment with several choices of time window.

Perkel *et al.* bounded the plot in a hexagon, hence the name snowflake plot. They used extensive computer simulations to show how different networks' firing patterns appeared on the snowflake plot. For instance, they showed that if the first neuron caused the second one to fire at a lag h with high probability (and the third neuron were firing independently of the other two), then there would be a dense set of points along a line perpendicular to the first axis at a distance h from the origin. They showed the results of other networks that included inhibition, feedback loops, and coincidence detectors. They also noted that with just these synaptic types, there were over seven hundred distinct networks of three neurons. One limitation of this method is that there is not a one-to-one correspondence between a particular network and its resulting snowflake plot: several distinct networks can give similar plots. Therefore, they recommended its use for screening purposes; that is, to identify a class of possible networks from a plot rather than to expect to identify a particular network.

The extension of this graphical procedure to p neurons yielding correspondingly high dimensional data is straightforward in some respects but presents considerable challenges. The relation among the cyclic differences above reduces the dimension of the data by only one to $(p-1)$. Thus, the problem here is that of detecting structure in high-dimensional data that is noisy. Projection pursuit is one methodology with this specific aim: see, for instance, Huber (1985) for further discussion and references.

One aim of projection pursuit is to identify interesting low dimensional projections of data. Here, the word "interesting" is often construed to mean "far from Gaussian" [Huber (1985, p. 443)], using some measure of entropy. An important issue is the computational feasibility of this procedure; that is, one major concern is the number of projections needed to be assured that a thorough search has been done. But in the problem of multiple spike trains, several directions are distinguished from our prior substantive knowledge. That is, candidates for interesting one-dimensional projections would look for evidence of pairwise excitation or inhibition. Similarly, candidates for interesting two-dimensional projections could be based on our knowledge of the various kinds of higher order interactions. Thus, the candidates for interesting projections are at least partly well defined; in short, this type of data set is well suited for exploratory data analysis using an interactive projection pursuit algorithm. The implementation of this more elaborate graphical aid, along with a careful study of its theoretical properties remains to be done.

The other graphical approach that we describe here is related to gravitational clustering, which is used in the pattern recognition literature, and was applied to this problem by Gerstein, Perkel, and Dayhoff (1985). The idea behind this approach is the following. First, the activity of neurons is mapped into motions of particles in Euclidean space of appropriate dimension. The forces exerted on particles by others are due to "charges" that represent interactions between the corresponding neurons. [Even though a charge is used here, indicating electrical attraction and repulsion, we use the term gravitational clustering because of earlier work on pattern recognition, that used the term: see Wright (1977).] The particles are allowed to move about until they begin to cluster as a result of the charges, and the resulting aggregation of the particles into smaller subgroups then presumably represent the functionally related, or cooperative, subgroups that are sought. Of course, detailed knowledge about the nature of the connection is not available from this technique.

Gravitational clustering is an attractive approach to such problems, but a number of cautionary remarks are in order. First, there is no evidence yet that the gravitational clustering approach models a biophysical mechanism. It is a formal mathematical tool to try to detect functional connections among neurons. Gerstein *et al.* recognize this limitation, saying that they are not interested in "dynamic realism but are using these dynamics only as a means to allow appropriate particles to aggregate" (p. 884). It is important, therefore, to consider variations of this model to see if the substantive conclusions about the functional connections change much when the details of the gravitational algorithms are changed. Next, this clustering procedure has the same difficulty that any clustering algorithm has: namely, without some external validation, it is not possible to say with any confidence that the derived clusters are indeed correct, rather than being artifacts of the clustering process. A related point here is that there are no guidelines on deciding when to stop the aggregation process; when a certain connection between two neurons is strong, the algorithm should quickly join them into one cluster; more extensive simulation studies are needed to understand the quantitative relationship between the time to joining and the strength of connection. This problem is similar to the difficulty in hierarchical clustering algorithms, where the choice of the appropriate time to stop splitting the sub-clusters is not clear, especially when there is no external validation.

28.3.5   Parametric Methods

When there is extensive information about a particular network, many detailed questions are bound to arise. In that case, it is often fruitful to propose parametric models to try to answer such questions.

One example comes from Brillinger's (1988) analysis of data from a network of neurons from *Aplysia*. Let the spike trains of two neurons, say

A and B be modeled as the point processes $A(t)$ (the number of spikes by A up to time t) and $B(t)$, respectively. Suppose that the effect of A on B is of interest. If A fires at time τ, let the postsynaptic effect on B be $a(t-\tau)$; a is called the summation function. If $\gamma(t)$ is the time since B last fired, then the potential at the trigger zone for B is

$$V(t) = \int_0^{\gamma(t)} a(u)dA(t-u) = \sum a(t - \tau_j).$$

Brillinger allows the firing threshold V_f to be the random function of t, $V_f(t) = V_f + \epsilon(t)$, where $\epsilon(t)$ is zero mean noise used to represent the contributions of unmeasured neurons influencing B. In order to include B's own effect on its internal potential, spontaneous firings, and its refractoriness, Brillinger suggested adding a recovery term $\theta_1\gamma(t) + \theta_2\gamma(t)^2 + \theta_3\gamma(t)^3$ to the membrane potential. Then the probability that B fires at time t given its past is given by the probit model

$$p_t = \Phi \left(\sum_{u=0}^{\gamma(t)-1} a(t)A(t-u) + \theta_1\gamma(t) + \theta_2\gamma(t)^2 + \theta_3\gamma(t)^3 \right);$$

hence, the likelihood given the firing times of A and B is

$$\prod_t p_t^{B(t)}(1 - p_t)^{1-B(t)}.$$

Brillinger suggests that the parameters of this model be estimated by maximum likelihood, which can be done using standard GLIM routines.

For the *Aplysia* data, Brillinger showed that the estimates of θ_i were all significant, and also provided an estimate of the summation function. Thus, a detailed study of the neuron's recovery is possible from this analysis. This method in principle extends to large numbers of neurons; it can also be modified to include other effects such as adaptation. See also Brillinger and Villa (1994) for a recent example of the use of semiparametric modeling. The use of a linear summation function can be extended to include higher order nonlinear terms. Such terms better model presynaptic signals that are bursty, regular, or very fast. Such methods are the point process versions of the Volterra expansions for Gaussian processes. For an account of the theory, see Brillinger (1975, 1992); and for examples of its use, see Brillinger, Bryant, and Segundo (1976).

Another example of a parametric model is motivated by Gerstein and Mandelbrot (1964), who proposed the first random walk model for the firing of a single neuron. Their model can be derived using the following argument. Let $\{N_i^E(t) : i = 1, \ldots, n_E\}$ and $\{N_j^I(t) : j = 1, \ldots, n_I\}$ represent the excitatory and inhibitory input Poisson processes, respectively. Suppose that N_i^E and N_j^I have rates λ_i^E and λ_j^I, and magnitudes α_i^E and α_j^I, respectively. Then the stochastic differential equation for $V(t)$ is

$$dV = \sum_{i=1}^{n_E} \alpha_i^E dN_i^E - \sum_{j=1}^{n_I} \alpha_j^I dN_j^I.$$

In the limit that the intensities tend to infinity and the magnitudes of the inputs tend to zero appropriately, the Poisson inputs are well approximated by white noise dW. Thus, $dV = \mu dt + \sigma dW$, where μ is the mean input per unit time, and σ is the standard deviation of that noisy input. Thus, $V(t) = V_0 + \mu t + \sigma W(t)$ is a Brownian motion starting at V_0 with drift μ and diffusion or variance σ^2. For a constant firing threshold V_f, the random time to firing is $T = \inf\{t > 0 : V(t) = V_f\}$. And when the net input is excitatory, $\mu > 0$ and T is a proper random variable with the inverse Gaussian density

$$f(t; d, \nu) = \frac{d}{\sqrt{2\pi t^3}} \exp\left[-\frac{(d - \nu t)^2}{2t}\right],$$

where $\nu = \mu/\sigma$ is the standardized drift and $d = (V_f - V_0)/\sigma$ is the standardized distance from the initial and firing potentials. For an account of other random walk or diffusion models of single neuron activity, see Ricciardi (1994).

Iyengar (1985) extended this work to a "Zeitgeber" model, in which an external stimulus (the "time-giver") drives two other neurons. This model assumes that the noise affecting each neuron is the sum of two components: a shared component and an independent specific component. The other assumptions for the model are similar to those of Gerstein and Mandelbrot above, so that the parameters of this model have similar physical interpretations. These assumptions lead to a correlated two-dimensional Brownian motion, $(X_1(t), X_2(t))$, where the correlation is related to the noise variance ratios thus. Let the noise variances be σ^2 for the shared noise and σ_i^2 for the noise particular to the ith neuron. Then,

$$\text{Corr}(X_1(t), X_2(t)) = \left[\left(1 + \frac{\sigma_1^2}{\sigma^2}\right)\left(1 + \frac{\sigma_2^2}{\sigma^2}\right)\right]^{-1/2}.$$

Here, the firing time distributions are given by the joint distribution of (τ_1, τ_2), where $\tau_i = \inf\{t : X_i(t) = a_i\}$ and a_i is the distance between the resting potential and the firing threshold for the ith neuron. The derivation of the joint distribution of (τ_1, τ_2) involves the solution of the heat equation inside a wedge in $\Re^2$, with specified initial and boundary conditions, and the use of the strong Markov property. As an aside, recall that the inverse Gaussian distribution has statistical properties that mirror those of the Gaussian distribution, and admit easy estimation and testing procedures. The statistical properties of this model for a bivariate inverse Gaussian have not been studied or compared with other proposals that are in the literature. Furthermore, extensions of this model to more neurons leading to higher dimensional analogs of this model also have not been done.

28.4 DISCUSSION

Some of the methods above, such as the use of the cross-intensity function, are rather generic in that they are used for point processes arising in many areas, not just neuroscience. Other methods, such as Brillinger's use of a recovery function, are tailored to the biophysical details of the neuron. Still other methods, such as Utikal's interacting counting process approach are flexible enough to model both neurophysiological phenomena and other phenomena (in this case, reliability of components). All of these methods have a role in neurophysiology. The generic methods play an important role during the initial phase of an investigation when little is known about the network. As more information is gathered and initial hypotheses are refined, parametric approaches come to the fore to assess them more precisely.

The routine gathering of such data is still relatively recent; therefore, there are many outstanding problems. For example, the graphical methods that have been proposed have both proved themselves to be useful screening devices, but they do not have adequate theoretical support. That is, a careful study of the properties of these methods to find out when they work well and what their limitations are, along with computational procedures to store and process the huge data sets have yet to be done.

Acknowledgment This paper was written while the author was visiting the Centre for Mathematics and its Applications at the Australian National University.

REFERENCES

Aersten, A. and Gerstein, G. L. (1985). Evaluation of neuronal connectivity: sensitivity of cross-correlation, *Brain Research*, **340**, 341–345.

Bartlett, M. S. (1963). The spectral analysis of point processes, *Journal of the Royal Statistical Society, Series B*, **25**, 264–280.

Borisyuk, G. N., Borisyuk, R. N., Kirillov, A. B., Kovalenko, E. I. and Kryukov, V. I. (1985). A new statistical method for identifying interconnections between neuronal network elements, *Biological Cybernetics*, **52**, 301–306.

Brillinger, D. R. (1975). The identification of point process systems, *Annals of Probability*, **3**, 909–929.

Brillinger, D. R. (1976). Estimation of the second order intensities of a bivariate stationary process, *Journal of the Royal Statistical Society, Series B*, **38**, 60–66.

Brillinger, D. R. (1988). Maximum likelihood analysis of spike trains of interacting nerve cells, *Biological Cybernetics*, **59**, 189–200.

Brillinger, D. R. (1992). Nerve cell spike train data analysis: a progression of technique, *Journal of the American Statistical Association*, **87**, 260–271.

Brillinger, D. R., Bryant, H. L. and Segundo, J. P. (1976). Identification of synaptic interactions, *Biological Cybernetics*, **22**, 213–228.

Brillinger, D. R. and Villa, A. E. P. (1994) Examples of the investigation of neural information processing by point process analysis, In *Advanced Methods of Physiological System Modeling*, Volume 3 (Ed., V. Z. Marmarelis), pp. 111–127, Plenum Press, New York.

Chornoboy, E. S., Schramm, L. P. and Karr, A. F. (1988). Maximum likelihood identification of neural point process systems, *Biological Cybernetics*, **59**, 265–275.

Cox, D. R. and Lewis, P. A. W. (1972). Multivariate point processes, In *Proceedings of the Sixth Berkeley Symposium on Mathematical Statistics and Probability*, **3**, 401–448.

Daley, D. J. and Vere-Jones, D. (1988). *An Introduction to the Theory of Point Processes*, Springer–Verlag, New York.

Doss, H. (1989). On estimating the dependence between two point processes, *Annals of Statistics*, **17**, 749–763.

Fee, M., Mitra, P. and Kleinfeld, D. (1996). Automatic sorting of multiple unit neuronal signals in the presence of anisotropic and non-Gaussian variability, *Journal of Neuroscience Methods*, **69**, 175–188.

Gerstein, G. L. and Mandelbrot, B. (1964). Random walk models for the spike activity of a single neuron, *Biophysical Journal*, **4**, 41–68.

Gerstein, G. L., Perkel, D. H. and Dayhoff, J. E. (1985). Cooperative firing activity in simultaneously recorded populations of neurons: detection and measurement, *Journal of Neuroscience*, **5**, 881–889.

Gerstein, G. L., Perkel, D. and Subramanian, K. (1978). Identification of functionally related neural assemblies, *Brain Research*, **140**, 43–62.

Huber, P. (1985). Projection pursuit, *Annals of Statistics*, **13**, 435–525.

Iyengar, S. (1985). Hitting lines with two-dimensional Brownian motion, *SIAM Journal of Applied Mathematics*, **43**, 583–589.

Karr, A. F. (1991). *Point Processes and their Statistical Inference*, Marcel Dekker, New York.

Kirkwood, P. A. (1979). On the use and interpretation of cross-correlation measurements in the mammalian central nervous system, *Journal of Neuroscience Methods*, **21**, 201–224.

Knox, C. K. (1974). Cross-correlation functions for a neuronal model, *Biophysical Journal*, **14**, 567–582.

Levitan, I. B. and Kaczmarek, L. K. (1991). *The Neuron: Cell and Molecular Biology*, Oxford University Press, New York.

MacGregor, R. J. (1987). *Neural and Brain Modeling*, Academic Press, New York.

Moore, G. P., Segundo, J. P., Perkel, D. H. and Levitan, H. (1970). Statistical signs of synaptic interaction in neurons, *Biophysical Journal*, **10**, 876–900.

Perkel, D. H., Gerstein, G. L., Smith, M. and Tatton, W. (1975). Nerve Impulse Patterns: A Quantitative Display Technique for Three Neurons, *Brain Research*, **100**, 271–295.

Ricciardi, L. M. (1994). Diffusion models of single neurons, In *Neural Modeling and Neural Networks* (Ed., F. Ventriglia), pp. 129–162, Pergamon Press, Oxford.

Ripley, B. D. (1976). The second order analysis of stationary point processes, *Journal of Applied Probability*, **13**, 255–266.

Ripley, B. D. (1977). Modelling spatial patterns (with discussion), *Journal of the Royal Statistical Society, Series B*, **39**, 172–212.

Rosenberg, J. R., Amjad, A. M., Breeze, P., Brillinger, D. R. and Halliday, D. M. (1989). The Fourier approach to the identification of functional coupling between neuronal spike trains, *Progress in Biophysics and Molecular Biology*, **53**, 1–31.

Shepherd, G. M. (1994). *Neurobiology*, Oxford University Press, New York.

Snyder, D. L. and Miller, M. I. (1991). *Random Point Processes in Time and Space*, Second edition, Springer–Verlag, New York.

Tuckwell, H. C. (1988). *Introduction to Theoretical Neurobiology*, Volumes 1 and 2, Cambridge University Press, New York.

Tuckwell, H. C. (1989). *Stochastic Processes in the Neurosciences*, SIAM, Philadelphia.

Utikal, K. (1997a). Nonparametric inference for Markovian interval processes, *Stochastic Processes and their Applications*, **67**, 1–23.

Utikal, K. (1997b). A new method for detecting neural interconnectivity, *Biological Cybernetics*, **76**, 459–470.

Wright, W. (1977). Gravitational clustering, *Pattern Recognition*, **9**, 151–166.

CHAPTER 29

SOME STATISTICAL ISSUES INVOLVING MULTIGENERATION CYTONUCLEAR DATA

SUSMITA DATTA

Georgia State University, Atlanta, GA

Abstract: In recent years, there has been increasing attention on studying the association or interaction between a nuclear gene or genotype and maternally inherited cytoplasmic components such as mitochondria. Questions have arisen concerning whether or not these associations can be explained without invoking neutral selection. A novel application of the dynamics of allelic or genotypic linkage disequilibria is in the construction of statistical tests for testing the null hypothesis of a neutral model. In this paper, such tests are considered by comparing the paths of a collection of disequilibria measures with their (conditional given the previous generation) expected trajectory. Since these tests are based on the complex nature of the interaction between genes or genotypes, they are expected to be more capable of capturing differences in behavior from that expected under the neutral theory than tests based on a single gene (haplotype) frequency. Also such tests are applicable even if the population is *not in existence for a long time* (so that an assumption of equilibrium is not reasonable) provided it is *observed over a few generations* which is possible in a controlled experimental setting.

As an alternative to a neutral or random drift model, researchers have considered different types of selection models involving fertility or viability selection. In such models it is of fundamental interest to estimate the selection parameters. In this paper we formulate an estimation scheme using multi-generation genotypic counts for a multiplicative fertility selection. Once the selection coefficients are estimated they can be used to test any statistical hypothesis of interest that can be formulated in terms of the selection coefficients. In particular, a test for the neutrality or no selection can be constructed based on such estimates.

Keywords and phrases: Nuclear gene, mitochondria, neutral selection, genotypic linkage disequilibria, viability selection, multiplicative fertility selection

29.1　INTRODUCTION

One of the simplest and most powerful hypotheses in population genetics is that most DNA markers are neutral [Kimura (1983)]. Under the neutral theory, such as a random drift model or drift with mutation or migration, precise predictions about the dynamics and equilibrium behaviors of individual DNA markers can be made. The bulk of the neutral theory concerns prediction about properties of individual genes like heterozygosity and rates of evolution [Nei (1987) and Li and Graur (1991)]. In selectionist theories, on the other hand, there is an emphasis on the epistatic interactions of genes and their resultant associations in genomes and populations [Wright (1969), Dobzhansky (1970) and Lewontin (1974)]. With the advent of new genome technologies it has now become feasible to test predictions about the dynamics and equilibrium behavior of many markers simultaneously to distinguish between neutral and selectionist hypotheses. In particular, studies (theoretical or simulation based) of the dynamics of various disequilibria measures are becoming increasingly important. A genotypic/allelic disequilibrium measures association between two (or more) genotypes/alleles in a population. Such measures provide new inferential tools in analyzing both a single cytonuclear or nuclear system as well as hybrid zone data.

Statistical properties such as the dynamics of the first two moments of a gametic linkage disequilibria in a nuclear system under random union of gametes (RUG) have been studied by Robertson (1952), Hill and Robertson (1966), Hill and Robertson (1968), Hill and Weir (1988), among others. Some form of an overall disequilibria in a multi-loci nuclear system has been considered in Hill (1975), Takahata (1982) etc. Effect of selection can be particularly important in such systems since it can produce strong correlation among the distribution of alleles or genotypes at different loci even without strong epistatic interactions [Franklin and Lewontin (1970)]. Hedrick and Thomson (1986) used an overall disequilibria as a test statistic for neutrality under the assumption that the system has reached an equilibrium.

The basic setup of the present paper differs from the above mentioned papers in an important way in that a cytonuclear setup is considered; for such setup it can be argued [Fu and Arnold (1992)] that a random union of zygote (RUZ) is more appropriate [Watterson (1972)]. However, the many of the methods in this paper can be useful in a RUG setup also with appropriate modification.

TABLE 29.1 Frequencies in a cytonuclear system

	Nuclear Genotype			
Cytoplasm	AA	Aa	aa	Total
M	p_1	p_2	p_3	q
m	p_4	p_5	p_6	$1-q$
Total	u	v	w	1

We consider a nuclear locus with possible alleles A and a and a cytoplasmic locus with alleles M and m. The frequencies of six cytonuclear genotypes are denoted $p_1, \ldots, p_6$ as in Table 29.1.

Asmussen *et al.* (1987) introduced three disequilibria measures D_1, D_2 and D_3 to measure association between a cytoplasmic gene and a nuclear genotype as given in the following table.

TABLE 29.2 Genotypic disequilibria in a cytonuclear system

	Nuclear genotype			Total
Cytoplasm	AA	Aa	aa	Total
M	$uq + D_1$	$vq + D_2$	$wq + D_3$	q
m	$u(1-q) - D_1$	$v(1-q) - D_2$	$w(1-q) - D_3$	$1-q$
Total	u	v	w	1

Only two of these three disequilibria can change independently since they satisfy $\sum_{i=1}^{3} D_i = 0$.

29.2 NEUTRALITY OR SELECTION?

One of the central tenets of evolutionary genetics is that most DNA markers distinguishing individuals and species are neutral and have little effect on individual fitness [Kimura (1983)]. Under this hypothesis the action of genetic drift or genetic drift in combination with mutation or migration can be used to describe the evolution of most DNA markers. If these "neutrality hypotheses" are correct, then one of the consequences (predictions) is the existence of a "molecular clock" in which differences between two DNA sequences descended from a common ancestral DNA sequence accumulate linearity with time on average. As a consequence, DNA sequence difference between extant species have been used to reconstruct the history of life.

Evidence supporting the neutral theory of molecular evolution has usually come from comparisons of empirical data to certain stationary equilibrium properties of neutral models. Such asymptotic properties of neutral populations include expected heterozygosity, variance in heterozygosity, vanishing allelic disequilibria etc. are seen to be in reasonable agreement with predictions from the neutral model [Fuerst, Chakraborty and Nei (1977), Chakraborty, Fuerst and Nei (1980), Dykhuizen and Hartl (1980) and Asmussen *et al.* (1989) etc.]

However such approaches of agreements based on the equilibrium behavior under the neutral theory have been questioned in recent years. For example, Gillespie (1979) has shown that the infinite allele neutral model and his model of selection in a random environment have the same stationary distribution and therefore the agreement between empirical observations and that predicted by the infinite allelic model noted by Fuerst, Chakraborty and Nei (1977) can be used with equal strength to support Gillespie's model of neutral selection. As another example, consider the tests by Watterson (1977) based on the sampling theory of Ewens' (1972). Rothman and Templeton (1980) showed that under some departure from the model assumption the underlying neutral model can yield frequency spectra and homozygosity similar to those expected from heterosis.

Besides the above theoretical developments, a number of researchers have recently designed experiments to test the neutrality of mtDNA markers [Clark and Lyckegaard (1988), MacRae and Anderson (1988), Fos *et al.* 1990, Nigro and Prout (1990), Pollak (1991), Arnason (1991), Kambhampati *et al.* (1992), Scribner and Avise (1994a,b), Hutter and Rand (1995), etc.] Singh and Hale (1990) suggested that the apparent "non-neutral" behavior may be also caused by mating preference and that any attempt to understand the role of selection on mtDNA variants one should first begin with simpler conspecific variants rather than with interspecific variants; however see McRae and Anderson (1990), Jenkins *et al.* (1996). Multi-locus empirical comparisons have been undertaken by Karl and Avise (1992) [also see McDonald (1996)], Berry and Kreitman (1993), McDonald (1994).

To remedy the criticisms of using a single generation test (assuming that the population has reached an equilibrium), in recent years, several tests of the neutral theory have been devised which use observations of allelic or gametic frequencies over several generations. Tests utilizing temporal data include Fisher and Ford (1947), Lewontin and Krakauer (1973), Schaffer, Yardley and Anderson (1977), Wilson (1980), Watterson (1982), Wilson et al. (1982) Williams *et al.* (1990). These papers considered tests based on the expected dynamics of allelic frequencies at a given locus. A novel approach based on the interactions of a nuclear genotype and a cytoplasmic marker in testing the neutrality hypothesis has recently been presented in Datta and Arnold (1996). This approach is based on the comparison of the observed cytonuclear disequilibria with those expected under a neutral model of random drift. Disequilibria between an mtDNA and a nuclear marker may prove to be more sensitive to departures from a neutrality hypothesis than the statistics derived from a single marker and a test based on an extensive description for the dynamics of cytonuclear disequilibria would be more powerful than a test based only on the equilibrium behavior.

The test of Datta and Arnold (1996) was applied to a vertebrate cage experiment involving two species of mosquito fish [Scribner and Avise (1994a,b)]. In this experiment an artificial hybrid zone composed of two competing species of mosquito fish that interbreed was established. Frequencies of cytonuclear genotypes and associated cytonuclear disequilibria were monitored over time and compared with their expectations under random drift.

In general, for experiments of the type described above there are two potential sources of variation in cytonuclear frequencies, namely, *statistical sampling variation* and *genetic sampling variation* [Weir (1990)]. Statistical sampling variation arises from sampling individuals from a population and using the estimated cytonuclear disequilibria from the sample. Genetic sampling variation arises from genetic drift, the sampling of gametes from a finite breeding pool of individual in nature to constitute the next generation. In the experiment of Scribner and Avise (1994a,b) in every generation, all individuals from an entire population were sampled and as a consequence, statistical sampling variation was eliminated. However more general cage experiment, as in Kiparsky (unpublished), would result in both sources of variation. In Datta *et al.* (1996), test statistics based on cytonuclear disequilibria were constructed which can take both sources of variation into account. The formulation in this case is considerably more involved. The sampling schemes are described below.

29.2.1 Sampling Schemes for Multi-Generation Data

We consider a general sampling scheme for the experiments for collecting relative frequencies of genotypic counts over a number of generations.

Sometimes in kitty-pool experiments [Scribner and Avise (1994a,b)], several independent populations are allowed to propagate in time and at each time point one of these populations is completely sacrificed to obtain the genotypic counts.

Usually, a complete census of the population is difficult and expensive.

We therefore consider a broad sampling scheme for the experiment in which at each generation only a portion (a random sample) of the population is sacrificed to obtain the counts. Also a common population is allowed to propagate in time.

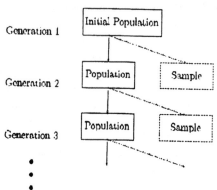

FIGURE 29.1 Format of the experiment considered in Datta *et al.* (1996)

More specifically, at each generation a portion of the adult population is collected by simple random sampling and sent for analysis after they form the next generation eggs by random mating. The eggs are then collected and placed in a cage to form the next generation. Thus the counts and hence the disequilibria measures in this case are only based on the sample and not on the population and therefore subject to the additional source of sampling variation. The scheme is described in Figure 29.1.

29.2.2 An Omnibus Test

With reference to the above experiments calculate the genotypic disequilibria $\widehat{D}_i(t)$, $i = 1, 2$, at different time points t where a hat signifies that these may be based only on the sample counts (in the second sampling scheme). Note that these can be thought of as estimates of the expected genotypic disequilibria $\mu_i(t) = E\, D_i(t)$, where this expected value will be calculated under a neutral model (e.g., random drift). Datta and Arnold (1996) and Datta *et al.* (1996) considered a chi-squared goodness of fit type test statistic of the form

$$T = (\widehat{D} - \mu)^T \widehat{\Sigma}^{-1} (\widehat{D} - \mu), \qquad (29.2.1)$$

under the sampling schemes mentioned above. Here $\widehat{D}$ is the vector of $\widehat{D}_1(t)$, $\widehat{D}_2(t)$ for all the generations under study and $\widehat{\Sigma}$ is its estimated variance-covariance matrix.

It was shown in the above mentioned papers that asymptotically, as both genetic sample size (the effective population size) as well as the statistical sample size increase, $\hat{D}$ is asymptotically bi-variate normal and therefore T has an approximate chi-square distribution under the null hypothesis of random drift.

29.2.3   Application to Gambusia Data

In Scribner and Avise (1994a,b), several lines of evidence were forwarded as demonstrations that genetic changes within replicate experimental *Gambusia* hybrid zones did not result from random drift. Consistency in the pattern and direction of change in mitochondrial and nuclear allele frequencies, consistency in reduction in population genetic diversity owing to loss of *G. affins* allele and lack of appreciable levels of inter-replicate genetic variance implicated the importance of non-random evolutionary forces. They compared the observed cytonuclear gene (and genotype) frequencies from their expectations under random drift; however no explicit overall test for cytonuclear drift was available at that time. Such a test is necessary to correctly control the overall type 1 error probability.

TABLE 29.3 Results for the Gambusia data [from Scribner *et al.* (1998)]

Locus	Value of test statistic T	P-value
Pep-A	21.18	0.048
Ada	24.90	0.015
Mdh-1	21.67	0.041
Aat-1	20.90	0.052
Ah-1	21.10	0.049

In Datta and Arnold (1996) the above test (1) was applied to the Scribner and Avise (1994) data at one locus; more complete results were given in Scribner *et al.* (1998). The RUZ model with random drift alone was rejected (at 5% level) for four of the five nuclear loci. Thus the simple genetic drift model does not explain the temporal changes in composite cytonuclear frequencies. Frequencies of parental *G. holbrooki* mitochondrial alleles and nuclear genotypes exceeded expected values during most time periods implying some selective advantage of offspring produced by *G. holbrooki* females.

29.2.4   Application to Drosophila Melanogaster Data

The above test procedure was applied to data from a cage experiment on
the fruit fly *Drosophila Melanogaster* conducted by M. Kiparsky. A PCR/4-
cutter method was developed to simultaneously score individual flies for
mtDNA haplotype and the three diploid genotypes at each of two second
chromosome loci. The significance of the above test in the context of this
data is that the decay of cytonuclear disequilibria rejects a random drift
model. See Datta *et al.* (1996) for the details. This non-neutral decay
of disequilibria was believed to be due to a preferential transmission in
heterozygotes of the nuclear chromosome that were from the same strain
as the mtDNA haplotype. The results for the DPP locus (with generation
2 sample missing) are given below.

TABLE 29.4   Results for the Drosophila data [from Datta *et al.* (1996)]

Gen.	Gen. Size	Sample Size	$\widehat{D}_1$	$\widehat{D}_2$	$\widehat{p}(0)$	$\widehat{q}(0)$	$\widehat{D}(0)$	T
0		85			0.51	0.22	0.03	
1	1826	69	-0.0256	0.0078				
2	694							
3	1621	77	0.0587	-0.0712				18.5*

T: Test statistic. * Significant at 5%

29.2.5   Tests Against a Specific Selection Model

The above test is omnibus since it works against any alternative to the
neutrality hypothesis. The above applications suggest that the researcher
may have an alternative model to explain the data. In such a case it would
be desirable to see if the data shows a significant inclination towards the
alternative model. However the omnibus test does not incorporate the
specific model structure in the alternative hypothesis. In fact, it is the
lack of the use of the alternative hypothesis in the formulation of the test
statistic that may make the power of the test not very good (it is fairly
typical of omnibus test). In this subsection, we would like to address the
question 'how to construct such a test statistic for this problem which will
perform better than the omnibus tests of Datta and Arnold (1996) and
Datta *et al.* (1996) in detecting a specific type of interesting selectionist
alternatives?'

TABLE 29.5   Selection coefficients in a cytonuclear system

	AA	Aa	aa
M	w_1	w_2	w_3
m	w_4	w_5	w_6

Although a number of interesting alternative selectionist hypotheses can be considered the following general approach may be attempted in all case. For the discussion here, we consider a viability selection mechanism.. Let us consider selection followed by random union of zygote (RUZ) in a cytonuclear systems. Let w_i, $i = 1, \ldots, 6$ denote the selection coefficients for the six genotypes (Table 29.5).

We assume that w_is measure the relative fitness of the genotypes for survival from birth to adulthood (viable selection coefficient). Thus prior to random mating the relative frequencies after selection of the six genotypes are $p'_i = w_i p_i / \overline{w}$, $i = 1, \ldots, 6$, with $\overline{w} = \sum w_i p_i$. Since it is only possible to measure the relative fitness in this formulation there are at the most five independent parameters in w_i and the usual convention (parametrization) is to let $w^* = \max(w_i) = 1$. In some cases the number of parameters is further restricted by assuming a parametric model of the form $w_i = w_i(\theta)$, where θ is a parameter of dimension four or less. The advantage of this is of course a gain in efficiency in the inference procedure provided the model is correctly specified. Below we consider three such models.

Examples of selection schemes

Consider a cytonuclear system in which selection is determined by phenotypes. Assume further that the mitochondrial allele M is at an induced selective advantage over m. This gives rise to a selection scheme as follows:

	AA	Aa	aa
M	1	1	$1 - s$
m	1	$1 - sh$	$1 - s$

A similar model in a nuclear system was considered in Ewens (1970, p. 27). In this model we may let $\theta = (s, h) \in [0, 1]^2$.

Next consider a situation in which there is no dominance in fitness for each mtDNA locus and that selection acts in opposite direction so that we have the following selection coefficients:

	AA	Aa	aa
M	1	$1 - \frac{1}{2}s_M$	$1 - s_M$
m	$1 - s_m$	$1 - \frac{1}{2}s_m$	1

When dominance is introduced, the selection scheme looks like

	AA	Aa	aa
M	1	$1 - h_M s_M$	$1 - s_M$
m	$1 - s_m$	$1 - h_m s_m$	1

In this model $\theta = (h_M, h_m, s_M, s_m)$ is a four dimensional parameter. See Ewens (1970, p. 38) for such a scheme.

As a last example of a parametric viability selection model consider hitchhiking of a neutral mitochondrial marker. Under this model, selection operates only on the nuclear locus. The following selection model can be used in this situation

	AA	Aa	aa
M	$1 - s_1$	1	$1 - s_2$
m	$1 - s_1$	1	$1 - s_2$

Such models have been studied by many authors such as Thomson (1977), Asmussen and Clegg (1981), Clark (1984), Asmussen (1986). This model is parametrized by a two dimensional $\theta = (s_1, s_2)$.

Construction of a test statistic

Consider viable selection followed by random union of zygotes. It can be shown that the expected cytonuclear disequilibria in the $(t+1)$th generation given the genotypic counts $X(t)$ in the tth generation are given by

$$
\begin{aligned}
E(D_1(t+1)|\boldsymbol{X}(t)) &= \frac{(N_{t+1} - 1)}{N_{t+1}} D(t, \theta) p(t, \theta) \\
&= f_1(\theta, p(t), N_{t+1}), \text{ say} \qquad (29.2.2)
\end{aligned}
$$

$$
\begin{aligned}
E(D_2(t+1)|\boldsymbol{X}(t)) &= \frac{(N_{t+1} - 1)}{N_{t+1}} D(t, \theta)(1 - 2p(t, \theta)) \\
&= f_2(\theta, p(t), N_{t+1}), \text{ say} \qquad (29.2.3)
\end{aligned}
$$

where N_{t+1} is the population size at generation $t + 1$, $w_i = w_i(\theta)$ are the selection coefficients,

$$
p(t, \theta) = \sum_{i=1}^{6} w_i(\theta) p_i / \overline{w},
$$

and

$$
\begin{aligned}
D(t, \theta) = \{ & w_1(\theta)p_1 + 0.5 w_2(\theta)p_2 \\
& - p(t, \theta)(w_1(\theta)p_1 + 0.5 w_2(\theta)p_2 + w_4(\theta)p_4 + 0.5 w_5(\theta)p_5) \} / \overline{w}.
\end{aligned}
$$

Owing to formulas (29.2.2) and (29.2.3) one can form an objective function measuring the departure of the disequilibria from their one-step expected value under selection given by

$$Q(\theta) = \sum_{t=1}^{m-1} \sum_{i=1}^{2} \left\{ D_i(t+1) - f_i(\theta, p(t), N_{t+1}) \right\}^2,$$

where m denote the number of generations for which data were collected. In practice we need to change the population statistics such as D and p to their sample estimates $\widehat{D}$ and $\widehat{p}$ which are obtained by replacing the population zygote frequencies by the corresponding sample frequencies. Let

$$\widehat{Q}(\theta) = \sum_{t=1}^{m-1} \sum_{i=1}^{2} \left\{ \widehat{D}_i(t+1) - f_i(\theta, \widehat{p}(t), N_{t+1}) \right\}^2,$$

The minimum value of Q over all possible selection coefficient combination under the model represents the "error sum of squares"

$$\widehat{Q}_* := \min_{\theta} \widehat{Q}(\theta).$$

The corresponding quantity under the null hypothesis of neutrality or "no selection" (which corresponds to $w_i \equiv 1$) is given by

$$\widehat{Q}_0 := \widehat{Q}(\theta_0)$$

where θ_0 represents the parameter value for which $w_i(\theta) \equiv 1$. In fact,

$$\widehat{Q}_0 = \sum_{t=1}^{m-1} \left\{ \left(\widehat{D}_1(t+1) - \frac{N_{t+1} - 1}{N_{t+1}} \widehat{D}(t)\widehat{p}(t) \right)^2 \right.$$
$$\left. + \left(\widehat{D}_2(t+1) - \frac{N_{t+1} - 1}{N_{t+1}} \widehat{D}(t)(1 - 2\widehat{p}(t)) \right)^2 \right\}$$

where $D(t)$ and $p(t)$ are the gametic disequilibrium corresponding to A/M and p is the frequency of M in generation t. The test statistic we propose to use to test the null hypothesis of "no selection" is given by

$$T = \widehat{Q}_0 - \widehat{Q}_*. \tag{29.2.4}$$

Note that $T \geq 0$ and larger values would indicate stronger evidence of selection. It resembles the "model sum of squares" or "model deviance" [McCullah and Nelder (1989)]. However due to the nonlinear nature of the formulas for expected disequilibria its large sample distribution is not likely to be a multiple of chi-square. Thus instead of using the large sample theory we would use a resampling procedure (parametric bootstrap; see below) to calculate the approximate P-value for the test.

Calculation of P-value using bootstrap

We are going to describe a resampling scheme to carry out a parametric bootstrap procedure [Efron (1979)]. It can be implemented in any language (C++, Fortan, Pascal) or package (S- Plus, SAS, Mathematica) which has a random number generator. We assume that the more general sampling scheme 2 was used in the original experiment.

Given our multi-generation data (to be referred to as the original sample or original data in the sequel) the resampling scheme mimics the genetic and statistical sampling procedures under the null hypothesis of random drift (or no selection) in the computer simulation. Here is how to generate the genotypic counts for a multi-level bootstrap sample. It is a customary notation to denote bootstrap quantities with asterics. Start the bootstrap chain with an initial population of genotypic frequencies as the estimated ones from the original sample, i.e., let $p_i^*(1) = \widehat{p}_i(1)$, $i = 1, \ldots, 6$. Next obtain the bootstrap genotypic counts at the sample level from the first generation bootstrap population via multinomial sampling $(n_1 \widehat{p}_1^*(1), \cdots, n_1 \widehat{p}_6^*(1)) \sim$ Multinomial$(n_1, \widehat{p}_1(1), \cdots, \widehat{p}_6(1))$, where n_t is the sample size at generation t of the original data.

TABLE 29.6 α_{fmk}

f	m	k 1	2	3	4	5	6
1 or 2							
	1	1	0	0	0	0	0
	2	0	0	0	1	0	0
	3	0	1	0	0	0	0
	4	0	0	0	0	1	0
3 or 4							
	1	0	1	0	0	0	0
	2	0	0	0	0	1	0
	3	0	0	1	0	0	0
	4	0	0	0	0	0	1

In general, having obtained the bootstrap genotypic frequencies at both the population and sample levels for generations $1, \ldots, t$, form the next generation of bootstrap genotypic counts via multinomial sampling under a RUZ model with drift alone. That is, letting $X_{fm}^*(t+1)$, $1 \leq m, f \leq 4$, denote the number of offspring with gametic type m from mother and type f from father, generate the vector of counts $(X_{11}^*(t+1), X_{12}^*(t+$

1), ..., $X_{44}^*(t+1)$) from a multinomial distribution of total size N_{t+1} and cell probabilities $(e_1^*(t)e_1^*(t),\ e_1^*(t)e_2^*(t),\ \ldots,\ e_4^*(t)e_4^*(t))$, where

$$e_i^*(t) = \sum_k \beta_{ik} p_k^*(t)$$

are the gametic frequencies in the bootstrap population at generation t obtained from the genotypic frequencies at the same population. The genotypic frequencies are then obtained as $p_k^*(t+1) = \sum_{f,m} \alpha_{fmk} X_{fm}^*(t+1)$, $1 \le k \le 6$. The coefficients α and β are given in Tables 29.6 and Table 29.7, respectively. This completes the (recursive) description of the genetic sampling for the generation $t+1$ for the bootstrap data. Next generate the bootstrap genotypic counts at the sample level from the $t+1$ generation bootstrap population via multinomial sampling $(n_{t+1}\widehat{p}_1^*(t+1), \ldots, n_{t+1}\widehat{p}_6^*(t+1)) \sim$ Multinomial$(n_{t+1}, p_1^*(t+1), \ldots, p_6^*(t+1))$. Denote the value of the test statistic T in (29.2.4) calculated at the bootstrap data by T^*.

TABLE 29.7 β_{ki}

k	i			
	1	2	3	4
1	1	0	0	0
2	1/2	0	1/2	0
3	0	0	1	0
4	0	1	0	0
5	0	1/2	0	1/2
6	0	0	0	1

The bootstrap resampling procedure consists of replicating the above steps a large number B (say 5000) of times and calculating the values of the test statistic T at each bootstrap replicate. These would result in B values of T^*, denoted $T_1^*, \ldots, T_B^*$. Finally, the bootstrap approximation to the P-value of the test is given by the proportion of times T_i^* exceeds T_{obs}, the value of the test statistic T calculated from the original multilevel sample. Of course, a 5% level test would be to reject the null hypothesis of "no selection" if and only if P-value < 0.05.

Power and sample size

All the traditional tests of neutrality hypothesis using multi-generation data are based on a single gene frequencies. The numerical studies in Kilpatrick and Rand (1995) reveals overall unsatisfactory power of such tests. The idea in Datta and Arnold (1996) or Datta *et al.* (1996) was to use cytonuclear data on genotypic frequencies through the use of cytonuclear disequilibria in order to gain power in detecting a departure from random drift. Indeed their simulation results show that under a migration alternative the power of the omnibus test can be at respectable level. For example a power of about 85% is achieved for a population size (genetic sampling) of 200 at a mutation rate of 15% or greater. However since this test did not make use of the alternative hypothesis in its construction it is possible to construct selection schemes under which this test can have undesirably low power. As shown in Datta *et al.* (1996), in a heterozygous viability selection model the power curve test remains rather flat for small to moderate level of selection before it picks up the effect of selection. The present test is developed within a specific selection model in which the null hypothesis is embedded (i.e., $w_i \equiv 1$). Therefore this test is expected to have a better power performance in detecting selection provided a correct (or nearly correct) premises for selection is made. However, a determination of this fact (theoretically or by simulation) is yet to be made.

29.3 INFERENCE FOR THE SELECTION COEFFICIENTS

Suppose one is faced with a situation where random drift alone cannot explain the dynamics of the multi-generation cytonuclear data and a selectionist model needs to be invoked for satisfactory description of the system. In such a situation, it is of fundamental interest to estimate the parameter involved in the selection coefficients of the model. Estimation of a selection coefficient in a simpler context was considered, among others, by Wright (1969), Dykhuizen and Hartl (1980) and Watterson (1982). For example Watterson (1982) considered selection on a single gene frequency and considered an approximate Gaussian likelihood for the conditional distribution of the gene frequency in the following generation given the data in the current generation where the effect of selection on the variance is ignored. Here we consider estimating the selection coefficients through the use of an approximate likelihood based on multi-generation cytonuclear genotypic counts. The likelihood is efficient in that it captures the genetic variation fully (which is affected by the selection coefficients). It is approximate because we choose to use the sample genotypic frequencies instead of the population genotypic frequencies. This seems to be a reasonable approach since the selection is not involved at the statistical sampling level. We explain the estimation procedure in detail in the next subsection.

29.3.1 A Multiplicative Fertility Selection Model

We consider a selection model in which the selection effect appears due to difference in fertility of gametic combinations. Suppose a population is observed over discrete generations $t = 1, 2, \ldots$, and let $X_{fm}(t)$ be the number of individuals in generation t receiving gamete type f from the father and gamete type m from the mother, $1 \leq f, m \leq 4$. Under the RUZ model [Watterson (1970)], along with fertility selection, the probability distribution of the counts $X(t+1) = (X_{11}(t+1), \ldots, X_{44}(t+1))$ in generation $t+1$, given the gametic combination counts up to time t, $\mathcal{H}_{\sqcup}$, is multinomial and is given by

$$
\begin{aligned}
Pr(\boldsymbol{X}(t+1) &= x(t+1)|\mathcal{H}_{\sqcup}) \\
&= \frac{N_{t+1}!}{x_{11}(t+1)! \cdots x_{44}(t+1)!} \prod_{f,m} \left(\frac{w_{fm}e_f(t)e_m(t)}{W} \right)^{x_{fm}(t+1)}
\end{aligned}
$$

$$(29.3.5)$$

where $N_{t+1} = \sum_{f,m} x_{fm}(t+1)$, and $W = \sum_{f,m} w_{fm}e_f(t)e_m(t)$. Here e's denote the (relative) frequencies of the various gametes given by $e_i(t) = \sum_k \beta_{ik}p_k(t)$ and p's are the genotypic (relative) frequencies $p_k(t+1) = \sum_{f,m} \alpha_{fmk}X_{fm}(t+1)$, where the constants α and β are given in Tables 29.6 and 29.7, respectively. The coefficients w_{fm} denote the (relative) fertility of the various gametic combinations. We would specialize to the case when the fertility selection coefficients are of the multiplicate form $w_{fm} = w_f w_m$, $1 \leq f, m \leq 4$; we may assume further that $\sum_i w_i = 1$, which makes them identifiable.

29.3.2 An Approximate Likelihood

Let us consider cytonuclear data collected over m generation following the sampling scheme in Figure 29.1. The experiment starts with an initial base population. Generations are discrete and nonoverlapping. In each generation, the individuals were allowed to mate and after the eggs were collected the adults were frozen for genotyping. Eggs were allowed to form the next generation. A random sample of n_t individuals from the frozen adult population of generation t were taken and genotyped to give the sample genotypic frequencies.

Let $\boldsymbol{w} = (w_1, \ldots, w_4)$ be the selection parameters described in the multiplicative fertility selection model described in Section 29.3.1. Under this model, it can be shown that the log-likelihood function of the population genotypic counts over m successive generations given the initial ones is given by

$$l_n(\boldsymbol{w}) = \sum_{t=1}^{m-1} \sum_{k=1}^{6} N_{t+1} p_k(t+1) \log(L_k(\boldsymbol{w},t)),$$

where

$$
\begin{aligned}
L_1(\boldsymbol{w},t) &= e_1^2(\boldsymbol{w},t) + e_1(\boldsymbol{w},t)e_2(\boldsymbol{w},t), \\
L_2(\boldsymbol{w},t) &= e_1(\boldsymbol{w},t)e_3(\boldsymbol{w},t) + e_2(\boldsymbol{w},t)e_3(\boldsymbol{w},t) + e_1(\boldsymbol{w},t)e_3(\boldsymbol{w},t) \\
&\quad + e_1(\boldsymbol{w},t)e_4(\boldsymbol{w},t), \\
L_3(\boldsymbol{w},t) &= e_3^2(\boldsymbol{w},t) + e_3(\boldsymbol{w},t)e_4(\boldsymbol{w},t), \\
L_4(\boldsymbol{w},t) &= e_2^2(\boldsymbol{w},t) + e_1(\boldsymbol{w},t)e_2(\boldsymbol{w},t), \\
L_5(\boldsymbol{w},t) &= e_1(\boldsymbol{w},t)e_4(\boldsymbol{w},t) + e_2(\boldsymbol{w},t)e_4(\boldsymbol{w},t) + e_2(\boldsymbol{w},t)e_3(\boldsymbol{w},t) \\
&\quad + e_2(\boldsymbol{w},t)e_4(\boldsymbol{w},t), \\
L_6(\boldsymbol{w},t) &= e_4^2(\boldsymbol{w},t) + e_3(\boldsymbol{w},t)e_4(\boldsymbol{w},t),
\end{aligned}
$$

and

$$e_i(\boldsymbol{w},t) = e_i w_i / \left(\sum_{j=1}^{4} e_i w_i \right), \qquad 1 \le i \le 4.$$

Note however, that the population genotypic relative frequencies p_k (and their linear combinations e_i, the gametic relative frequencies are not observable and hence the above likelihood is not usable. We propose to replace them by their sample estimates $\widehat{p}_k$ (which is just the corresponding sample proportion) to produce an approximate likelihood $\hat{l}_n(w)$ that is "consistent" for the true likelihood

$$\hat{l}_n(w) = \sum_{t=1}^{m-1} \sum_{k=1}^{6} N\,\widehat{p}_k(t+1)\,\log(\widehat{L}_k(w,t)). \qquad (29.3.6)$$

The formulas for $\widehat{L}_k$ are those for L_k with the e replaced by $\widehat{e}$ in them. Our proposal is to estimate w by the maximizer of the approximate likelihood, i.e., $\widehat{w} = \operatorname{argmax}\hat{l}_n(w)$, where the maximization is to be carried out subject to $\sum_i w_i = 1$.

We recommend estimating the variance-covariance matrix of the selection parameter estimates by parametric bootstrap. Note that a bootstrap sample can be generated following the same sampling scheme described above (Figure 29.1) with the initial cytonuclear genotypic relative frequencies and the selection coefficients replaced by their estimates.

A number of simulation studies were conducted to examine the finite (moderate) sample behavior of the approximate MLE. They show excellent performance both in terms of bias and variance over a wide range of parameter values. Furthermore, they reveal the asymptotic multivariate normality of the estimator. The details are available in Datta (1999).

29.3.3   Application to Hypotheses Testing

The parametric estimates can be used to construct Wald type test statistic [Serfling (1980)] for testing hypotheses involving the selection coefficients. Many interesting null hypotheses, when formulated in terms of w, will be of the form $H_0 : Hw = c$, where H and c are known matrices and constants. As a special case, one may consider a simple null hypothesis $H_0 : w = w_0$. In particular if $w_0 = (1/4, 1/4, 1/4, 1/4)$, then the resulting test would be a test of a neutral model. A test statistic for the general linear hypothesis will be of the form $T = (H\widehat{w}-c)^T(H\widehat{\Sigma}H^T)^{-1}(H\widehat{w}-c)$, where $\widehat{\Sigma}$ is an estimated variance covariance matrix of $\widehat{w}$. A simpler test statistic for the simple null hypothesis $H_0 : w = w_0$ that does not require numerical maximization is given by the approximate Rao type score test statistic $T = \dot{l}_n(w_0)$. Its P-value can be computed via a similar parametric bootstrap as indicated earlier with a viability selection model and a different test statistic.

REFERENCES

Arnason, E. (1991). Perturbation-reperturbation test of selection vs. hitch-hiking of the two major alleles of *Esterase-5* in *Drosophila pseudoobscura*, *Genetics*, **129**, 145–168.

Asmussen, M. A. (1986). The dynamics of interlocus associations in the three-locus hitchhiking model, *Journal of Mathematical Biology*, **24**, 361–380.

Asmussen, M. A. and Clegg, M. T. (1981). Dynamics of the linkage disequilibrium function under models of gene-frequency hitchhiking, *Genetics*, **99**, 337–356.

Asmussen, M. A., Arnold, J. and Avise, J. C. (1989). The effects of assortative mating and migration on cytonuclear disequilibria in hybrid zones, *Genetics*, **112**, 923–934.

Berry, A. J. and Kreitman, M. (1993). Molecular analysis of an allozyme cline: alchol dehydrogenase in Drosophila melanogaster on the east coast of North America, *Genetics*, **134**, 869–893.

Chakraborty, R., Fuerst, P. A. and Nei, M. (1980). Statistical studies on protein polymorphism in natural populations. III. Distribution of allele frequencies and the number of alleles per locus, *Genetics*, **94** 1039–1063.

Clark, A. G. (1984). Natural selection with nuclear and cytoplasmic transmission. I. A deterministic model, *Genetics*, **107**, 679–701.

Clark, A. G. and Lyckegaard, E. M.S. (1988). Natural selection with nuclear and cytoplasmic transmission. III. Joint analysis of segregation and mtDNA in Drosophila melanogaster, *Genetics*, **118**, 471–481.

Datta, S. (1999). Hypotheses testing for different selection models using multi-generation cytonuclear data, *Proceedings of American Statistical Association, Biometrics Section* (to appear).

Datta, S. and Arnold, J. (1996). Diagnostics and a Statistical test of neutrality hypotheses using the dynamics of cytonuclear disequilibria, *Biometrics*, **52**, 1042–1054.

Datta, S., Fu, Y. X. and Arnold, J. (1996). Dynamics and equilibrium behavior of cytonuclear disequilibria under genetic drift. *Theoretical Population Biology*, **50**, 298–324.

Datta, S., Kiparsky, M., Rand, D. M. and Arnold, J. (1996). A statistical test of a neutral model using the dynamics of cytonuclear disequilibria, *Genetics*, **144**, 1985–1992.

Dobzhansky, T. (1970). *Genetics of the Evolutionary Process*, Columbia University Press, New York.

Dykhuizen, D. and Hartl, D. L. (1980). Selective neutrality of 6PGD allozymes in *E. Coli* and the effects of genetic background, *Genetics*, **96**, 801–817.

Efron, B. (1979). Bootstrap Methods: Another look at the jackknife, *Annals of Statistics*, **7**, 1–26.

Ewens, W. J. (1970). *Mathematical Population Genetics*, Biomathematics, Vol. 9, Springer-Verlag, New York.

Ewens, W. J. (1972). The sampling theory of selectively neutral alleles, *Theoretical Population Biology*, **3**, 87–112.

Fisher, R. A. and Ford, E. B. (1947). The spread of a gene in natural conditions in a colony of the moth *Panaxia dominula L. Heredity*, **1**, 143–174.

Fos, M., Dominguez, M. A., Latorre, A. and Moya, A. (1990). Mitochondrial DNA evolution in experimental populations of *Drosophila pseudoobscura*, *Proceedings of the National Academy of Sciences, USA*, **87**, 4198–4201.

Franklin, I. and Lewontin, R. C. (1970). Is the gene the unit of selection? *Genetics*, **65**, 701–734.

Fu, Y. X. and Arnold, J. (1992). Dynamics of cytonuclear disequilibria in finite populations and a comparison with a two-locus nuclear system, *Theoretical Population Biology*, **41**, 1–25.

Fuerst, P. A., Chakraborty, R., and Nei, M. (1977). Statistical studies on protein polymorphism in natural populations. I. Distribution of single locus heterozygosity, *Genetics*, **86**, 455–483.

Gillespie, J. H. (1979). Molecular evolution and polymorphism in a random environment, *Genetics*, **74**, 175–195.

Griffiths, R. C. (1981). Neutral two-locus multiple allele models with recombination, *Theoretical Population Biology*, **19**, 169–186.

Hedrick, P. W. and Thomson, G. (1986). A two-locus neutrality test: Applications to Humans, *E. Coli* and Lodgepole pine, *Genetics*, **112**, 135–156.

Hill, W. G. (1975). Linkage disequilibrium among multiple neutral alleles produced by mutation in finite population, *Theoretical Population Biology*, **8**, 117–126.

Hill, W. G. and Weir, B. S. (1988). Variances and covariances of squared linkage disequilibria in finite populations, *Theoretical Population Biology*, **33**, 54–78.

Hill, W. G. and Roberston, A. (1966). The effect of linkage on limits to artificial selection, *Genetic Research*, **8**, 269–294.

Hill, W. G. and Roberston, A. (1968). Linkage disequilibrium in finite populations, *Theoretical and Applied Genetics*, **38**, 226–231.

Hutter, C. M. and Rand, D. M. (1995). Competition between mitochondrial haplotypes in distinct nuclear genetic environments: *Drosophila pseudoobscura* vs *Drosophila persimilis*, *Genetics*, **140**, 537–548.

Jenkins, T. M., Babcook, C., Geiser, D. M. and Anderson, W. W. (1996). Cytoplasmic incompatibility and mating preference in Colombian *Drosophila pseudoobscura*, *Genetics*, **142**, 189–194.

Karl, S. A. and Avise, J. C. (1992). Balancing selection at allozyme loci in oysters: implications from nuclear RFLPs, *Science*, **256**, 100–102.

Kilpatrick, S. T. and Rand, D. M. (1995). Conditional hitchhiking of mitochondrial DNA: frequency shifts of *Drosophila melanogaster* mtDNA variants depend on neutral genetic background, *Genetics*, **141**, 1113–1124.

Kimura, M. (1983). *The Neutral Theory of Molecular Evolution*, Cambridge University Press, New York.

Kambhampati, S., Rai, K. S. and Verleye, D. M. (1992). Frequencies of Mitochondrial DNA haplotypes in laboratory cage populations of the mosquito, *Aedes albopictus*, *Genetics*, **132**, 205–209.

Lewontin, R. C. (1974). *Genetic Basis of Evolutionary Changes*. Columbia University Press, New York.

Lewontin, R. C. and Krakauer, J. (1973). Distribution of gene frequency as a test of the theory of selective neutrality of polymorphisms, *Genetics*, **74**, 175–195.

Macrae, A. and Anderson, W. W. (1988). Evidence of non-neutrality of mitochondrial DNA haplotypes in *Drosophila pseudoobscura*, *Genetics*, **120**, 485–494.

Macrae, A. and Anderson, W. W. (1990). Can mating preference explain changes in mtDNA haplotype frequency? *Genetics*, **124**, 999–1001.

McCullagh, P. A and Nelder, J. A. (1989). *Generalized Linear Models*, Second edition, Chapman and Hall, London.

McDonald, J. H. (1994). Detecting natural selection by comparing geographic variation in protein and DNA polymorphisms, In *Non-Neutral Evolution* (Ed., B. Golding), pp. 88–100, Chapman and Hall, New York.

McDonald, J. H. (1996). Lack of geographic variation in anonymous polymorphisms in the American Oyster *Crassostrea virginica*, *Molecular Biology and Evolution*, **13**, 1114–1118.

Nei, M. (1987). *Molecular Evolutionary Genetics*, Columbia University Press, New York.

Nigro, L. and Prout, T. (1990). Is there selection on RFLP differences in mitochondrial DNA? *Genetics*, **125**, 551–555.

Li, W. H. and Graur, D. (1991). *Fundamentals of Molecular Sequence Data*, Sinauer, Sunderland, MA.

Pollak, P. E. (1991). Cytoplasmic effects on components of fitness in tobacco hybrids, *Evolution*, **45**, 785–790.

Roberston, A. (1952). The effect of inbreeding on the variation due to recessive genes, *Genetics*, **37**, 189–207.

Rothman, E. D. and Templeton, A. R. (1980). A class of models of selectively neutral alleles, *Theoretical Population Biology*, **18**, 135–150.

Schaffer, H. E., Yardley, D. and Anderson, W. W. (1977). Drift or selection: test of gene frequency variation over generations, *Genetics*, **87**, 371–379.

Scribner, K. T. and Avise, J. C. (1994a). Population cage experiments with a vertebrate: genetics of hybridization in *Gambusia* fishes, *Evolution*, **48**, 155–171.

Scribner, K. T., and Avise, J. C. (1994b). Cytonuclear genetics of experimental fish hybrid zone inside biosphere 2, *Proccedings of the National Academy of Sciences*, **91**, 5066–5069.

Scribner, K. T., Datta, S., Arnold, J. and Avise, J. C. (1998). Temporal changes in cytonuclear disequilibrium in experimental *Gambusia* hybrid zones: deviations from expectations under genetic drift and tests for tests for neutrality of mtDNA variants, *Genetica*, **105**, 101–108.

Serfling, R. J. (1980). *Approximation Theorems of Mathematical Statistics*, John Wiley & Sons, New York.

Singh, R. S., and Hale, L. R. (1990). Are mitochondrial DNA variants selectively non-neutral? *Genetics*, **124**, 995–997.

Takahata, N. (1982). Linkage disequilibrium, genetic distance and evolutionary distance under a general model of linked genes or a part of the genome, *Genetic Research*, **39**, 63–77.

Thomson, G. (1977). The effect of a selected locus on linked neutral loci, *Genetics*, **85**, 753–788.

Watterson, G. A. (1970). The effect of linkage in finite random-mating population, *Theoretical Population Biology*, **1**, 72–87.

Watterson, G. A. (1972). Errata, *Theoretical Population Biology*, **3**, 117.

Watterson, G. A. (1977). Heterosis or neutrality? *Genetics*, **85**, 789–814.

Watterson, G. A. (1982). Testing selection at a single locus, *Biometrics*, **38**, 323–331.

Weir, B. S. (1990). *Genetic Data Analysis*, Sinauer Associates, Sunderland, MA.

Wright, S. (1969). *Evolution and Genetics of Populations, Vol II. The Theory of Gene Frequencies*, University Press of Chicago.

Williams, C. J., Anderson, W. W. and Arnold, J. (1990). Generalized linear modeling methods for selection component experiments, *Theoretical Population Biology*, **37**, 389–423.

Wilson, S. R. (1980). Analyzing gene-frequency data when the effective population size is finite, *Genetics*, **95**, 489–502.

Wilson, S. R., Oakeshott, J. G., Gibson, J. B. and Anderson, P. R. (1982). Measuring selection coefficients affecting the *Alchol dehydrogenase* polymorphism in *Drosophila pseudoobscura*, *Genetics*, **100**, 113–126.

CHAPTER 30

THE PERFORMANCE OF ESTIMATION PROCEDURES FOR COST-EFFECTIVENESS RATIOS

JOSEPH C. GARDINER   ALKA INDURKHYA
ZHEHUI LUO

Michigan State University, East Lansing, MI

30.1 INTRODUCTION

The cost-effectiveness ratio (CER) is a widely used summary statistic for comparing competing health care programs relative to their cost and benefit. The CER is defined as the ratio of the incremental cost of the test program to the incremental benefit, using the next best alternative the referent program as the comparator. The CER is most useful when the test intervention costs more and produces extra health benefit. When cost is measured in dollars and effectiveness in life years the CER is then the additional cost of the test intervention for each additional unit of health benefit. With data on cost and benefit from samples of patients from the two interventions, the CER can be estimated and where feasible, a confidence interval (CI) constructed. Three parametric approaches to obtaining a CI have been proposed. We refer to these as the Fieller, symmetric and Bonferroni intervals. Additionally, resampling methods based on the bootstrap have also been advocated [Chaudhary and Stearns (1996)].

The CER estimator is a ratio estimator, which often makes its distribution skewed. Confidence intervals such as the Fieller and bootstrap

547

intervals account for this skewness and might be expected to have better performance characteristics with respect to coverage probability. They may also be preferable in practice to the symmetric and Bonferroni intervals when assessing power and sample size requirements for tests of hypothesis on the CER, when the confidence interval is used to formulate the test procedure [Gardiner *et al.* (1999)]. The Bonferroni interval ignores the likely correlation between cost and health benefit. Its interval is wider than the comparable Fieller interval [Laska, Meissner and Seigel (1997)] which in turn is wider than the corresponding symmetric interval [Gardiner, Bradley and Huebner (2000)] whenever this comparison can be made. Existence of a finite length Fieller interval of confidence level $1 - \alpha$ is guaranteed when the effectiveness difference is statistically significant at level α [Gardiner, Bradley and Huebner (2000)]. The symmetric interval exists whenever the estimate of incremental effectiveness is non zero.

In this article we examine through extensive Monte Carlo simulation the performance of these three parametric CIs for the CER. Our primary criterion is the coverage probability. We also comment on use of these intervals for hypothesis testing on the CER and power assessments for cost-effectiveness studies.

30.2   CONFIDENCE INTERVALS FOR CER

The form of the CER will depend on the measures of health benefit and cost at the individual level, and the type of cost and health outcome data available. For example, in longitudinal studies with life expectancy as a measure of benefit and non sampled resource utilization, life-expectancy restricted to a finite time horizon and present value of all resource use can be defined in terms of the underling survival distribution [Gardiner *et al.* (1995)]. In this paper we take a simpler view in which mean values for costs and effectiveness obtained from independent samples (C_{0j}, B_{0j} : $1 \leq j \leq n_0$) on the referent intervention and (C_{1j}, B_{1j} : $1 \leq j \leq n_1$) on test intervention are used to estimate the CER. The incremental cost is estimated by the difference in sample means, $\bar{C}_1 - \bar{C}_0$ and for the incremental benefit by $\bar{B}_1 - \bar{B}_0$ which then yields the estimate $\hat{\theta} = (\bar{C}_1 - \bar{C}_0)/(\bar{B}_1 - \bar{B}_0)$ for the CER($= \theta$). The population means, variances and correlations of (C_{0j}, B_{0j}) and (C_{1j}, B_{1j}) are in first two rows of Table 30.1 and appearing in the third row, with $n_1 = kn_0$ are the theoretical mean and standard deviation for the estimated incremental cost $\hat{\mu}_c = \bar{C}_1 \bar{C}_0$ and estimated incremental benefit $\hat{\mu}_c = \bar{B}_1 \bar{B}_0$ and their correlation. We assume that $\mu_c \neq 0$.

TABLE 30.1 Parameters in joint distributions of cost and effectiveness in test and referent interventions

Intervention	N	Cost	
		Mean	SD
Test	n_1	μ_{1c}	ω_1
Referent	n_0	μ_{0c}	ω_0
Difference		$\mu_c =$ $\mu_{1c} - \mu_{0c}$	$\sigma_c =$ $\dfrac{(\omega_0^2 + k^{-1}\omega_1^2)^{1/2}}{\sqrt{n_0}}$

Intervention	N	Effectiveness		Correlation
		Mean	SD	
Test	n_1	μ_{1e}	τ_1	ρ_1
Referent	n_0	μ_{0e}	τ_0	ρ_0
Difference		$\mu_e =$ $\mu_{1e} - \mu_{0e}$	$\sigma_e =$ $\dfrac{(\tau_0^2 + k^{-1}\tau_1^2)^{1/2}}{\sqrt{n_0}}$	$\rho =$ $\dfrac{(\rho_0\omega_0\tau_0 + k^{-1}\rho_1\omega_1\tau_1)}{n_0\sigma_c\sigma_e}$

Fieller interval

A finite length $100(1 - \alpha)\%$ CI for θ exists if, and only if the effectiveness difference $\bar{B}_1 - \bar{B}_0$ is significant at level α [Gardiner, Bradley and Huebner (2000)]. Moreover, if the cost difference $\bar{C}_1 - \bar{C}_0$ is significant at level α the CI for θ is displaced to the left or right of zero according as $\hat{\theta} < 0$ or $\hat{\theta} > 0$. If the cost difference is not significant then the CI contains zero. The confidence limits are

$$L_f, U_f = (1 - y)^{-1}\left\{ (\hat{\theta} - y\rho\sigma)c/\sigma_e) \pm (1 - \rho^2)^{1/2}(\sigma_c/\sigma_e)\sqrt{y(1 - y_v)} \right\} \tag{30.2.1}$$

where z is the $(1 - 1/2\alpha)$-th percentile of standard normal distribution,

$$y = z^2\sigma_e^2/\hat{\mu}_e^2 \quad \text{and}$$

$$v = \left(\frac{\hat{\mu}_c/\sigma_c}{\hat{\mu}_e/\sigma_e} - \rho\right)^2 \Big/ (1 - \rho^2). \tag{30.2.2}$$

In (30.2.1) and (30.2.2) the parameters σ_c, σ_e and ρ will be replaced by estimates obtained from sample variances and correlations for ω_i, τ_i and ρ_i $(i = 0, 1)$. Note that because of the assumed significance of the incremental effectiveness at level α we have $y < 1$.

Symmetric interval

The confidence limits are of the form $\hat{\theta} \pm z \times (AVar(\hat{\theta}))^{1/2}$, with an estimate of the asymptotic variance $AVar(\hat{\theta})$ derived from the asymptotic variance of $\bar{C}_1 - \bar{C}_0 - \theta(\bar{B}_1 - \bar{B}_0)$ and the consistency of $\bar{B}_1 - \bar{B}_0$. The confidence limits are

$$L_s, U_s = \hat{\theta} \pm (1 - \rho^2)^{1/2}(\sigma_c/\sigma_e)\sqrt{y(1 + v)} \qquad (30.2.3)$$

with the unknown parameters σ_c, σ_e and ρ replaced by estimates. It can be shown [Gardiner, Bradley and Huebner (2000)] that the Fieller interval is almost surely wider than the symmetric interval. However, if the difference in effectiveness is highly significant then the two CIs are practically the same. The symmetric interval has also been referred to as the Taylor series interval [O'Brien *et al.* 1994)] because $AVar(\hat{\theta})$ may be derived from a Taylor series expansion of $(\hat{C}_1 - \hat{C}_0)/(\hat{B}_1 - \hat{B}_0)$ at (μ_c, μ_e).

Bonferroni interval

As the name implies the CI for θ is derived by applying the Bonferroni inequality to $P[L_c < \mu_c < U_c, L_e < \mu_e < U_e]$ where (L_c, U_c) and (L_e, U_e) are separate CIs for μ_c at level $1 - \alpha_c$ and μ_e at $1 - \alpha_e$ respectively, with $\alpha_c + \alpha_e \leq \alpha$. The confidence limits for θ are functions of (L_c, U_c, L_e, U_e) and will depend on their signs [Laska, Meissner and Siegel (1997) and Wakker and Lkassen (1995)]. If the incremental effectiveness is not significant a finite interval does not exist. When $L_c > 0$ and $L_e > 0$, the CI may be taken as

$$(L_b, U_b) = \left(\frac{\hat{\mu}_c - z'\sigma_c}{\hat{\mu}_e + z'\sigma_e} \ , \ \frac{\hat{\mu}_c + z'\sigma_c}{\hat{\mu}_e - z'\sigma_e} \right) \qquad (30.2.4)$$

where $\hat{\mu}_c = \bar{C}_1 - \bar{C}_0$, $\hat{\mu}_e = \bar{B}_1 - \bar{B}_0$ and z' is the $(1 - 1/4\alpha)$-th percentile of the standard normal distribution. For this case both cost and effectiveness are significantly higher for the test intervention than in the referent intervention. Moreover, the corresponding Fieller interval is wholly contained in the Bonferroni interval [Laska, Meissner and Siegel (1997)] .

30.3   COMPARISON OF INTERVALS

The symmetric interval (30.2.3) for the CER always exists provided the estimated difference in effectiveness $\hat{\mu}_e$ is not zero. However, a finite width $1 - \alpha$ level CI under the Fieller method obtains if and only if the difference in effectiveness is significant. This means that the test of $H_{0e} : \mu_e = 0$ is significant at level α based on the rule that rejects H_{0e} if $|\hat{\mu}_e/\sigma_e| > z$. Because σ_e would generally be unknown it must be replaced by the sample standard deviation. If the α level test of H_{0e} is not significant then the

Fieller interval is an unbounded interval of the form $(-\infty, L) \cup (U, \infty)$ or the entire real line. As noted earlier the width of the Fieller interval is almost surely wider than that of the symmetric interval, and the Bonferroni interval wholly contains the Fieller interval.

When H_{0e} is highly significant in the sense that $y = z^2 \sigma_e^2 / \hat{\mu}_e^2 << 1$ the Fieller and symmetric intervals are approximately the same. Since the Bonferroni interval ignores the correlation between costs and effectiveness we may compare it to the symmetric interval when its width is greatest with respect to ρ. From (30.2.3) this half-width is proportional to the square root of $1 - 2\rho\hat{\theta}(\sigma_e/\sigma_c) + (\hat{\theta}\sigma_e/\sigma_c)^2$. Therefore when $\rho = -1$ we obtain

$$L_s, U_s = \hat{\theta} \pm z \left(\frac{\sigma_c}{\hat{\mu}_e} \right) \left(1 + \hat{\theta}\frac{\sigma_e}{\sigma_c} \right). \tag{30.3.5}$$

For the Bonferroni interval (30.2.4) assuming y is small we get approximately the same limits in (30.3.5) with z replaced by z'. Because $z' > z$ the Bonferroni CI is still wider than the widest symmetric CI.

The central limit theorem for sample averages $\bar{C}_1$, $\bar{C}_2$, $\bar{B}_1$ and $\bar{B}_0$ ensures the validity of the CIs (30.2.1), (30.2.3) and (30.2.4) with respect to the stipulated coverage probability $1 - \alpha$. From the consistency of sample means we obtain the consistency of $\hat{\theta}$. The Bonferroni interval will be conservative guaranteeing coverage of at least $1 - \alpha$. For large sample sizes n_1, n_0 the Fieller and symmetric intervals should be very similar. Although the asymptotic distribution of $\hat{\theta}$ is used to obtain $\mathrm{SD}(\hat{\theta})$ for constructing the symmetric CI, the bias of $\hat{\theta}$, $E(\hat{\theta}) - \theta$ does not exist. A more interesting quantity is $E(\hat{\theta}|\hat{\mu}_e > a) - \theta$ where $a > 0$ is a specified level of minimum incremental effectiveness. The uniform integrability of sample means ensures that $E(\hat{\theta}|\hat{\mu}_e > a)$ converges to θ, provided $\mu_e > a$. For normally distributed samples this conditional expectation is

$$\rho(\sigma_c/\sigma_e) + (\mu_c - \rho(\sigma_c/\sigma_e)\mu_e) \left(\int_b^\infty \frac{\phi(x)}{\mu_e + \sigma_e x} dx \right) \Big/ (1 - \Phi(b))$$

where $b = (a - \mu_e)/\sigma_e$. Therefore in this situation we have

$$E(\hat{\theta}|\hat{\mu}_e > a) - \theta$$
$$= \{\theta - \rho(\sigma_c/\sigma_e)\} \left(-1 + \frac{1}{1 - \Phi(b)} \int_b^\infty \frac{\phi(x)}{1 + (\sigma_e/\mu_e)x} dx \right) \tag{30.3.6}$$

which shows that the left hand side has the opposite sign of $\theta - \rho(\sigma_c/\sigma_e)$.

30.4 SIMULATION STUDIES

Specifying distributions and parameters

The parameters $\mu_{ic}, \mu_{ie}, \omega_i, \tau_i$ and ρ_i $(i = 0, 1)$ in Table 30.1 will be fixed for the different distributions of cost and effectiveness. The three distributions that we will consider make the following transformed cost and effectiveness measures bivariate normal for both the referent $(i = 0)$ and test interventions $(i = 1)$: [1] (C_i, B_i), [2] $(\log C_i, B_i)$, and [3] $(\log C_i, \log B_i)$. Therefore in [2] and [3] the costs are log normally distributed, and in [3] the effectiveness is also log normal. In [2] the parameters of the bivariate distribution may be specified using Lemma 30.4.1.

Lemma 30.4.1 *Suppose* $\log C_i \sim N(\lambda_i, \zeta_i^2)$ *and* $Corr(\log C_i, B_i) = \gamma_i$. *Then*

$$\zeta_i^2 = \log(1 + \omega_i^2/\mu_{ic}^2)$$

$$\lambda_i = \log \mu_{ic} - \frac{1}{2}\zeta_i^2 \qquad and$$

$$\gamma_i = \rho_i \omega_i / \mu_{ic} \zeta_i.$$

PROOF The first two equations follow immediately from expressions for the mean and variance of the log normal distribution, that is $\mu_{ic} = \exp(\lambda_i + \frac{1}{2}\zeta_i^2)$ and $\omega_i^2 = \mu_{ic}^2(\exp(\zeta_i^2) - 1)$.

For a bivariate normal pair (X, Y) with zero means, unit variances and correlation r we have $E(e^{tX}Y) = rte^{t^2/2}$ for any real number t. This follows by evaluating the expectation by first conditioning on Y and using the fact that the conditional distribution of X given Y is $N(rY, 1 - r^2)$. Then $\rho_i = Corr(e^{\log C_i}, B_i) = e^{\lambda_i}\gamma_i\zeta_i e^{\zeta_i^2/2}/\omega_i$ and the third equation obtains. $\square$

Similarly the parameters in [3] may be specified using the following Lemma, the proof of which is straightforward.

Lemma 30.4.2 *Suppose* $\log C_i \sim N(\lambda_{ic}, \zeta_{ic}^2)$, $\log B_i \sim N(\lambda_{ie}, \zeta_{ie}^2)$ *and* $Corr(\log C_i, \log B_i) = \gamma_i'$. *Then*

$$\zeta_{ic}^2 = \log(1 + \omega_i^2/\mu_{ic}^2), \quad \lambda_{ic} = \log \mu_{ic} - \frac{1}{2}\zeta_{ic}^2$$

$$\zeta_{ie}^2 = \log(1 + \tau_i^2/\mu_{ie}^2), \quad \lambda_{ie} = \log \mu_{ie} - \frac{1}{2}\zeta_{ie}^2$$

and

$$\gamma_i' = \log\left(1 + \frac{\rho_i \omega_i \tau_i}{\mu_{ic}\mu_{ie}}\right) \bigg/ \zeta_{ic}\zeta_{ie}.$$

To generate the data for our simulations we need specify the underlying means μ_{ic}, μ_{ie} the correlations ρ_i and coefficients of variation (CV) for cost ω_i/μ_{ic}, and for effectiveness τ_i/μ_{ie}. In practice we would expect greater variation in costs than in effectiveness measures. Coefficients of variation of 10-30% for costs are not uncommon. On the other hand in studies such as randomized trials designed to demonstrate a difference in the effectiveness of two interventions, we would expect the variation in effectiveness to be much smaller, with greater variation in the test intervention than in the referent (standard).

30.5 RESULTS

Each simulation was based on 1000 replications from samples drawn from the three sets of distributions for costs and effectiveness. The means and coefficients of variations were specified for the parent distributions and Lemmas 30.4.1 and 30.4.2 were used to generate the data to conform to these specifications. The sample sizes n_0, n_1 for the referent and test interventions were the same: either 250, 100 or 50. In all cases $(\mu_{0c}, \mu_{0e}) = (30,5)$ and $(\mu_{1c}, \mu_{1e}) = (40,6)$. If costs are in thousands of dollars and effectiveness in quality-adjusted life years (QALY5) the true CER is \$10,000/QALY. For each sample 95% confidence limits were computed by (1) for the Fieller interval, by (30.2.3) for the symmetric interval and by (30.2.4) for the Bonferroni interval, as appropriate. For the latter different expressions for the confidence limits obtain that depend of the direction and significance of the cost difference [Laska and Meissner (1997)]. Both the Fieller and Bonferroni CIs are unbounded when the effectiveness difference is not significant. The empirical coverage is the proportion of samples in which the generated bounded CI contained the theoretical CER (=10).

Tables 30.2, 30.3 and 30.4 give the results of sample sizes of 250. Here we expect the asymptotic properties of our estimates to clearly hold. With normally distributed samples (Table 30.2) the coverage by the Fieller and symmetric intervals are essentially the same across the different levels of the CV of costs and effectiveness. The correlations ρ_0, ρ_1 have a greater influence on the standard deviation of the CER estimate than on the coverage probability. As expected at negative correlations there is greater variation in the CER estimates and corresponding wider confidence intervals. There is even greater variation in the estimates with log normal distributions for either costs or effectiveness (Tables 30.3 and 30.4). In these simulations because of the large sample size of 250 for all samples the effectiveness difference was significant at the 5% level. Therefore a finite Fieller interval was guaranteed. Also the cost difference was positive and significant in which case the Bonferroni interval is given by (30.2.4).

TABLE 30.2 Normal distributions for cost and effectiveness
$(n_0 = n_1 = 250; (\mu_{0c}, \mu_{0e}) = (30, 5); (\mu_{1c}, \mu_{1e}) = (40, 6))$

CV (costs effectiveness)		Correlation	Coverage probability			Distribution of CER		
Referent	Test	ρ_0, ρ_1	Fieller	Symmetric	Bonferroni	Mean	SD	Range
.2, .02	.2, .02	.5, .5	.944	.945	.995	10.06	.59	8.3–11.9
.2, .02	.2, .02	.7, .7	.959	.958	.996	10.02	.58	8.0–11.6
.2, .02	.2, .02	0,0	.959	.960	.992	10.02	.65	7.8–11.9
.2, .02	.3, .05	.3, .7	.957	.961	.999	10.0	.69	7.5–12.0
.3, .1	.3, .1	.3, .3	.961	.959	1.00	9.95	.90	7.2–13.1
.3, .1	.1, .1	.3, .5	.950	.949	.999	9.99	.67	7.4–12.1
.3, .1	.3, .1	−.2, −.3	.956	.952	.998	9.99	.88	6.3–12.6

TABLE 30.3 Log normal cost and normal effectiveness distributions
$(n_0 = n_1 = 250; (\mu_{0c}, \mu_{0e}) = (30, 5); (\mu_{1c}, \mu_{1e}) = (40, 6))$

CV (costs effectiveness)		Correlation	Coverage probability			Distribution of CER		
Referent	Test	ρ_0, ρ_1	Fieller	Symmetric	Bonferroni	Mean	SD	Range
.2, .02	.2, .02	.5, .5	.945	.946	.995	10.06	.59	8.3–12.0
.2, .02	.2, .02	.7, .7	.951	.950	.997	10.02	.57	8.1–11.6
.2, .02	.2, .02	0,0	.952	.952	.989	10.02	.65	7.9–11.8
.2, .02	.3, .05	.3, .7	.948	.949	1.00	10.002	.76	7.9–12.4
.3, .1	.3, .1	.3, .3	.949	.951	1.00	10.03	.96	7.2–12.9
.3, .2	.3, .2	.3, .5	.948	.955	1.00	10.00	1.09	5.6–13.9
.2, .02	.2, .02	−.3, −.2	.953	.954	.988	10.03	.67	8.1–11.9
.3, .2	.3, .2	−.3, −.2	.946	.951	.995	10.18	1.59	5.7–16.6
.9, .2	.9, .2	.3, .5	.946	.953	.998	9.91	2.60	(3.3)–17.6

Negative values denoted by parentheses ()

TABLE 30.4 Log normal cost and normal effectiveness distributions
$(n_0 = n_1 = 250; (\mu_{0c}, \mu_{0e}) = (30, 5); (\mu_{1c}, \mu_{1e}) = (40, 6))$

CV (costs effectiveness)		Correlation	Coverage probability			Distribution of CER		
Referent	Test	ρ_0, ρ_1	Fieller	Symmetric	Bonferroni	Mean	SD	Range
.3, .1	.5, .25	.5, .5	.945	.951	1.00	10.04	1.29	6.4–14.5
.3, .1	.5, .25	0, 0	.945	.951	.997	10.13	1.80	5.4–17.2
.3, .1	.5, .25	−.5, −5	.942	.951	.981	10.23	2.24	4.7–11.6
.3, .2	.3, .05	.5, .5	.951	.951	1.00	10.02	.90	5.3–13.1
.3, .2	.3, .05	.25, .25	.953	.955	.999	10.04	1.03	7.2–14.1
.3, .2	.3, .05	−.5, −.5	.943	.943	.989	10.01	1.37	6.5–14.6

When the coefficient of variation for cost is high it is more likely that in some samples the difference in cost between the interventions will not be significant. A confidence interval for the CER in this case would include zero, and the distribution of the estimated CERs could still be quite skewed. This was seen in one set of simulations (Table 30.3) where the CER estimates ranged from (−3.3) to 17.6. Although the mean of 9.91 was still close to the theoretical value of 10, the average standard deviation was relatively high. Both the Fieller and symmetric intervals provided coverage near the targeted value of 95%.

With a smaller sample size it is more likely that in some samples the effectiveness difference or cost difference might not be significant. When the effectiveness difference is not significant the Fieller and Bonferroni intervals will be unbounded but a finite symmetric interval can still be computed. Some simulation results for sample sizes of 50 are given in Table 30.5, for log normal cost and normal effectiveness, and in Table 30.6 when both measures are normally distributed. The distribution of the CER is now markedly skewed, especially if the coefficients of variation are high. The Fieller method gave unbounded intervals in less than 0.5% of the 1000 simulation runs. Unbounded intervals appeared in about 1% of the time for the Bonferroni method. All methods gave very wide confidence intervals for the CER with the Bonferroni intervals being by far the very widest.

TABLE 30.5 Log normal cost and normal effectiveness distributions
$(n_0 = n_1 = 50; (\mu_{0c}, \mu_{0e}) = (30, 5); (\mu_{1c}, \mu_{1e}) = (40, 6))$

CV (costs effectiveness)		Correlation	Coverage probability			Distribution of CER		
Referent	Test	ρ_0, ρ_1	Fieller	Symmetric	Bonferroni	Mean	SD	Range
.2, .1	.5, .25	.5, .5	.953	.957	.985	10.4	2.93	3.8–27.7
.2, .1	.5, .25	0, 0	.930	.926	.985	10.4	4.49	0.9–56.5
.3, .2	.3, .2	.3, .3	.940	.941	.988	10.3	3.12	2.7–37.1
.8, .2	.8, .2	.3, .3	.941	.951	.990	10.1	6.02	(9.9)–38.4
.3, .2	.3, .1	−.5, −.5	.947	.951	.992	10.5	3.53	3.3–31.4
.2, .1	.5, .25	−.5, −.5	.934	.903	.973	10.7	5.62	1.0–67.8

Negative values denoted by parentheses ()

TABLE 30.6 Normal cost and normal effectiveness distributions
$(n_0 = n_1 = 50; (\mu_{0c}, \mu_{0e}) = (30, 5); (\mu_{1c}, \mu_{1e}) = (40, 6))$

CV (costs effectiveness)		Correlation	Coverage probability			Distribution of CER		
Referent	Test	ρ_0, ρ_1	Fieller	Symmetric	Bonferroni	Mean	SD	Range
.2, .1	.3, .05	.5, .5	.950	.958	1.00	10.0	1.73	4.3–15.8
.2, .1	.5, .25	.5, .5	.936	.961	.991	10.1	2.89	2.1–26.2
.8, .2	.8, .2	.3, .3	.944	.969	.989	10.1	5.8	(9.7)–44.9
.3, .2	.3, .1	−.5, −.5	.953	.947	.990	10.4	3.45	3.5–28.7
.5, .2	.5, .2	0,0	.947	.948	.990	10.5	4.77	(0.38)–54.3

Negative values denoted by parentheses ()

Additional simulations carried out for sample size of $n_0 = n_1 = 100$ revealed differences that were intermediate between the two cases discussed here (Table 30.7). Figures 30.1 and 30.2 show the distribution of the CER for the first and one but last entries in Table 30.7. The greater skewness is seen with a negative correlation.

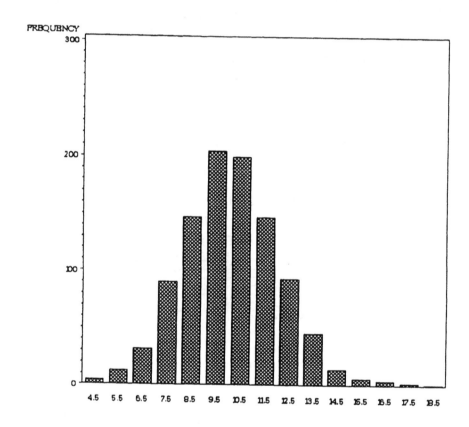

FIGURE 30.1 Distribution of CER
(normal cost, normal effectiveness, correlation = .5)

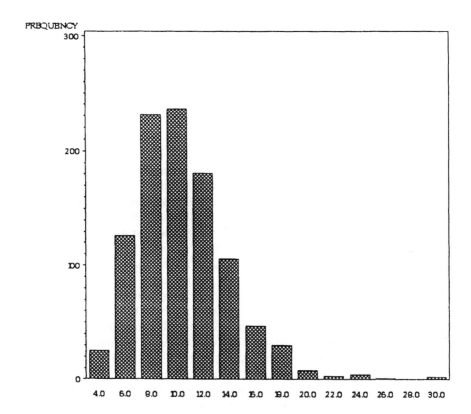

FIGURE 30.2 Distribution of CER
(normal cost, normal effectiveness, correlation $= -.5$)

TABLE 30.7 $(n_0 = n_1 = 100; (\mu_{0c}, \mu_{0e}) = (30, 5); (\mu_{1c}, \mu_{1e}) = (40, 6))$

Costs, effectiveness		Correlation	Coverage probability			Distribution of CER		
CV	Distribution	ρ_0, ρ_1	Fieller	Symmetric	Bonferroni	Mean	SD	Range
.2, .1	N, N	.5, .5	.956	.965	1.00	10.1	1.97	4.4–18.6
.5, .25	LN, N		.946	.939	1.00	10.0	1.99	4.5–19.7
	LN, LN		.928	.938	.999	10.1	2.08	4.3–19.0
.2, .1	N, N	0,0	.956	.955	1.00	10.3	2.76	3.3–24.3
.5, .25	LN, N		.941	.941	.999	10.2	2.82	3.8–26.5
	LN, LN		.933	.942	.997	10.3	2.95	2.9–24.2
.2, .1	N, N	−.5, −.5	.959	.941	.993	10.5	3.39	2.7–25.5
.5,.25	LN, N		.942	.938	.989	10.4	3.51	3.5–29.7
	LN, LN		.944	.934	.984	10.5	3.66	2.1–29.9

CV: Referent on first line, test on second line; N = Normal, LN = log Normal

30.6 RECOMMENDATIONS

Our theoretical and empirical investigation generally favor the Fieller method for constructing a confidence interval for the CER based on random samples of cost and effectiveness data from two independent samples. The Fieller method performs the best across the different scenarios we tested. It correctly accounts for the correlation between the cost and effectiveness measures, and provides coverage close to the desired level. The symmetric interval, which is easier to compute than the Fieller interval, could be used when the effectiveness difference is highly significant. Generally, these two methods yield similar coverage probability. Although the Bonferroni interval is the easiest to compute among the three intervals it is generally very wide and the coverage far too conservative. The Bonferroni method should not be used when there is a strong positive correlation between costs and effectiveness. For sample size and power calculations for tests of hypotheses on the CER in cost-effectiveness studies, accounting for the positive correlation leads to higher statistical power for given sample size,[7] than when this correlation is ignored [Gardiner, Bradley and Heubner (2000) and Briggs and Gray (1998)]. Conversely, to achieve a specified power our sample size requirements will be smaller by accounting for this correlation. The computation of the bias in the normal case in (30.3.6) points to decreasing bias with increasing correlation.

The influence of the assumed distributions of cost and effectiveness on the performance of the three methods was not discernable, at least in the cases we considered. This is especially true for coverage probability. With a log normal distribution for cost, the distribution of the CER was more skewed. Theoretically we would expect the methods to perform well when the distributions are assumed to be normal. Our study shows that the effects of the variation in costs and effectiveness can have a pronounced impact on the precision of the confidence interval when the sample size is small.

When a specific parametric distribution is assumed for the cost and effectiveness measures would a more specific estimation procedure for the CER yield improved estimates? For instance if the distribution of cost is skewed, a distribution such as the log normal could be considered appropriate. The underlying parameters of the log normal cost distribution and those of the distribution of effectiveness in Lemmas 30.4.1 and 30.4.2 can be estimated [see, for example, Zhou (1998), Zhou, Melfi and Hui (1997a) and Zhou, Gao and Hui (1997b)] by using sample means, variances and covariances on the transformed scale. The CER is then a function of these underlying parameters. For example, if costs are log normal, that is, $\log C_i \sim N(\lambda_{ic}, \zeta_{ic}^2)$, the numerator of the CER is $\exp(\lambda_{1c} + \frac{1}{2}$. If effectiveness is normally distributed in each sample, the difference in mean effectiveness is $\mu_{1e} - \mu_{0e}$. Therefore the CER is a function of these six parameters which would be estimated by substituting the appropriate estimates of these parameters. The distribution of this CER estimate would involve in addition the two correlations. Because of the assumed distributions the estimated covariance matrix of these parameter estimates will have several zero entries. Asymptotic theory may be invoked to derive the distribution of the CER. It is unclear whether this fully parametric approach has an appreciable advantage in addressing the precision of the CER estimate compared to that based on sample means and covariances. Our simulations studies and those of other researchers [see, for example, Chaudhary (1996) and Polsky et al. (1997)] point to the robustness of the Fieller method.

Acknowledgment Research supported in part by the Agency of Health Care Policy & Research under grant 1 R01 HS 0943.

REFERENCES

Briggs, A. H. and Gray, A. M. (1998). Power and sample size calculations for stochastic cost-effectiveness analysis, *Meducak Decision Making*, **18**, suppl: S81–S92.

Chaudhary, M. A. and Stearns, S. C. (1996). Estimating confidence intervals for cost-effectiveness ratios: An example from a randomized trial, *Statistics in Medicine*, **15**, 1447–1458.

Gardiner, J. C., Bradley, C. J. and Huebner, M. (1999). The cost-effectiveness ratio in the analysis of health care programs, In *Handbook of Statistics–18* (Eds., C. R. Rao and P. K. Sen), North-Holland, Amsterdam.

Gardiner, J. C., Hogan, A., Holmes-Rovner, M., Rovner, D., Griffith, L. and Kupersmith, J. (1995). Confidence intervals for cost-effectiveness ratios, *Medical Decision Making*, **15**, 254–263.

Gardiner, J. C., Huebner, M., Jetton, J, and Bradley, C. J. (1999). Power and sample size assessments for tests of hypotheses on cost-effectiveness ratios, *Submitted for publication*.

Laska, E. M., Meissner, M. and Siegel, C. (1997). Statistical inference for cost-effectiveness ratios, *Health Economics*, **6**, 229–242.

O'Brien, B. J., Drummond, M. F., Labelle, R. J. and Willan, A. (1994). In search of power and significance: issues in the design and analysis of stochastic cost-effec-tiveness studies in health care, *Medical Care*, **32**, 150–163.

Polsky, D., Glick, H. A., Willke, R. and Schulman, K. (1997). Confidence intervals for cost-effectiveness ratios: A comparison of four methods, *Health Economics*, **6**, 243–252.

Wakker, P. and Klassen, M. P. (1995). Confidence intervals for cost/effectiveness ratios, *Health Economics*, **4**, 373–381.

Zhou, X. H. (1998). Estimation of the log-normal mean, *Statistics in Medicine*, **17**, 2251–2264.

Zhou, X. H., Melfi, C. A. and Hui, S. L. (1997a). Methods for comparison of cost data, *Annals of Internal Medicine*, **127**, 752–756.

Zhou, X. H., Gao, S. J. and Hui, S. L. (1997b). Methods for comparing the means of two independent log-normal samples, *Biometrics*, **53**, 129–1135. Correction **53**, 1566.

CHAPTER 31

MODELING TIME-TO-EVENT DATA USING FLOWGRAPH MODELS

APARNA V. HUZURBAZAR

University of New Mexico, Albuquerque, NM

Abstract: Flowgraph models provide an innovative approach for the analysis of time-to-event data. Time-to-event data is especially important in two major areas of statistics: reliability and survival analysis. This article introduces flowgraph models the context of system reliability. A flowgraph is a graphical representation of a set of relations describing a stochastic system. It consists of a set of nodes connected by directed line segments (branches) that model the dependence of an output variable on an input variable. A practical application consists of using flowgraph methods to access the system moment generating function. This moment generating can be inverted either exactly or approximately by using methods such as saddlepoint approximations to give estimated densities, cumulative distribution functions, reliability or survival, and hazard functions. We demonstrate the viability of these methods by considering systems in series, parallel, and with feedback loops. We also extend these methods to phase type distributions. An application to modeling HIV/AIDS in survival analysis using flowgraphs is also presented.

Keywords and phrases: Reliability, survival analysis, HIV/AIDS, hydraulic pump

31.1 INTRODUCTION

Flowgraph models are useful for a wide variety of problems. The focus of this article is on illustrating the use of flowgraph models for time-to-event data. Time-to-event data is especially important in both reliability and survival analysis and this article will present one application from each area. Flowgraph methods have recently been developed for Bayesian survival analysis by Butler and Huzurbazar (1997). The focus of Butler and Huzurbazar (1997) is on Bayesian prediction using flowgraphs and the methodology presented includes models of disease progression for kidney failure, cancer, and HIV/AIDS. Situations involving left, right, and interval censored data are illustrated using AIDS data from the San Francisco Men's Health Study.

Flowgraph models also provide useful extensions to phase type (PH) distributions. PH distributions are defined to be distributions of absorption times in Markov processes. In survival analysis, these distributions are used to model stages of disease progression that lead to an absorbing end state, usually death. Aalen (1995) presents many different PH distributions for modeling survival times in situations where the overall survival time involves progression through several stages. Analysis of these models is quite complicated and requires a number of simplifying assumptions. The most common of these is the Markovian assumption which necessitates exponential waiting times between stages of the disease. While this may be a reasonable assumption for many situations, it is also quite restrictive. Flowgraph modeling extends these to generalized PH distributions by allowing the use of any waiting time distribution with a tractable moment generating function (MGF). Such extensions of PH distributions using flowgraph models are discussed in Huzurbazar (1999).

A flowgraph is a graphical representation of a set of relations describing a stochastic system. It consists of a set of nodes connected by directed line segments (branches) that model the dependence of an output variable on an input variable. In a typical flowgraph analysis, our interest is in accessing the system (MGF) for a particular problem. This MGF can be inverted either exactly or by using approximation methods such as saddlepoint approximations [cf. Daniels (1954)]. This gives estimated densities, cumulative distribution functions (CDFs), reliability or survival functions, and hazard functions. The focus of this article is on illustrating the use of flowgraph models in accessing the MGF of the quantities of interest. Section 31.2 provides a detailed introduction to flowgraphs. Section 31.3 reanalyzes data from Limnios (1992) using a flowgraph model. Section 31.4 proposes a feedforward model for HIV/AIDS with an exact inversion of the MGF from the resulting flowgraph model. This an alternative to a PH analysis of this model and the analysis presented here is more general than a PH analysis. For illustrative clarity, in this article, we focus on examples of flowgraph models where exact inversion of the MGF is possible. Saddlepoint methods, which are generally used in conjunction with flowgraph models for approximate inversion of the MGFs, are beyond the scope of this paper. For a discussion on saddlepoint methods for problems in survival analysis see Huzurbazar and Huzurbazar (1999).

31.2 INTRODUCTION TO FLOWGRAPH MODELING

Consider a hydraulic system with 2 pumps physically in parallel and with appropriate valving so that the system can operate with only one pump if necessary. Limnios (1992) analyzes this system for various quantities of interest including mean time to total failure. The analysis is quite complicated and requires a number of simplifying assumptions, namely exponential waiting times so that Markov models can be used for the analysis. Figure 31.1 is the flowgraph of the hydraulic system. For now we assume that the pumps operate independently and that the system can operate with only 1 pump if necessary. This is an example of a series system with feedback. State 0 represents zero failed pumps, state 1 represents 1 failed pump, and state 2 is the state with 2 failed pumps. The backward transition from state 1 to state 0 represents the repair of a failed pump. We will use portions of this flowgraph to illustrate the basics of solving flowgraph models before analyzing the full hydraulic system problem.

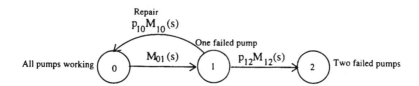

FIGURE 31.1 Flowgraph model for hydraulic pump system

31.2.1 Flowgraph Models for Series Systems

First, we consider the system of Figure 31.1 *without* feedback, i.e. removing the transition from $1 \rightarrow 0$. Flowgraphs are simplified by solving them. In Figure 31.1 let Y_1 be the random variable representing the time for passage from state 0 to state 1 with MGF $M_{01}(s)$ and Y_2 represent time for passage from state 1 to state 2 with MGF $M_{12}(s)$. The passage time from state 0 to state 2 is the sum of two independent random variables Y_1 and Y_2. The MGF of the total waiting time for pump failure is $M_{01}(s)M_{12}(s)$. We can now replace the flowgraph in Figure 31.1 with an equivalent flowgraph consisting of only two nodes, 0 and 2, and one branch labeled $M_{01}(s)M_{12}(s)$, the *equivalent transmittance*. For finite systems where passage from input to output is certain to occur, the is equivalent transmittance is the overall system MGF. This process of reducing a larger flowgraph to a smaller one is called "solving" a flowgraph. This procedure is also known as block diagram reduction in engineering. However, in engineering, probability distributions and random variables are not included in the modeling.

Series models lead to convolutions of random variables and hence are quite straightforward to analyze using flowgraphs. For example a series model with n states has an equivalent transmittance given by the product of the MGFs of the $n - 1$ independent waiting times involved in the convolution. This property has been amply exploited in reliability analysis for independent and identically distributed random variables where the sum is tractable and leads to tractable functional forms for expressions of system reliability. With a flowgraph model, the random variables need only be conditionally independent, not identically distributed.

31.2.2   Flowgraph Models for Parallel Systems

Figure 31.2 shows the flowgraph of a parallel system. These structures are familiar to statisticians in the form of the finite mixture distributions that they yield. Suppose state 1 represents the testing station for a component. The component is prone to one of two types of failures which moves the system to state 2 or state 3. Each branch is labeled with the probability of taking that branch and the MGF of the waiting time. This quantity, *probability* × *MGF*, is defined as the *branch transmittance*. The probability of transition from $1 \rightarrow 2$ is p_{12} and the MGF of the waiting time for passage is $M_{12}(s)$ which gives $p_{12}M_{12}(s)$ as the branch transmittance. The transition probability from $1 \rightarrow 3$ is $p_{13} = 1 - p_{12}$ with $M_{13}(s)$ as the MGF of the waiting time for passage. The MGF of the overall waiting time is a finite mixture distribution: with probability p_{12} it is $M_{12}(s)$, the MGF of the waiting time distribution for $1 \rightarrow 2$ and with probability p_{13} it is $M_{13}(s)$, the MGF of the waiting time distribution for $1 \rightarrow 3$. Therefore, the MGF of the waiting time for first passage to either state 2 or state 3 is given by $p_{12}M_{12}(s) + p_{13}M_{13}(s)$. If Y_1 represents the waiting time to state 2 and Y_2 represents the waiting time to state 3, this flowgraph describes the waiting time for the occurrence of $\min\{Y_1, Y_2\}$. If distributional assumptions about Y_1 and Y_2 were made then $p_{12} = P(Y_1 < Y_2)$. Note that $M_{12}(s)$ is not the MGF of Y_1 but rather the MGF of the actual "competitive waiting time", the "competing risks" distribution in survival analysis. If the competitive waiting time from $1 \rightarrow 2$ were inverse Gaussian and from $1 \rightarrow 3$ were gamma, the overall waiting time distribution would be the mixture of these with the mixing parameter determined by the probabilities.

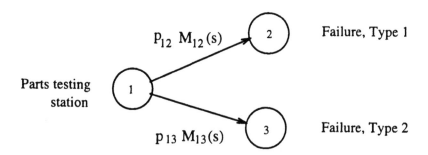

FIGURE 31.2 Flowgraph model for a parallel system

Figure 31.2 is an example of a flowgraph with two "paths". A *path* from beginning to end is any possible sequence of nodes from input to output that does not pass through any intermediate node more than once. In this case, the paths are $1 \rightarrow 2$ and $1 \rightarrow 3$ with no intermediate nodes. Series and parallel components can be combined to form larger systems.

31.2.3 Flowgraph Models with Feedback

The feedback loop, shown in Figure 31.3, is a generalization of the testing system above where we allow a part to be repaired when the system is in state 1. This model is useful for modeling the number of times the part fails before passing inspection or how long it takes for the part to successfully pass the test. From the initial state, 1, the probability of staying in state 1, i.e. failing the inspection, is p. The waiting time for remaining in state 1 has MGF $M_{11}(s)$, the waiting time for inspection and repair. Transition from state 1 to state 2 occurs with probability $1 - p$ and waiting time MGF $M_{12}(s)$. The feedback loop can be solved for the equivalent transmittance $1 \rightarrow 2$. The process starts over again whenever one remains in state 1, so the equivalent transmittance is $T_{12}(s) = pM_{11}(s)T_{12}(s) + (1 - p)M_{12}(s)$, which yields

$$T_{12}(s) = \frac{(1 - p)M_{12}(s)}{1 - pM_{11}(s)}.$$

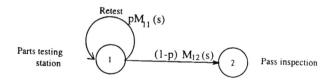

FIGURE 31.3 Flowgraph model for a feedback loop

More complex systems such as those with numerous types of paths and loops require the use of techniques such as Mason's rule [see Mason (1953)]. Mason's Rule is an algebraic procedure that allows computation of the equivalent transmittance of the solved flowgraph from any node A to any node B. This requires that we identify all of the paths and loops of the system. Loops are identified by their *order*. A *first-order* loop is any closed path that returns to the starting node of the loop without passing through any node more than once. The transmittance of a first-order loop is the product of the individual transmittances involved in its passage. Higher order loops are defined as follows: a jth-order loop consists of j non-touching first-order loops. Its transmittance is the product of the transmittances of the first order loops it contains.

The general form of Mason's rule gives the equivalent transmittance, i.e., the MGF, from input to output as

$$T_E(s) = \frac{\sum_i P_i(s)[1 + \sum_j (-1)^j L_j^i(s)]}{1 + \sum_j (-1)^j L_j(s)}, \qquad (31.2.1)$$

where $P_i(s)$ is the transmittance for the ith path, $L_j(s)$ in the denominator is the sum of the transmittances over the jth-order loops, and $L_j^i(s)$ is the sum of the transmittances over jth-order loops sharing no common nodes with the i path, i.e., loops not touching the path.

31.3   RELIABILITY APPLICATION: HYDRAULIC PUMP SYSTEM

We consider the hydraulic pump system with feedback of Limnios (1992). In the physical system, the pumps are in parallel. The system is designed to work with 1 or 2 pumps. In Section 31.2.1 we considered this as an equivalent series system with a flowgraph model in which the states represent the number of failed (or functioning) pumps. In addition, we now allow for a failed pump to be repaired which makes this a series system with feedback. The equivalent flowgraph model is given in Figure 31.1. Limnios assumes that each pump operates independently and fails according to an $exp(\lambda)$ waiting time distribution. Once failed, a pump is repaired at an $exp(\mu)$ waiting time distribution. In state 0 we have two functioning pumps and waiting time to failure is the minimum of two independent $exp(\lambda)$ random variables, $exp(2\lambda)$. In state 1, we have one working pump and one failed pump. The transition from state 1 occurs to state 2 if the working pump fails before the failed pump is fixed and transition occurs to state 0 if the failed pump is repaired before the good pump fails. Therefore, we observe the minimum of an $exp(\lambda)$ with an $exp(\mu)$, which is $exp(\lambda + \mu)$. The corresponding transition probabilities and waiting time MGFs are

$$p_{01} = 1, \quad M_{01}(s) = \frac{2\lambda}{2\lambda - s}$$

$$p_{10} = \frac{\mu}{\lambda + \mu}, \quad M_{10}(s) = \frac{\mu + \lambda}{\mu + \lambda - s}$$

$$p_{12} = \frac{\lambda}{\lambda + \mu}, \quad M_{12}(s) = \frac{\mu + \lambda}{\mu + \lambda - s}$$

Using (31.2.1) in conjunction with the flowgraph model readily gives the MGF of the waiting time to total failure (TTF) as

$$M(s) = \frac{\left(\frac{\lambda}{\lambda+\mu}\right)\left(\frac{2\lambda}{2\lambda-s}\right)\left(\frac{\lambda+\mu}{\lambda+\mu-s}\right)}{1 - \left(\frac{\mu}{\lambda+\mu}\right)\left(\frac{\mu+\lambda}{\mu+\lambda-s}\right)\left(\frac{2\lambda}{2\lambda-s}\right)} \tag{31.3.2}$$

Limnios found system parameter estimates of $\hat{\lambda} = 1/1000$ and $\hat{\mu} = 1/100$, and after much detailed algebra, a mean time to total failure (MTTF) of 6800. The flowgraph analysis of the same model is much simpler and gives the correct MTTF $M'(0) = 6500$. While the MTTF is a quantity of interest in systems analysis and derivable from well known methods for Markovian systems, the flowgraph model provides this in a very simple form and gives much more than the MTTF. Flowgraph models provide the *entire* waiting time distribution of the TTF as well as the system reliability and hazard functions. We rewrite (31.3.2) as

$$M(s) = \frac{2\lambda^2}{s^2 - (3\lambda + \mu)s + 2\lambda^2} \tag{31.3.3}$$

$$= 2\lambda^2 \left[\frac{1}{(\beta - \alpha)(\alpha - s)} + \frac{1}{(\alpha - \beta)(\beta - s)} \right]$$

$$\text{for } s < min(\alpha, \beta), \ \alpha \neq \beta \tag{31.3.4}$$

where $\alpha = \alpha(\lambda, \mu)$ and $\beta = \beta(\lambda, \mu)$ are the factors of the denominator of (31.3.3). The estimated parameter values give $\hat{\alpha} = 0.012844$ and $\hat{\beta} = 0.00015571$. Each term in (31.3.4) is the MGF of an exponential distribution. Inverting, gives the density of T, the time to total failure as the mixture of exponentials,

$$f(t) = \frac{2\lambda^2}{\alpha - \beta}[e^{-\beta t} - e^{-\alpha t}] \quad \text{for } \lambda > 0, \ \mu > 0, \ t > 0. \qquad (31.3.5)$$

Figure 31.4 gives the maximum likelihood estimates for the density, $\hat{f}(t)$, and the hazard function, $\hat{h}(t)$. If these waiting times were not exponential, a situation where methodology for Markovian systems does not work, the flowgraph model is still applicable. See Huzurbazar (1998) for development of flowgraph models for complex engineering systems such as cellular telephone networks.

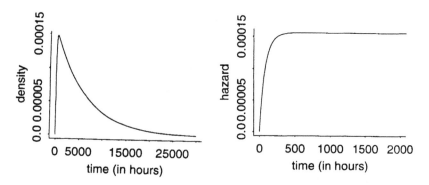

FIGURE 31.4 Density and hazard function for hydraulic pump application

31.4 SURVIVAL ANALYSIS APPLICATION: A FEED FORWARD MODEL FOR HIV

Another type of feed forward model is useful in modeling the incubation time for AIDS. The Markov model was suggested by Brookmeyer and Liao (1992). Aalen (1995) proposes a phase type (PH) distribution model for this problem.

Figure 31.5 shows the flowgraph model used to generalize Aalen's PH formulation. This model is similar to the series model used in Longini (1989) but with an additional feed forward to account for treatment. The

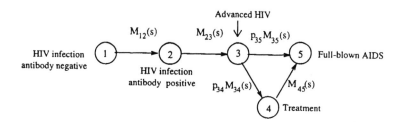

FIGURE 31.5 Flowgraph model for HIV/AIDS

overall waiting time MGF from the equivalent transmittance is

$$M(s) = p_{35}M_{12}(s)M_{23}(s)M_{35}(s) + p_{34}M_{12}(s)M_{23}(s)M_{34}(s)M_{45}(s)$$
(31.4.6)

and $p_{34} + p_{35} = 1$. We use a Markov model with waiting time distributions $Exp(\alpha)$ for $1 \to 2$, $Exp(\beta)$ for $2 \to 3$, $Exp(\lambda)$ for $3 \to 5$, $Exp(\gamma)$ for $3 \to 4$, and $Exp(\theta\lambda)$ for $4 \to 5$. Note that we use exponential distributions only to retain comparability with previously suggested models. We can use any other distribution with a tractable MGF in the flowgraph model of (31.4.6). The parameter γ represents the rate at which treatment can be offered. The parameter θ represents the factor with the progression of AIDS is slowed down as an effect of treatment. The MGF of the survival time distribution of the time to incubation is

$$
\begin{aligned}
M(s) = {} & \left(\frac{\lambda}{\lambda+\gamma}\right)\left(\frac{\lambda+\gamma}{\lambda+\gamma-s}\right)\left(\frac{\alpha}{\alpha-s}\right)\left(\frac{\beta}{\beta-s}\right) \\
& + \left(\frac{\gamma}{\lambda+\gamma}\right)\left(\frac{\alpha}{\alpha-s}\right)\left(\frac{\beta}{\beta-s}\right)\left(\frac{\lambda+\gamma}{\lambda+\gamma-s}\right)\left(\frac{\theta\lambda}{\theta\lambda-s}\right)
\end{aligned}
$$
(31.4.7)

Using a partial fraction expansion, the overall waiting time density is

$$
\begin{aligned}
f(t) = {} & \frac{\lambda^2\beta\theta - \alpha\beta\lambda + \gamma\beta\theta\lambda}{(\lambda+\gamma-\alpha)(\beta-\alpha)(\theta\lambda-\alpha)}\alpha e^{-\alpha t} \\
& + \frac{\lambda^2\beta\theta - \alpha\beta\lambda + \gamma\alpha\theta\lambda}{(\lambda+\gamma-\beta)(\alpha-\beta)(\theta\lambda-\beta)}\beta e^{-\beta t} \\
& + \frac{\alpha\beta\lambda(\theta-1)}{(\alpha-\lambda-\gamma)(\beta-\lambda-\gamma)(\lambda\theta-\gamma-\lambda)}(\lambda+\gamma)e^{-(\lambda+\gamma)t} \\
& + \frac{\alpha\beta\gamma}{(\alpha-\theta\lambda)(\beta-\theta\lambda)(\lambda+\gamma-\theta\lambda)}\theta\lambda e^{-\theta\lambda t},
\end{aligned}
$$
(31.4.8)

a finite mixture of exponentials. This can be used to acquire the corresponding CDF, survival, and hazard functions. The form in (31.4.8) is equivalent to the overall survival function for this model given as in Aalen (1995). Alternatively, symbolic algebra or numerical inversion can be used in conjunction with the flowgraph. More generally, for a non-Markov model with some non-exponential waiting times, analytic inversion of the MGF using symbolic algebra can be cumbersome or even impossible. Saddlepoint approximations remain a viable inversion method. The flowgraph model requires only that the waiting time MGFs exist. See Huzurbazar (1999) for generalizations of other PH models.

31.5 CONCLUSION

Flowgraph models provide an innovative approach to the analysis of time-to-event data. Computational aspects begin with determining a flowgraph model for a given system. In the reliability context this is determined by the physical system. However, in the (medical) survival analysis setting this flowgraph must come from the subject matter specialists. Available data on the whole or partial system must be incorporated into the model. Estimation based on the flowgraph model can be performed in the Bayesian framework or via maximum likelihood. If the likelihood is intractable, flowgraph models can be used to reconstruct likelihood [see Butler and Huzurbazar (1997)]. This situation occurs when data is incomplete, as often happens in survival analysis. MGFs of relevant quantities of interest are then computed from the flowgraph model. Inversion methods such as saddlepoint approximations can be used to convert this into a density, CDF, reliability or survival, and hazard functions. For complex systems we strongly recommend the use of a symbolic algebra package such as MAPLE. This paper focused on discussing the methodology when the system MGF was exactly invertible. Flowgraph models handle the traditional analysis of Markovian systems but also allow the use of non-exponential waiting times thus extending the modeling to semi-Markov processes. They also allow analysis of systems with feedforward and feedback loops in the presence of general waiting time distributions. These systems are intractable by standard methods even under the Markovian assumption of exponential waiting times. The only constraint is that the MGF of the waiting time distributions used at each stage be tractable. This includes a wide class of distributions commonly used to model time-to-event data such as the exponential, gamma, Weibull, inverse Gaussian, the compound exponential, and the Gompertz. In such cases, the overall MGF can be quite complicated and much be inverted approximately to recover the estimated density, CDF, reliability or survival, and hazard functions. Flowgraph models are applicable to a wide variety of situations. In the medical setting, flowgraphs are useful as models for disease progression or models of phase type. In the engineering setting, they can be used for systems analysis and design of complex systems such as cellular telephone networks.

REFERENCES

Aalen, O. O. (1995). Phase type distributions in survival analysis, *The Scandinavian Journal of Statistics*, **22**, 447–463.

Brookmeyer, R. and Liao, J. (1992). Statistical methods for reconstructing infection curves, In *AIDS Epidemiology: Methodological Issues* (Eds., N. P. Jewell, K. Dietz, and V. T. Farewell), pp. 39–60, Birkhäuser, Basel.

Butler, R. W. and Huzurbazar, A. V. (1997). Stochastic network models for survival analysis, *Journal of the American Statistical Association*, **92**, 246–257.

Daniels, H. (1954). Saddlepoint approximations in statistics, *Annals of Mathematical Statistics*, **25**, 631–650.

Huzurbazar, A. V. (2000). Analysis of engineering systems data using flowgraph models, *Technometrics* (to appear).

Huzurbazar, A. V. (1999). Flowgraph models for generalized phase type distributions with non-exponential waiting times, *The Scandinavian Journal of Statistics*, **26**, 145–157.

Huzurbazar, S. and Huzurbazar, A. V. (1999). Survival and hazard functions for progressive diseases using saddlepoint approximations, *Biometrics*, **55**, 203–210.

Limnios, N. (1992). Throughput availability in Markov systems, *IEEE Transactions on Reliability*, **41**, 219–224.

Longini, I., Clark, W., Byers, R., Ward, J., Darrow, W., Lemp, G. and Hethcote, H. (1989). Statistical analysis of the stages of HIV infection using a Markov model, *Statistics in Medicine*, **8**, 831–843.

Mason, S. J. (1953). Feedback theory—some properties of signal flow graphs, *Proceedings of the Institute of Radio Engineers*, **41**, 1144–1156.

Part IX

Applications to Economics and Management

CHAPTER 32

INFORMATION MATRIX TESTS FOR THE COMPOSED ERROR FRONTIER MODEL

ANIL K. BERA

University of Illinois at Urbana-Champaign, Champaign, IL

NARESH C. MALLICK

Alabama Agricultural & Mechanical University, Normal, AL

Abstract: Estimation of composed error frontier models is generally conducted under certain strict assumptions. In practice, however, these assumptions are not tested thoroughly. This is probably because simple workable tests are not yet available for these models. This paper develops easily computable specification tests for half-normal composed error frontier models. The tests are based on the information matrix (IM) and moment test principles. These tests are applied to the well-known Cowing (1970) steam-electric data set. Our tests reveal no serious misspecification of the cost model, while for the output model the null hypothesis of correct specification is rejected strongly.

Keywords and phrases: Frontier model, composed error, information matrix test, moment conditions, parameter variation, heteroskedasticity, nonnormality

32.1 INTRODUCTION

Given the technology of production and inputs bundle, a production frontier refers to the maximum output obtainable. Its dual is the cost frontier

575

which refers to the minimum cost required to produce a given level of output. The distance by which a firm lies below its output frontier or above its cost frontier is a measure of the firm's inefficiency in the frontier literature. The total inefficiency can be broken down into two parts: allocative and technical inefficiencies. If a firm's given inputs allocation is optimal for some higher level of output than the observed one (i.e., a firm operates on the expansion path), the firm is said to be technically inefficient. On the other hand, if it operates at any point on the isoquant for the observed output other than its optimal one, the firm is said to be allocatively inefficient. Operation at any point other than these two points mentioned above and the optimal point for the observed output results in both technical and allocative inefficiencies. It is obvious from these definitions that inefficiencies of any sort causes a firm to deviate from its either frontiers and is costly to the firm.

Substantial research has been done following the pioneering work of Farrell (1957) focusing on the analysis of these inefficiencies. The research on the frontier model received further attention following the independent formulation of the composed error stochastic frontier model by Aigner, Lovell and Schmidt (1977) and by Meeusen and Van den Broeck (1977). A large number of models, both for cross-sectional and panel data, have been fitted under different distributional assumptions for the asymmetric error term. However, so far, not much attention has been directed toward testing the specifications of these models. Lee (1983) tested the half-normal and truncated normal distributional assumptions separately for the asymmetric error term against a Pearson family of distributions. Schmidt and Lin (1984) performed a test of the normality of residuals. Rao's score statistic does not exist in this particular case because the scores vanish under the null hypothesis. As an alternative solution, they suggested the use of the skewness coefficient $\sqrt{b_1}$ to test for the normality of composed error against positively skewed distribution and rejected the null for the well-known steam-electric data set previously used by Schmidt and Lovell (1979). To overcome the vanishing score problem, Lee and Chesher (1986) developed an extremum test procedure based on the second-order derivatives of the log-likelihood function. Kopp and Mullahy (1990) also considered the test of symmetry of the error term and conducted tests of overidentifying restrictions as a by product of their generalized method of moments (GMM) estimation. A test of symmetry is not a specification test for the frontier model, it is a test for the presence of inefficiency. The GMM overidentification test is a valuable tool for checking the validity of the maintained assumptions of the frontier model. However, the selection of moment conditions is somewhat arbitrary. Our objective here is to develop straightforward tests for the frontier models that could be routinely used in practice. The tests are based on the White information matrix (IM) test. These are also moment tests, but now the moments are based on a well-defined and objective principle. Under the null hypothesis of correct specification, the IM equality holds and the standard maximum likelihood approach leads to an asymptotically valid inference. However, under misspecification, the IM equality fails, and the moment conditions of the IM test are based on the difference between the two estimates of the information matrix. Basically, we are interested in detecting the misspecifications that lead to invalid inference. From this point of view, the IM test principle provides a natural selection of moment conditions. Recent applications of the IM test to standard regression and other econometric models have resulted in some simple and interesting specification tests, for example, see Hall (1987), Bera and Lee (1993), White (1994), and Bera and Zuo (1996).

The plan of this paper is as follows. In Section 32.2, we find the moment conditions on the basis of the IM test both for the output and cost stochastic composed error models. In Section 32.3, we empirically test the moment conditions jointly and separately for both the models using the Cowing (1970) steam-electric data set. Finally, in Section 32.4, we offer some concluding remarks.

32.2 INFORMATION MATRIX TESTS FOR FRONTIER MODELS

32.2.1 The Elements of the IM Test for the Output Model

The composed error stochastic output frontier model (in logarithm form) for Cobb-Douglas production technology $y = A \prod_{j=1}^{m} x_j^{\alpha_j}$ is given by

$$
\begin{aligned}
\ln y_i &= \ln A + \sum_{j=1}^{m} \alpha_j \ln x_{ij} + u_i + v_i, \quad i = 1, 2, \ldots, n \\
&= k + \sum_{j=1}^{m} \alpha_j \ln x_{ij} + \varepsilon_i
\end{aligned}
\tag{32.2.1}
$$

where y and x's denote output and inputs, respectively, $k = \ln A$ and $\varepsilon_i = u_i + v_i$ is the composed error term. In the composed error, the symmetric part $v_i \in (-\infty\, \infty)$ is assumed to be distributed as $N(0, \sigma_v^2)$ while the asymmetric part $u_i \in (-\infty\, 0]$ (the logarithm of technical efficiency) is assumed to be distributed as half-normal $|N(0, \sigma_u^2)|$ and is independent of v_i. Under these assumptions, the probability distribution of ε_i is given by [see Aigner *et al.* (1977)]

$$g(\varepsilon_i, \theta) = \frac{2}{\sigma} h\left(\frac{\varepsilon_i}{\sigma}\right)\left[1 - F\left(\frac{\lambda \varepsilon_i}{\sigma}\right)\right], \qquad (32.2.2)$$

where $\theta = (\sigma^2, \lambda, \beta')'$, with $\beta = (k, \alpha_1, \alpha_2, \ldots, \alpha_m)'$ and $\lambda = \sigma_u/\sigma_v$, $\sigma^2 = \sigma_u^2 + \sigma_v^2$, $h(\varepsilon_i/\sigma) = (1/\sqrt{2\pi})e^{-\varepsilon_i^2/(2\sigma^2)}$, and $F(\lambda \varepsilon_i/\sigma)$ denotes standard normal distribution function evaluated at $(\lambda \varepsilon_i/\sigma)$. From (32.2.2), the log-density function for the i-th observation is given by

$$l_i(\theta) = \text{constant} - \frac{1}{2}\ln\sigma^2 - \frac{\varepsilon_i^2}{2\sigma^2} + \ln\left[1 - F\left(\frac{\lambda \varepsilon_i}{\sigma}\right)\right]. \qquad (32.2.3)$$

As mentioned earlier, the frontier models are not generally tested for their specification validity. This may be due to the fact that simple workable tests are not yet available for these models. In the econometric literature Newey (1985) and Tauchen (1985) type moment tests have been found to be very useful. Under their framework, the procedure involves testing moment conditions, like $E(\nu_{(t,i)}(\theta)) = 0$, $t = 1, 2, \ldots, N$ (the number of moment conditions). These moment conditions are true when the model is correctly specified. However, the selection of these moment functions $\nu_{(t,i)}(\theta)$ is a difficult task. For our choice of $\nu_{(t,i)}(\theta)$, we utilize White's (1982) information matrix (IM) test principle. This test exploits the IM equality that under the correct specification of the model $-E\left(\partial^2 l_i(\theta)/\partial\theta\partial\theta'\right) = E\left[\left(\partial l_i(\theta)/\partial\theta\right)\cdot\left(\partial l_i(\theta)/\partial\theta'\right)\right]$ holds. When this equality holds,

$$D = E\left[\left(\frac{\partial^2 l_i(\theta)}{\partial\theta\partial\theta'}\right) + \left(\frac{\partial l_i(\theta)}{\partial\theta}\right)\cdot\left(\frac{\partial l_i(\theta)}{\partial l_i(\theta)'}\right)\right]$$

will be a null matrix, otherwise it will have some nonzero elements. Each element of the symmetric matrix D gives us a separate moment condition, and there are $p(p+1)/2(= N)$ distinct moment conditions embedded in D, where $p(= m + 3)$ denotes the number of parameters in (32.2.2). To derive the expression for the IM test, we stack all N distinct moment conditions $[\nu_{(1,i)}(\theta), \nu_{(2,i)}(\theta), \ldots, \nu_{(N,i)}(\theta)]$ of D into a single vector $\mu_i(\theta)$ of dimension $(N \times 1)$ which is given by

$$\mu_i(\theta) = \text{vech}\left(\frac{\partial^2 l_i(\theta)}{\partial\theta\partial\theta'} + \frac{\partial l_i(\theta)}{\partial\theta}\cdot\frac{\partial l_i(\theta)}{\partial\theta'}\right), \qquad (32.2.4)$$

where 'vech' stacks the upper triangular elements of the symmetric matrix D in a column vector. To derive the IM test, let us define the following

$$M(\theta) = \frac{1}{n}\sum_{i=1}^{n}\frac{\partial\mu_i(\theta)}{\partial\theta'}, \quad d_i(\theta) = \frac{\partial l_i(\theta)}{\partial\theta}, \quad \text{and} \quad \bar{\mu}(\theta) = \frac{1}{n}\sum_{i=1}^{n}\mu_i(\theta).$$

If $\hat{\theta}$ denotes MLE, the IM test statistic is given by

$$\text{IM} = n\bar{\mu}(\hat{\theta})'\Omega(\hat{\theta})^{-1}\bar{\mu}(\hat{\theta}), \tag{32.2.5}$$

where

$$\Omega(\hat{\theta}) = \frac{1}{n}\sum_{i=1}^{n}[\mu_i(\hat{\theta}) - M(\hat{\theta})\,K_i(\hat{\theta})^{-1}\,d_i(\hat{\theta})]\,[\mu_i(\hat{\theta}) - M(\hat{\theta})\,K_i(\hat{\theta})^{-1}\,d_i(\hat{\theta})]'$$

with $K_i(\hat{\theta}) = -[\partial^2 l_i(\hat{\theta})/\partial\theta\partial\theta']$. Although the above formulation of the IM test appears to be somewhat complicated, involving the third derivatives of the likelihood function, the test can be implemented easily. One way is to run a regression of unit vector on $d_i(\hat{\theta})$ and $\mu_i(\hat{\theta})$ and compute the nR^2 where R^2 is the uncentered coefficient of determination. Under the null hypothesis of correct specification, asymptotically nR^2 is distributed as χ_N^2. An alternative method is to run a multivariate regression of $\mu_i(\hat{\theta})$ on $d_i(\hat{\theta})$ with intercepts, and test the significance of the intercept vector. In our case $\mu_i(\theta)$ has six broad elements corresponding to $\theta = (\sigma^2, \lambda, \beta')'$. The elements are given below (for detail derivation, see Equations (32.4.33)–(32.4.38) in the Appendix A):

$$\frac{\partial^2 l_i(\theta)}{\partial(\sigma^2)^2} + \left(\frac{\partial l_i(\theta)}{\partial\sigma^2}\right)^2 = \frac{1}{4\sigma^4}[e_i^4 - 6e_i^2 + 3$$
$$+ \lambda(\lambda^2 + 2)z_i e_i^3 - 5\lambda z_i e_i] \tag{32.2.6}$$

$$\frac{\partial^2 l_i(\theta)}{\partial\sigma^2\partial\lambda} + \left(\frac{\partial l_i(\theta)}{\partial\sigma^2}\right)\left(\frac{\partial l_i(\theta)}{\partial\lambda}\right) = \frac{1}{2\sigma^2}[2z_i e_i - (1 + \lambda^2)z_i e_i^3] \tag{32.2.7}$$

$$\frac{\partial^2 l_i(\theta)}{\partial\sigma^2\partial\beta'} + \left(\frac{\partial l_i(\theta)}{\partial\sigma^2}\right)\cdot\left(\frac{\partial l_i(\theta)}{\partial\beta'}\right) = \frac{1}{2\sigma^3}[e_i^3 - 3e_i - 2\lambda z_i$$
$$+ \lambda(\lambda^2 + 2)z_i e_i^2]s_i' \tag{32.2.8}$$

$$\frac{\partial^2 l_i(\theta)}{\partial\lambda^2} + \left(\frac{\partial l_i(\theta)}{\partial\lambda}\right)^2 = \lambda z_i e_i^3 \tag{32.2.9}$$

$$\frac{\partial^2 l_i(\theta)}{\partial\lambda\partial\beta'} + \left(\frac{\partial l_i(\theta)}{\partial\lambda}\right)\cdot\left(\frac{\partial l_i(\theta)}{\partial\beta'}\right) = \frac{1}{\sigma}[z_i - (1+\lambda^2)z_i e_i^2]s_i'$$

(32.2.10)

$$\frac{\partial^2 l_i(\theta)}{\partial\beta\partial\beta'} + \left(\frac{\partial l_i(\theta)}{\partial\beta}\right)\cdot\left(\frac{\partial l_i(\theta)}{\partial\beta'}\right) = \frac{1}{\sigma^2}[e_i^2 - 1$$
$$+\lambda(\lambda^2 + 2)e_i z_i]s_i s_i', \quad (32.2.11)$$

where $s_i = (1, \ln x_{i1}, \ln x_{i2}, \ldots, \ln x_{im})'$, $z_i = f(\lambda\varepsilon_i/\sigma)/(1 - F(\lambda\varepsilon_i/\sigma))$ (the output hazard function), and $e_i = \varepsilon_i/\sigma$.

The first thing to note is that if we put $\lambda = 0$, all our moment conditions reduce to those in Hall (1987), who applied the IM test to the standard regression model. When $\lambda = 0$, we essentially have $u_i = 0$ for all i, i.e., there is no inefficiency component. Although, the separate explanations of each moment condition are difficult, we shall make an attempt to explain the important ones. Under the null hypothesis of correct specification, the moment condition in (32.2.9) tests $E(z_i e_i^3) = 0$. This could be viewed as a test for skewness. We know $E(\varepsilon) = -\sqrt{2/\pi}\,\sigma_u$ and $\text{Var}(\varepsilon) = \text{Var}(v) + \text{Var}(u) = \sigma_v^2 + [(\pi-2)/\pi]\sigma_u^2 = \sigma^2 - [E(\varepsilon)]^2$. Therefore, $E(\varepsilon^2) = \sigma^2$ implies $E(\varepsilon_i/\sigma)^2 = E(e_i^2) = 1$, this together with the implication of the test with reference to the parameter λ, test indicator (32.2.6) vanishes when $E(e_i^4) = 3$, and $E(z_i e_i) = 0$. Therefore, the IM test component corresponding to σ^2 in (32.2.6), is testing for kurtosis. The indicator corresponding to (32.2.11) tests moment conditions with respect to the production function parameters vector β, and is a symmetric matrix of dimension $(m+1) \times (m+1)$, having $(m+1)(m+2)/2$ distinct elements. A typical element of (32.2.11) is given by

$$\frac{1}{\sigma^2}[(e_i^2 - 1) + \lambda(\lambda^2 + 2)z_i e_i]\ln x_{ij}\ln x_{ij'},$$

and this provides a test for heteroskedasticity in the frontier model. If we put $\lambda = 0$, we obtain White's (1980) heteroskedasticity test for the traditional linear regression model.

Lee (1983) tested the existence of half-normal and truncated normal distributional assumptions for the asymmetric error term u against truncated Pearson family of distributions:

$$f(u) = \frac{\exp\left(\int_u^0 \frac{a+t}{b_0+b_1 t+b_2 t^2}dt\right)}{\int_{-\infty}^0 \exp\left(\int_s^0 \frac{a+t}{b_0+b_1 t+b_2 t^2}dt\right)ds}.$$

To test the null hypothesis that the correct density function for u is half-normal, i. e.,

$$f(u) = \sqrt{2/(\pi\sigma_u^2)}e^{-\frac{u^2}{2\sigma_u^2}}, \quad u \le 0,$$

Lee (1983) tested the null $H_0 : a = 0$, $b_1 = 0$, and, $b_2 = 0$, using Rao's score test. Under the null, the three nonzero elements of the score vector corresponding to a, b_1, and b_2 use the differences between the conditional (given ε) and unconditional moments of the asymmetric error term u. When these three nonzero elements of the score vector are simplified in our notations, for i^{th} observation, they can be written as

$$h_1 = \frac{1}{1+\lambda^2}[\sigma^2 z_i - \lambda^2 \sigma e_i - \sqrt{(2/\pi)}(1+\lambda^2)\sigma_u],$$

$$h_2 = \frac{\sigma^2\lambda^2}{(1+\lambda^2)^3}[2(\sigma^2 z_i - \sqrt{(2/\pi)}\lambda^2\sigma_u - 3\sigma\lambda^2 e_i + \sigma^2\lambda^2 e_i^2 - \sigma\lambda^4 e_i^3],$$

$$h_3 = \frac{\sigma^4\lambda^4}{(1+\lambda^2)^4}[\lambda^4 e_i^4 + 6\lambda^2 e_i^2 + 3 - \sigma\lambda^2 z_i e_i^3 - 5\sigma z_i e_i - 3(1+\lambda^2)^2].$$

The score principle verifies the zero expectations of the above indicators. Two of these indicators, h_2 and h_3, can be closely linked with our moment conditions (32.2.8) and (32.2.6), respectively. These two indicators test skewness and kurtosis of the model. The first component (h_1) tests the first raw moment condition for the standardized composed error e_i. By construction the tests based on h_1, h_2, and h_3 are concerned only with the distributional assumption of the composed error. On the other hand, the IM test principle, tests the *overall* specification of the model.

We now get the theoretical components of the IM test for the steam-electric data set which has three inputs ($m = 3$); capital (K), fuel (F), and labor (L). The vector of production function parameters β has four elements ($k, \alpha_K, \alpha_F, \alpha_L$), and the parameter vector θ of (32.2.2) has a total of six elements ($p = 6$) i. e., $\theta = (\sigma^2, \lambda, k, \alpha_K, \alpha_F, \alpha_L)'$. This indicates, for our data set, the information matrix IM will have twenty-one distinct moment conditions. To avoid some notational confusion, let us number the relative position of the elements of θ by $(1, 2, \ldots, 6)$. With these notations, the twenty-one elements of $\mu_i(\theta)$ of (32.2.4) for i-th observation are given in the following upper triangular matrix

$$\begin{bmatrix} \nu_{(1,i)}(1,1) & \nu_{(2,i)}(1,2) & \nu_{(3,i)}(1,3) & \nu_{(4,i)}(1,4) & \nu_{(5,i)}(1,5) & \nu_{(6,i)}(1,6) \\ & \nu_{(7,i)}(2,2) & \nu_{(8,i)}(2,3) & \nu_{(9,i)}(2,4) & \nu_{(10,i)}(2,5) & \nu_{(11,i)}(2,6) \\ & & \nu_{(12,i)}(3,3) & \nu_{(13,i)}(3,4) & \nu_{(14,i)}(3,5) & \nu_{(15,i)}(3,6) \\ & & & \nu_{(16,i)}(4,4) & \nu_{(17,i)}(4,5) & \nu_{(18,i)}(4,6) \\ & & & & \nu_{(19,i)}(5,5) & \nu_{(20,i)}(5,6) \\ & & & & & \nu_{(21,i)}(6,6) \end{bmatrix},$$

where $\nu_{(t,i)}(l,q)$ denotes the t-th $(t = 1, 2, \ldots, 21)$ population moment condition between the parameters l and q with $l, q = 1, 2, \ldots, 6$.

Under the null of correct specification of true data generating process (DGP), the expected value of these twenty-one moment conditions are separately zero. Tauchen (1985), suggested carrying out any moment test by running a standard regression of the test criteria (the moment condition) on $d_i(\theta)$ and an intercept term. In our case

$$d_i(\theta) = \left[\frac{\partial l_i(\theta)}{\partial \sigma^2}, \frac{\partial l_i(\theta)}{\partial \lambda}, \frac{\partial l_i(\theta)}{\partial k}, \frac{\partial l_i(\theta)}{\partial \alpha_K}, \frac{\partial l_i(\theta)}{\partial \alpha_F}, \frac{\partial l_i(\theta)}{\partial \alpha_L} \right]'.$$

The test is performed by checking the significance of the intercept term. A typical regression equation for $t-th$ population moment condition $\nu_{(t,i)}(l,q)$ is given by

$$\nu_{(t,i)}(l,q) = \delta_t + \gamma' d_i(\hat{\theta}) + \zeta_i, \qquad (32.2.12)$$

where $\gamma = (\gamma_1, \gamma_2, \ldots, \gamma_6)'$ and ζ_i is error term of this auxiliary regression. Note that $d(\bar{\theta}) = \sum_{i=1}^{n} d_i(\hat{\theta})/n = 0$, and OLS estimate of δ_t is $\hat{\delta}_t = \bar{\nu}_t - \hat{\gamma}' d(\bar{\theta}) = \bar{\nu}_t$, where $\bar{\nu}_t = \sum_{i=1}^{n} \nu_{(t,i)}/n$. Therefore, testing the significance of δ_t equal to zero is equivalent to testing $E(\nu_{(t,i)}) = 0$. To test the hypothesis of no misspecification, we run all twenty-one moment conditions simultaneously using seemingly unrelated regression equation (SURE) technique and perform a joint test that $\delta_t = 0$ for all t. To test each moment condition separately, we run OLS of each moment condition $\nu_{(t,i)}(l,q)$ on $d_i(\hat{\theta})'$ and an intercept term and test intercept (δ_t) equals zero. The same procedure is carried out to test the moment conditions for the cost frontier model. The moment conditions for the cost model is derived in the following section.

32.2.2   The Elements of the IM Test for the Cost Model

The composed error cost frontier model is given by

$$\begin{aligned} C_i &= C_{y_i}^o e^{u_i + v_i} \\ &= ra^{-\frac{1}{r}} \prod_{j=1}^{m} \left(\frac{p_{ij}}{\alpha_j} \right)^{\frac{\alpha_j}{r}} y_i^{\frac{1}{r}} e^{u_i + v_i}. \end{aligned} \qquad (32.2.13)$$

Therefore, the corresponding regression model in logarithm form for the cost model is

$$\ln C_i = \ln r - \frac{1}{r}(k + \sum_{j=1}^{m} \alpha_j \ln \alpha_j - \sum_{j=1}^{m} \alpha_j \ln p_{ij} - \ln y_i) + \omega_i. \qquad (32.2.14)$$

In this formulation of the composed error cost frontier model (32.2.13), $C_{y_i}^o = ra^{-1/r} \prod_{j=1}^m (p_{ij}/\alpha_j)^{\alpha_j/r} y_i^{1/r}$ denotes the frontier (optimal) cost for the observed output y_i with $r = \sum_{j=1}^m \alpha_j$, the returns to scale parameter and p_{ij} is the price of j-th input for i-th firm ($i = 1, 2, \ldots, n$). In the composed error $\omega_i(= u_i + v_i \in (-\infty\,\infty))$, the symmetric component $v_i \in (-\infty\,\infty)$ is distributed as $N(0, \sigma_v^2)$ while the asymmetric part $u_i \in [0\,\infty)$ (denotes the negative logarithm of cost efficiency) is distributed as $|N(0, \sigma_u^2)|$ and is independent of v_i. Under these assumptions, the density function of ω_i is given by

$$g(\omega_i, \theta) = \frac{2}{\sigma} h\left(\frac{\omega_i}{\sigma}\right) F\left(\frac{\lambda\omega_i}{\sigma}\right), \qquad (32.2.15)$$

where $h(\omega_i/\sigma) = 1/\sqrt{2\pi}e^{-\omega_i^2/(2\sigma^2)}$, $\sigma^2 = \sigma_u^2 + \sigma_v^2 = E(\omega_i^2)$, $\lambda = \sigma_u/\sigma_v$ and $F(\lambda\omega_i/\sigma)$ is the standard normal distribution function. The log-density function for the i-th firm is given by

$$l_i(\theta) = \text{constant} - \frac{1}{2}\ln\sigma^2 - \frac{\omega_i^2}{2\sigma^2} + \ln F\left(\frac{\lambda\omega_i}{\sigma}\right). \qquad (32.2.16)$$

Note that we use the same notations λ, σ_u^2, σ_v^2 and σ^2 for both the cost and output models but they are model specific. The elements of IM test for the cost model are (for detail derivations, see Equations (32.4.52)–(32.4.57), Appendix B):

$$\frac{\partial^2 l_i(\theta)}{\partial(\sigma^2)^2} + \left(\frac{\partial l_i(\theta)}{\partial\sigma^2}\right)^2 = \frac{1}{4\sigma^4}[e_i^4 - 6e_i^2 + 3$$
$$-\lambda(\lambda^2 + 2)e_i^3 t_i + 5\lambda e_i t_i] \qquad (32.2.17)$$

$$\frac{\partial^2 l_i(\theta)}{\partial\sigma^2\partial\lambda} + \left(\frac{\partial l_i(\theta)}{\partial\sigma^2}\right)\left(\frac{\partial l_i(\theta)}{\partial\lambda}\right) = \frac{1}{2\sigma^2}[(1 + \lambda^2)e_i^2 - 2]e_i t_i \qquad (32.2.18)$$

$$\frac{\partial^2 l_i(\theta)}{\partial\sigma^2\partial\beta'} + \left(\frac{\partial l_i(\theta)}{\partial\sigma^2}\right)\cdot\left(\frac{\partial l_i(\theta)}{\partial\beta'}\right) = \frac{1}{2r\sigma^3}[e_i^3 - 3e_i + 2\lambda t_i$$
$$-\lambda(\lambda^2 + 2)e_i^2 t_i]a_i' \qquad (32.2.19)$$

$$\frac{\partial^2 l_i(\theta)}{\partial\lambda^2} + \left(\frac{\partial l_i(\theta)}{\partial\lambda}\right)^2 = -\lambda e_i^3 t_i \qquad (32.2.20)$$

$$\frac{\partial^2 l_i(\theta)}{\partial \lambda \partial \beta'} + \left(\frac{\partial l_i(\theta)}{\partial \lambda}\right) \cdot \left(\frac{\partial l_i(\theta)}{\partial \beta'}\right) = \frac{1}{r\sigma}[(1 + \lambda^2)e_i^2 - 1]t_i a_i'$$

$$(32.2.21)$$

$$\frac{\partial^2 l_i(\theta)}{\partial \beta \partial \beta'} + \left(\frac{\partial l_i(\theta)}{\partial \beta}\right) \cdot \left(\frac{\partial l_i(\theta)}{\partial \beta'}\right) = \frac{1}{r^2\sigma^2}[e_i^2 - 1$$
$$-\lambda(\lambda^2 + 2)e_i t_i]a_i a_i' \quad (32.2.22)$$

where $l_{ij} = \ln p_{ij} - \ln \alpha_j$, $e_i = \omega_i/\sigma$, $t_i = f(\lambda\omega_i/\sigma)/F(\lambda\omega_i/\sigma)$ (the cost hazard function), and $a_i = (-1, d_i/r + l_{i1}, d_i/r + l_{i2}, \ldots, d_i/r + l_{im})$, with $d_i = k + \sum_{j=1}^m \alpha_j \ln \alpha_j - \sum_{j=1}^m \alpha_j \ln p_{ij} - \ln y_i$.   Here a_i is a vector of confounded regressors and is a function of y_i, input quantities' price vector, $p_i = (p_{i1}, p_{i2}, \ldots, p_{im})'$, and β. If we compare the cost frontier model's moment conditions (32.2.17)–(32.2.22) with the output moment conditions (32.2.6)–(32.2.11), we see that they are identical except that z_i is replaced by $-t_i$ and s_i is replaced by (a_i/r). In the output model s_i denotes $\partial \ln y_i/\partial \beta$ while (a_i/r) denotes $\partial \ln C_{y_i}^o/\partial \beta$ in the cost model. The interpretations of the moment conditions are also similar to those of the output model.

32.3   EMPIRICAL RESULTS

32.3.1   Output Model Estimation

Estimation of the output model (32.2.1) maximizing the log-likelihood function (32.2.3), for the U.S. steam electric data set with three inputs; capital (K), fuel (F) and labor (L), consisting of 111 observations on output, input quantities and their prices using LIMDEP software package is given in Table 32.1.

TABLE 32.1

parameter	k	λ	α_K	α_F	α_L	σ^2
estimate	-4.207	4.957	0.045	1.070	0.005	0.026
t-ratio	-19.886	2.273	3.007	60.13	0.214	

The total sum of squares (SST), sum of squares of errors (SSE), and sum of residuals (TSR) for this output model at this parameters estimate, respectively, are 108.19, 1.52, and −9.61. The estimates of the variances for the asymmetric and symmetric errors are, respectively $\hat{\sigma}_u^2 = 0.025$ and

$\hat{\sigma}_v^2 = 0.001$. These estimates provide 123.51 as the value of the log-likelihood function (LF).

We further report, at these estimates the mean value of $\hat{e}_i^2 = \hat{\varepsilon}_i^2/\hat{\sigma}^2$ is 0.53, far less than the theoretical expected value of one. The vector of average scores at these estimates provided by LIMDEP is $(-8.620, -0.005, 3.994, 67.296, 64.139, 46.401)$.

The MLE requires zero average value of each score at the estimates. The average scores vector at these estimates is far from being a null vector. Therefore, use of these estimates for testing the moment conditions will not be appropriate. In a quest to find better estimates, we used the simple genetic algorithm (SGA). We would like to report here that the use of IMSL package yielded a slope vector also far from being null, no better than what LIMDEP provided. The estimates provided by SGA are relatively much better in terms of the slopes and the value of the objective function (the log-likelihood function). The estimates provided by SGA are given in Table 32.2.

TABLE 32.2

parameter	k	λ	α_K	α_F	α_L	σ^2
estimate	-4.187	4.015	0.047	1.069	0.003	0.015
t-ratio	-1.715	7.267	0.379	8.008	0.013	

At these estimates, SSE and TSR, respectively, are 1.67 and -10.42. The value of the log-likelihood function (LF) is 147.42, relatively much higher than 123.51, LIMDEP provides. The estimated mean of $\hat{e}_i^2 = \hat{\varepsilon}_i^2/\hat{\sigma}^2$ is 1.0 as compared with 0.53 given by LIMDEP, which is far below the theoretical value of 1.0. These estimates provide a discernible improvement in the estimate of average vector of scores, which is $(0.087, -0.001, 0.004, -0.012, 0.189, 0.007)$. Comparing the values of the log-likelihood function, the vector of scores, and mean of $\hat{e}_i^2$ at these estimates with that of provided by LIMDEP, it is evident that SGA estimates are better.

32.3.2 Moments Test for the Output Model

Here, we test all twenty-one different possible moment conditions listed in (32.2.6)–(32.2.11) with respect to different elements of θ, using the estimates provided by SGA. To test that they are jointly zero for all twenty-one conditions, we run these twenty-one regressions by SURE technique and test that all intercepts are zero. The numerical value of Wald's statistic is 20,794.08. This high value of the test statistic indicates that the hypothesis of correct specification of the model does not hold. Since the value of the

Wald's statistic is so high, we tested the zero intercept of each moment condition on the basis of separate regressions. The numerical values of *t*-*ratios* for testing the intercept term for these regressions are presented in the following upper triangular matrix

$$
\begin{array}{c}
\sigma^2 \\
\lambda \\
k \\
\alpha_K \\
\alpha_F \\
\alpha_L
\end{array}
\left(
\begin{array}{cccccc}
\sigma^2 & \lambda & k & \alpha_K & \alpha_F & \alpha_L \\
1.272 & 0.470 & -1.284 & -1.359 & -1.290 & -1.309 \\
 & -10.557* & 0.186 & 0.218 & 0.217 & 0.162 \\
 & & 0.308 & -2.331* & -3.082* & -1.500 \\
 & & & -2.310 & -3.032* & -2.425* \\
 & & & & -3.066* & -2.679* \\
 & & & & & -1.480
\end{array}
\right).
$$

Elements with an asterisk (*) are significant at 5% level of significance. Focusing our attention on the elements of the matrix, all the *t*-*ratios* of the first row are insignificant, indicating that we cannot reject the moment conditions (32.2.6)–(32.2.8). The second diagonal element representing the *t*-*ratio* with respect to the parameter λ, tests the moment condition (32.2.9) and is highly significant, indicating the skewness of the model. The elements of rows 3–6 are the *t*-*ratios* corresponding to the moment condition (32.2.11). These moment conditions are related to the different parameters of the production technology and therefore, are of prime importance. These indicators (32.2.11) test the heteroskedasticity of the composed error term. Of these ten *t*-*ratios* only two are insignificant. This indicates that the proposition of constant variance of the composed error term for this model is not true. On the basis of joint and separate tests of all the moment conditions, we may conclude that the null hypothesis of no misspecification cannot be accepted for the output composed error model for this data set. There is another way to interpret the IM test results. Chesher (1983) demonstrated that the IM test is equivalent to testing the randomness of the parameters. Since the IM test components corresponding to the regression parameters are significant, it implies that the parameters, in particular α_K and α_F, cannot be taken as fixed across firms. Similarly, the strong significance of the indicator relating to the parameter λ, reveals that the inefficiency parameter λ could also be varying across firms. There has been recent attempt in the literature to capture the varying nature of inefficiency [for example, see, Battese and Coelli (1988) and Kumbhakar (1990)] and heteroscedasticity [Caudill, Ford and Gropper (1995)].

32.3.3 Cost Model Estimation

The observed cost model (32.2.14), has the parameters restriction that the sum of the coefficients of $\ln p_{ij}$ is unity. This restriction imposes a problem in the estimation of the model by maximizing the log-likelihood function (15). To avoid this restriction, Schmidt and Lovell (1979) addressed the problem a little differently. The frontier minimum cost $C_y^o = r a^{-1/r} \prod_{j=1}^m (p_j/\alpha_j)^{\alpha_j/r} y^{1/r}$ is a homogeneous function of degree one in input prices. Therefore, in deriving the cost frontier model, they scaled both sides of the deterministic part of equation (13) by the price of any one input (price of labor for this steam electric data set). Then, the revised cost frontier model becomes

$$\ln(C_i/p_{im}) = K + \sum_{j=1}^{m-1} \phi_j \ln p_{ij} + \frac{1}{r} \ln y_i + \omega_i, \qquad (32.3.23)$$

where $K = \ln r - 1/r(k + \sum_{j=1}^m \alpha_j \ln \alpha_j)$ and $\phi_j = \alpha_j/r$. They estimated this reduced form of the cost model (32.3.23). It is to be noted that the above model (32.3.23), still has the parameters restriction of $\sum_{j=1}^{m-1} \phi_j < 1$. The LIMDEP estimate of the composed error cost frontier model (32.3.23) reconfirms the results of Schmidt and Lovell (1979) for this data set. Since the model (32.3.23) is a reduced form model and still has the parameters restriction, and the slope functions are not easily derivable, we estimated the model (32.2.14) incorporating all restrictions by SGA. The parameters estimate together with their respective t-ratio are given in Table 32.3.

TABLE 32.3

parameter	k	λ	α_K	α_F	α_L	σ^2
estimate	-5.913	5.668	0.168	0.962	0.121	0.027
t-ratio	-1.668	0.260	0.479	2.453	0.366	

The total sum of squares (SST), residual sum of squares (SSE), and sum of residuals (TRS) are, respectively, 82.42, 3.04, and 14.15. The log-likelihood function (LF) evaluated at this estimate is 119.49. The mean of $\hat{e}_i^2 = \hat{\omega}_i^2/\hat{\sigma}^2$ is 1.0. The estimate of the average scores vector of the log-likelihood function for this cost model evaluated at these estimates is (0.041, −0.0, 0.008, −0.06, −0.01, 0.037).

32.3.4 Moments Test for the Cost Model

Using the parameters estimate of Table 32.3, we test all twenty-one moment conditions (32.2.17)–(32.2.22), jointly as well as separately, following the same method as in the output model. The joint test of all intercepts terms are zero also give a high value of Wald test statistic χ^2, rejecting the null of no misspecification. However, the separate test of each moment condition gives somewhat encouraging results toward accepting the null hypothesis of no misspecification. The *t-ratios* for testing the null hypothesis of zero intercept terms of all twenty-one distinct regressions are provided in the following upper triangular matrix

$$
\begin{array}{c}
\sigma^2 \\ \lambda \\ k \\ \alpha_K \\ \alpha_F \\ \alpha_L
\end{array}
\left(
\begin{array}{cccccc}
\sigma^2 & \lambda & k & \alpha_K & \alpha_F & \alpha_L \\
2.486^* & 0.491 & -1.107 & -0.641 & -0.6550 & -0.357 \\
 & -0.491_* & -0.665 & -0.764 & -0.783 & -0.807 \\
 & & 4.820_* & 1.234 & 1.480 & 0.581 \\
 & & & 1.304 & 1.477 & 0.942 \\
 & & & & 1.571 & 1.129 \\
 & & & & & 0.742
\end{array}
\right).
$$

Based on the entries in the matrix, we see all the moment conditions are accepted except the two identified by an asterisk (*). These two *t-ratios* test the moment conditions (32.2.17) and (32.2.22) with respect to the parameter σ^2 and k are significant. We see that for the same data set the parameters estimates derived from the cost and the output models differ markedly and the test conclusions are also are quite different.

Comparing the production function parameter estimates derived from the output and cost frontier models estimations listed, respectively, in Table 32.2 and Table 32.3, we see, for the same data set, the estimates differ substantially. Also our tests conclusions on the null hypothesis of no misspecifications of the cost and output models, are quite different. Schmidt and Lovell (1979) find that the parameter estimates derived from the cost model estimation is economically more meaningful than its counterpart obtained from the output model. Our empirical results also substantiate their findings. But these findings are not in a clear agreement with the well-known *Shephard's duality* theorem which states that the estimates derived from the estimation of the cost and output models should be identical for the Cobb-Douglas production function. In this regards, Chung (1994) reports, "The rationale for this modern approach to the theory of production is that a production technology is identically represented by either the production function or the corresponding cost function." All the empirical findings in the frontier literature do not substantiate this important theorem and seem to advocate the cost model over its output counterpart. We would like to offer possible explanations of this issue. The formulation of output frontier model assumes that the observed output y is allocatively efficient and may be only technically inefficient (i. e., a firm knows and operates on the expansion path). Under this strict assumption of no allocative inefficiency (no human error), $f(= A \prod_{j=1}^{m} x_j^{\alpha_j} \geq y)$ is a technically (hence perfectly) efficient output for the observed cost $C = \sum_{j=1}^{m} p_j x_j$. If there is no allocative inefficiency, the observed input $x_j = x_j(p, \beta, C)$ for all $j = 1, 2, \ldots, m$ [see Chung (1994, pp. 201–202)] and substitution of which makes $f = f(p, \beta, C)$. Therefore, use of $f = A \prod_{j=1}^{m} x_j^{\alpha_j}$ or its counterpart $f(p, \beta, C)$ should render identical parameter estimates otherwise the assumption of allocative efficiency is not appropriate. On the other hand, derivation of the cost composed error frontier model takes the endogenously determined optimal cost $C = C(p, \beta, y)$ and takes care of both inefficiencies (allocative and technical) and therefore, yields more meaningful parameter estimates. Mallick (1995), considering both allocative and technical inefficiencies of a production process, formulates the alternative cost and output frontier models using $f = f(p, \beta, C)$ and finds the identical parameter estimates for both the models substantiating the *duality* theorem. In summary, we can conclude that formulation of frontier model of any type (cost or output) uses the optimal cost or output as the case may be and the at the optimal level the bivariate nature between cost and output cannot be ignored on the ground that price vector p and cost C cannot be used as exogenous variables in formulating the output frontier model.

32.4   CONCLUSION

Following the Aigner *et al.* (1977) formulation of the frontier models, we tested the model assuming the half-normal distribution for the asymmetric error term. The test for no misspecification of frontier models is carried out using White's IM test principle. We perform the test following Tauchen's (1985) suggestion, by running joint regressions of all the moment conditions on the score vector and an intercept term by SURE technique. We also ran OLS of each moment condition on the same set of explanatory variables. In each case, we test the null hypothesis of zero intercept term to test the overall null hypothesis of no misspecification of the models. Our empirical evidence of this test on the Cowing (1970) steam electric data both for the cost and output composed error frontier models leads to contradictory inferences. The null hypothesis of no misspecification is rejected for both the models on the basis of the joint test of all intercept terms. Based on testing individual moment conditions, the correct specification of the cost model was found to be acceptable, while the correct specification of the output frontier model was rejected strongly. Our test reveals some other features. We strongly suspect that the assumption of fixed parameters set for all the firms in the data set may not be true. We should note that the results of this

paper are all based on asymptotic tests, and, therefore, their finite sample
properties cannot be guaranteed from their asymptotic distributions. Finally, the tests we have developed are specific to the particular models and
distributions that we have considered; different information matrix test will
result under different model and distributional assumptions.

APPENDIX A

Derivations of the elements of IM test for the output model

The log-likelihood (log-density) function for i-th observation is

$$l_i(\theta) = \text{constant} - \frac{1}{2}\ln\sigma^2 - \frac{\varepsilon_i^2}{2\sigma^2} + \ln\left(1 - \mathrm{F}\left(\frac{\lambda\varepsilon_i}{\sigma}\right)\right).$$

For a better understanding and quick derivations of the components of IM
test mentioned in Section 32.2, we derive the following results first that we
will refer to repeatedly:

$$\frac{\partial F_i}{\partial\lambda} = f_i e_i, \quad \frac{\partial F_i}{\partial\sigma^2} = -\frac{\lambda}{2\sigma^2}f_i e_i, \quad \frac{\partial F_i}{\partial\beta'} = -\frac{\lambda}{\sigma}f_i s_i',$$

$$\frac{\partial f_i}{\partial\lambda} = -\lambda f_i e_i^2, \quad \frac{\partial f_i}{\partial\sigma^2} = \frac{\lambda^2}{2\sigma^2}f_i e_i^2, \quad \text{and} \quad \frac{\partial f_i}{\partial\beta'} = \frac{\lambda^2}{\sigma}f_i e_i s_i',$$

where $s_i = (1, \ln x_{i1}, \ln x_{i2}, \ldots, \ln x_{im})'$, f_i is standard normal density evaluated at $\lambda\varepsilon_i/\sigma$, and $e_i = \varepsilon_i/\sigma$. To derive the second-order derivatives we
shall also use the following results:

$$\frac{\partial z_i}{\partial\lambda} = -z_i e_i(\lambda e_i - z_i)$$

$$\frac{\partial z_i}{\partial\sigma^2} = \frac{\lambda}{2\sigma^2}(\lambda e_i - z_i)z_i e_i$$

$$\frac{\partial z_i}{\partial\beta'} = \frac{\lambda}{\sigma}(\lambda e_i - z_i)z_i s_i',$$

where $z_i = f(\lambda\varepsilon_i/\sigma)/(1 - \mathrm{F}(\lambda\varepsilon_i/\sigma))$. Using these, the first-order derivatives (scores $d(\theta)$) with respect to each element of the parameters vector
$(\sigma^2, \lambda, \beta')$ are given by

$$\frac{\partial l_i(\theta)}{\partial\sigma^2} = \frac{1}{2\sigma^2}(e_i^2 + \lambda z_i e_i - 1) \tag{32.4.24}$$

$$\frac{\partial l_i(\theta)}{\partial\lambda} = -z_i e_i \tag{32.4.25}$$

$$\frac{\partial l_i(\theta)}{\partial\beta'} = \frac{1}{\sigma}(e_i + \lambda z_i) s_i'. \tag{32.4.26}$$

The second-order derivatives are

$$\frac{\partial^2 l_i(\theta)}{\partial(\sigma^2)^2} = \frac{1}{4\sigma^4}(2 - 4e_i^2 - 3\lambda z_i e_i - \lambda^2 z_i^2 e_i^2 + \lambda^3 z_i e_i^3) \quad (32.4.27)$$

$$\frac{\partial^2 l_i(\theta)}{\partial\sigma^2\partial\lambda} = \frac{1}{2\sigma^2}(1 - \lambda^2 e_i^2 + \lambda z_i e_i)e_i z_i \quad (32.4.28)$$

$$\frac{\partial^2 l_i(\theta)}{\partial\sigma^2\partial\beta'} = \frac{1}{2\sigma^3}(\lambda^3 z_i e_i^2 - \lambda^2 z_i^2 e_i - \lambda z_i - 2e_i)s_i' \quad (32.4.29)$$

$$\frac{\partial^2 l_i(\theta)}{\partial\lambda^2} = z_i e_i^2(\lambda e_i - z_i) \quad (32.4.30)$$

$$\frac{\partial^2 l_i(\theta)}{\partial\lambda\partial\beta'} = \frac{1}{\sigma}[1 - \lambda e_i(\lambda e_i - z_i)]z_i s_i' \quad (32.4.31)$$

$$\frac{\partial^2 l_i(\theta)}{\partial\beta\partial\beta'} = \frac{1}{\sigma^2}[\lambda^2(\lambda e_i - z_i)z_i - 1]s_i s_i'. \quad (32.4.32)$$

Using these results, the elements for the IM (the moment functions or indicators) test can be written as

$$\frac{\partial^2 l_i(\theta)}{\partial(\sigma^2)^2} + \left(\frac{\partial l_i(\theta)}{\partial\sigma^2}\right)^2 = \frac{1}{4\sigma^4}(e_i^4 - 6e_i^2 + 3$$
$$+\lambda(\lambda^2 + 2)z_i e_i^3 - 5\lambda z_i e_i) \quad (32.4.33)$$

$$\frac{\partial^2 l_i(\theta)}{\partial\sigma^2\partial\lambda} + \frac{\partial l_i(\theta)}{\partial\sigma^2}\frac{\partial l_i(\theta)}{\partial\lambda} = \frac{1}{2\sigma^2}[2z_i e_i - (1 + \lambda^2)z_i e_i^3] \quad (32.4.34)$$

$$\frac{\partial^2 l_i(\theta)}{\partial\sigma^2\partial\beta'} + \frac{\partial l_i(\theta)}{\partial\sigma^2}\frac{\partial l_i(\theta)}{\partial\beta'} = \frac{1}{2\sigma^3}[e_i^3 - 3e_i - 2\lambda z_i$$
$$+\lambda(\lambda^2 + 2)z_i e_i^2]s_i' \quad (32.4.35)$$

$$\frac{\partial^2 l_i(\theta)}{\partial\lambda^2} + \left(\frac{\partial l_i(\theta)}{\partial\lambda}\right)^2 = \lambda z_i e_i^3 \quad (32.4.36)$$

$$\frac{\partial^2 l_i(\theta)}{\partial\lambda\partial\beta'} + \frac{\partial l_i(\theta)}{\partial\lambda}\frac{\partial l_i(\theta)}{\partial\beta'} = \frac{1}{\sigma}[z_i - (1 + \lambda^2)z_i e_i^2]s_i' \quad (32.4.37)$$

$$\frac{\partial^2 l_i(\theta)}{\partial\beta\partial\beta'} + \frac{\partial l_i(\theta)}{\partial\beta}\frac{\partial l_i(\theta)}{\partial\beta'} = \frac{1}{\sigma^2}[e_i^2 - 1 + \lambda(\lambda^2 + 2)e_i z_i]s_i s_i'. $$
$$(32.4.38)$$

APPENDIX B

Derivations of the elements of the IM test for the cost model

The observed cost model in logarithm form that incorporates the overall efficiency is

$$\ln C_i \;=\; \ln r - \frac{1}{r}\Big(k + \sum_{j=1}^{m} \alpha_j \ln \alpha_j - \sum_{j=1}^{m} \alpha_j \ln p_{ij} - \ln y_i\Big) + \omega_i$$

$$\;=\; \ln r - \frac{d_i}{r} + \omega_i \,,$$

where $d_i = k + \sum_{j=1}^{m} \alpha_j \ln \alpha_j - \sum_{j=1}^{m} \alpha_j \ln p_{ij} - \ln y_i$. The probability density function of ω_i and its log-likelihood function are, respectively,

$$g(\omega_i, \theta) \;=\; \frac{2}{\sigma} h\Big(\frac{\omega_i}{\sigma}\Big) F\Big(\frac{\lambda \omega_i}{\sigma}\Big) \quad \text{and}$$

$$l_i(\theta) \;=\; \text{constant} - \frac{1}{2}\ln \sigma^2 - \frac{\omega_i^2}{2\sigma^2} + \ln F\Big(\frac{\lambda \omega_i}{\sigma}\Big).$$

To find the first-order condition of the log-likelihood function we shall use the following results:

$$\frac{\partial \omega_i}{\partial k} = \frac{1}{r}, \quad \frac{\partial \omega_i}{\partial \alpha_j} = -\frac{1}{r}(d_i/r + l_{ij}), \quad \frac{\partial d_i}{\partial k} = 1,$$

$$\frac{\partial d_i}{\partial \alpha_j} = 1 - l_{ij}, \quad \frac{\partial l_{ij}}{\partial \alpha_j} = -\frac{1}{\alpha_j}, \quad \frac{\partial f_i}{\partial \lambda} = -\lambda f_i e_i^2,$$

$$\frac{\partial f_i}{\partial \sigma^2} = \frac{\lambda^2}{2\sigma^2} f_i e_i^2, \quad \frac{\partial f_i}{\partial k} = -\frac{\lambda^2}{r\sigma} f_i e_i, \quad \frac{\partial f_i}{\partial \alpha_j} = \frac{\lambda^2}{r\sigma}(d_i/r + l_{ij}) f_i e_i,$$

$$\frac{\partial F_i}{\partial \lambda} = f_i e_i, \quad \frac{\partial F_i}{\partial \sigma^2} = -\frac{\lambda}{2\sigma^2} f_i e_i,$$

$$\frac{\partial F_i}{\partial k} = \frac{\lambda}{r\sigma} f_i, \quad \text{and} \quad \frac{\partial F_i}{\partial \alpha_j} = -\frac{\lambda}{r\sigma}(d_i/r + l_{ij}) f_i,$$

where f_i is the standard normal density evaluated at $(\lambda \omega_i / \sigma)$, $l_{ij} = \ln p_{ij} - \ln \alpha_j$, and $e_i = \omega_i / \sigma$. The score functions (first-order conditions) for the cost model are:

$$\frac{\partial l_i(\theta)}{\partial \sigma^2} \;=\; \frac{1}{2\sigma^2}(e_i^2 - \lambda e_i t_i - 1) \tag{32.4.39}$$

$$\frac{\partial l_i(\theta)}{\partial \lambda} \;=\; e_i t_i \tag{32.4.40}$$

$$\frac{\partial l_i(\theta)}{\partial \beta'} \;=\; \frac{1}{r\sigma}(e_i - \lambda t_i) a_i' \,, \tag{32.4.41}$$

where $a_i = [-1, (d_i/r + l_{i1}), (d_i/r + l_{i2}), \ldots, (d_i/r + l_{im})]'$ and $t_i = \frac{f(\lambda e_i)}{F(\lambda e_i)}$ (the cost hazard function). To derive the second-order derivatives of the log-likelihood function for the cost model we make use of the following results:

$$\frac{\partial t_i}{\partial \sigma^2} = \frac{\lambda}{2\sigma^2}(\lambda e_i + t_i)e_i t_i,$$

$$\frac{\partial t_i}{\partial \lambda} = -(\lambda e_i + t_i)e_i t_i, \quad \text{and}$$

$$\frac{\partial t_i}{\partial \beta'} = \frac{\lambda}{r\sigma}(\lambda e_i + t_i)t_i a_i'.$$

The second-order derivatives of the log-likelihood function are:

$$\frac{\partial^2 l_i(\theta)}{\partial(\sigma^2)^2} = \frac{1}{4\sigma^4}(2 - 4e_i^2 + 3\lambda e_i t_i - \lambda^2 e_i^2 t_i^2 - \lambda^3 e_i^3 t_i) \tag{32.4.42}$$

$$\frac{\partial^2 l_i(\theta)}{\partial \sigma^2 \partial \lambda} = \frac{1}{2\sigma^2}(\lambda^2 e_i^2 + \lambda e_i t_i - 1)e_i t_i \tag{32.4.43}$$

$$\frac{\partial^2 l_i(\theta)}{\partial \sigma^2 \partial \beta'} = \frac{1}{2r\sigma^3}(-\lambda^3 e_i^2 t_i - \lambda^2 e_i t_i^2 + \lambda t_i - 2e_i)a_i' \tag{32.4.44}$$

$$\frac{\partial^2 l_i(\theta)}{\partial \lambda^2} = -e_i^2 t_i(\lambda e_i + t_i) \tag{32.4.45}$$

$$\frac{\partial^2 l_i(\theta)}{\partial \lambda \partial \beta'} = \frac{1}{r\sigma}[\lambda e_i(\lambda e_i + t_i) - 1]t_i a_i' \tag{32.4.46}$$

$$\frac{\partial^2 l_i(\theta)}{\partial k^2} = -\frac{1}{r^2\sigma^2}(\lambda^3 e_i t_i + \lambda^2 t_i^2 + 1) \tag{32.4.47}$$

$$\frac{\partial^2 l_i(\theta)}{\partial k \partial \alpha_j} = \frac{1}{r^2\sigma^2}[(d_i/r + l_{ij})(1 + \lambda^3 e_i t_i + \lambda^2 t_i^2)] \tag{32.4.48}$$

$$\frac{\partial^2 l_i(\theta)}{\partial \alpha_j^2} = \frac{1}{r^2\sigma^2}[(d_i/r + l_{ij})^2(-\lambda^3 e_i t_i - \lambda^2 t_i^2 - 1)]$$
$$+ \frac{1}{r^2\sigma}(e_i - \lambda t_i)[1 - \frac{r}{\alpha_j} - 2(d_i/r + l_{ij})] \tag{32.4.49}$$

$$\frac{\partial^2 l_i(\theta)}{\partial \alpha_j \partial \alpha_s} = \frac{1}{r^2\sigma^2}[(d_i/r + l_{ij})(d_i/r + l_{is})(-\lambda^3 e_i t_i - \lambda^2 t_i^2 - 1)]$$
$$+ \frac{1}{r^2\sigma}(e_i - \lambda t_i)[1 - \frac{2d_i}{r} - (l_{ij} + l_{is})] \text{ for } j \neq s. \tag{32.4.50}$$

Now if we take the expectation of all the second-order derivatives, the expectations of (32.4.49) and (32.4.50) simplify further. Using the first-order conditions, (32.4.39)–(32.4.41), the second term of (32.4.49) and (32.4.50) can be, respectively, written as:

$$\frac{1}{r^2\sigma}(e_i - \lambda t_i)[1 - r/\alpha_j - 2(d_i/r + l_{ij})]$$

$$= -\frac{1}{r}\left[\left(1 - \frac{r}{\alpha_j}\right)\frac{\partial l_i(\theta)}{\partial k} + 2\frac{\partial l_i(\theta)}{\partial \alpha_j}\right] \quad \text{and}$$

$$\frac{1}{r^2\sigma}(e_i - \lambda t_i)[1 - (d_i/r + l_{ij}) - (d_i/r + l_{is})]$$

$$= -\frac{1}{r}\left[\frac{\partial l_i(\theta)}{\partial k} + \frac{\partial l_i(\theta)}{\partial \alpha_j} + \frac{\partial l_i(\theta)}{\partial \alpha_s}\right].$$

The expectations of these two terms evaluated at MLE are zero. Using these results, (32.4.47)–(32.4.50) can be written in a matrix notation as

$$\frac{\partial^2 l_i(\theta)}{\partial \beta \partial \beta'} = -\frac{1}{r^2\sigma^2}[\lambda^2(\lambda e_i + t_i)t_i + 1]a_i a_i'. \tag{32.4.51}$$

Using the above results, the elements for the IM test (the moment functions or indicators) are:

$$\left(\frac{\partial l_i(\theta)}{\partial \sigma^2}\right)^2 + \frac{\partial^2 l_i(\theta)}{\partial(\sigma^2)^2} = \frac{1}{4\sigma^4}[e_i^4 - 6e_i^2 + 3$$
$$-\lambda(\lambda^2 + 2)e_i^3 t_i + 5\lambda e_i t_i] \tag{32.4.52}$$

$$\left(\frac{\partial l_i(\theta)}{\partial \sigma^2}\right)\left(\frac{\partial l_i(\theta)}{\partial \lambda}\right) + \frac{\partial^2 l_i(\theta)}{\partial \sigma^2 \partial \lambda} = \frac{1}{2\sigma^2}[(1 + \lambda^2)e_i^2 - 2]e_i t_i \tag{32.4.53}$$

$$\left(\frac{\partial l_i(\theta)}{\partial \sigma^2}\right)\left(\frac{\partial l_i(\theta)}{\partial \beta'}\right) + \frac{\partial^2 l_i(\theta)}{\partial \sigma^2 \partial \beta'} = \frac{1}{2r\sigma^3}[e_i^3 - 3e_i + 2\lambda t_i$$
$$-\lambda(\lambda^2 + 2)e_i^2 t_i]a_i' \tag{32.4.54}$$

$$\left(\frac{\partial l_i(\theta)}{\partial \lambda}\right)^2 + \frac{\partial^2 l_i(\theta)}{\partial \lambda^2} = -\lambda e_i^3 t_i \tag{32.4.55}$$

$$\left(\frac{\partial l_i(\theta)}{\partial \lambda}\right)\left(\frac{\partial l_i(\theta)}{\partial \beta'}\right) + \frac{\partial^2 l_i(\theta)}{\partial \lambda \partial \beta'} = \frac{1}{r\sigma}[(1 + \lambda^2)e_i^2 - 1]t_i a_i' \tag{32.4.56}$$

$$\left(\frac{\partial l_i(\theta)}{\partial \beta}\right) \cdot \left(\frac{\partial l_i(\theta)}{\partial \beta'}\right) + \frac{\partial^2 l_i(\theta)}{\partial \beta \partial \beta'} = \frac{1}{r^2\sigma^2}[e_i^2 - 1$$
$$-\lambda(\lambda^2 + 2)e_i t_i]a_i a_i'. \tag{32.4.57}$$

Acknowledgements We would like to thank Janet Fitch, Philip Garcia, Subal Kumbhakar, and Robin Sickles for their helpful comments and suggestions. Thanks are also due to Peter Schmidt for his help and for providing us with the data set used in this paper. However, we retain the responsibility of any other errors. The first author would like to acknowledge financial support from the Bureau of Economic and Business Research and the Research Board of the University of Illinois at Urbana-Champaign.

REFERENCES

Aigner, D., Lovell, C. A. K. and Schmidt, P. (1977). Formulation and estimation of stochastic frontier functions models, *Journal of Econometrics*, **6**, 21–37.

Battese, G. E. and Coelli (1988). Prediction of firm-level technical efficiencies with generalized frontier production function and panel data, *Journal of Econometrics*, **38**, 387–399.

Bera, A. K. and Lee, S. (1993). Information matrix test, parameter heterogeneity and ARCH: A synthesis, *Review of Economic Studies*, **60**, 229–240.

Bera, A. K. and Zuo, X–L. (1996). Specification test for linear regression model with ARCH process, *Journal of Statistical Planning and Inference*, **50**, 283–308.

Caudill, S. B., Ford, J. F. and Gropper, D. M. (1995). Frontier estimation and firm specific inefficiency measures in the presence of heteroscedasticity, *Journal of Business and Economic Statistics*, **13**, 105–111.

Chung, J. W. (1994). *Utility and Production Functions: Theory and Applications*, Blackwell Publisher, USA.

Cowing, T. J. (1970). Technical change in steam-electric generation: An engineering approach, *Ph.D. Dissertation*, University of California, Berkeley, CA.

Farrell, M. J. (1957) The measurement of productive efficiency, *Journal of Royal Statistical Society, Series A*, **120**, 253–281.

Hall, A. (1987). The information matrix test for linear model, *Review of Economic Studies*, **54**, 257–263.

Kopp, R. J. and Mullahy, J. (1990). Moment-based estimation and testing of stochastic frontier models, *Journal of Econometrics*, **46**, 165–183.

Kumbhakar, S. C. (1990). Production frontiers, panel data, time-varying technical inefficiency, *Journal of Econometrics*, **45**, 201–211.

Lee, L-F. (1983). A test for distributional assumptions for stochastic frontier models, *Journal of Econometrics*, **22**, 245–267.

Lee, L-F. and Chesher, A. (1986). Specification testing when score test statistics are identically zero, *Journal of Econometrics*, **31**, 121–149.

Mallick, N. C. (1995). Specification tests and a reformulation of frontier models: An alternative approach to efficiency estimation, *Ph.D. Dissertation*, University of Illinois at Urbana–Champaign, Champaign, Illinois.

Meeusen, W. and Van den Broeck, J. (1977). Efficiency estimation from Cobb-Douglas production functions with composed error, *International Economic Review*, **18**, 435–443.

Schmidt, P. and Lovell, C. A. K. (1979). Estimating technical and allocative inefficiencies relative to stochastic production and cost frontiers, *Journal of Econometrics*, **9**, 343–366.

Schmidt, P. and Lin, T-F. (1984). Simple tests of alternative specifications in stochastic frontier models, *Journal of Econometrics*, **24**, 349–361.

Tauchen, G. (1985). Diagnostic testing and evaluation of maximum likelihood models, *Journal of Econometrics*, **30**, 415–443.

White, H. (1982), Maximum likelihood estimation of misspecified models, *Econometrica*, **50**, 1–25.

White, H. (1994). *Estimation, Inference and Specification Analysis*, Cambridge University Press, Cambridge, England.

CHAPTER 33

GENERALIZED ESTIMATING EQUATIONS FOR PANEL DATA AND MANAGERIAL MONITORING IN ELECTRIC UTILITIES

H. D. VINOD   R. R. GEDDES

Fordham University, Bronx, NY

Abstract: Vinod (1997, 1998) discuss the Godambe-Durbin theory of estimating functions (EFs) and its potential in econometrics. Here we consider a popular application of EFs called generalized estimating equations (GEE). It is typically applied to panel data, where the heteroscedasticity is analytically related to β, the regression parameter, and where the dependent variable is binary. Geddes (1997) studies panel data on regulated electric utilities with exclusive geographic franchises, and the turnover of the chief executive officer (CEO) on the job. Our GEE estimates reverse his somewhat counterintuitive result that firm performance variables do not affect the turnover of the CEO. We test the empirical validity of predictions of (i) regulatory slack, (ii) rent seeking, and (iii) political pressure hypotheses, and reject the first.

Keywords and phrases: Estimating functions, panel data logits, econometrics, GLM, managerial turnover

33.1 THE INTRODUCTION AND MOTIVATION

Vinod (1997, 1998) discuss applications of Godambe-Durbin EFs in econometrics. This paper provides a new panel data application of EFs called

597

GEE, which is popular in biostatistics [Dunlop (1994), Diggle *et al.* (1994) and Liang and Zeger (1995)]. We provide an explanation of why GEE is popular by showing that it is simpler and theoretically superior to its competition: least squares (LS) and maximum likelihood (ML). Since econometricians rarely use anything other than LS or ML, this explanation is novel. Although the underlying results are known in the EF literature [Godambe and Kale (1991) and Heyde (1997)], their application to the panel data case clarifies and highlights the advantages of EFs.

We consider a typical logit-type specification and apply GEE to panel data with limited (binary) dependent variables. This application of EF theory will explain why EFs yield simpler and superior estimators here. Consider T real variables y_i $(i = 1, 2, \ldots, T)$:

$$y_i \sim \text{IND}(\mu_i(\beta), \sigma^2 \nu_i(\beta)), \text{ where } \beta \text{ is } p \times 1 \qquad (33.1.1)$$

where IND suggests an independently (not necessarily identically) distributed random variable (r.v.) with mean $\mu_i(\beta)$, variance $\sigma^2 \nu_i(\beta)$ and σ^2 does not depend on β. Let $y = y_i$ be a $T \times 1$ vector and $V = Var(y) = \sigma^2 Var(\mu(\beta))$ denote the $T \times T$ covariance matrix. The IND assumption implies that $V(\mu)$ is a diagonal matrix depending only on the i-th component μ_i of the $T \times 1$ vector μ.

The common parameter vector of interest is β that measures how μ depends on covariates x. The heteroscedastic variances $\nu_i(\beta)$ are somewhat unusual. We emphasize that $\mu_i(\beta)$ and $\nu_i(\beta)$ in (33.1.1) are functionally related to each other through β, implying a "special" kind of heteroscedasticity. If y_i are discrete stochastic processes, (time series data) then μ_i and ν_i are conditional on past data. The usual log-likelihood is:

$$LnL = -(T/2)ln2\pi - (T/2)(ln\sigma^2) - S_1 - S_2 \qquad (33.1.2)$$

where $S_1 = (1/2)\Sigma_{i=1}^{T} ln\nu_i$, and $S_2 = \Sigma_{i=1}^{T}[y_i - \mu_i(\beta)]^2/[2\sigma^2\nu_i(\beta)]$. The first order condition (FOC) for generalized LS (or GLS) is $\partial(S_2)/\partial\beta = 0$. The FOC for maximizing the LnL (ML estimator) is

$$\partial(S_1 + S_2)/\partial\beta = [\partial S_2/\partial\mu_i][\partial\mu_i/\partial\beta] + [\partial(S_1 + S_2)/\partial\nu_i][\partial\nu_i/\partial\beta] = 0,$$

using $\partial S_1/\partial\mu_i = 0$. Thus

$$
\begin{aligned}
\partial LnL/\partial\beta =\ & \Sigma_{i=1}^{T}(y_i - \mu_i)(\partial\mu_i/\partial\beta)/(\sigma^2\nu_i) \\
& -\Sigma_{i=1}^{T}(\partial\nu_i/\partial\beta)/(2\nu_i) \\
& +\Sigma_{i=1}^{T}(y_i - \mu_i)^2(\partial\nu_i/\partial\beta)/(2\sigma^2\nu_i^2).
\end{aligned}
$$

In our context, the quasi score function (QSF) equals the first term

$$[\partial S_2/\partial\mu_i][\partial\mu_i/\partial\beta].$$

Its expectation,

$$E(QSF) = \Sigma_{i=1}^{T}(y_i - \mu_i)(\partial\mu_i/\partial\beta)/(\sigma^2\nu_i) = 0,$$

since $\nu_i > 0$ and $Ey_i = \mu_i$ are assumed. Thus the QSF defined this way alone yields an unbiased EF. Since $(\partial\nu_i/\partial\beta) \neq 0$ is assumed, the inclusion of the remaining two terms of the FOC for ML would obviously lead to a biased EF.

Wedderburn (1974) was motivated by applications to the generalized linear model (GLM), where one is unwilling to specify any more than mean and variance properties. His quasi-likelihood function (QLF) is a hypothetical integral of the QSF. The true integral (i.e., the likelihood function) can fail to exist when the "integrability condition" of symmetric partials is violated, McCullagh and Nelder (1989, p. 333). The EFs are defined as functions of data and parameters, $g(y, \beta)$. Unbiased EFs satisfy $E(g) = 0$. Godambe's (1960) optimal EFs minimize $[Var(g)]/(E\partial g/\partial\beta)^2$.

Godambe (1985) proved that the optimal EF is the quasi-score function (QSF). The optimal EFs (QSFs) are computed from the means and variances, without assuming further knowledge of higher moments (skewness, kurtosis) or the form of the density. The methods based on QLFs are generally regarded as "more robust." For example, Liang et al. (1992, p. 11) show that the traditional likelihood requires additional restrictions.

In matrix notation write (33.1.1) as: $y = \mu + \epsilon$, $E\epsilon = 0$, $E\epsilon\epsilon' = \sigma^2 V = \sigma^2 diag(\nu_i)$. If $D = \{\partial\mu_i/\partial\beta_j\}$ is a $T \times p$ matrix, McCullagh and Nelder (1989, p. 327) show that the $QSF(\mu, \nu)$ is:

$$QSF(\mu, \nu) = D'V^{-1}(y - \mu)/\sigma^2 = \Sigma_{i=1}^{T}(y_i - \mu_i)(\partial\mu_i/\partial\beta)/(\sigma^2\nu_i).$$
$$(33.1.3)$$

The optimal EF estimator of β is obtained by solving the (nonlinear) equation $QSF = 0$ for β. The following three key properties of the QSFs lead to optimality of EF estimators:

(i) Since $E(y - \mu) = 0, E(QSF) = 0$, implying that QSF is an unbiased EF.

(ii) $Cov(QSF) = D'\Omega^{-1}D/\sigma^2 = I_F$, the Fisher information matrix.

(iii) Since $-E(\partial QSF/\partial\beta) = Cov(QSF) = I_F$, its variance reaches the Cramer-Rao lower bound.

These statements do not require V to be diagonal as in (33.1.3), only that V be symmetric positive definite having known functions of β. Vinod

(1997) gives examples where the EF estimator coincides with the least squares (LS) and maximum likelihood (ML) estimators. Our panel data logit example here is more interesting, because the EF estimator is distinct from both LS and ML estimators. While the LS and ML solve FOCs similar to (33.1.4) below, the EF estimator solves $QSF = 0$. The chain rule on the FOC requires a second term involving $(\partial \nu_i / \partial \beta)$, which is nonzero from (33.1.1) due to the special heteroscedasticity. Hence the FOC's of LS and ML are unnecessarily complicated. Our arguments in favor of EFs are (a) that FOCs can be biased EFs and (b) that LS and ML can fail to reach the Cramer-Rao bound, i.e., property (iii) above. We summarize this as:

Result 33.1.1 The first order conditions for ML imply a superfluous second term in:

$$\partial(S_1 + S_2)/\partial\beta = QSF(\mu, \nu) + [\partial(S_1 + S_2)/\partial\nu_i][\partial\nu_i/\partial\beta] \qquad (33.1.4)$$

where $QSF(\mu, \nu)$ is from (33.1.3). The FOCs for GLS are similar to (33.1.4), except that the S_1 term is absent. For both ML and GLS, the second term of (33.1.4) is nonzero under special heteroscedasticity conditions. Only when $(\partial\nu_i/\partial\beta) = 0$, i.e., when the heteroscedasticity does not depend on β, FOCs lead to unbiased EFs, proved to be desirable in EF theory, Heyde (1997).

Depending on how complicated $(\partial\nu_i/\partial\beta)$ is, the second term in (33.1.4) obviously complicates the derivation of ML (normal) equations. Our discussion surrounding Result 33.2.1 of the following section will explain why similarly complicated second terms are present in the so-called 'panel logit/probit' models in econometrics. Econometric literature surveyed in Hajivassiliou and Ruud (1994) uses ingenuity and simulations to surmount the complications. Unfortunately, these attempts ignore the deeper fault of the first order conditions causing biased and/or inefficient equations. We emphasize that $QSF = 0$ is always unbiased and its variance always reaches the Cramer-Rao bound. By contrast the first order conditions defining the ML (or GLS) estimators can be biased equations and their variance may not reach the lower bound whenever $\partial\nu_i/\partial\beta$ values are nonzero.

A lesson of the EF-theory is that biased estimators can be acceptable but biased and inefficient EFs should be avoided. This is why the EF estimator obtained by solving $QSF = 0$ cannot be worse than the full-blown ML estimator. Although counter-intuitive, the simpler $QSF = 0$ is actually superior to ML whenever the heteroscedastic variance $\nu_i(\beta)$ depends on β, in light of (33.1.4) above. The following section provides further details regarding the GEE model for panel data logits and probits.

33.2 GLM, GEE & PANEL LOGIT/PROBIT (LDV) MODELS

This section provides an introduction to general linear model (GLM) and to the EF literature leading to GEE models. It may be skipped by statisticians familiar with the GLM and the GEE. We include it because even recent econometric literature dealing with logit, probit and limited dependent variable (LDV) models continues to ignore GLM and GEE models. For example, Baltagi (1995), Hajivassiliou and Ruud (1994) and Bertschek and Lechner (1998)and their references do not even mention GLM or GEE.

The econometric context of this paper is a limited dependent variable model for panel data (time series of cross sections) typically estimated by the logit or probit. These LDV models are well known in biostatistics since the 1930s. GEE models generalize the LDV models by incorporating time dependence among repeated measurements for an individual subject. When the biometric panel is of laboratory animals having a common heritage, the time dependence is sometimes called the "litter effect." The GEE models incorporate different kinds of litter effects characterized by serial correlation matrices $R(\phi)$ defined later as functions of a parameter vector ϕ. This section ends with a statement of the formulas for the GEE estimator and its variance. We have included a limited discussion of the economic issues regarding our application to CEO turnover, although the details are postponed till the next section.

In light of Result 33.1.1 above, to show that a quasi-ML (or GEE) estimator for panel data logit models is superior to the ML and LS, we have to establish that it has a special kind of heteroscedasticity. This is done in Result 33.2.1 of this section. In preparation for that result and for a better understanding of GEE, we include some discussion of the generalized linear model (GLM) literature, McCullagh and Nelder (1989). This literature shows that the logit is not merely convenient, but implies a "canonical link" for which a "sufficient" statistic exists. Since EF theory and GLM modeling terminology is not well known in econometrics, we begin by placing this material in the familiar context of a regression model with T ($t = 1, \ldots, T$) observations and p regressors:

$$y = X\beta + \epsilon, \quad E(\epsilon) = 0, \quad E\epsilon\epsilon' = \sigma^2\Omega. \tag{33.2.5}$$

The generalized least squares estimator (GLS) minimizes the error sum of squares. If $E(\epsilon\epsilon') = \sigma^2\Omega$ is a known diagonal matrix which is not a function of β, GLS is obtained by solving the normal equations: $g(y, X, \Omega) = X'\Omega^{-1}X\beta - X'\Omega^{-1}y = 0$, for β. Under normality of errors, the GLS coincides with the maximum likelihood estimator. Here the EF estimator defined by $QSF = 0$ leads to the same normal equations. If the diagonals of Ω are functions of β as in panel logit models, the $QSF = 0$ equations are called GEE and are developed as (33.2.10), (33.2.15) and (33.2.16) after we explain the link functions of GLM.

Remark 33.2.1 The GLS is extended into the general linear model (GLM) in three steps, McCullagh and Nelder (1989).

(i) Instead of $y \sim N(\mu, \sigma^2\Omega)$ we allow non-normal distributions with various relations between mean and variance functions. Non-normality permits the expectation $E(y) = \mu$ to take on values only in a meaningful restricted range (e.g., nonnegative integer counts or binary outcomes).

(ii) Define the systematic component $\eta = X\beta = \Sigma_{j=1}^{p}x_j\beta_j$, where $\eta \in (-\infty, \infty)$, is a linear predictor.

(iii) A monotonic differentiable link function $\eta = h(\mu)$ relates $E(y)$ to the systematic component $X\beta$. The t-th observation satisfies $\eta_t = h(\mu_t)$. For GLS, the link function is identity, or $\eta = \mu$, since $y \in (-\infty, \infty)$. When y data are counts of something, we need a link function which makes sure that $X\beta = \mu > 0$. Similarly, for y as binary (dummy variable) outcomes, $y \in [0, 1]$, we need a link function $h(\mu)$ which maps the interval $[0, 1]$ for y on $(-\infty, \infty)$ for $X\beta$. In the CEO example below, we use a binary dummy dependent variable.

Remark 33.2.2 To obtain generality, the normal distribution is often replaced by a member of the exponential family of distributions, which includes Poisson, binomial, gamma, inverse-Gaussian, etc. It is well known that "sufficient statistics" are available for the exponential family. In our context, $X'y$ which is a $p \times 1$ vector similar to β, is a sufficient statistic. A "canonical" link function is one for which a sufficient statistic of $p \times 1$ dimension exists. Some well known canonical link functions for distributions in the exponential family are: $h(\mu) = \mu$ for Normal, $h(\mu) = log\mu$ for Poisson, $h(\mu) = log[\mu/(1 - \mu)]$ for Binomial, and $h(\mu) = -1/\mu$ is negative for gamma distributions.

Remark 33.2.3 Since $h(\mu) = -1/\mu$, based on the gamma distribution, is rarely used in econometrics, it is useful to remark on the special features of this link function. The gamma density is:

$$f(x) = (1/\Gamma(\alpha))e^{-x}\theta x^{\alpha-1}\theta^{\alpha}; \text{ where}, x \geq 0, \theta > 0, \alpha > 0. \quad (33.2.6)$$

Its mean is α/θ, variance is α/θ^2, and the coefficient of variation defined as the (standard deviation)/mean is $\alpha^{-0.5}$, which is a constant, since α is a constant parameter. Thus in applications where the variance increases with the mean, keeping the coefficient of variation constant, the gamma distribution with a fixed α is attractive. Since the support of the gamma density is $[0, \infty)$, rather than the $(-\infty, \infty)$, this is restrictive. However, for many economic variables, including our CEO example in the following section, this may be a desirable restriction. A competitor of the gamma model is the Log-Normal. Firth (1988) supports the gamma over the Log-Normal under mutually reciprocal misspecifications.

Now we state and prove the known result that when y is binary, heteroscedasticity measured by $Var(\epsilon_t)$, the variance of ϵ_t, depends on the regression coefficients β. This dependence result also holds true for the more general case, where y is a categorical variable (e.g., poor, good and excellent as three categories) and to panel data where we have a time series of cross sections. The general cases tend to be tedious and are discussed in the EF literature.

Result 33.2.1 The heteroscedastic $Var(\epsilon_t)$ is a function of the regression coefficients β for a special case where y_t is a binary (dummy) variable from time series (or cross sectional) data (up to a possibly unknown scale parameter).

PROOF.   Let P_t denote the probability that $y_t = 1$. Our interest is in relating this probability to various regressors at time t, or X_t. If the binary dependent variable y_t in (33.2.5) can assume only two values (1 or 0), then regression errors ϵ_t also can and must assume only two values: $1 - X_t\beta$ or $-X_t\beta$. The corresponding probabilities are: P_t and $(1 - P_t)$ respectively, which can be viewed as realizations of a binomial process. Note that

$$E(\epsilon_t) = P_t(1 - X_t\beta) + (1 - P_t)(-X_t\beta) = P_t - X_t\beta. \qquad (33.2.7)$$

Hence the assumption that $E(\epsilon_t) = 0$ itself implies that $P_t = X_t\beta$. Thus we have the result that P_t is a function of the regression parameters β. Since $E(\epsilon_t) = 0$, the $Var(\epsilon_t)$ is simply the square of the two values of ϵ_t weighted by the corresponding probabilities. After some algebra, thanks to certain cancelations, we have $Var(\epsilon_t) = P_t(1 - P_t) = X_t\beta(1 - X_t\beta)$. This proves the key result that both the mean and variance depend on β, where EFs have superior properties. $\square$

We can extend the above result to other situations with limited dependent variables. In econometrics, the canonical link function terminology of Remark 33.2.2 is rarely used. Econometricians typically replace y_t by unobservable (latent) variables and write the regression model as:

$$y_t^* = X_t\beta + \epsilon_t,$$

where the observable

$$y_t = 1 \text{ if } y_t^* > 0, \text{ and } y_t = 0 \text{ if } y_t^* \leq 0. \tag{33.2.8}$$

Now write $P_t = \Pr(y_t = 1) = \Pr(y_t^* > 0) = \Pr(X_t\beta + \epsilon_t > 0)$, which implies that:

$$P_t = \Pr(\epsilon_t > -X_t\beta) = 1 - \Pr(\epsilon_t \leq X_t\beta) = 1 - \int_{-\infty}^{X_t\beta} f(\epsilon_t)d\epsilon_t, \tag{33.2.9}$$

where we have used the fact that ϵ_t is a symmetric random variable defined over an infinite range with density $f(\epsilon_t)$. In terms of cumulative distribution functions (CDF) we can write the last integral in (33.2.9) as $F(X_t\beta) \in [0,1]$. Hence $P_t \in [0,1]$ is guaranteed. It is obvious that if we choose a density which has an analytic CDF, the P_t expressions will be convenient. For example, $F(X_t\beta) = [1 + exp(-X_t\beta)]^{-1}$ is the analytic CDF of the standard logistic distribution. From this, econometric texts derive the logit link function $h(P_t) = log[P_t/(1 - P_t)]$ somewhat arduously. Since $P_t/(1 - P_t)$ is the ratio of the odds of $y_t = 1$ to the odds of $y_t = 0$, the practical implication of the logit link function is to regress the log odds ratio on X_t. Clearly, as P_t in $[0,1]$, the logit is defined by $h(P) \in (-\infty, \infty)$. The probit model is similar and also popular in econometrics. It was first used for bioassay in 1935 and uses the inverse of the CDF of the unit normal distribution as the link function: $h(P_t) = \Phi^{-1}(P_t)$.

Remark 33.2.4 The normality assumption is obviously unrealistic when the variable assumes only a few values, or when the researcher is unwilling to assume precise knowledge about skewness, kurtosis, etc. Recall that QSF of (33.1.3) is the optimum EF and satisfies three key properties. Econometricians generally use a "feasible GLS" estimator, where the heteroscedasticity problem is solved by simply replacing $Var(\epsilon_t)$ by its sample estimates. In the present context of binary data, $Var(\epsilon_t)$ is a function of β and minimizing the S_2 with respect to (wrt) β would have to allow for the dependence of $Var(\epsilon_t)$ on β. See (33.1.4) and Result 33.1.1 above. In the 1970's some biostatisticians simply ignored such dependence on β for computational convenience. The EF-theory proves the surprising result that it would be suboptimal to incorporate the dependence of $Var(\epsilon_t)$ on β by including the extra term in the FOCs of (33.1.4). An initial appeal of EF-theory in biostatistics was that it provided a formal justification for the quasi-ML estimator used since the 1970's. We shall see that GEE goes beyond quasi-ML by offering more flexible correlation structures for panel data.

As in McCullagh and Nelder (1989), we denote the log of the quasi-likelihood by $Q(\mu; y)$ for μ based on the data y. For the Normal distribution $Q(\mu; y) = -0.5(y-\mu)^2$, the variance function $Var(\mu) = 1$ and the canonical link is $h(\mu) = \mu$. For the binomial, $Q(\mu; y) = ylog[\mu/(1 - \mu)] + log(1 - \mu), Var(\mu) = \mu(1 - \mu), h(\mu) = log[\mu/(1 - \mu)]$. For the gamma, $Q(\mu; y) = -y/\mu - log\mu, Var(\mu) = \mu^2$ and $h(\mu) = -1/\mu$. Since the link function of the gamma has a negative sign, the signs of all regression coefficients are reversed if the gamma distribution is used. The quasi-score functions (QSFs) become our EFs as in (33.1.3):

$$\partial Q/\partial \beta = D'\Omega^{-1}(y - \mu) = 0 \qquad (33.2.10)$$

where $\mu = h^{-1}(X\beta)$ and $D = \{\partial \mu_t/\partial \beta_j\}$ is a $T \times p$ matrix of partials and Ω is $T \times T$ diagonal matrix with entries $Var(\mu_t)$ as noted above. The GLM estimate of β is given by solving (33.2.10) for β. Thus the complication arising from a binary (or limited range) dependent variable is solved by using the GLM method.

33.2.1 GLM for Panel Data

The panel data involve an additional complication from three possible subscripts i, j and t. There are $(i = 1, \ldots, N)$ individuals about which cross sectional data are available in addition to the time series over $(t = 1, \ldots, T)$ on the dependent variable y_{it} and p regressors x_{ijt}, with $j = 1, \cdots p$. We avoid subscript j by defining x_{it} as a $p \times 1$ vector. Geddes (1997) estimates a logit model for panel data from electric utilities focusing on the tenure of the chief executive officer (CEO) relating it to age, salary, job performance, price charged for electricity, etc. His logit model estimates suggest the somewhat counterintuitive result that CEO job performance variables do not have a statistically significant effect on the survival of the CEO. This paper reviews that result from the GEE perspective.

Let y_{it} represent a binary choice variable such that $y_{it} = 1$, if the CEO is removed and $y_{it} = 0$, otherwise. Let $P_{i,t}$ denote the probability of turnover of ith CEO $(i = 1, \ldots, N)$ at time t $(t = 1, \ldots, T)$ and note that:

$$E(y_{it}) = 1P_{i,t} + 0(1 - P_{i,t}) = P_{i,t}. \qquad (33.2.11)$$

Now, we remove the time subscript by collecting elements of $P_{i,t}$ and y_{it} into $T \times 1$ vectors and write $E(y_i) = P_i$, as a vector of turnover probabilities for the i-th individual CEO. Let X_i be a $T \times p$ matrix of data on regressors for ith individual. As before, let β be a $p \times 1$ vector of regression parameters. If the method of latent variables is used, the decision to remove a CEO is assumed to be based on latent unobservable positive dissatisfaction y_{it}^* by the board of directors with the CEO's performance. Thus

$$
\begin{aligned}
y_{it} &= 1, \text{ if } y_{it}^* > 0 \text{ or} \\
y_{it} &= 0, \text{ if } y_{it}^* \leq 0
\end{aligned}
\qquad (33.2.12)
$$

where $y_{it} = 1$ if the board is dissatisfied with the CEO and $y_{it} = 0$ if the board is satisfied.

Following the GLM terminology of link functions, we may write the panel data model as:

$$
\begin{aligned}
h(P_i) &= X_i \beta_i + \epsilon_i, \\
E(\epsilon_i) &= 0, \qquad E\epsilon_i \epsilon_i' = \sigma^2 \Omega_i \text{ for } i = 1, \ldots N.
\end{aligned}
\qquad (33.2.13)
$$

Now the logit link has $h(P_i) = log[P_i/(1 - P_i)]$, probit link has $h(P_i) = \Phi^{-1}(P_i)$ and the gamma density of (33.2.6) implies reciprocal link $h(\mu) = -1/\mu$.

33.2.2   Random Effects Model from Econometrics

Instead of N separate β_i parameters for each CEO as in (33.2.13), econometricians often pool the data for all CEOs and split the errors as $\epsilon_{it} = M_i + \nu_{it}$, where ν_{it} represents "random effects" and M_i denotes the "individual effects." Using the logit link, the log-odds ratio in a so-called random effects model is written as:

$$
log(P_{i,t}/(1 - P_{i,t})) = x_{it}' \beta + M_i + \nu_{it},
\qquad (33.2.14)
$$

The random effects model also assumes that $M_i \sim IID(0, \sigma_M^2)$ and $\nu_{it} \sim IID(0, \sigma_\nu^2)$ are independent of each other and also independent of the regressors x_{it}. It is explained in the panel data literature, Baltagi (1995, p. 178), that these individual effects complicate matters significantly. Note that under the random effects assumptions in (33.2.13), covariance over time is nonzero, $E(\epsilon_{it} \epsilon_{is}) = \sigma_M^2$. Hence independence is lost and the joint likelihood (probability) cannot be rewritten as a product of marginal likelihoods (probabilities). Since the only feasible maximum likelihood implementation involves numerical integration, we may consider a less realistic "fixed effects" model where the likelihood function is a product of marginals. Unfortunately, the fixed effects model still faces the so-called "problem of incidental parameters" (the number of parameters M_i increases indefinitely as $N \to \infty$). Some other solutions from the econometrics literature referenced by Baltagi include Chamberlain's (1980) suggestion to maximize a conditional likelihood function. These ML or LS methods continue to suffer from unnecessary complications arising from the extra term (See Eq.. 33.1.4), which would make their FOCs (See Eq.. 33.1.3) biased and inefficient.

33.2.3 Derivation of GEE, the Estimator for β and Standard Errors

Next, we describe how panel data GEE methods can avoid the difficult and inefficient LS or ML solutions in the econometrics literature. We shall write a quasi score function justified by the EF-theory as our GEE. We achieve a fully flexible choice of error covariance structures by using link functions of the GLM. Since GEE is based on the QSF (See Eq.. 33.1.3), only the mean and variance are assumed to be known. The distribution itself can be any member of the exponential family with almost arbitrary skewness and kurtosis - not just the normal distribution assumed in the literature. Denoting the log likelihood for i-th individual by L_i we construct a $T \times 1$ vector $\partial L_i/\partial \beta$. Similarly, we construct y_i and $\mu_i = h^{-1}(X_i'\beta)$ as $T \times 1$ vectors and suppress the time subscripts. Denote a $T \times p$ matrix of partial derivatives by $D_i = \{\partial \mu_i/\partial \beta_j\}$ for $j = 1, \cdots p$. When there is heteroscedasticity but no autocorrelation, $Var(y_i) = \Omega_i = diag(\Omega_t)$ is $T \times T$ diagonal matrix of variances of y_i over time. Using these notations, the i-th QSF similar to (33.2.10) above is:

$$\partial L_i/\partial \beta = D_i'\Omega_i^{-1}(y_i - \mu_i) = 0. \tag{33.2.15}$$

When panel data are available with repeated N measurements over T time units, GEE methods view this as an opportunity to allow for both autocorrelation and heteroscedasticity. The pooling over i leads to an aggregate QSF from (33.2.15) called generalized estimating equation (GEE):

$$\Sigma_{i=1}^N D_i'V_i^{-1}(y_i - \mu_i) = 0, \text{ where } V_i = \Omega_i^{0.5}R(\phi)\,\Omega_i^{0.5}, \tag{33.2.16}$$

where $R(\phi)$ is a $T \times T$ matrix of serial correlations viewed as a function of a vector of parameters ϕ. The sandwiching of $R(\phi)$ autocorrelations between two matrices of (heteroscedasticity) standard deviations in (33.2.16) makes V_i a proper covariance matrix. The GEE user can simply specify the general nature of the autocorrelations by choosing $R(\phi)$ from the following list, stated in increasing order of flexibility. The list contains common abbreviations used by authors of software.

(i) 'Independence' means $R(\phi)$ is the identity matrix.

(ii) 'Exchangeable' $R(\phi)$ means that all intertemporal correlations defined by $corr(y_{it}, y_{is}) = \phi$, are constant.

(iii) 'AR(1)' or first order autoregressive model implies that $R(\phi)$ or $corr(y_{it}, y_{is})$ simply equals $\phi^{|t-s|}$.

(iv) 'Unstructured' correlations in $R(\phi)$ means that $corr(y_{it}, y_{is}) = \phi_{ts}$ with $T(T-1)/2$ distinct values for all pairwise correlations.

Finally, solving (33.2.16) for β gives the GEE estimator, which is usually iteratively estimated. Liang and Zeger (1986) suggest a "modified Fisher scoring" algorithm for these iterations. The initial choice of $R(\phi)$ is usually the identity matrix and standard GLM is first estimated. The GEE algorithm then estimates $R(\phi)$ from the residuals of the GLM and iterates until convergence. We use Smith (1996) software in S-PLUS language on an IBM compatible computer. The theoretical justification for iterations exploits the property that a QML estimate is consistent even if $R(\phi)$ is misspecified, [Zeger and Liang (1986) and McCullagh and Nelder (1989, p. 333)]. Denoting by $\hat{R}$ the estimates of R, the asymptotic covariance matrix of GEE estimator is:

$$Var(\hat{\beta}_{gee}) = \sigma^2 A^{-1} B A^{-1}, \quad \text{with} \quad A = \Sigma_{i=1}^{N} D_{i'} \hat{V}_i^{-1} D_i$$

and

$$B = \Sigma_{i=1}^{N} D_{i'} \hat{R}_i^{-1} R_i \hat{R}_i^{-1} D_i. \tag{33.2.17}$$

This expression yields the robust standard errors reported in our numerical work in the next section. See Zeger and Liang (1986) and Dunlop (1994) for further discussion and references.

In the following section we use the gamma family to fix the relation between the mean and variance, instead of the traditional binomial or Poisson family. This is mainly because the gamma family gives better fits as measured by the lowest residual sum of squares (RSS) than other families of distributions. Remark 33.2.3 above notes other reasons why the gamma family and its canonical link may be appropriate. McCullagh and Nelder (1989, p. 290) suggest a deviance function as the difference between two log likelihoods instead of RSS. We do not use deviances, since they need artificial truncation to avoid computing log of a near zero number for the gamma family. A proof of consistency of the GEE estimator is given by Li (1997). Heyde (1997, p. 89) gives the necessary and sufficient conditions under which GEE are fully efficient asymptotically. Lipsitz *et al.* (1994) report simulations showing that GEE are more efficient than ordinary logistic regressions. In conclusion, this section has shown that the GEE estimator is practical with attractive properties for the type of data studied here.

33.3 GEE ESTIMATION OF CEO TURNOVER AND THREE HYPOTHESES

In this section, we discuss the motivation for examining managerial turnover using GEE. An important way to align the interests of managers with those of owners is by linking managerial turnover to firm performance. Removal of a manager by a board of directors for performance reasons is a negative signal to the managerial labor market. If boards remove managers when a firm performs poorly, an inverse relationship between managerial turnover and performance will result. Using a variety of data sets, performance measures and empirical techniques, researchers have confirmed this relationship in many industries. Salancik and Pfeffer (1980) report that when outsiders own stock there is a positive and significant correlation between profit margin and managerial tenure. Warner, Watts, and Wruck (1988) and Coughlan and Schmidt (1985) find that the probability of managerial turnover is inversely related to abnormal stock price performance. Weisbach (1988) finds that, given the behavior of stock returns, the probability of managerial turnover is negatively related to accounting performance. Barro and Barro (1990) report a negative relationship between turnover and performance for bank CEOs. The evidence is thus strongly supportive of the hypothesis that performance and managerial turnover are negatively related, consistent with an incentive-alignment view of CEO turnover. There are a number of reasons to believe that the performance-turnover relationship described above may be affected by utility regulation. Investor-owned utilities are typically regulated by state public utility commissions, which administer rate-of-return regulation under exclusive geographic franchises. Two fundamental managerial functions, investment and financing, are determined through the regulatory process. Hence regulation is crucial for managerial decisions.

Regulation supplants decisions normally made by firm owners and their managers with the administrative process. The significance of this control has been the subject of considerable debate. Some researchers suggest that rate-of-return regulation allows managers to incur high "agency costs" with little fear of removal by owners. After all, the return to managerial efficiency, given that the maximum rate-of-return is achieved, is zero. In their classic article, Alchian and Kessel (1962) state: "If regulated monopolists are able to earn more than the permissible pecuniary rate of return, then "inefficiency" is a free good, because the alternative to inefficiency is the same pecuniary income and no "inefficiency." This is essentially a "regulatory slack" view of regulation, which takes regulation as exogenous: if the firm is earning at least the allowed rate-of-return then managers need not be concerned with operating in owners' interests, and managerial inefficiency has zero opportunity cost in the alternative use of maximizing firm value. Regulation leads to a situation where monitoring by owners has a low return, and owners engage in little of it.

Others suggest that regulatory "rent-seeking" behavior will lead to managerial monitoring by owners. For example, Crain and Zardkoohi (1978, 1980) rely on a rent-seeking hypothesis to arrive at the conclusion that managerial control mechanisms operate in regulated firms. They submit that there are potential monopoly rents available through regulation, which can be obtained via rent-seeking activity by managers, and that private owners monitor on this basis. Firm resources, rather than being devoted to profit-enhancing activities such as product development and marketing, are directed at influencing the allowed rate-of-return and other regulated variables that affect economic profits, making regulatory outcomes endogenous to managerial behavior. Such rent-seeking activities include political contributions and public relations programs, as well as payments to lawyers and consultants. Here managerial inefficiency carries a non-zero opportunity cost, providing an incentive for owners to monitor managers.

In a third view, commentators suggest that the regulatory process exposes managerial behavior to political forces. In Stigler's (1971) immortal words, the regulatory process "automatically admits powerful outsiders to industry's councils," or as Joskow, Rose and Shepard (1993) state, "Economic regulation imposes political outcomes in place of some private decisions or market outcomes." This view implies that the regulatory process provides organized pressure groups with a mechanism for translating their interests into outcomes. One important party likely to obtain greater control over the regulatory process is the consumer. In many regulatory processes, consumers are granted avenues by which they can organize and are given a special say in proceedings. The political power of consumers relative to shareholders is expected to increase under a "political pressure" hypothesis. The variable that best measures consumer wealth is the real price of electricity, and we expect consumers to monitor managers on that basis. As applied to turnover, this implies that the political forum in which regulated CEOs operate results in turnover that responds positively to increases in electricity prices.

Our data set, described below, allows us to compare the predictions of

(i) regulatory-slack, (ii) rent-seeking, and (iii) political pressure hypotheses in the case of US electric utilities. In testing these alternative hypotheses, we are able to show how the GEE approach represents an improvement over standard limited dependent variable techniques. Below, we describe our data sources, variables, and present summary statistics. We show how our data allow tests of these hypotheses. We then present estimates of managerial turnover using GEE.

33.3.1 Description of Data

A sample of 95 investor-owned electric utilities (IOUs) was taken from those listed in the Statistics of Privately Owned Electric Utilities in the United States, for which financial and managerial data were available. The test years run from 1966 through 1988. We used numerous data sources to compile a large data set on managerial turnover and variables potentially affecting turnover; see Geddes (1997) for details. The variables in the data set are briefly described below. Due to our need for data on CEO age, salary, and performance measures, the ultimate number of observations in the data set was 790 in variable groups discussed below.

Turnover measure

The focus of this study is on the probability of a change in the senior manager of an electric utility. Senior managers are defined as the president or CEO of an IOU. If a change in the president or CEO was observed but actually the individual moved into the chairman's position this was not counted as a turnover. Our measure of managerial turnover, TURN=1 if the CEO left the firm, and zero otherwise. Data on reasons for departure (firing, quits or illness) are unavailable.

Managerial characteristics

We control for three important managerial characteristics. AGE, the age of the senior manager, is expected to affect turnover positively since managers are more likely to change as they approach retirement age. This effect is more explicitly studied by a dummy variable: RETIRE=1, if the manager is aged 63 through 66. A non-trivial number of managerial changes occur around normal retirement age, and are likely to be unrelated to performance. TENURE is the number of years served as the CEO, and is also expected to positively affect turnover. SALARY is the annual real compensation including bonus of the CEO and we expect that it will be negatively related to turnover, if higher paid CEOs are less likely to leave.

Firm characteristics

We include SIZE as measured by the annual real sales of the firm in dollars. It is often alleged that larger firms have a higher rate of CEO turnover [Warner, Watts, and Wruck (1988)]. We also examine the responsiveness of CEO turnover at regulated firms to shareholder wealth, as measured by accounting returns. Despite the problems with using earnings data to measure economic profits [Fisher and McGowan (1983)], accounting returns do measure short-term profits, rather than the discounted present value of the expected future cash flows of the firm, as measured by the stock price. Since stock prices are forward-looking, they incorporate the possibility that the board will remove the CEO after poor performance [Weisbach (1988)]. The use of stock prices may therefore understate the effects of managerial monitoring. Also, Joskow, Rose and Shepard (1996) note that accounting returns are likely to be relatively more important in electric utilities, which are regulated. Changes in accounting returns are thus the best available measure of changes in owner wealth when examining managerial turnover in electricals. The two measures used here are ROA and ROE. ROA is the firm's realized return-on-assets = (gross income)/(total assets). ROE is the realized return on equity = (gross income)/(total stockholder's equity).

Regional variables

These dummy variables control for the area or region of the country in which the firm operates. They are a proxy for such diverse factors as different federal air pollution standards, availability of cheap hydroelectric power, age of the capital stock, different population densities, weather, etc. To create these variables, the country was divided into seven regions. Northeast (NEDUM), Mid-Atlantic (MADUM), Southeast (SEDUM), South-central (SCDUM), Northwest (NWDUM), Midwest (MWDUM), and the Southwest (SWDUM). The suffix DUM is for dummy variable, i.e., they equal unity if the electric utility operated in that region and zero otherwise. The omitted category was the Southwest. Table 33.1 summarizes the predictions of the three hypotheses and the variables used to test them. Table 33.2 reports the GEE estimation results using the gamma density and assuming that the correlation structure over time is unrestricted [compare to Table 2 of Geddes (1997)]. Since the likely sign of the coefficient is known from the theories discussed above, we use one-sided tests.

33.3.2   Shareholder and Consumer Wealth Variables for Hypothesis Testing

Shareholder wealth

Shareholder wealth comes from accounting returns. Since it is unlikely that managers would respond to the level of shareholder wealth, the year-to-year changes in ROA and ROE were used to construct two new variables, ΔROA and ΔROE. These variables measure the change in shareholder wealth prior to a particular observed firm-year. The regulatory slack hypothesis predicts that managerial turnover will be unrelated to changes in returns, while the rent-seeking hypothesis predicts that turnover will be negatively related to changes in returns. An alternative test of the rent-seeking hypothesis involves "allowed" returns. Since regulatory rent-seeking activity includes attempts by managers to influence the return allowed by the regulatory commission it may result in a greater deviation of the realized rate-of-return from that allowed by the commission. That is, managers may devote resources to maximizing the deviation between the realized and allowed return, and may be monitored by owners on that basis.

The rent-seeking hypothesis is further tested by relating managerial turnover in IOUs to differences between the actual and allowed returns, and to changes in allowed returns. Two new variables were created, ΔDEVROA and ΔDEVROE, which are the year-to-year deviations between the return allowed by the regulatory commission and the realized return for ROA and ROE respectively. The rent-seeking hypothesis predicts that the probability of turnover will decrease as ΔDEVROA and ΔDEVROE increase.

Customer wealth

To test the political pressure hypothesis, a measure of the change in customer wealth was developed. A variable called ΔPRICE was created, which is the year-to-year change in the real price of electricity sold by the firm. If consumers exercise power via the political process, then managerial turnover will be positively related to changes in price. That is, managers will be removed when the real price of electricity rises, and conversely.

TABLE 33.1 Tested hypotheses and predictions

Hypothesis	Variable	Predicted Effect on Turnover
Regulatory slack	ΔROA, ΔROE	No effect
Rent-seeking	ΔROA, ΔROE, ΔDEVROA, ΔDEVROE	Negative
Political pressure	ΔPRICE	Positive

33.3.3 Empirical Results

Firm performance measures

Table 33.2 presents GEE estimates of the effects of firm performance on managerial turnover. It is important to note that the gamma function reverses the signs of the coefficients relative to the logit. In Table 33.2, most variables have the expected effect on turnover. For example, RETIRE, TENURE and ln(SIZE) significantly increase the probability of managerial change. With retirement effects controlled for by RETIRE, it does not appear that AGE affects CEO turnover. SALARY significantly decreases turnover, as predicted. Most regional variables have little effect, with the exception of the Southeast dummy, which decreases the turnover probability. Importantly, both ΔROA and ΔROE decrease the probability of managerial turnover, with ΔROE significant at the 2 percent level. This is strongly at odds with logit estimates reported in Table 33.2 of Geddes (1997, p. 275). These GEE estimates provide support for the rent-seeking hypothesis over the regulatory slack hypothesis, as discussed above.

Turnover and allowed returns

The rent-seeking hypothesis can also be tested by examining the effect managers have on differences between allowed and realized rates of return. That is, managers may be monitored by owners on the basis of their ability to achieve a return higher than that allowed by the regulatory commission. Here, the rent-seeking hypothesis predicts that the probability of turnover will be negatively related to the deviation between actual and allowed returns, ΔDEVROA and ΔDEVROE. GEE parameter estimates including ΔDEVROA and ΔDEVROE are reported in Columns 2 and 3 of Table 33.3.

TABLE 33.2 Generalized estimating equations estimates of the effects of firm performance on managerial turnover in investor-owned utilities

Regressor	(1)	(2)
Constant	73.207 (1.797)**	83.243 (2.147)**
RETIRE	-.806 (-.172)	-.173 (-.044)
AGE	.032 (.076)	-.077 (-.235)
TENURE	-.984 (-3.527)**	-.968 (-3.523)**
SALARY	.042 (1.704)**	.0475 (2.028)**
Ln(SIZE)	-4.672 (-1.911)**	-5.065 (-2.083)**
ΔROA	21.378 (1.452)*	–
ΔROE	–	.478 (2.086)**
NEDUM	2.378 (.502)	2.156 (.491)
SEDUM	28.340 (1.304)*	28.339 (1.298)*
MADUM	-.046 (-.010)	.585 (.128)
NWDUM	24.628 (.570)	24.869 (.578)
MWDUM	.422 (.127)	1.072 (.376)
SCDUM	-.068 (-.019)	.199 (.061)
RSS	47.158	46.928
N	790	790

Note: Robust Z-statistics are in parenthesis. * Significant at the .10 level; ** Significant at the .05 level, one-tailed test. The Variance-to-mean relation is gamma. The link function is reciprocal. Signs are reversed when using gamma.

The reduced number of observations available here is due to lack of data on allowed returns for certain firm-years. Here, ΔDEVROA is significant at approximately the 6 percent level, with greater deviations of the realized return from the allowed decreasing the probability of managerial turnover. These tests do not provide support for a regulatory slack hypothesis, but are consistent with a rent-seeking hypothesis. They suggest that owners do monitor managers on the basis of deviations from allowed returns.

Managerial turnover and electricity price changes

GEE estimates incorporating ΔPRICE are reported in column 1 of Table 33.3. These estimates support an important conclusion: managerial turnover in IOUs is sensitive to increases in price. This is consistent with estimates reported in Geddes (1997), but levels of significance are higher here. It appears that managers in electric utilities are also monitored on the basis of price changes, consistent with a "political pressure" hypothesis.

TABLE 33.3 GEE estimates of the effects of output price and allowed returns on managerial turnover in investor-owned utilities

Regressor	(1)	(2)	(3)
Constant	59.122 (1.312)*	76.509 (1.676)**	68.388 (1.583)*
RETIRE	2.694 (-0.547)	1.805 (0.418)	-1.46 (-0.301)
AGE	0.113 (0.253)	-0.045 (-0.09)	0.0003 (0.001)
TENURE	-0.994 (-3.07)**	-1.362(3.401)**	-1.053 (-2.906)**
SALARY	0.0298 (1.131)	0.042 (1.285)	0.0334 (1.251)
ln(SIZE)	-3.566(-1.372)*	-4.244(-1.372)*	-3.846 (-1.594)*
*Delta*PRICE	-4.065 (-2.51)**	——	——
*Delta*DEVROA	——	0.623 (1.522)*	——
*Delta*DEVROE	——	——	0.247 (0.635)
NEDUM	1.783 (0.379)	3.336 (0.577)	0.142 (0.027)
SEDUM	25.03 (1.213)	41.015 (1.143)	40.646 (1.112)
MADUM	-1.955 (-0.458)	0.250 (0.049)	-0.76 (-0.15)
NWDUM	23.397 (0.541)	19.496 (0.498)	20.242 (0.515)
MWDUM	-0.802 (-0.201)	-0.388 (-0.079)	-0.564 (-0.125)
SCDUM	-0.36 (-0.091)	-0.829 (-0.176)	-1.239 (-0.288)
RSS	48.067	61.923	40.506
N	776	685	692

Note: Robust Z-statistics are in parenthesis. * Significant at the .10 level; ** Significant at the .05 level, one-tailed test. The Variance-to-mean relation is gamma. The link function is reciprocal. Signs are reversed when using gamma.

33.4 CONCLUDING REMARKS

This paper reviews recent developments in the estimating function literature and explains why estimation problems involving limited dependent variables are particularly promising for applications of the EF theory. Our Result 33.1.1 shows that whenever heteroscedasticity is related to β the traditional LS or ML estimators have an unnecessary extra term leading to biased and inefficient EFs. We note that recent econometric literature, surveyed in Hajivassiliou and Ruud (1994) or Baltagi (1995) while ignoring the simpler GEE methods, is suggesting computer intensive nonparametric and simulation based method of moments estimators. These estimators are obviously suboptimal, since they fail to remove the extra term mentioned above. Our Result 33.2.1 shows why binary dependent variables have such heteroscedasticity.

The panel data GEE estimator in (33.2.16) is implemented by Liang and Zeger's (1986) "modified Fisher scoring" algorithm with variances given in (33.2.17). The flexibility of the GEE estimator arises from its ability to specify the matrix of autocorrelations $R(\phi)$ as a function of a set of parameters ϕ. We use unstructured $R(\phi)$ with minimum prior assumptions to achieve robustness and use a "canonical link" function satisfying "sufficiency" properties available for all distributions from the exponential family. It is well known that this family includes many of the familiar distributions including Normal, binomial, Poisson, exponential, gamma, etc. Our Remark 33.2.3 explains the advantages of the gamma family with its canonical link used here, which is almost never used in econometrics.

The regression results reported here are consistent with the "rent-seeking" hypothesis regarding the intensity of managerial monitoring of firm managers by owners in the electric utility industry. The GEE estimates imply rejection of the "monopoly slack" view. This reverses the conclusions suggested by logit tests in Geddes (1997). There is also evidence that managers are monitored by consumers on the basis of price changes, which is consistent with a "political pressure" hypothesis. Overall, this suggests that the GEE estimation technique represents an improvement over standard binary dependent variable techniques, especially for panel data.

REFERENCES

Alchian, A. A. and Kessel, R. A. (1962). Competition, Monopoly, and the Pursuit of Money, In *Aspects of Labor Economics* (Eds., H. Gregg Lewis), pp. 157-175, Princeton University Press, Princeton.

Baltagi, B. H. (1995). *Econometric Analysis of Panel Data*, John Wiley & Sons, New York.

Barro, J. R. and Barro, R. J. (1990). Pay, performance, and turnover of bank CEOs, *Journal of Labor Economics*, **8**, 448–481.

Bertschek, I. and Lechner, M. (1998). Convenient estimators for panel data probit model *Journal of Econometrics*, **87**, 329–371.

Chamberlain, G. (1980). Analysis of covariance with qualitative data, *Review of Economic Studies*, **47**, 225–238.

Coughlan, A. T. and Schmidt, R. M. (1985). Executive compensation, management turnover, and firm performance, *Journal of Accounting and Economics*, **7**, 43–66.

Crain, W. M. and Zardkoohi, A. (1978). A test of the property-rights theory of the firm: Water utilities in the United States, *Journal of Law and Economics*, **21**, 395–408.

Crain, W. M. and Zardkoohi, A. (1980). X-Inefficiency and nonpecuniary rewards in a rent-seeking society: A neglected issue in the property rights theory of the firm, *American Economic Review*, **70**, 784–792.

Diggle, P. J., Liang, K-Y. and Zeger, S. L. (1994). *Analysis of Longitudinal Data*, Clarendon Press, Oxford.

Dunlop, D. D. (1994). Regression for longitudinal data: A bridge from least squares regression, *The American Statistician*, **48**, 299–303.

Firth, D. (1988). Multiplicative errors: log-normal or gamma?, *Journal of the Royal Statistical Society, Series B*, **50**, 266–268.

Fisher, F. M. and McGowan. J. J. (1983). On the misuse of accounting rates of return to infer monopoly profits, *American Economic Review*, **73**, 82–97.

Geddes, R. R. (1997). Ownership, regulation, and managerial monitoring in the electric utility industry, *Journal of Law and Economics*, **40**, 261–288.

Godambe, V. P. (1960). An optimum property of regular maximum likelihood estimation, *Annals of Mathematical Statistics*, **31** 1208–1212.

Godambe, V. P. (1985). The foundations of finite sample estimation in stochastic processes, *Biometrika*, **72**, 419–428.

Godambe, V. P. and Kale, B. K. (1991). Estimating functions: an overview. In *Estimating Functions* (Ed., V. P. Godambe), Chapter 1, Clarendon Press, Oxford.

Hajivassiliou, V. A. and Ruud, P. A. (1994). Classical estimation methods for LDV models using simulation, In *Handbook of Econometrics*, Vol. 4 (Eds., R. F. Engle and D. L. McFadden), Chapter 40, Elsevier, Amsterdam.

Heyde, C. C. (1997). *Quasi-Likelihood and Its Applications*, Springer-Verlag, New York.

Jensen, M. C. and Murphy, K. J. (1990). Performance pay and top management incentives, *Journal of Political Economy*, **98**, 225–264.

Joskow, P., Rose, N. and Shepard, A. (1993). Regulatory constraints on CEO compensation, *Brookings Papers on Economic Activity: Microeconomics*, 1–58.

Joskow, P., Rose, N. and Wolfram, C. (1996). Political constraints on executive compensation: Evidence from the electric power industry, *Rand Journal of Economics and Management Science*, **27**,165–182.

Li, B. (1997). On consistency of generalized estimating equations, In *Selected proceedings of the Symposium on Estimating Functions* (Eds., I. Basawa, V. P. Godambe and R. L. Taylor), IMS Lecture Notes-Monograph Series, Vol. 32, pp. 115–136.

Liang, K. and Zeger, S. L. (1986). Longitudinal data analysis using generalized linear models, *Biometrika*, **73**, 13–22.

Liang, K. and Zeger, S. L. (1995). Inference based on estimating functions in the presence of nuisance parameters, *Statistical Science*, **10**, 158–173.

Liang, K., Quaqish, B. and Zeger, S. L. (1992). Multivariate regression analysis for categorical data (with discussion), *Journal of the Royal Statistical Society, Series B*, **54**, 3–40.

Lipsitz, S. R., Fitzmaurice, G. M., Orav, E. J. and Laird, N. M. (1994). Performance of generalized estimating equations in practical situations, *Biometrics*, **50**, 270–278.

McCullagh, P. and Nelder, J. A. (1989). *Generalized Linear Models*, Second Edition, Chapman and Hall, London.

Salancik, G. R. and Pfeffer, J. (1980). Effects of ownership and performance on executive tenure, *US Corporations, Academy of Management Journal*, **23**, 653–664.

Stigler, G. J. (1971). The theory of economic regulation, *Bell Journal of Economics and Management Science*, **2**, 3–21.

Smith, D. M. (1996). Oswald: Object-oriented software for the analysis of longitudinal data in S, http://www.maths.lancs.ac.uk/Software/Oswald, See also references to Vincent Carey's implementation of GEE.

Vinod, H. D. (1997). Using Godambe-Durbin estimating functions in econometrics, In *Selected Proceedings of the Symposium on Estimating Functions* (Eds., I. Basawa, V. P. Godambe and R. L. Taylor), IMS Lecture Notes-Monograph Series, Vol. 32, pp. 215–237, Hayward, California.

Vinod, H. D. (1998). Foundations of statistical inference based on numerical roots of robust pivot functions (Fellow's Corner), *Journal of Econometrics*, **86**, 387–396.

Warner, J. B., Watts, R. L. and Wruck, K. H. (1988). Stock prices and top management changes, *Journal of Financial Economics*, **20**, 461–492.

Weisbach, M. (1988). Outside directors and CEO turnover, *Journal of Financial Economics*, **29**, 431–460.

Wedderburn, R. W. M. (1974). Quasi-likelihood functions, generalized linear models and the Gaussian method, *Biometrika,* **61**, 439–447.

Zeger, S. L. and Liang, K-Y. (1986). Longitudinal data analysis for discrete and continuous outcomes, *Biometrics*, **42**, 121–130.